The Falsification of Science
Our Distorted Reality

by

John Hamer

Edited by, and with contributions from

Shannon Rowan

The Falsification of Science

John Hamer

Paperback Edition First Published in Great Britain in 2021
by John Hamer

eBook Edition First Published in Great Britain in 2021
by John Hamer

Copyright © John Hamer 2021

John Hamer has asserted his rights under 'the Copyright Designs and Patents Act 1988' to be identified as the author of this work.

All rights reserved.

No part of this document may be reproduced or transmitted in any form or by any means, electronic, mechanical, photocopying, recording, or otherwise, without prior written permission of the Author.

ISBN: 9798703758427

Acknowledgements and Thanks

I should like to take the opportunity to offer my thanks and gratitude to Shannon Rowan, my very good friend and co-author on other projects, for her many contributions to this book.

As well as providing some of the actual words to be found within this volume, her profound knowledge of all things geopolitical, her wisdom, her in-depth, expert knowledge of photography and alternative health, and indeed her entire worldview has also been my inspiration for several of the ideas and topics to be found within these pages.

And if all that was not enough, she has also acted as both my editor and 'eagle-eyed' proof-reader.

Thank you again Shannon, for your valued input – and advice. It is all very much appreciated.

I should also like to thank my publisher, Nicola Mackin at aSys Publishing for her diligence, hard work, help and advice – as always. Nothing is ever too much trouble and she is never anything other than an absolute delight to work with.

Thank you, Nicola.

My thanks also go to Tracey Northern who (just in time) corrected some of my misconceptions in the Health, Food and Medicines chapter!

More Books by the Same Author…

Roots Detective. 2008 (Canonbury Publishing)

The Falsification of History. 2012 (Rossendale Publishing)

RMS Olympic. 2014 (Rossendale Publishing)

Titanic's Last Secret. 2015 (aSys Publishing)

Behind the Curtain (volumes 1 & 2). 2016 (aSys Publishing)

JFK – A Very British Coup. 2019 (aSys Publishing)

Dedication

This book is dedicated to the memory of my 'baby brother,' Paul who sadly passed away on the 31st of January 2021 after suffering so much during his final few years.

May you rest in peace forever Paul, now that all your earthly worries are over. Goodnight and god bless until we meet again. I will never forget you and all the fun we had together in our younger days.

Contents

The Prologue 1

Chapter 1 In the Beginning 17
    The Big Bang Theory 17
        Dark Matter and Dark Energy 26
    The Electric Universe 29
        Redshift 41
        The Sun 43
        The Planets 46
        Gravity 48
    The Theory of Relativity 56
    The Holographic Universe 59

Chapter 2 Evolution 67
        Dinosaurs 95

Chapter 3 Ancient Technology and Knowledge 112
    High-tech Stonework 113
    Ancient Astronomers 119
    Flexible Glass 122
    The Antikythera Mechanism 125
    The 1200 Year-old Telephone 127
    Greek Fire 130
    The 300 Million-Year Old Screw 130
    Han Purple and Egyptian Blue 132
    The Baghdad Battery 135
    Viking Compasses 136
    The Lycurgus Cup 138

Chapter 4 Freemasonry and the Royal Society 140

Freemasonry 140
The Royal Society 153

Chapter 5 Flat Earth 165

Airplane Flights in the Southern Hemisphere 183
Solar Eclipses 185
Spinning Earth? 188
More 'Proofs' of a Flat Earth 189
The Bedford Level Experiments 195
The Bishop Experiment and other examples of 'Impossible' Viewpoints 198
The Michelson-Morley Experiment 204
Earth 'Photographs' courtesy of NASA 206

Chapter 6 NASA and Outer Space 209

Jack Parsons 211
Aleister Crowley 214
L. Ron Hubbard 215
Lysenkoism 219
The Challenger 'Disaster' 227
Hubble Space Telescope 229
The Apollo Project 230

Chapter 7 The Nuclear Weapons Hoax 291

Chapter 8 Health, Food and Medicine 333

Health 333
    Vaccines 344
    Spanish Flu 1918/9 367
    ADHD and ADD 396
    Cancer 408

    Autism 418

    Dental Amalgam 421

    Energy-Saving Lightbulbs 424

    Sunscreen 427

    Invented Diseases 432

    Cholesterol 433

Food and Nutrition 440

    Aspartame 442

    Genetically Modified Organisms (GMOs) 447

    High Fructose Corn Syrup (HFCS) 458

    Monosodium Glutamate (MSG) 459

    Fluoride 461

    Contaminated Drinking Water 469

    Junk Food 472

    Codex Alimentarius 475

Medicine 480

    The Rockefeller Medical Paradigm 480

Chapter 9 The Environment 492

    The Great 'Climate Change' Hoax 492

    HAARP – The High Frequency Active Auroral Research Program 515

    Chemtrails 530

    Fracking 535

    Renewable Energy 543

    5G and 'Smart' Technology 555

Chapter 10 Suppressed Technology 572

    Nikola Tesla 573

    Direct Energy Weapons 575

    Wireless Electricity 584

    Radiant Energy 585

- Motionless Electromagnetic Generator 586
- Mechanical Heater 586
- Water Powered Car 587
- Implosion Engine 588
- Cold Fusion 590
- Antigravity Devices 592
- The OTC-X1 593
- Conclusion 595

Chapter 11 The 'Unexplained' Human Mind 602
- Extrasensory Perception 603
  - Clairaudience 604
  - Clairvoyance 604
  - Mediumship 605
  - Psychometry 605
  - Precognition 606
  - Remote Viewing 606
  - Retrocognition 607
  - Telepathy 607
  - Telekinesis/Psychokinesis 608
- Astral Projection 608
  - Meditation 608
  - Out of Body Experiences 609
- Automatic Writing 610
- Déjà Vu et al 612
  - Déjà Vu 612
  - Déjà Vécu 612
  - Déjà Visité 612
  - Déjà Senti 613
  - Jamais Vu 613
  - Presque Vu 614

 L'esprit de l'escalier 614
 Capgras Delusion 614
Spirituality 615
Mainstream Religion 619
Reincarnation and 'Life between Lives' 623

Chapter 12 The Great Global Reset 2020 629

Agenda 21/2030 629
COVID-19 – The Real Truth 639
 Mandatory Vaccinations 646
 A Financial Crash? 650
 1984 is here at last! 651
 The Internet of Things and YOU 667
 The 'Spanish Flu' Pandemic of 1918-1919 – Reprise! 669
 A Pandemic of Testing 673
 The Trouble with RNA Tests 674
 'Hidden in Plain Sight' 678
 Robot, Good--- Human, Bad 680
 Humans are a Virus 681
 'We Live in Zoom Now' 682
 A New Type of War 683
 Depopulation Agenda 684
 The Real Death Toll 687
 Talk of 'cases' is totally misleading 688
 A Coronavirus is Nothing to Fear 688
 Big Brother is Watching 690

The Epilogue 698

The Prologue

"Those who are able to see beyond the shadows and lies of their culture will never be understood, let alone believed, by the masses." Plato

"I know that most men, including those at ease with problems of the greatest complexity, can seldom accept even the simplest and most obvious truth, if it would be such as would oblige them to admit the falsity of conclusions which they have delighted in explaining to colleagues, which they have proudly taught to others, and which they have woven, thread by thread, into the fabric of their lives." Leo Tolstoy, 20th Century Russian dissident writer

"To be a scientist is to be naïve. We are so focused on the truth; we fail to consider how few actually want us to find it. But it is always there whether we see it or not, whether we choose to or not. The truth doesn't care about our needs or wants, it doesn't care about our governments, our ideologies, our religions. It will lie in wait for all time and this at last is the gift of Chernobyl. I once feared the cost of truth but now I only ask, what is the cost of lies?" Valeriy Legasov, the Soviet nuclear physicist who deeply regretted his part in the elaborate Soviet cover-up of the Chernobyl disaster and who took his own life by hanging at 1.23 am on the 26th April 1988, exactly two years (to the exact minute) after the event

----

Unlike the subject of this book's predecessor and sister volume, *The Falsification of History*, science is usually systematically falsified for two quite distinct reasons...

Firstly, as with history it may be falsified for political expediency, that is for the purpose of maintaining and propounding a false paradigm which supports the ongoing political agenda, but secondly and most commonly, science is often falsified for financial gain, usually directly benefiting the ruling Elite corporatocracy in some way.

'But hold on a minute…' I hear you say, '…how can this possibly be true? Surely there would be a huge number of people 'whistle-blowing' and exposing these perpetrators as the criminals and frauds they surely are?' And you would be right. There ARE indeed a huge number of people aware of and attempting to expose the perpetrators, but as with anything else that contradicts an 'official diktat,' the overwhelming power and influence of those in whose best interests it is to deceive us, and who control most of the world, including the media in all its forms, ensures that only small snippets of information escape into the mainstream. However, this is more than enough to 'paint a picture' of reality for those of us who are broadly aware of the ongoing deceit, and this allows us to expose the sordid truth embedded in all the many establishment lies we are able to dissect.

The 'scientific worldview' is immensely influential because the sciences have been so successful in many respects. They touch all our lives through technology and through modern medicine – although not always necessarily for the better. Our intellectual world has been transformed by an immense expansion of knowledge, deep down into the most microscopic particles of matter and far out into the vastness of space, with hundreds of billions of galaxies in an ever-expanding universe – allegedly.

Yet here, in the third decade of the twenty-first century, and when science and technology seem to be at the peak of their power, when their influence has spread all over the world and when their triumphal proclamations on all matters scientific appear indisputable, unexpected problems are disrupting all of the sciences from within. Most scientists take it for granted that these problems will eventually be solved by more research along conventional lines, but many people, including myself, believe that they are symptoms of a much deeper malaise.

The Falsification of Science

"Today's scientists have substituted mathematics for experiments, and they wander off through equation after equation and eventually build a structure which has no relation to reality." Nikola Tesla

In this book's 'sister' volume, *The Falsification of History*, I argued that science is being held back and indeed being deliberately deceitful, by promoting centuries-old assumptions that have hardened into dogmas that have become the de facto 'truth.' The biggest scientific delusion of all is that science already knows all the answers. The finer details may still need working out but, in principle, the fundamental questions are settled. 'Twas ever thus.

Indeed, contemporary science is based on the claim that all reality is material or physical and that there is no reality but material reality, and consciousness is merely a by-product of the physical activity of the brain. It would have us believe that matter is wholly unconscious, and that a divine intelligence exists only as an idea in human minds and hence within human heads.

These beliefs are powerful, not because most scientists think about them critically but actually because they do not. They believe, or would have us believe that the *facts* of science are real enough as are the techniques that scientists use, and the technologies based on them, but in actual fact, the belief system that governs conventional scientific thinking is an act of faith, grounded in a nineteenth-century ideology.

But please do not mistake my motivations, I am pro-science, but I would prefer the sciences to be less dogmatic and assumptive, and more, well...*scientific* and I believe that the sciences will be wholly regenerated once they are liberated from the dogmas and assumptions that currently constrict and distort them.

Here are the ten core beliefs that all of mainstream science erroneously takes for granted...

1. Everything is essentially mechanical. Animals, for example, are complex mechanisms, rather than living organisms with goals of their own. Even people are machines, 'lumbering robots,' in

zoologist Richard Dawkins' vivid, but in my view wildly inaccurate phrase, with brains that are in effect genetically programmed computers.

2. All matter is unconscious. It has no inner life or subjectivity, or point of view and even human consciousness is an illusion produced by the material activities of brains.

3. The total amount of matter and energy always remains constant (with the exception of the 'Big Bang,' when all the matter and energy of the universe suddenly appeared from nowhere).

4. The laws of nature are fixed. They are the same today as they were at the beginning, and they will stay the same forever.

5. Nature is purposeless, and evolution is a hard, proven fact and not an insubstantial theory.

6. All biological inheritance is material, carried in the genetic material, DNA and in other material structures.

7. Minds are inside heads and are nothing but the activities of brains. When we observe a tree, the image of the tree we see is not 'out there,' where it appears to be, but inside our brain.

8. Memories are stored as material traces in brains and are erased at death.

9. Unexplained phenomena such as telepathy, ESP, levitation, teleportation, and memories of past lives are all ridiculous fantasies.

10. Mechanistic, allopathic (Rockefeller) medicine is the only kind that really works.

Together, these beliefs make up the philosophy or ideology of materialism, whose central assumption is that everything is essentially material or physical – even human minds.

The Falsification of Science

This belief system became dominant within science in the late nineteenth century under extreme Freemasonic influence (as I will expand further upon, in this volume) and is now taken for granted. Many scientists are unaware that materialism is actually nothing more substantial than an assumption; they simply think of it as science, or the scientific view of reality, or the scientific worldview. They are taught about it but are never allowed the opportunity to discuss it objectively whilst absorbing it by a kind of intellectual osmosis from the all-encompassing, existing, and prevailing agenda laid down by vested interests of various kinds.

In its everyday usage, 'materialism' simply refers to a way of life devoted entirely to material interests, a preoccupation with wealth, possessions, and luxury. These attitudes are very much encouraged by its close relation, the scientific materialist philosophy, which denies the existence of any spiritual realities or non-materialistic goals, but propounds a philosophy totally governed by Elite-engineered dogma which absolutely denies the possibility of the existence of any phenomena other than those detectable by the five generally accepted human senses. As a sidenote, more freethinking scientists have estimated that the human senses can only detect something like 0.05% (or one two thousandth) of the entire spectrum of reality. If true, this means that our so-called 'science' concerns itself only with that tiny percentage of knowledge, whilst blatantly disregarding the rest of creation in all its glory.

For more than two hundred years, materialists have promised that science will eventually explain everything in terms of physics and chemistry. Science will prove that living organisms are complex machines, minds are simply brain activity and nature is purposeless and believers of this particular 'faith,' for such it is, are sustained by the further belief that scientific discoveries will eventually justify their often flimsy theories. Karl Popper, the 20th century British-Austrian 'philosopher of science,' referred to this stance as 'promissory materialism' because it depends in effect, on issuing promissory notes for discoveries not yet made. Maybe unsurprisingly, despite all the achievements of science and technology, materialism is now facing a credibility crisis that was unimaginable in the twentieth century.

"How many times have you heard someone proudly proclaim, 'I believe in science!?' The statement is generally meant to silence and ridicule anyone who dares question orthodox science and portray these 'lay persons' as ignorant, superstitious and religious. The irony of this statement, admitting to 'belief' in something that is constantly touted as 'factual' and thereby presented as not requiring belief or faith, seems to be forever lost on these scientific adherents – forever blind to their own dogmatic positions." Shannon Rowan, geopolitical researcher, and author

Francis Crick (the English discoverer of the double helix shape of the DNA molecule along with the American James Watson, in the early 1950s) and Sydney Brenner were avid materialists and Crick was also a staunch atheist. They believed that there were two major unsolved problems in biology; development and consciousness and they had not been solved because the people who worked on them were not molecular biologists – or even very bright. Crick and Brenner pledged that they were going to find the answers within ten years, or maybe twenty, maximum. Brenner would take developmental biology, and Crick, consciousness.

Both tried their level best. Brenner was awarded the Nobel Prize in 2002 for his work on the development of a tiny worm, *Caenorhabdytis elegans* and Crick corrected the manuscript of his final paper on the brain the day before he died in 2004. At his funeral, his son Michael said that what motivated him was *"not the desire to be famous, wealthy, or popular, but to knock the final nail into the coffin of vitalism."* Vitalism is the theory that living organisms are truly alive, and not explicable in terms of physics and chemistry alone. But such dedication to 'prove' one's own particular belief system is in fact endemic amongst scientists and it is this more than anything else that creates and facilitates a severe distortion of the truth in so many scientific disciplines as facts are 'bent' to suit a personal agenda rather than a truly objective, open-minded approach to research being applied from the outset.

But the simple truth anyway, is that both Crick and Brenner failed abjectly. The problems of development and consciousness remain unsolved to this day and many details have been discovered, dozens of genomes have been sequenced and brain scans are ever more precise, but there is still no proof whatsoever that life and minds can be explained by physics and chemistry alone.

Mainstream science is often defended by its disciples in the utilisation of the deceptive phrase 'the scientific method,' as though this is some catch-all, get-out clause that somehow exonerates fraud and misrepresentation. It is often invoked in defence of criminal and insidious activities such as the mass prescription drugging of the population, vaccines, EMF radiation, GMOs, and geo-engineering. In even more extreme cases, science is deployed as a weapon to serve insane agendas such as the eugenics-based programmes of surreptitious sterilisation and depopulation of 'undesirables.' Science itself as a discipline, is not to blame for this; my issue is only with those pushing dangerous and wicked agendas under the rubric of science and yet, most unthinking people will blindly bow in reverence to these white-coated 'priests' of science without further research or investigation. (I refer you back to the previous quote by Shannon Rowan). And indeed, this is how all the great hoaxes of our present times are facilitated.

How can we trust science when it decrees with zero evidence that vaccines save lives, despite the fact that they provably injure, paralyse and kill with impunity?

How can we trust 'the scientific method,' when there is such a massive amount of fraudulent scientific research, as exposed by former Big Pharma directors, doctors, medical journal editors, insiders, and other so-called 'experts?'

How can we trust the 'peer review,' process when it has been proven to fail so often and is totally open to fraudulent practices, nepotism and cronyism?

"Peer review is the evaluation of work by one or more people with similar competences as the producers of the work (peers). It functions as a form of self-regulation by qualified members of a profession within the relevant field. Peer review methods are used to maintain quality standards, improve performance, and provide credibility. In academia, scholarly peer review is often used to determine an academic paper's suitability for

publication. Peer review can be categorized by the type of activity and by the field or profession in which the activity occurs, e.g., medical peer review." Wikipedia™

Interesting, and perhaps true to a point, but science and society in general's absolute insistence that this process is infallible and therefore scrupulously reliable and valid as a basis for the scientific evaluation of anything at all, is sadly, at best completely wrong and at worst nothing more than a convenient basis for the large-scale deception of we, the masses.

Peer review is at the very heart of all the advances in all of the scientific disciplines. It is the method by which substantial grants are allocated, new 'facts' are decided upon and published, academics are promoted, and Nobel prizes are awarded, yet it is extremely difficult to define. It has until recently gone completely un-investigated and unstudied and its defects are far easier to identify than its positive attributes, yet it shows no sign of ever being replaced as a basis for the 'advancement' of science. In fact, it is analogous in many ways to democracy; a system replete with issues of many guises, but maybe the 'least worst' we have. Although even that much is highly debatable.

Whenever conclusions from any research are peer reviewed their integrity is automatically regarded as beyond question and even journalists recognise this. For example, some years ago, when the *British Medical Journal* published a highly controversial paper that argued that a new 'disease,' a form of female sexual dysfunction, had actually been 'created' artificially by pharmaceutical companies, the first question asked by medical science was of course, 'is this paper peer reviewed?' The implication being that if it had then it was certainly an indisputable fact and had it not been, then it was not. Simple. In fact, as it transpired it had not been peer reviewed as the authors of the paper had been unable to elicit any kind of support from their peers due to fears of a huge backlash from 'Big Pharma' and therefore it was not regarded as a serious piece of research and was ultimately discarded because of this, despite the possible validity of the research undertaken.

The Falsification of Science

As stated, peer review is actually impossible to define in operational terms and indeed how would a 'peer' be defined? If it is someone undertaking exactly the same kind of research, then he or she is probably a direct competitor and unlikely to be unbiased in any assessment or judgement of the validity of the research. Or could it be someone in the same broad discipline or someone who is merely an expert in methodology? And what exactly is defined as a review? Is it someone simply saying 'that looks fine to me,' which is sadly what peer reviews often tend to be? Or alternatively is it someone checking in great detail throughout the entire paper, asking for the raw data, repeating analyses, checking all the references, and making detailed suggestions for improvement? Such a review is extremely rare to non-existent in reality.

But what is abundantly clear is that all forms of peer review are protean; extremely variable, undefined, and changeable. It is completely safe to assume that the systems of every journal and every grant-giving body are different in at least some detail; and some systems are very different indeed. There are even some scientific journals that use the following methodology...the editor simply looks at the title of the paper and sends it to two friends whom he/she thinks know something about the subject and if they both give it the cursory, metaphorical 'nod' then the editor approves it for publication. And if both advise against publication the editor then rejects the paper. However, if the reviewers disagree, the editor will then send it to a third reviewer and act accordingly upon whatever he or she advises. This is actually akin to the tossing of a coin in effect, because the level of agreement between reviewers on whether a paper should be published is no better or more reliable than pure chance.

Robbie Fox, the former editor of the *Lancet*, who was no admirer of peer review, once speculated as to whether anybody would notice if he were to exchange the piles of papers marked 'publish' and 'reject.' He also once joked that the *Lancet* had adopted a system of throwing a pile of papers down the stairs and only publishing those that reached the bottom! And he was in fact once challenged by two well respected British researchers to publish an issue of the journal comprised only of papers that had failed peer review and to see if anyone even noticed. He wrote back, *"How do you know I haven't already done that!?"*

A recent systematic review of all the available evidence on peer review concluded that *...the practice of peer review is based on faith in its effects, rather than on facts."*

But what is the actual purpose of peer review? One answer may be that it is a method by which the best grant applications for funding and the best papers for publishing in a scientific journal may be selected. It is extremely difficult, if not downright impossible to test this premise because there is no agreed definition of what constitutes a 'good' paper or a 'good' research proposal. In addition, what should peer review be tested against? Pure chance or a much simpler process?

Stephen Lock, when editor of the *British Medical Journal* conducted an experiment in which he alone decided which of a consecutive series of papers submitted to the journal would be accepted for publication. He also submitted *all* the papers received, to the usual process of selection for peer review and the result was that there was little to no difference between the papers he chose and those selected after the full process of peer review. Even this small study suggested that an elaborate process is unnecessary. But it would be a bold journal editor that permanently steered away from the sacred path of peer review. The severe and unbridled wrath of all of the scientific hierarchy would almost certainly befall him or her.

Significantly, peer review is also extremely susceptible to fraudulent activities. One major, respected scientific journal recently conducted several studies where major errors were deliberately inserted into papers that were then sent to many 'independent' reviewers. No-one spotted all of the errors, some reviewers did not spot any at all, and most reviewers spotted only about twenty-five percent of them. Peer review occasionally does detect fraud however, but only by chance, and does not even come close to being a reliable method for detecting fraud because it works purely on 'trust' and it is therefore an absolutely simple process to fake results.

The Falsification of Science

And there are several ways in which it is possible to abuse the process of peer review. Ideas may be stolen and presented as original and it is also possible to produce an unjustly harsh review to block or at least slow down the publication of the ideas of a competitor. And perhaps more importantly and relevant to the title of this book, it is absolutely possible to fake an entire succession of various 'studies' in order to hide or in some way completely diminish the dangers of a particular technology, for financial gain or to further a particular agenda.

In addition, anything that is 'peer-reviewed,' depending on whether it confirms or denies supporting evidence for a particular agenda, is either heavily promoted, heavily criticised, or even ignored. The value of peer reviews are therefore questionable at best and are subject only to be either used as propaganda to support a particular premise or are propagandised against, as the case may be.

How can we trust the much-vaunted scientific 'hierarchy of evidence,' when it is such a simple matter to perpetrate a fraud, whether it be outright omission, distortion by changing data sets, publishing deliberately falsified data, and many of the other tricks used to substitute 'real' science for corporate junk science?

The 'hierarchy of evidence' is often referred-to by scientists and doctors as an integral component of the scientific method. It is the purported 'gold standard' by which we can arrive at the truth, and by which we can compare and contrast different pieces of evidence and determine which should be given more credence and which should be given less. The hierarchy of evidence is often raised in debates by those pushing various agendas such as the vaccine, GMO, and fluoride agendas as a way of proving that the 'science' behind those agendas is sound. However, there is a huge problem with invoking hierarchy of evidence to prove a case, but firstly, I should define what is meant by the 'hierarchy of evidence.'

In a nutshell, the hierarchy of evidence, although different people and institutions construe it slightly differently, assigns most credence to RCTs (Randomised Controlled Trials) then secondly to other types of controlled trials, thirdly to other studies such as cohort and case-control studies, and lastly to expert opinion, case reports and anecdotal evidence. Wikipedia™ uses the writings of Greenhalgh to frame the hierarchy of evidence as follows...

- Systematic reviews and meta-analyses of RCTs with definitive results.'

- Randomised controlled trials with definitive results.

- Randomised controlled trials with non-definitive results.

- Cohort studies

- Case-control studies

- Cross sectional surveys

- Case reports

- Evidence from a systematic review of all relevant RCTs, or evidence-based clinical practice guidelines based on systematic reviews of RCTs

- Evidence obtained from at least one well-designed RCT

- Evidence obtained from well-designed controlled trials without randomisation, quasi-experimental

- Evidence from well-designed case-control and cohort studies

- Evidence from systematic reviews of descriptive and qualitative studies

- Evidence from a single descriptive or qualitative study

- Evidence from the opinion of authorities and/or reports of expert committees

Without becoming embroiled in the details and specifics of all the above, what it actually means in simple, plain English is that if a corporation is for example, attempting to bring a certain product to the open market and it is known by themselves (and others) to be harmful, this matters not a jot. All that is needed is to obtain enough favourable RCT reviews and then it may be *legally and scientifically* claimed that the product is safe. Once this is achieved then there will be no shortage of scientists, doctors, politicians, and public institutions willing to either propound this point of view, or at least defend it – for a price of course. They do not even need to be privy to the secret, because they have been conditioned and propagandised to implicitly trust the false 'hierarchy of evidence' paradigm.

And so how would a devious corporation, cartel or cabal achieve this?

There is a well-known maxim in law that *fraud vitiates all,'* meaning that if fraud is involved in any way in any agreement, then it totally invalidates the entire document, not just the small part that may be affected. It is not possible for there to be 'partial fraud' and to then consider the rest of any contract to be trustworthy. If someone has lied, omitted relevant information, or deceived in any way, the entire premise collapses and such is the case with so much of so-called 'science.' Fraud has rendered it as nothing more than corporate junk science, a pale imitation of what 'real' science is meant to represent.

"It is simply no longer possible to believe much of the clinical research that is published, or to rely on the judgment of trusted physicians or authoritative medical guidelines . . . I take no pleasure in this conclusion, which I reached slowly and reluctantly over my two decades as an editor of The New England Journal of Medicine." Marcia Angell, former editor of said, *The New England Journal of Medicine*

So, you may well ask, what is the point of the 'hierarchy of evidence' when scientists can simply omit and/or distort data whenever it suits their purposes? Or to put it another way, how can we believe it when some government agency or huge corporation tells us that their product is safe because the science has settled it? How do we know it is not just more junk science or corporate science disguising itself as real science? Unfortunately, the sad truth is that we cannot assume that a product is safe simply because it is legal or allegedly scientifically 'proven' to be harmless.

Apart from the possibility of rampant fraud, the hierarchy of evidence has another major problem and that is that it disregards real stories of suffering as 'low level evidence.' This means that in cases where for example, vaccines, Big Pharma drugs, radiotherapy or chemotherapy, 5G, mobile (cell) phones, wi-fi, or indeed any other industrial or medical product or service deemed safe, actually kills or seriously harms people, the stories are treated as unprovable, isolated incidents, coincidences or unimportant anecdotes and swept under the carpet – all in the name of protecting reputations and corporate profits.

It is saddening to hear stories of how many real people have been harmed by corporate science and its slavish devotion to the 'hierarchy of evidence,' but it will always argue that those harmed are basically within 'acceptable parameters' because we 'need to consider how many (hypothetical) people were saved.' How is it possible though to prove how many were saved? It is not possible and therefore the 'hierarchy of evidence' remains a faith-based belief-system, and not one based upon true scientific principles.

Allopathic doctors or 'Rockefeller medics' once laboured under horrendous misconceptions, such as the practices of electric shock therapy, blood-letting as a medical and surgical technique and advocating the smoking of tobacco as being 'good for you.' Was any of this real science and if not, how do we know that today's science is any more 'real' than those examples? I strongly suggest that we do not.

Although it presents itself as a rational force for the advancement of knowledge, much of mainstream science, is often simply dogma or a pseudo-religion propounding insubstantial 'theories' that somehow become 'facts' in the minds of the masses. In truth, science in general, loves to blur the boundaries between hard facts and theories and in this way it is able to present a completely fictitious view of reality, benefitting those unseen forces who control us all. For example, 'germ theory' is (obviously) just a theory regarded as fact, 'relativity'

The Falsification of Science

is a theory, the 'Big Bang' is a theory and 'evolution' is only an insubstantial theory too, and none of these are proven, indisputable facts as claimed deceptively by scientists of all disciplines. But no doubt these theories will all be replaced by newer, more politically expedient ones in the fullness of time, as has been the case from time immemorial.

And speaking of 'facts,' it is extremely doubtful as to whether scientific dogma is any more substantial than religious dogma. Science demands that the universe and all it contains, is represented in a certain manner and then simply labels anyone whose worldview falls outside of these set parameters as a 'denier,' a 'heretic' or even as 'deluded' or that old favourite, a crazy 'conspiracy theorist.' Ultimately, how much can we really trust mainstream, materialistic science when it officially denies the existence of anything that the severely limited range of human senses cannot detect? As stated earlier in this introduction, more, shall we say, open-minded scientists, those who have no financial 'axe to grind,' or corporate master to serve, estimate that the five human senses of sight, sound, smell, touch and taste are only able to discern a mere 0.05% of reality. Even that great champion of mainstream edicts, Albert Einstein, said that, *"the field is the sole governing agency of the particle."* In other words, consciousness is the very basis of matter and life in the Universe, and yet science in general refuses to even accept its existence, let alone investigate the vast possibilities it presents.

I would also like to make it clear that I am not formally scientifically trained other than advanced (A) level physics at grammar school around fifty years ago. However, this is not an issue as within these pages I am attempting to impart reasons for the scientific deceptions that occur on a regular basis, and not take much more than a cursory glance into the deep scientific issues themselves. However, of necessity, I have had to include a modicum of scientific 'stuff' but have endeavoured to keep this to an absolute minimum. In short, this is not a science book meant only for the scientific community, merely an introduction to the murky world of corporate/establishment science and the deception that holds us all in its thrall.

John Hamer

Here is a very apposite quote from this book's sister volume, 'The Falsification of History'...

"Science will never answer our questions as long as it is run by a profit-focused, Elite-controlled scientific community that has no real interest in truth-seeking and really only exists in any case, to serve the interests of the Elite families in the generating of obscene profits and to maintain the false reality they have carefully created and nurtured and not as they would have you believe, for the furtherance of human knowledge or to create a 'better world' for us all." John Hamer, *'The Falsification of History'*

"The devotion to truth at all costs has gradually given place to the blind pursuit of the superficially plausible; the direction towards the most seductive, in which advance has been easiest, has been taken without regard to preservation of contact with the base, which is the truth of experience and reason." Herbert Dingle, Professor Emeritus of History and Philosophy of Science, University of London

So, to conclude this introduction, it is my self-assigned brief within the pages of this book, to attempt to identify all the many and various ways in which science deliberately deceives us and to demonstrate how and perhaps more importantly, why, many of these often-heinous deceptions are perpetrated.

I would ask you now to continue, but please prepare to be shocked by what you are about to read...

John Hamer

Chapter 1

In the Beginning...

Firstly, please be reassured. This first chapter does admittedly present some maybe, 'difficult' concepts for the non-scientific, but I am sure that you will find the remaining chapters to be much less 'technical' in nature.

The Big Bang Theory

The first and most pertinent question we should ask, is a simple one. How was the universe, our solar system and perhaps most pertinently of all, the planet, which is our home, created?

And the equally simple answer is that no-one knows. Not I...no-one. Not even mainstream science, despite their disingenuous, dogmatic, even arrogant claims to the contrary.

We are told confidently, but without any justification or proof by scientists, that around fourteen billion Earth years ago there was 'nothing,' and that this 'nothing' suddenly exploded in a 'Big Bang' and created an unimaginably vast universe of billions of galaxies, stars, solar systems and planets and on many of these planets, including of course, our own, life gradually evolved –– also from nothing and eventually produced sentient, (allegedly) super-intelligent beings which we refer to as 'the human race.' These so-called facts are repeated ad nauseum by the scientific establishment and its apologists the controlled media, until we poor dupes cannot even conceive that this may be inaccurate. It is constantly presented to us as 'settled science' and 'absolute fact' – and we, for the most part, absolutely believe it.

However, if we begin to analyse all of the above, does it not seem to be a rather glib conclusion despite the very questionable science assigned to it in order to furnish a plausible explanation? Surely, an explosion that creates something 'from nothing' would seem to immediately contradict one of mainstream science's great tenets; the First Law of Thermodynamics, which clearly states that matter cannot be created nor destroyed, it can only be changed from one form to another.

Matter cannot simply create itself and in the 'real world' cannot spontaneously arise from nothing, whether or not the instigating event is an explosion. Within the bounds of natural law all effects must have a cause and because of this fact, the spontaneous appearance of hydrogen atoms out of nothing (*ex nihilo* creation) is a definite breach of the aforementioned First Law of Thermodynamics which asserts that matter, under *natural* circumstances, can neither be created nor destroyed. Therefore, since it is not a natural event, it is by definition a supernatural event, otherwise known as 'a miracle.' So, science obviously now believes in miracles, does it? Seriously though, this is a rather weak starting point for a totally unrealistic and materialistic scenario.

But please do not worry, science has a get-out clause for this apparent paradox. Several physicists have recently proposed that the laws of physics are 'different' elsewhere in the universe. However, this assertion is not supported by even the smallest shred of scientific evidence, it is just yet another flimsy 'theory' and this amply demonstrates the lengths to which some will go in a futile attempt to 'prove' a finite beginning for the universe. Indeed, science is replete with such 'logic patches' (as I refer to them), which patch and conceal holes in insubstantial, unproven and in many cases, totally inadequate, even fraudulent theories.

However, please do not mistake my meaning, I do not have an answer to this conundrum, myself – the difference being though, that I do not feel that I must invent one to fill a gaping hole in the knowledge base. But it would seem that science does. It must protect its norms from being assailed by the truth – for reasons that will become clearer as we progress.

There have been many and varied explanations proposed, for the beginning of the universe down the millennia and centuries, ranging from the plausible to the utterly ridiculous and all points in-between, but as already stated, the plain and simple fact is that *NO-ONE KNOWS THE REAL TRUTH.*

The calculated acceleration of distant galaxies and galaxy clusters is just one of several serious dilemmas for the Big Bang theory. For example, what is the means of propulsion and how could a galaxy accelerate through the universe? There are no theories because the entire concept is nonsense, but it is a conclusion that one must reach if the 'Redshift' theory is true. Redshift (or the Doppler effect) is purportedly the phenomenon of the displacement of the spectrum of an astronomical object toward longer (red) wavelengths. It is a change in wavelength that occurs when a given source of waves (e.g., light, sound, or radio waves) and an observer are in rapid motion in opposite directions to each other. But obviously the 'Redshift' theory is also wrong as there is absolutely no known way in which a galaxy may increase its alleged acceleration through the cosmos. This contradicts all of mainstream science's edicts.

Following the current laws of physics, the surrounding mass from other galaxies, galaxy clusters, black matter, light, and unknown particles would have the effect of slowing the expansion. Masses produce a gravitational field that pulls them together – they do not push away from each other. Expansion at an accelerating rate is impossible. To reiterate, the Redshift theory is plainly wrong.

Einstein's General Theory of Relativity states that the universe must be contracting. He originally believed that the universe was static and introduced a cosmological constant as a correction factor, but he later referred to the constant as, *"my biggest blunder"* after accepting the Redshift theory, that the universe is expanding. But it now appears that his 'biggest blunder' was actually the acceptance of his second biggest blunder. Einstein's General Theory of Relativity does *not* support an accelerating expansion.

There are now dozens of more logical theories that have empirical evidence behind them, that the 'Big Bang' lacks. And yet, were you to ask a professor of particle physics about this, he will respond with pretentious physics jargon that threatens to verbally slap you into submission for the heinous crime of desecrating the holiness of the Big Bang theory.

So, how much more proof do we need to demonstrate that the Big Bang is pure fiction? The following debunks of the Big Bang are only the tip of a very large iceberg…

Direct observation of galaxies shows that every mature galaxy has either a super-massive black hole in the centre or an ion jet. These black holes 'suck-in' and destroy millions of stars. Conversely, ion jets spin out millions of stars from their super-brilliant column of ionic energy. Stars and solar systems are created by ion jets and destroyed by super-massive black holes. There is no other observation that needs be made to disprove the Big Bang.

With the Hubble telescope, 200 billion galaxies have allegedly been mapped and many have been analysed to find the consistent pattern of oscillating galactic ion jets and super-massive black holes. The order of galactic creation does not actually need an ejaculatory explosion theory to explain what Mother Nature does through rhythm oscillations between creation and destruction (ion jets and black holes).

CERN (The European Council for Nuclear Research) is a science machine/weapon whose adherents claim that it can produce the same atmosphere that existed during the Big Bang. And if that hubris is not enough, CERN also claims that it can go 'beyond' matter and has the capacity to open a doorway between dimensions. CERN scientists even plan to try to push antimatter through this doorway to 'see what happens.' It has become the church and altar of secular humanist scientists, who are spiritless materialists that worship the god-entity they call the Big Bang. Physical, tangible matter is the only god for these scientists who all believe without question that they are annihilated and become nothing more than dust, after death.

However, the only thing consistent about CERN is its almost child-like insistence that it is equal to or greater than the creator, whoever or whatever that may be. The clear problem with this argument is that physicists do not believe in a 'moment of creation' and the reason that CERN uses references to 'the divine' is to create complex brain patterns called narrative networks.

Reference to spiritual concepts creates higher order narrative networks within our brains. These narrative networks of the divine are used to brainwash and indoctrinate 'believers' in all-matters of science. Please be under no illusions, the science propounded by the CERN scientists is just another form of religion requiring belief and faith. So far, none of its experiments, or rather 'scientific' pronouncements, have been replicated by any other 'collider' scientists in the world.

Here are some more considerations that I would suggest that physicists consider before asking taxpayers to pay another penny for CERN's fraudulent pseudo-science. I may not be a formally trained scientist, and neither am I a meteorologist, but in reality I only need to look out of the window to tell whether or not it is raining, or indeed whether or not it appears *likely* to rain..

- *Static universe models fit observational data better than expanding universe models.*

The Big Bang can match each of the critical observations, but only with adjustable parameters, one of which (the cosmic deceleration parameter) requires mutually exclusive values to match different tests. Without ad hoc theorising, this point alone falsifies the Big Bang. (adapted from *Meta Research Bulletin 11*, 13-6-2002)

- *The microwave 'background' makes more sense as the limiting temperature of space heated by starlight, than as the remnant of a fireball.*

Intergalactic matter is like a 'fog,' and would therefore provide a simpler explanation for the microwave radiation, including its black body-shaped spectrum. None of the predictions of the background temperature based on the Big Bang were close enough to qualify as successes and the Big Bang offers no explanation for the kind of intensity variations with wavelength seen in radio galaxies.

- *Element abundance predictions using the Big Bang require too many adjustable parameters to make them work.*

The best the Big Bang can claim is consistency with observations using the various ad hoc models to explain the data for each light element.

- *The universe has too much large-scale structure (interspersed 'walls' and voids) to form in a time as short as 10-20 billion years.*

To get around this problem, scientists propose that galaxy speeds were initially much higher and have slowed due to some sort of 'viscosity' of space.

- *The average luminosity of quasars must decrease with time in just the right way so that their average apparent brightness is the same as all redshifts, which is exceedingly unlikely.*

It is not as if the Big Bang provides a reason as to why quasars should evolve in just this exact, magical way. But that is required to explain the observations using the Big Bang interpretation of the redshift of quasars as a measure of cosmological distance.

- *The ages of globular clusters appear older than the universe.*

Astronomers have studied this for the past decade and more but resist the 'observational error' explanation because that would almost certainly push the Hubble age older, which creates several new problems for the Big Bang theory.

- *The local streaming motions of galaxies are too high for a finite universe that is supposed to be uniform.*

The Falsification of Science

The only Big Bang alternative to the apparent result of large-scale streaming of galaxies is that the microwave radiation is in motion relative to us. Either way, this results in a death sentence for the Big Bang theory.

- *Invisible Dark Matter of an unknown nature must be the dominant ingredient of the entire universe.*

The Big Bang requires sprinkling galaxies, clusters, superclusters, and the universe with ever-increasing amounts of this invisible, as-yet-undetected, theoretical 'Dark Matter' to maintain the theory's viability. Overall, over 90% of the universe must be made of something never yet detected in any way.

- *The most distant galaxies in the Hubble Deep Field show insufficient evidence of evolution, with some of them having higher redshifts than the highest-redshift quasars.*

The Big Bang theory requires that stars, quasars, and galaxies in the early universe are 'primitive,' meaning mostly metal-free, because it requires many generations of supernovae to build up metal content in stars. But the latest evidence suggests lots of metal in the earliest quasars and galaxies.

- *If the open universe we see today is extrapolated back near the beginning, the ratio of the actual density of matter in the universe to the critical density must differ from unity.*

Inflation failed to achieve its goal when many observations went against it so to maintain consistency and salvage inflation, the Big Bang theory has now been modified to introduce two new 'adjustable' parameters... the cosmological constant, and 'Dark Energy.'

- *The Big Bang predicts that equal amounts of matter and antimatter were created in the initial explosion.*

Experiments are being conducted, searching for evidence of this asymmetry, so far with zero success. Another example of desperately trying to find evidence for an insubstantial theory rather than starting from a standpoint of zero knowledge and working with known 'facts.'

- *The Big Bang theory violates the first law of thermodynamics, that energy cannot be either created or destroyed, by requiring that new space filled with 'zero-point energy' be continually created between the galaxies.*

- *Elliptical galaxies supposedly bulge along the axis of the most recent galaxy merger.*

However, the angular velocities of stars at different distances from the centre are all different, making an elliptical shape formed in that way, unstable. Such velocities would shear the elliptical shape until it was smoothed into a circular disc. Where are all the galaxies in the process of being sheared?

- *Redshifts are quantised for both galaxies and quasars. So are other properties of galaxies.*

This should not happen under Big Bang premises.

- *Measurements of the two-point correlation function for optically selected galaxies follow an almost perfect power law over nearly three orders of magnitude in separation.*

However, this result disagrees with all the Big Bang's various modifications.

- *The absorption lines of damped Lyman-alpha systems are seen in quasars.*

The relative abundances have surprising uniformity, completely unexplained by the Big Bang theory.

- *The fundamental question of why it is that at early cosmological times, bound aggregates of 100,000 stars (globular clusters) were able to form, remains unsolved in the Big Bang.*

The Falsification of Science

It is no mystery in infinite universe models such as the 'Electric Universe' theory.

- *Blue galaxy counts show an excess of faint blue galaxies by a factor of 10 at magnitude 28.*

This implies that the volume of space is larger than in the Big Bang, whereas it should get smaller the further one regresses in time.

These are just a few of the anomalies that prove that the Big Bang theory is a complete nonsense. It is surely clear to anyone – especially physicists one would think – that the entire premise is completely untenable. And yet they continue to propound what has become a laughably flimsy, totally unsubstantiated theory whilst simultaneously, desperately searching for a non-existent 'proof' and are perpetually plugging the 'logic gap' with fantastic inventions such as 'Dark Matter' and 'Dark Energy' in order to paper over the abundant cracks.

"This time it is the creed of scientism and the pseudo-religion of the big bang that stands in the way of progress. In truth, we have no real understanding of matter, light, magnetism, gravity, quantum behaviour, subatomic particles, stars, galaxies, or... need I go on? Stories of creation and what did and did not happen in the universe over the past 13.7 billion years are crackpot schemes by celebrities of less talent than many in the Natural Philosophy Alliance but of greater prestige. We have too much information and too little real understanding. Many of the things we are taught 'just ain't so.' This realisation frees the mind to view everything afresh. It is the spark required to rekindle enthusiasm for science and drive progress. There is so much to be discovered." Wal Thornhill, September 2011

The Big Bang is just a myth, a hoax, a scam, call it what you will. The overall general 'scientific' argument in support of its authenticity as a fact, is akin to an adult saying *"because I said so!"* to a four-year old child who questions one of their statements.

John Hamer

Dark Matter and Dark Energy

"Dark Matter is a hypothetical form of matter that is thought to account for approximately 85% of the matter in the universe, and about a quarter of its total energy density. The majority of Dark Matter is thought to be non-baryonic in nature, possibly being composed of some as yet undiscovered subatomic particles. Its presence is implied in a variety of astrophysical observations, including gravitational effects that cannot be explained unless more matter is present than can be seen." Wikipedia™

I think the above quote from the establishment-controlled Wikipedia, tells us all we need to know about the deceptions taking place. But let us analyse it in more detail...

"*...a hypothetical form of matter*" 'Hypothetical' indeed!

"*...thought to account for...*" 'Thought to?' It gets better.

"*...85% of the matter in the Universe*" 85%?!!! Is this an admission that they only 'know' what 15% of the universe consists of, perchance?

"*...thought to be non-baryonic in nature...*" There is an awful lot of 'thinking' going on here and not much 'knowing.'

"*...possibly being composed of some as-yet undiscovered subatomic particles*" 'Possibly?' Are they being serious?! I am afraid they are. And as for 'as yet undiscovered subatomic particles'? No further comment needed!!

"*...its presence is implied*" Again, implication is not fact.

"*...effects that cannot be explained unless more matter is present than can be seen.*"

Words simply just fail me, really.

Similarly, here is Wikipedia's™ definition of 'Dark Energy'...

"*In physical cosmology and astronomy, Dark Energy is an unknown form of energy which is hypothesized to permeate all of space, tending to accelerate the expansion of the universe. Dark Energy is the most accepted*

The Falsification of Science

hypothesis to explain the observations since the 1990s indicating that the universe is expanding at an accelerating rate."

Again, I would ask you to note the implicit assumptions, disguised as fact, hidden within the above words. There is no doubt in my mind that the scientific community has an agenda which the pushing of the Big Bang theory serves well, and so to make it fit they must come up with all kinds of crazy hypotheses and 'sub-theories' that 'patch it up.' Instead of wasting all that time, money, and energy on attempting to make a totally inadequate theory plausible against all odds, would it not be much, much easier to actually spend time searching for the 'real' solution to the question of how the universe began? Yes, it would, but of course that does not suit their purposes, and nor would it enhance their agenda in any way.

Dark Matter and Dark Energy MUST therefore be contrived to exist in order to plug the gaping holes in the Big Bang theory, according to mainstream science. Does this not run somewhat contrary to the so-called 'scientific principle' of letting the known facts dictate the theory? It seems to me that what tends to happen all too often in these dark days for 'science,' is that the theory is plucked from thin air in an attempt to fulfil a particular agenda and then thousands upon thousands of scientists spend decades trying vainly to come up with the science to 'prove' the hypothesis. And in doing so, provide their insubstantial theories with so many allegedly 'supporting' theories, all of which tend to be even more preposterous than the initial premise itself.

But what does NASA itself, that veritable bastion of truth and genuine science, have to say about Dark Matter and Dark Energy?

Direct from the NASA website, nasa.gov...

John Hamer

"This NASA / Hubble telescope image (above) shows the distribution of Dark Matter... Dark Matter is an invisible form of matter. It accounts for 85% of the mass of the universe, but scientists still do not know exactly what it is."

No, please do not laugh. This is what they are actually telling us. So, the picture above, according to NASA, *shows* the distribution of Dark Matter, yet it is '*invisible*,' eh? But surely if it has mass, then it must be visible according to the laws of physics? Or is this just yet another case of adjusting fixed, immovable laws to suit their circumstances?

It then goes on to tell us that, *"Dark Matter and Dark Energy are elusive, invisible phenomena... There are some things in the universe that humans are not meant to understand."*

All patronising waffle of course, which I personally read as '*Don't worry about it you non-scientifically-trained ignoramuses. It's a mystery alright, but just leave the science to we professionals, OK?*'

In truth, the universe is actually made-up of electromagnetic particles, otherwise known as 'plasma.' But NASA and mainstream science does not 'like' plasma or anything that suggests an electric universe because this does not fit the pseudo-scientific model, as decreed by their puppet-masters. We live in a constant electrical field surrounded by electricity in all its forms and indeed the electromagnetic force is 1,000 trillion, trillion, trillion times stronger than the (alleged) gravitational force which according to mainstream science is the foundation stone of the entire universe.

So let us in future, refer to Dark Matter and Dark Energy as what they actually are ... plasma.

The Electric Universe

There are now two major theories of the origin of the universe in existence. One is the 'official' Big Bang fairy tale as outlined above and the other, very few have ever heard of, or can even speculate as to its existence. As always tends to be the case, the unofficial one is virtually an 'underground' movement, strongly refuted by the mainstream, consisting mainly of books, blogs, DVDs, and active discussion groups on the internet. However, both are advocated by people with science, cosmology, physics, electrical engineering, or electronics degrees, as well as there being Nobel prize winners on both sides of the debate. The unofficial position is certainly not mysticism or shrouded in New Age spiritualism but based on solid experimental data and proven observation, yet one is considered legitimate by the mainstream and the other to be advocated by total 'crazies,' or 'conspiracy theorists.' The 'official' story based on the Big Bang theory, proposes the existence of Black Holes, one of which allegedly has the mass of a billion suns as well as advocating a warped, eleven-dimensional space, and physically impossible Neutron stars. As previously outlined, the 'official' theory is totally replete with such 'logic patches' in a vain attempt to resolve the otherwise insoluble mathematical equations.

In addition, the official fairy story proposes that 80-90% of the matter (depending on which source you believe) in the universe is made up of Dark Energy, as well as Dark Matter that emits no light and is invisible. But of course, no one knows the composition of Dark Matter or Dark Energy and there are no ideas even forthcoming on possible candidates for its composition. Furthermore, there is no experimental evidence for the proposed, yet elusive graviton (gravity particle) or the Higgs-Boson particle, despite all the multi-billion dollar efforts of the scientists who operate the Large Hadron Collider, in Geneva, Switzerland at the CERN research establishment. It is still impossible to unify quantum mechanics with gravity. Gravity waves have still not been detected after two years and building two 4-kilometre machines underground, one in Louisiana, USA and the other in Washington, to detect them (the LIGO project). Indeed, Albert Einstein died, still trying to reconcile gravity with quantum theories.

Those supporting the unofficial position, as always, are constantly denied a mainstream platform and are being denied publication in peer-reviewed scientific journals, which is the very lifeblood of a serious scientific career. The unofficial story states that there are no Black Holes, there was never a Big Bang and in short that the Universe is electric, that it is based on electromagnetic forces that are 10^{39} times stronger than the forces of gravity. It maintains that gravity, an extremely weak force, cannot possibly explain the formation of galaxies, nor the planets. This all makes perfect sense to me.

The official version of events states that the Universe is 13.7 billion years old but it has been demonstrated that the Universe is riddled with vast galaxies that could not reasonably have formed naturally, under current scientific laws in less than 200 billion years whereas the Electric Universe theory can explain scientifically, many of the gamma ray and X-ray phenomena as viewed from the Hubble telescope, floating in space above the surface of the Earth with a much-improved resolution of images of stars and galaxies.

The Electric Universe also offers a totally different perspective on the energy of the Sun and its sources. This theory also questions the redshift of the Doppler effect, as codified in 'Hubble's Law,' and which is used to calculate the distance of galaxies from our own solar system and thus the alleged age of the Universe. It also questions the fact that the galaxies are stated to be accelerating away from us and that this rate is also accelerating, the result of which would be that in the long term, the sky will be totally dark at night as the stars will all have moved away, and the only visible galaxy will be Andromeda. For this last discovery, a US-born Australian citizen Brian Schmidt and his colleagues were awarded the Nobel Prize in Physics in 2011, once again putting an official stamp on the unknowable and elusive Dark Matter.

The Electric Universe theory, quite rightly, maintains that the present, universally accepted Big Bang theory of the universe was derived from mammoth leaps of abstract thinking using mathematics, without any verification from direct experimentation or observation, but accepted simply because the equations could be solved mathematically. This method is known as 'deductive,' in that it deduces the universe's make-up, ignoring observational data and experimentation. I would conclude from this that the sciences of astronomy and astrophysics have been totally compromised, ending in a total mathematical cul-de-sac.

Advocates of the Electric Universe theory do not question the existence of gravity but simply believe that it plays a much lesser, indeed very minor role behind the much greater force of electromagnetism. Newton's law of gravity explains the motions of falling bodies on Earth and placing satellites into orbit, however, inside the nucleus of an atom, Newton's laws do not apply. Newton's laws also fail to explain the way that galaxies rotate because it is outside its domain of validity. These are simply more examples of so-called immutable scientific 'laws' being selective.

The official story is ubiquitous. Even the British science fiction writer Terry Pratchett parodied the Big Bang and the Bible in one of his books, stating that, *"In the beginning there was nothing – which exploded."* Strangely, the Big Bang theory has recently been given a massive boost by the Catholic Church and religion in general. This is because the Big Bang fits neatly into a schema where God created the Universe in a Big Bang, thereby proving that God must exist and neatly circumventing whatever may have happened before the Big Bang, what precipitated this unexplainable event and who or what created God in the first place. The Big Bang indeed, gives the idea of a creator a certain 'scientific' validity and in addition, it also provides an apocalyptic view of the universe, doomed to decay in a final annihilation, a universe hostile to human endeavour, which is a very apt vision for the Catholic Church.

The idea of the Big Bang was first proposed in 1927 by a Belgian Catholic priest, Abbe Georges Lemaitre, who was also a scientist, lecturing at the Catholic University of Louvain. Lemaitre proposed that the Universe had expanded from an initial point, which he called the 'Primeval Atom.' He also asked, who else could have created something out of nothing but God? Fred Hoyle, a British scientist, who at first scoffed at the notion, eventually gave it its commonly accepted name, the 'Big Bang.' Another eminent plasma cosmologist, Hannes Alfven, said *"I was there when Abbe Georges Lemaitre first proposed this theory,"* he recalled. *"Lemaitre said in private that this theory was a way to reconcile science with St. Thomas Aquinas' theological dictum of creatio ex nihilo or creation out of nothing."*

The Vatican has unsurprisingly provided large sums of money to astrophysicists who adhere to this official story. The Vatican Observatory (Specola Vaticana) is an astronomical research and educational institution supported by the Holy See, so, obviously totally independent, and impartial, then?! Originally based in Rome, it now has its headquarters and laboratory at the summer residence of the Pope in Castel Gandolfo, Italy, and an observatory at the Mount Graham International Observatory in the United States. Dr Guy Consolmango is one of a team of twelve astronomers, all of them Jesuits, working for the Vatican. The group specialises in galaxy formation and inflationary universes, which is also a key element of the Big Bang theory.

So, it can be observed, that both mainstream science and religion now present a united front on the 'science' behind the Big Bang and its associated 'patch' theories. This is quite a turnaround from the historical position of more than a century ago when the two factions were at loggerheads over the science. I believe this to be the Jesuit influence – which incidentally I do not cover in this book, except in passing, but I would certainly advise anyone interested in this aspect, to do the research themselves.

The Falsification of Science

Paul Davies a well-known physicist and popular science writer has attended many conferences hosted by the Vatican and has written books, titled *'The Mind of God,' 'The Last Three Minutes,'* and *'God and the New Physics.'* He was awarded the Templeton Prize for Progress in Religion, in 1995 for showing 'extraordinary originality' in advancing humanity's understanding of God or spirituality. Previous winners included Alexander Solzhenitsyn, Mother Teresa, the evangelist Billy Graham and Charles Colson, infamous as the Watergate burglar, who on his release from prison, 'found God.'

Davies maintains that, *"science offers a surer path to God than religion,"* and succeeded in injecting mysticism and religion into science. The Pontifical Academy of Science, of which Stephen Hawking was a former member, kept the senior cardinals and the Pope up to date with the latest scientific developments. Even Hawking bought into mainstream religion for a while, but eventually changed his mind.

Marxism also rears its ugly head on the topic. Why would Marxism be interested in the origin of the Universe you may ask? Well, as it is a very apposite topic and Marxism increasingly takes centre stage in many aspects of modern life, it is perhaps unsurprising. Cultural Marxism is after all, becoming all-pervasive in today's society, with its extremist, gay and general minority rights bordering on supremacy, its relentless pushing of the transgender agenda, lobbying for the legalisation of paedophilia, mass immigration into Europe resulting in white genocide, the increasingly compulsory sacrifice of individual rights, freedoms and freedom of speech and last but not least the onslaught of 'political correctness' to allegedly serve the 'greater good' and the general, extreme so-called 'liberalisation' of the entire world.

Marx and Engels' view was that the universe was infinite, that it has always existed, that order evolves out of chaos (Ordo ab Chao – the illuminist creed) and that it is always evolving, will continue to evolve with no limits imposed, and that it has no beginning and will have no end. They also believed that as life began on Earth, it would also have begun on other planets by a similar mechanism. Of course, they were writing at the very beginning of the age of mass scientific discoveries, when it was still thought that the universe was wrapped up in an aether, through which light travelled. Dividing up the tasks, in true socialist fashion, Engels took-on the study of science and nature to explain its relation to philosophy and dialectics, whilst Marx concentrated on the economic questions and societal implications. But they were never in disagreement over either project, fully collaborating to the end. Engels was never able to finish his major work, *'The Dialectics of Nature,'* as he was too busy editing Marx's *'Das Kapital'* after his colleague's untimely death.

The adherents of the Electric Universe theory are almost totally marginalised by the adherents of 'Big Bang' dogma. The two factions have no official dialogue between them, do not attend each other's respective conferences and do not publish in the same journals. How could astrophysics reach this state of affairs? It is extremely sad and shameful that two contrasting points of view cannot be debated in an open, friendly, and respectful manner. The exchanges on the internet are vitriolic and poisonous, with one recent post criticised as, *"A blind man in a dark room, looking for a black hole,"* followed by a totally black picture. Surely a truly scientific approach would involve sensible adult discussion, properly funded research into alternatives, and friendly debate and not the sheer intransigence exhibited by the followers of the 'Big Bang cult.' This fact alone, smacks of a sinister agenda at play.

One group of scientists are in effect excommunicated, silenced, ignored, maligned, and ostracised, and receive very little, if any, funding whilst a tiny few, the 'believers of the faith,' have access to billions of dollars for the purposes of their research. However, it is perhaps no surprise that it is the deeply entrenched interests of officially sanctioned scientists, which controls the peer-reviewed publications, and that has the strongest influence in distributing the vast research funds. There are enormous vested interests supporting the present, incumbent Big Bang theory, indeed, many scientific academics have staked their entire careers on it, including the famous figures, Stephen Hawking, George Gamow, Brian Greene, Paul Davies, Alan Guth, Sir Martin Rees and John Wheeler who first mooted the existence of Black Holes, besides thousands of other lesser known mathematicians. It is much easier to toy with complex mathematical equations, which cannot be understood by 99.99% of the population than to make direct observations of the universe and to understand the complex relationships of plasma, electric charges, and magnetism – as Nikola Tesla correctly pointed-out.

In his 1991 book *'The Big Bang Never Happened; A startling refutation of the dominant theory of the Origin of the Universe,'* Eric J. Lerner was possibly the first to alert the general public to the possibility that the Big Bang was not a 'set-in-stone' scientific premise, but merely an insubstantial theory. He linked the current, appalling state of cosmology to the dwindling research funds for science in the USA because of the exorbitant cost of the Vietnam War and the downturn of the Apollo projects run by NASA. Theoretical cosmologists need no funding at all, he postulated, just pen and paper, manipulating equations and running programmes all day on desk-top computers. He regarded it all as research 'on the cheap' and this enormous growth of the theoretical side created an imbalance with the practical collection of observational data and Lerner also believed that science and society inevitably influenced each other through events occurring in society in general. He wrote, *"the faltering universe of the Big Bang became a metaphor for the faltering economy – both equally inevitable processes, beyond the control of mere mortals."* More importantly, the collapse of the Soviet Union no longer made advances in space exploration necessary for propaganda purposes, to prove capitalism could achieve a consumer utopia for the masses better than communism ever would.

NASA has an annual budget of $15 billion and has stated openly that it simply will not fund any cosmological research that is adverse to the Big Bang, which for them has assumed an almost religious devotion. More than anything, this situation is a damning indictment of the peer review scientific publication system, which will automatically fail to promote new hypotheses as the entrenched editors and reviewers have invested their careers in the protection of existing establishment ideas. The situation is so drastic that a letter from cosmology dissidents was finally accepted in the *'New Scientist'* in the 22nd May 2004 edition, complaining bitterly about the bias and discrimination against non-establishment thinking. It was signed by noted astronomers, Herman Bondi, Thomas Gold and Eric J. Lerner as well as another 218 astronomers at various institutions around the world and 187 independent researchers. Here is an extract of the letter...

"... *in cosmology today, doubt and dissent are not tolerated, and young scientists learn to remain silent if they have something negative to say about the standard Big Bang model. Those who doubt the Big Bang fear that saying so will cost them their funding. Even observations are now interpreted through this biased filter, judged right or wrong depending on whether or not they support the Big Bang. So discordant data on redshift, lithium and helium abundances, and galaxy distribution, among other topics, are ignored or ridiculed. This reflects a growing dogmatic mindset that is alien to the spirit of free scientific inquiry. Today, virtually all financial and experimental resources in cosmology are devoted to Big Bang studies. Funding comes from only a few sources, and all the peer-review committees that control them are dominated by supporters of the Big Bang. As a result, the dominance of the Big Bang within the field has become self-sustaining, irrespective of the scientific validity of the theory.*"

How very true.

The three most prominent pioneers of plasma cosmology are Kristian Birkeland, Irving Langmuir, and Hannes Alfven.

The Electric Universe theory states that electrical forces must be considered when studying the cosmos and it also states that 99% of the visible universe is charged due to the loss of electrons from atoms, leaving positively charged ions and negatively charged electrons. A molecular cloud of very cold gas and dust can be ionised by nearby radiating stars or cosmic rays, with the resulting ions and electrons assuming organised plasma characteristics. This then leaves a very tiny amount of material scattered meagrely throughout virtually empty space. In outer space, on average there is perhaps one particle per cubic centimetre, compared with 10^{13} per cc in Earth's atmosphere. This mixture of neutral and charged matter is called plasma, and it is suffused with electromagnetic fields and the proportion of ions is quantified by the degree of ionisation. The degree of ionisation of a plasma can vary from less than 0.01% up to 100%, but plasma behaviour will occur across this entire range due to the presence of the charged particles and the charge separation, so typical of plasma behaviour throughout the entirety of interstellar space. The first person to use the term 'plasma' to describe ionised gases, was the aforementioned Irving Langmuir.

This is a very different situation to that which we find on Earth, where according to conventional science, virtually 100% of matter exists only in its three major states of solid, liquid or gas. Earth is a cool, stable, and almost neutral planetary environment but this is a rarity in the universe. Both Newton and Einstein lived in the 0.001% of the universe that is not plasma, beneath the atmosphere of a rocky planet and so they could not be expected to understand plasma in 1687 and 1905, respectively. Plasma is a rare entity on Earth, however, flames, neon lights, electric arc welding and lightning are all examples of a plasma. Whenever electrical charges move, they generate magnetic fields and without moving electrical charges, magnetic fields cannot exist. Because of the ever-moving electrical current, present in space, a magnetic field is produced, which astrophysicists refer to as a gas, instead of its correct name – plasma but this can be extremely confusing as a plasma never acts as a gas and does not obey Boyle's laws, which specifically refer only to gases.

Magnetic fields around the Earth and the Sun are recognised by the scientific establishment but it totally denies their presence in plasma or electrical currents, stating that electrons will move at the speed of light to short-circuit any electric differential. If charged particles are moving, however, they are accompanied by magnetic fields and this changes the magnetic configuration of those fields. Changes in a magnetic field in turn, create electric fields and thereby affect currents themselves, so fields that start with moving particles represent very complex interactions, feedback-loops, and complex mathematics.

In the Big Bang theory, moving electrical charges and their effects causing magnetism, have been totally ignored. Changes in the direction of the Earth's magnetic field, for instance, cannot be explained by the official Big Bang theory. More than two hundred polar reverses have allegedly taken place in the last 65 million years here on Earth and at least four have occurred in the last four million years according to the mainstream version of 'history.' It is said that about 700,000 years ago the north magnetic pole was located in the Antarctic and vice versa and that we are currently in the process of a weakening of the Earth's magnetic field, which will eventually end in a new reversal. But also bear in mind that in fact the 'pole reversal' theory is not consistent with the flat Earth hypothesis either, so maybe that is all just more scientific propaganda? (this will be covered in chapter 5)

The Big Bang theory is based on two huge assumptions...

1. That gravity alone determines the structure and movement of stars and galaxies, and

2. The redshift of objects in space are a true reflection of their distance and that these objects are receding.

Dark Matter and Dark Energy had to be invented and added to the 'official theory,' as the movement of stars, galaxies and super-galaxies cannot be explained by the force of gravity alone, as it is far too weak. Dark Energy and Dark Matter are undetectable and invisible and are therefore totally unknowable, contrary to mainstream science's assertions.

The Electric Universe theory is partly based on the pioneering work of Norwegian scientist, Kristian Birkeland. In 1902-10, he was the first person to propose that electric currents come from the Sun, flow into the Earth's upper atmosphere and cause the Auroras, (Borealis and Australis, the northern and southern lights.) These currents pass through and excite the plasma high above the Earth's atmosphere to such a degree that it becomes visible, and glows. Birkeland constructed a magnetised metal sphere, suspended in a vacuum, and generated electrical discharges to the sphere, which he called a Terrella. Using this model, he was able to reproduce aurora-type displays, analogues of planetary rings, sunspots and other effects seen in the cosmos. Additionally, he risked his own life and those of his assistants, by measuring the electric field under auroras in the bleak 24-hour winter darkness and freezing winds of the Arctic snowfields.

The Falsification of Science

The electric field or the Coulomb force between an electron and a proton is 10^{39} times more powerful than the gravitational attraction between the two. (10^{39} is a 1 followed by 39 zeroes) In addition, gravity reduces inversely to the square of the distance, whereas electricity decreases only linearly with the distance between the forces. Therefore, Birkeland currents are far more effective than gravity for organising very thin gases and dust into stars and galaxies.

In order to demonstrate the huge distances involved, imagine the Sun and its nearest star Proxima Centauri, allegedly more than 4.2 light years away (40×10^{12} kms, or 40 quadrillion kilometres) being the size of just two dust particles (0.25mm). On this same scale, they would be over 6.4 km apart, so it is very easy to understand that the gravitational attraction is absolutely miniscule and that they could never randomly bump into each other.

Space probes, launched by NASA and other space agencies have allegedly discovered that space contains atoms, dust, ions and electrons and so although the density of matter in space is very low, it is not zero, but in space, gravity only becomes significant in those places where the electromagnetic forces are shielded or neutralised. As early as 1937, Hannes Alfven, proposed that our galaxy, the Milky Way, contains a large spiral, magnetic field and that charged particles move in spiral orbits within it, owing to forces exerted within the field. Plasma physicists can trace the evolution of observed galactic forms from basic electromagnetic principles, that can be repeated by experiments on Earth in the laboratory. For example, Alfven's student, Anthony Peratt, demonstrated in the laboratory that electric forces can organise spiral galaxies, as well as explain why they form in strings and why they rotate, without the necessity of resorting to the 'logic patch' of a massive Black Hole in the centre of a galaxy. Thus, galaxies are not merely groups of remote-from-each-other stars but consist primarily of plasma.

Alfven, the Swedish Nobel prize winner, who died in 1995, aged 86 years, was the founder of the modern field of plasma physics, the study of electrical discharges in low pressure gases. It is, however, somewhat ironic that he received the Nobel Prize in Physics in 1970 for a previous, totally incorrect theory, which he attempted to repudiate, during his acceptance speech. But nevertheless, now, astrophysicists use this 'wrong theory' on magnetism to justify their work, totally ignoring Alfven's subsequent repudiation. Alfven firmly believed that, *"astrophysics should be the extrapolation of laboratory physics, that we must begin from the present universe and work our way backwards to progressively more remote and uncertain epochs."*

Thus, the scientific method must be grounded in observation in the laboratory, from space probes and the Hubble telescope, leading to sustainable theories derived from observation, and not beginning with random, abstract theories and pure mathematics, designed to perpetuate an agenda and in the process inventing weird and wildly improbable sub-theories to account for anomalies.

Indeed, Alfven was quite a rebel. He was a politically active scientist who challenged the establishment, a very rare breed indeed. He locked horns with the Swedish government in the 1960s when he thwarted their plans to build nuclear reactors, pointing out their infeasibility as well as huge, technical errors in the specifications. He was subsequently threatened with loss of research funding (that old story again), and he therefore had to partially move his research to the US to survive this unjustifiable assault on his work. An interesting aside is that he wrote a political/scientific satire called *'The Great Computer'* under a pseudonym, which related a takeover of the planet by computers, now a recurring theme in contemporary science-fiction. Alfven used this novel as a vehicle not only to parody the growing obsession of governments and industry with the then fledgling computer industry, but also to pillory the Swedish establishment. He made it perfectly clear that it was the greed of the corporations, the short-sightedness of government bureaucrats and the hunger for power of the politicians that led to the future, he wryly outlined, as a utopia for automation. Of course, we still face similar problems today, on a global scale.

Redshift

Hydrogen is a very common element in the universe. Examination of the spectrum of ordinary light passing through hydrogen, produces a signature spectrum specific only to hydrogen with specific features at regular intervals. If this same pattern of intervals is observed in a spectrum from a distant source, say a star or a galaxy, but occurring at shifted wavelengths, it may also be identified as hydrogen, but it is said to be 'redshifted.' The dimmer the galaxy, the more its light is shifted towards the 'red end' of the spectrum. Redshift (and its opposite, blueshift) may be characterised by the relative difference between the observed and emitted wavelengths (or frequency) of an object. In astronomy, it is customary to refer to this change using a dimensionless quantity, as a redshift of 'z' and because of Edwin Hubble's work in 1929, the inference was drawn that a redshift implies distance, that there is a linear relationship and that the distance of the galaxy could therefore be accurately measured. He based the actual distance on Henrietta Leavitt's formulation of the period-luminosity relationship (for which she received no credit in her lifetime). Quite typically, no-one actually questioned the possibility that a galaxy may just simply be dimmer and relatively closer to Earth. The assumption that the star is moving away from us to explain the redshift was based on the well-known Doppler effect in sound waves, the pitch of a train whistle or a police car siren, increasing as it approaches and decreasing as it moves away from us. Thus, the 'fact' that redshift implies recessional velocity, as with many other theories and assumptions, became dogma in astronomy. This of course, is a major 'patch' to the Big Bang theory but is nevertheless totally discredited by the Electric Universe theory.

Edwin Hubble to his credit, pointed out that contrary to the establishment edict, redshift was probably not due to the Doppler effect. One of the main objections Hubble had to the Big Bang theory derived from his study of the brightness of stars. He maintained that if stars were receding at the rate indicated by their redshift, their brightness would appear to be diminished. Instead, he observed that there was no such diminishing of brightness. He recognised that the only real evidence directly supporting the Big Bang was the observed redshift, but sadly, Hubble's mental faculties began to deteriorate markedly in the early 1950s and he was therefore never able to prove the actual cause of redshift. Subsequent to his death at the relatively young age of 63 years in 1953, most scientists soon defaulted-to and accepted erroneously that Hubble's redshift *was* indeed caused by the Doppler effect.

Another influential figure in this drama, was Milton La Salle Humason, who began working at the Mount Wilson Observatory as a mule driver and janitor in the early 1900s during the time that Edwin Hubble began his research. Eventually Humason progressed to the position of secretary of the Mount Wilson and Palomar observatories and until his death in 1972, vehemently propounded the Big Bang theory over all other theories of 'creation.' This resulted in supporters of the Big Bang theory having complete freedom of access to the observational instruments controlled by Humason, whilst opponents of the Big Bang were denied all access.

Even Carl Sagan was candid enough to write in his book, *Cosmos*, in 1980 that, *"There is nevertheless a nagging suspicion among some astronomers that all may not be right with the deduction from the redshift of galaxies via the Doppler effect, that the universe is expanding. The astronomer Halton Arp has found enigmatic and disturbing cases where a galaxy and a quasar or a pair of galaxies, that are in apparent physical association have very different redshifts ... If Arp is right, the exotic mechanisms proposed to explain the energy source of distant quasars, supernova chain reactions, supermassive black holes, and the like – would prove unnecessary. Quasars need not then be very distant."*

Halton Arp indeed demonstrated that faint, highly red-shifted objects, such as quasars, are intrinsically faint because of their relatively young age and that it has nothing whatsoever to do with their distance from Earth. Quasars are 'born' episodically from the nucleus of active galaxies and they initially move very quickly along the spin-axis away from their parent. As they mature, they become brighter and slow down, as if gaining in mass. Finally, they evolve into companion galaxies.

The resulting decreasing quasar redshift occurs in discrete steps which points to a process whereby protons and electrons go through several small, quantised increases in mass as the electrical stress and power density within the quasar increases. The charge required comes via an electrical 'umbilical cord,' in the form of the parent galaxies' nuclear jet. So, they play an important role in the creation of new galaxies. Eventually, Arp came to realise that the standard, accepted model of astronomy is totally wrong, based on these observations.

He believed that redshift is entirely due to photons of light interacting or colliding with the electrons in the plasma of intergalactic space and thus losing energy. The more interactions they undergo, the more energy they lose and the lower their frequency therefore becomes. As the frequency reduces, the wavelength increases and thus the photons are redshifted, utilising the Mossbauer effect. Arp was one of those who were unjustly denied access to the Palomar telescope, simply for questioning the redshift assumptions, and was forced to move to Germany after being a staff astronomer for 29 years. He also experienced severe difficulties in having his papers published in American journals. This is yet another example of many, of the extreme bias exhibited by the establishment, against anyone or anything, not following their stated, unassailable decrees.

The Sun

According to the theories of Fred Hoyle and George Gamov, the energy of our Sun is produced by the continuous conversion of hydrogen into helium by a fusion reaction, deep in the interior and is eventually transported to the surface by radiation and convection, which is alleged to take 100,000 to 200,000 years. When all the hydrogen is converted to helium, after a total lifetime of 9½ billion years, then our Sun will collapse, and the Earth will be consumed in the resulting, massive explosion. Supposedly then, we have 'just' another 5 billion years left on Earth.

It is often stated that all the heavier metals we find on Earth were initially formed inside a star, much like our own Sun, meaning in effect that all the elements in our bodies were initially forged in a star. 'We are stardust' and all that jazz! But even after over eighty years of experimentation, no sustained controlled fusion reaction has ever been successfully performed in a physics laboratory. In addition, during the conversion of hydrogen to helium within stars, there should be an ejection of electron type neutrinos, but unfortunately for this at best, tenuous theory, the volume of neutrinos emitted by the Sun represents only one third of the predicted value, a fact which remains to this day, a total embarrassment to solar astronomers. The official model also offers no explanation for the existence of the corona, the plasmasphere or the solar wind.

Instead, in the Electric Universe model, the Sun is dominated by electrical and magnetic properties, implying that it possesses a massive, positive electrical charge. The Sun is a ball of plasma, charged positively by a gravitationally-induced flow of electrons towards the surface and it is the electrical repulsion which will prevent its collapse and not the conversion of hydrogen into helium, which is alleged to be the cause of it.

Again, in the Electric Universe model, stars may be described as giant spheres of slow-motion lightning and it is this simple hypothesis that best matches observational evidence. Fusion actually takes place on the surface of an electric star and not deep within its interior. The primary indicator for a star's behaviour is the current density at its surface and current flow from the solar wind can be observed on planets with magnetic fields which have polar 'cusps' or 'holes' that guide charged particles down and through the body, creating auroral displays in the upper atmosphere. Thus, the charged particles from the Sun are responsible for the auroras in both the Southern and Northern hemispheres.

"In the electric model, the Sun beneath the photosphere is simply a cool body, not at a temperature of a million degrees. The magnetic field of a Sunspot is due to a strong field aligned current, punching a hole through the photospheric plasma. This produces a Birkeland Current. The solar flares behave like lightning and are due to the electrification of the Sun's

atmosphere, analogous to electrification of storm clouds on Earth." Ralph Juergens.

Juergens was the first to describe the electric discharge of the Sun in the 1970s and one day, maybe, his genius and refusal to 'toe the party line,' will hopefully be fully recognised.

The key variable that determines the apparent size, brightness and colour of a star is electrical stress. The Sun is immersed in an extremely low-density plasma so those atoms which can be excited to emit visible light are those very close to the Sun in the corona, which is the glowing outer portion of the atmosphere of the sun. The corona is heated to two million degrees where oxygen atoms are ionised, lying above a much cooler surface. There are also ionised tornados, thousands of kilometres high and flecked with lightning, that provide the heat and visible light of the Sun.

This also means that the universe is connected through these voltage differentials and electrical phenomena. A charged body in plasma forms a bubble or sheath which provides a smooth transition between the differing electric potentials of the two plasma regions, due to the formation of the double layer, ie. the two distinct plasma regions. The Sun's plasma sheath is what the official model refers to as the heliospheric boundary. The current is carried throughout the solar system by a relatively low density of ionisation, where the planets orbit. The Sun is simply exhibiting the plasma glow discharge of a positively charged body in space in the same way as a glow discharge tube. The weak electrical field causes the acceleration of the solar wind in the inner solar system and a slow drift of electrons towards the Sun. Recently a satellite above the Earth's pole detected electrons streaming from the Earth towards the Sun as one would predict from the positively charged Sun. Hydrogen to helium fusion simply does not exist and cannot explain the reverse of the temperature gradient on the Sun.

All this means that the Sun could possibly continue forever, as long as it has a positive electric charge, created by galactic currents. However, there is downside to this. Conditions in a star can change very quickly and so the future of our Sun is not quite so certain as we may believe, and so we may not have even as long as 5 billion years left. We cannot possibly know whether or when the Birkeland current powering our Sun will experience a surge or alternatively a complete blackout.

The Planets

Similarly, all the planets in the solar system are charged bodies, including the moons, comets, and asteroids. Each planet is surrounded by its own plasma, the plasmasphere and a double layer separates the plasmasphere from the solar plasma. The planet Venus and Saturn's moon, Titan have little or no magnetic field but do have a large plasma sphere, thus a magnetosphere is not interchangeable with a plasmasphere. All four of the large moons of Jupiter lie within its plasmasphere and they are therefore electrically connected to the planet. Io, the innermost of these four moons, is presently experiencing electrical discharges from Jupiter. The famous volcanoes on Io cannot be volcanoes as we commonly know them, as they 'wander around' over distances of many miles, but electric arc discharges have a tendency to wander too, so it makes sense that these so-called 'volcanoes' should be regarded as an electric arc that offers astronomers the opportunity of studying electric plasma discharges. In addition, a 2008 press release from the Jet Propulsion Laboratory announced that Jupiter's rings are electrically charged too. The ions flow in an electric circuit to and from Jupiter to Io. When NASA launched the probe, *New Horizons* on a mission to study Pluto and Charon, the 'plumes' of Tvashtar, the gigantic 'volcano' on Io, were found to be filamentary in structure, indicating strongly that they are actually corona arc discharges from the electric 'hot spots' linking Io with Jupiter. Moving through Jupiter's intense magnetic field creates strong charge separation (voltage differential) and a resulting electrical current in a circuit of some 2 trillion watts of power flowing between Io and Jupiter's polar areas.

On another note, Jupiter has at least sixty-three moons and five of the smaller moons rotate in the opposite direction to Jupiter. Clearly, this indicates strongly that the moons could not have formed from the same 'accretion disc' as proposed by the official theory, as their angular momentums would all be in the same direction. A similar situation applies with Neptune and its moon, Triton.

Saturn actually emits more energy than it receives; 2.3 times more in fact, so it is obviously being powered by another, unknown source. There is good evidence that Saturn once existed as an independent body from the Sun and as such, it would have received even more energy in the recent past, its power source having since been usurped by the Sun. The plasmasphere of Saturn is an electrical environment, causing everything from dark-mode plasma discharges, to gigantic lightning bolts that flash across the ring plane. When the Cassini-Huygens spacecraft got close enough to finally begin observing Saturn, lightning of immense power, up to a million times more powerful than that on Earth, was discovered. Planets with magnetic fields can capture ionised particles to form a giant electrified magnetosphere. Enceladus, a small moon that orbits within Saturn's ring plane, actually causes Saturn's magnetosphere to bend. The effect is due to a flow of electrical charge that occurs when particles from Enceladus interact with the magnetosphere of Saturn. Thus, a demonstrable electrical effect is occurring between Saturn and Enceladus.

Cosmic Microwave Background Radiation (CMB) is often used as evidence to support the untenable Big Bang theory, even though this has absolutely nothing to do with the Big Bang. The CMB is just simply the temperature of the observable universe.

In fact, the Electric Universe fits very neatly with the basic tenets of the Marxist philosophy of dialectical materialism. And that is that matter is always in motion, always changing, according to its own specific laws. Matter cannot be created from nothing as the Big Bang would have us all believe, but it is always undergoing transformation and evolving. The Big Bang theory never did fit into this general philosophy and many Marxists had expressed their concerns about the Big Bang theory, even before the Electric Universe was mooted. Not that I am suggesting for one moment that dialectic materialism or indeed Marxism is an admirable philosophy in any other respect, very far from it indeed, it is merely an observation.

To those readers who are new to all this possibly confusing science, this information and re-interpretation of data may seem extraordinary and at times bizarre, but again, as always, I would urge you to undertake your own research on this topic and even if you are still not convinced afterwards, at the very least please remain sceptical of the proponents of the Big Bang theory, as the full facts are not being disclosed. There is obviously much more information available, especially using that most ubiquitous of tools, the Internet.

There are plenty of detailed explanations to be found online, as to why the official stories of Red Giants, Neutron stars, Supernovae, White Dwarfs, asteroids, and in particular the comets and the Hertzsprung-Russell diagram, which classifies stars according to temperature and cosmic rays, are completely wrong, whereas the Electric Universe theory can explain many aspects that defy explanation by the official, Big Bang theory and overall is a much more plausible hypothesis by comparison.

Today's astronomers and astrophysicists are similar to the Catholic cardinals and priests of the Middle Ages, who refused to look through Galileo's telescope. Their dishonest distortions and cavalier dismissals of the problems surrounding the theory of the Big Bang theory are facilitating the destruction of what little credibility still remains in mainstream science. There is, indeed, a veritable torrent of information and data, which utterly destroys the Big Bang hypothesis along with Black Holes, Dark Energy, Dark Matter, the Hubble constant, the age of the universe, accelerating galaxies and much, much more. There needs to be a paradigm shift in cosmology, on the scale of Galileo's epic struggle with the Catholic church, and once this happens, if it is ever allowed to, then many prominent astrophysicists and mathematicians will be thoroughly discredited, and their reputations will be in tatters.

Gravity

In 1687, the British Freemasonic Occultist and Magus, Sir Isaac Newton, published his major work *'Principia Mathematica,'* which hypothesises gravitation, and included all the efforts of Copernicus, Brahe, Kepler, and Galileo before him.

By the late 1800s, others had discovered many errors in Newton's theory, the most well-known of which are the errors with the movement of the planet Mercury. And by 1915, Einstein believed he had fixed the errors in Newton's work, with his Special Theory of Relativity. He theorised that gravity is what occurs when space and time is curved or warped around a mass, such as a star or a planet. Thus, a star or planet would cause a dip in space so that any other object that ventured too close would tend to 'fall into' the dip. Einstein basically explained how gravity is more than just a force, it is a curvature in the space-time continuum.

Both of Einstein's Theories of Relativity are still taught in colleges and universities today, as though they were totally factual and 'settled science.' They purport that in our own Solar System, not only does the Sun exert gravity on all the planets, keeping them in their orbits, but each planet exerts a force of gravity on the Sun, as well as all the other planets, too, all to varying degrees based on the mass and distance between the bodies. And it goes beyond just our Solar System, as actually, every object that has mass in the Universe attracts every other object that has mass – again, all to varying degrees based on mass and distance.

But this 'theory' of gravity is incomplete, the mathematics is incredibly complex, and no-one on Earth truly understands it.

"Gravity still remains one of the biggest mysteries of physics and the biggest obstacle to a universal theory that describes the functions of every interaction in the universe accurately. If we could fully understand the mechanics behind it, new opportunities in aeronautics and other fields would appear." 'Universe Today'

So in that case, why not accept that gravity is not what it is always portrayed to be and search for another, more logical explanation? That will never happen unfortunately, because science's vested interests wish to keep the same false paradigm in place. To change things in such a radical way and accept that gravity does not exist would mean a radical new rethink of the entire framework of physics, derailing far too many lucrative 'gravy trains' in the process.

Schools and NASA teach our impressionable children and older students alike, that the 1000 mph spinning of the ball Earth at the equator, and the 0 mph spinning of the ball Earth at the north pole, have almost nothing to do with gravity. The official line is that Earth spins because it has always spun, and because there is no friction to slow it down. But of course we should also consider that even the 'spinning of the ball Earth' is very likely a convenient myth too, but regardless, this alleged motion nevertheless still contradicts the 'theory' of gravity! (see the later chapter on 'flat Earth')

So, now we understand the mainstream view of the theory of gravity, let's tear it all apart, shall we? Yes? Okay then, strap in and here we go...

Remember, 'gravity' is a man-made label for an imaginary magical force, loaded with errors. There are even many mainstream 'dyed-in-the-wool' scientists who are beginning to question the existence of gravity.

With the rotation speed of the Earth (in ball terms) being much faster at the equator than further north or south, surely this affects the magical force of gravity?

'*Universe Today*' even admitted that there is only a very slight differential in gravity due to the rotation of the Earth... "*You might be surprised to know that the force of gravity on Earth actually changes depending on where you're standing on it. The first reason is because the Earth is rotating. This rotation is trying to spin you off into space, but don't worry, this force isn't much. The gravity of Earth at the equator is 9.789 m/s^2, while the force of gravity at the poles is 9.832 m/s^2. In other words, you weigh more at the poles than you do at the equator because of this centripetal force.*" Let us examine this part of the quotation for a moment... "*This rotation is trying to spin you off into space, but don't worry, this force isn't much.*" Take a wet tennis ball spinning and then multiply that by a million, give or take a little, and that is the force that '*Universe Today*' says "*...is not much.*"

The Falsification of Science

So, if we are standing on the equator at the widest point of the hypothetical globe, spinning at 1000 mph, our weight will be 80 kilos and if we then travel by plane to either north or south poles (this is not actually allowed, but never mind) at the narrowest point of the hypothetical globe, the rotation of the Earth would be reduced to around 150 mph. The gravitational pull, if it existed, would be so great that our weight would be around 3000 kilos and we would be squashed like an insect.

Mainstream science conveniently ignores the rotation speeds of the Earth at different places and states it has next to no effect on the magical force of gravity. This in itself should be a massive 'red flag.'

The Solar System is travelling at an average speed of 828,000 km/h (230 km/s) or 514,000 mph (143 miles per second) within its trajectory around the alleged galactic centre, which is about one 1300th of the speed of light. But this apparently has no effect on the magical force of gravity either. This again, is another huge 'red flag.'

Please note also, that our Moon does not fall into the Earth, the Earth does not fall into the Sun and satellites do not fall into the Earth, which would certainly be the case if we take gravitational theory as written. Mainstream science also proclaims that gravity is so strong that upside-down water in the southern hemisphere, at the surface of the ocean is being pulled to the centre of the Earth. But this force is simultaneously also not *that* strong, in that this water can move in any direction with currents and tides and be easily collected, splashed, or thrown by a human hand.

Also mainstream science tells us that gravity is so strong that water bends around the globe, yet we all know that water always remains flat and finds its own level. Note here that mainstream science states that the Earth's spin, and the speed of the solar system through space both have no effect on the oceans, and the magical force of gravity pulls water molecules to the centre of the Earth.

Science also claims that the Earth is surrounded by the vast vacuum of space and that we are moving through space at a speed of 514,000 miles per hour, yet the feeble pull of gravity on the atmosphere adheres it to the rapidly spinning globe. Even a child knows that a vacuum will suck air right into it, so why does the unimaginable vastness of space not pull our atmosphere into it? And have you ever thought about what happens at the boundary between the atmosphere and the vacuum? What scientific principle is there that explains that away?

So, to reiterate, why is gravity strong enough to hold people, buildings, and the oceans on the rapidly spinning ball-Earth and against the might of centrifugal force, but so weak as to allow balloons, birds, insects, flowers and even smoke to easily defeat its pull?

"Most people in England have either read, or heard, that Sir Isaac Newton's theory of gravitation was originated by his seeing an apple fall to the Earth from a tree in his garden. Persons gifted with ordinary common-sense would say that the apple fell down to the Earth because, bulk for bulk, it was heavier than the surrounding air, but if, instead of the apple, a fluffy feather had been detached from the tree, a breeze would probably have sent the feather floating away, and the feather would not reach the Earth until the surrounding air became so still that, by virtue of its own density, the feather would fall to the ground." Lady Blount, *'Clarion's Science Versus God's Truth'*

A famous 'flat-Earther' in the early 20th century, Wilbur Voliva, gave lectures all across America against Newtonian astronomy. He usually began his talks by walking on stage with a book, a balloon, a feather and a brick, and asking the audience, *"How is it that a law of gravitation can pull up a toy balloon and cannot pull up a brick? I throw up this book. Why doesn't it go on up? That book went up as far as the force behind it forced it and it fell because it was heavier than the air and that is the only reason. I cut the string of a toy balloon. It rises, gets to a certain height and then it begins to settle. I take this brick and a feather. I blow the feather. Yonder it goes. Finally, it begins to settle and comes down. This brick goes up as far as the force forces it and then it comes down because it is heavier than the air. That is all."*

"Any object which is heavier than the air, and which is unsupported, has a natural tendency to fall by its own weight. Newton's famous apple at

The Falsification of Science

Woolsthorpe, or any other apple when ripe, loses hold of its stalk, and being heavier than the air, drops as a matter of necessity, to the ground, totally irrespective of any attraction of the Earth. For, if such attraction existed, why does not the Earth attract the rising smoke which is not nearly so heavy as the apple? The answer is simple – because the smoke is lighter than the air, and, therefore, does not fall but ascends. Gravitation is only a subterfuge, employed by Newton in his attempt to prove that the Earth revolves round the Sun, and the quicker it is relegated to the tomb of all the Capulets, the better will it be for all classes of society." David Wardlaw Scott, 'Terra Firma'

"*The 'law of gravitation' is said by the advocates of the Newtonian system of astronomy, to be the greatest discovery of science, and the foundation of the whole of modern astronomy. If, therefore, it can be shown that gravitation is a pure assumption, and an imagination of the mind only, that it has no existence outside of the brains of its expounders and advocates, the whole of the hypotheses of this modern so-called science fall to the ground as flat as the surface of the ocean, and this 'most exact of all sciences,' this wonderful 'feat of the intellect' becomes at once the most ridiculous superstition and the most gigantic imposture to which ignorance and credulity could ever be exposed.*" Thomas Winship, 'Zetetic Cosmogeny'

"*That the Sun's path is an exact circle for only about four periods in a year, and then of only a few hours – at the equinoxes and solstices – completely disproves the 'might have been' of circular gravitation, and by consequence, of all gravitation ... If the Sun were of sufficient power to retain the Earth in its orbit when nearest the Sun, when the Earth arrived at that part of its elliptical path farthest from the Sun, the attractive force (unless very greatly increased) would be utterly incapable of preventing the Earth rushing away into space 'in a right line forever,' as astronomers say.*

On the other hand, it is equally clear that if the Sun's attraction were just sufficient to keep the Earth in its proper path when farthest from the Sun, and thus to prevent it rushing off into space; the same power of attraction when the Earth was nearest the Sun would be so much greater, that (unless the attraction were very greatly diminished) nothing would prevent the Earth rushing towards and being absorbed by the

Sun, there being no counterbalancing focus to prevent such a catastrophe! As astronomy makes no reference to the increase and diminution of the attractive force of the Sun, called gravitation, for the above necessary purposes, we are again forced to the conclusion that the great 'discovery' of which astronomers are so proud is absolutely non-existent." Thomas Winship, *'Zetetic Cosmogeny'*

"We are asked by the Newtonians to believe that the action of gravitation, which we can easily overcome by the slightest exercise of volition in raising a hand or a foot, is so overwhelmingly violent when we lose our balance and fall a distance of a few feet, that this force, which is imperceptible under usual conditions, may, under extraordinary circumstances, cause the fracture of every limb we possess? Common-sense must reject this interpretation. Gravitation does not furnish a satisfactory explanation of the phenomena here described, whereas the definition of weight already given does, for a body seeking in the readiest manner its level of stability would precisely produce the result experienced. If the influence which kept us securely attached to this Earth were identical with that which is powerful enough to disturb a distant planet in its orbit, we should be more immediately conscious of its masterful presence and potency; whereas this influence is so impotent in the very spot where it is supposed to be most dominant that we find an insurmountable difficulty in accepting the idea of its existence." N. Crossland, *'New Principia'*

"Gravity is simply density and buoyancy. People argue that things with different densities fall at the same rate through air. However, that is because both items have reached critical density in relation to its medium: air. If you were to change the medium from air to say, water or liquid mercury, the critical density to achieve the same rate of falling would increase significantly. Critical density is directly proportional to the medium density. Thus the denser the medium...the denser the objects would have to be in order to achieve the same rate of falling. A basketball and a rock might fall at the same speed in air. However...drop them in a thicker medium like water...and they will not fall at the same rate. That is because critical density has not been achieved by both the rock and the basketball, in water as its medium. In a vacuum...critical density is zero and is the reason why objects of any density fall at exactly the same rate. Any medium denser than a vacuum has a greater critical

density than zero... thus the reason why objects that haven't attained critical density fall slower in certain mediums. Critical density variation is 'gravity.'" Darrell Dragoo

"So, an object is dropped from a high altitude, this would not and does not, slow down as it gets closer to Earth. The object is free-falling and accelerating as it falls due to the laws of density, not because some magical force is 'pulling it down.' There is more oxygen closer to Earth than say twenty miles up, so how does one propose 'gravity' explains this and density does not? Oxygen is a denser gas than nitrogen and the other majority elements in our atmosphere. Where is the 'pulling force' pulling the oxygen down? The rate at which objects rise or fall depends on the resistance (or lack thereof) in the medium surrounding." Eric Dubay

A submarine can use compressed air to ascend in water.

A hot-air-balloon can use hot air to ascend in the atmosphere.

A plane can use pressure difference across its flying surfaces to gain lift. There is no pulling. Only falling and rising. No force 'pulls down.'

However, if as well as the rest of the universe, the Earth/Moon relationship is also governed by electromagnetism, this would explain why one face of the Moon is permanently presented towards the Earth and electromagnetism would also explain how the Moon affects the Earth's ocean tidal systems.

It is nonsense to believe that it is the Moon's gravitational attraction that 'pulls' the ocean tides around. As we have already discussed, the gravitational pull of the Earth is ridiculously weak and that of the Moon even more so. Any gravitational attraction generated by the Moon would be more than cancelled out in any case by that of the Earth. And if gravity is really a physical force, then the Earth's excess gravitational pull would have drawn-in the Moon many millions of years ago.

No, gravity was a theory postulated by the Freemason, Isaac Newton in the seventeenth century, to explain why we are not flung from the spinning Earth ball, that is after all, allegedly rotating at around 1000 mph+. In fact 'up/down' is a natural state of affairs. If anything is 'up' and unsupported, it will naturally fall 'down.' There is no need for the presence of a superfluous physical force such as gravity in order for this to occur.

The natural order of things is that lighter, less dense objects will float in water whilst heavier, denser objects will sink. Gravity is not responsible for this phenomenon; the laws of density and buoyancy are responsible. For example, an inflated beach ball held under water will immediately spring to the surface, ignoring the so-called 'law of gravity' but absolutely obeying the laws of density and buoyancy. Temperature and density determines what moves up or down and not the mythical gravity.

Many scientists are now starting to see massive errors in gravitational theory, and please note that Newton was in fact, an occultist first, and a scientist second. Even the BBC's very own 'useful idiot,' Professor Brian Cox, idolised by many for being the very 'last word' on physics, is now being ridiculed in some quarters for his immovable, establishment-decreed stance on gravity and other scientific, so-called 'truisms.' In my humble opinion, Cox should have stuck to what he was very average at – that is being an extremely mediocre keyboard player in a truly execrable, yet inexplicably popular teenage pop band.

So to summarise, gravity just simply does not exist and if you still want to insist that it does – bring me some and show it to me.

The Theory of Relativity

"Einstein's Relativity work is a magnificent mathematical garb, which fascinates, dazzles and makes people blind to the underlying errors. The theory is like a beggar clothed in purple whom ignorant people take for a king... its exponents are brilliant men, but they are meta-physicists rather than scientists." Nikola Tesla

In fact, the theories of both 'Special Relativity' and 'General Relativity' have now both been thoroughly debunked over and over again by numerous scientists. But of course, the theory is still presented as fact (as with so many others) because it does not suit the prevailing agenda to accept that it is thoroughly unreliable and disputed 'science.' The arguments against the sustainability of the theory tend to be quite technical, but I shall do my best to explain in simple terms. As stated in the introduction, I am no trained scientist, but you really do not need to be in order to understand this basic stuff.

At present, mainstream physicists seem to have fully accepted Einstein's Theory of Relativity and accept it as the foundation of modern physics because the theory appears perfectly logical and its predictions seem to be supported by numerous experiments and observations.

However, if one re-examines these experiments carefully and with an open mind, serious problems immediately become apparent. Many experiments are considered as evidence of relativistic effects but reveal that they either have null effects or are wrongly interpreted or calculated. For example, the behaviours of clocks in the Hafele-Keating experiment interpreted as the results of relativistic time dilation caused by the relative speed of an inertial reference frame are actually absolute and do not change with the change of inertial reference frames. In fact, the Hafele-Keating experiment indicates the existence of a medium in the space that can slow down the frequencies of atomic clocks when they have velocities relative to the medium, and the Fizeau experiment reveals that the existence of a medium called 'aether,' relative to which the speed of light is constant. Although it is possible that the medium to slow down atomic clocks may be different from aether as multiple media may co-exist in the space.

The existence of aether means that the two postulates of Relativity are wrong for light and electromagnetic waves because the speed of light and the electromagnetic wave equations should be valid only in the inertial reference frame moving with the local aether, just as the acoustic wave equation is valid only in the inertial reference frame, moving with the local air.

The Lorentz invariance of the clock time makes it possible to synchronise clocks in all inertial reference frames to produce the absolute and universal physical time as demonstrated in the universal synchronisation of all the satellite clocks and ground clocks of the Global Positioning System. Therefore, the time of Relativity is no longer the physical time measured with physical clocks.

Moreover, it is now proven that the Lorentz Transformation is the same in redefining time and space as functions of Galilean time and space to produce an artificially constant speed of light in all inertial reference frames. The relationship between the Relativity space-time and Galilean space-time has revealed that the time dilation and length contraction of Relativity in a moving inertial reference frame observed on the stationary inertial reference frame are just illusions. Using the relationship can also prove that the real speed of light measured with clocks still follows Newton's velocity addition formula, which directly renders false the postulate that the speed of light is constant in all inertial reference frames.

All these findings lead to the conclusion that Relativity is seriously flawed. In fact so flawed as to render it useless as a theory of physics. Thus, all relativistic spacetime model-based physics theories (Quantum Field theory, General Theory of Relativity, Big Bang theory, String theories, etc.) become questionable. Disproving Relativity and other related theories of physics will not lead to any kind of crisis (except in the minds of its staunch, immovable adherents) but instead will open up new opportunities for scientists to develop new theories for all the known and unknown physical phenomena. The dynamics of aether may lead to the discovery of new methods of super-fast travel, as the 'faster than light is impossible' speed limit imposed by Relativity is no longer valid.

The theory of Relativity is yet another attempt by an establishment scientist to make the science fit the manufactured model of reality. It has so many errors and misconceptions that as Tesla stated, above, it has blinded otherwise seemingly intelligent men and women to the truth of our entire existence.

The Falsification of Science

The Holographic Universe

Does objective reality really exist though?

In 1982 a remarkable event took place at the University of Paris. A research team led by the physicist Alain Aspect performed what may turn out to be one of the most important experiments of the 20th century. This of course was never reported in the mainstream media. In fact, unless you are in the habit of reading certain lesser-known scientific journals you probably have never even heard Aspect's name, although there are some who believe his discovery may change the face of science.

Aspect and his team discovered that under certain circumstances subatomic particles such as electrons are able to instantaneously communicate with each other regardless of the distance separating them. It does not matter whether they are ten feet or ten quadrillion miles apart.

Somehow, each particle always seems to 'know' what the other is doing. The only problem with this is that it violates Einstein's long-held tenet that nothing, not even communication can travel faster than the speed of light. Since travelling faster than the speed of light is tantamount to breaking the time barrier, this daunting prospect has caused some physicists to find elaborate ways to explain away Aspect's findings. But it has also inspired others to offer even more radical explanations.

A University of London physicist, David Bohm, for example, believes that Aspect's findings imply that objective reality does not exist and that despite its apparent solidity, the universe is nothing but a 'phantasm,' a gigantic and splendidly detailed hologram. In order to understand why Bohm makes this startling assertion, one must first understand a little about holograms. A hologram is a three-dimensional photograph made with the aid of a laser.

To construct a hologram, the subject is first bathed in the light of a laser beam and then a second laser beam is bounced off the reflected light of the first and the resulting interference pattern (the area where the two laser beams comingle) is captured on film. Then, when the film is developed, it resembles a meaningless swirl of light and dark lines but as soon as the developed film is illuminated by another laser beam, a three-dimensional image of the original object appears.

The three-dimensionality of such images is not the only remarkable characteristic of holograms. If a hologram of a rose is cut in half and then illuminated by a laser, each half will still be found to contain the entire image of the rose. Indeed, even if the halves are divided again and again, each snippet of film will always be found to contain a smaller but intact version of the original image. Unlike 'normal' photographs, every part of a hologram contains all the information possessed by the whole. The 'whole in every part' nature of a hologram provides us with an entirely new way of understanding organisation and order. For most of its history, western science has laboured under the bias that the best way to understand a physical phenomenon, whether a frog or an atom, is to dissect it and study its respective parts.

A hologram teaches us that some things in the universe may not lend themselves to this approach. If we try to take apart something constructed holographically, we will not get the pieces of which it is made, we will only get smaller 'wholes.'

This insight suggested to Bohm another way of understanding Aspect's discovery. Bohm believes that the reason that subatomic particles are able to remain in contact with each other regardless of the distance separating them, is not because they are sending some sort of mysterious signal back and forth, but because their separateness is actually an illusion. He argues that at some deeper level of reality such particles are not individual entities but are actually extensions of the same fundamental 'something.'

In order to enable us to better visualise what he meant; Bohm offered the following illustration...

Imagine an aquarium containing a fish. Imagine also that you are unable to see the aquarium directly and your knowledge about it and what it contains comes from two television cameras, one directed at the front of the aquarium and the other directed at its side. As you stare at the two television monitors, you might assume that the fish on each of the screens are separate entities. Because the cameras are set at different angles, each of the images will be slightly different. But as you continue to watch the two fish, you will eventually become aware that there is a certain relationship between them.

When one turns, the other also makes a slightly different but corresponding turn; when one faces the front, the other always faces toward the side. If you remain unaware of the full scope of the situation, you might even conclude that the fish must be instantaneously communicating with one another, but this is clearly not the case. But Bohm argues that this is precisely what is happening with subatomic particles in Aspect's experiment.

Also according to Bohm, the apparent faster-than-light connection between subatomic particles is really telling us that there is a deeper level of reality to which we are not privy, a more complex dimension beyond our own that is analogous to the aquarium. And, he adds, we view objects such as subatomic particles as separate from one another because we are seeing only a portion of their reality.

Such particles are not separate 'parts,' of the same entity, but facets of a deeper and more underlying unity that is ultimately as holographic and indivisible as the previously mentioned rose. And since everything in physical reality is comprised of these 'phantoms,' then the universe is itself a projection, a hologram.

In addition to its phantom-like nature, such a universe would possess other bizarre features. If the apparent separateness of subatomic particles is illusory, this consequently means that at a deeper level of reality, all things in the universe are infinitely interconnected. The electrons in a carbon atom in the human brain are connected to the subatomic particles that comprise every other living thing, every heart that beats, and every star that shimmers in the sky.

Everything interpenetrates everything else, and although human nature may seek to categorise, pigeonhole, and subdivide the various phenomena of the universe, all apportionments are of necessity artificial and all of nature is ultimately a seamless web.

In the holographic universe, even time and space are no longer regarded as fundamentals and because concepts such as location, break down in a universe in which nothing is truly separate from anything else, time and three-dimensional space, like the images of the fish on the TV monitors, would also have to be viewed as projections of this deeper order. At its deeper level, reality is a kind of super-hologram in which the past, present, and future all exist simultaneously. This suggests that given the proper tools it may even be possible someday, to reach into the super-holographic level of reality and physically extract scenes from the long-forgotten past.

But whatever else the super-hologram may contain is an open-ended question. Allowing, for the sake of argument, that the super-hologram is the matrix that has given birth to everything in our universe, at the very least it contains every subatomic particle that has been or will be – every configuration of matter and energy that is possible, from snowflakes to quasars, from blue whales to gamma rays. It should be seen as a sort of cosmic storehouse of 'all that is, was and ever will be.'

Although Bohm concedes that we have no way of knowing for certain what else might lie hidden in the super-hologram, he does venture to say that we have no reason to assume it does not contain more. Or as he puts it, perhaps the super-holographic level of reality is a *"mere stage"* beyond which lies *"an infinity of further development."*

However, Bohm is not the only researcher who has found evidence that the universe is a hologram. Working independently in the field of brain research, Stanford neurophysiologist Karl Pribram also became persuaded of the holographic nature of reality. Pribram was drawn to the holographic model by the puzzle of how and where memories are stored in the brain and for decades now, numerous studies have shown that rather than being confined to a specific location, memories are dispersed throughout the brain in some inexplicable way.

The Falsification of Science

In a series of landmark experiments in the 1920s, the brain scientist Karl Lashley found that no matter what portion of a rat's brain he removed, he was unable to eradicate its memory of how to perform complex tasks it had learned prior to surgery. The only problem was that no one was able to find a mechanism that might explain this curious 'whole in every part' nature of memory storage.

Then in the 1960s Pribram encountered the concept of holography and realised he had found the explanation for which brain scientists had been searching. Pribram, who died in 2015 at the grand old age of 96, then reached the conclusion that memories are encoded not in neurons, or small groupings of neurons, but in patterns of nerve impulses that criss-cross the entire brain in the same way that patterns of laser light interference criss-cross the entire area of a piece of film containing a holographic image. In other words, Pribram believed that the brain is itself a hologram.

Pribram's theory also explained how the human brain can store so many memories in so little space. It has been estimated that the human brain has the capacity to memorise something in the region of ten billion bits of information during the average human lifetime (or roughly the same amount of information contained in five sets of the Encyclopaedia Britannica).

Similarly, it has been discovered that in addition to their other capabilities, holograms possess an astounding capacity for information storage – simply by changing the angle at which the two lasers strike a piece of photographic film, it is possible to record many different images on the same surface. It has been demonstrated that one cubic centimetre of film can hold as many as ten billion bits of information.

Humans' incredible ability to quickly retrieve whatever information we need from the enormous store of our memories becomes more understandable if the brain functions according to holographic principles. If someone asks us to tell him what comes to mind when he says the word 'zebra,' we do not have to clumsily sort back through some gigantic, cerebral alphabetic file to arrive at an answer. Instead, associations such as 'striped,' 'horse-like,' and 'African animal,' all pop into our heads instantly.

Indeed, one of the most amazing things about the human thinking process is that every piece of information seems instantly cross-correlated with every other piece of information and this is another feature intrinsic to the hologram. Because every portion of a hologram is infinitely interconnected with every other portion, it is perhaps nature's supreme example of a cross-correlated system.

The storage of memory is not the only neurophysiological puzzle that becomes more tractable in light of Pribram's holographic model of the brain. Another is how the brain is able to translate the avalanche of frequencies it receives via the senses (light frequencies, sound frequencies, etc.) into the concrete world of our perceptions. Encoding and decoding frequencies is precisely what a hologram does best. Just as a hologram functions as a sort of lens, a translating device able to convert an apparently meaningless blur of frequencies into a coherent image, Pribram believed that the brain also comprises a lens and uses holographic principles to mathematically convert the frequencies it receives through the senses into the inner world of our perceptions.

Indeed, an impressive, ever-expanding body of evidence suggests that the brain uses holographic principles to perform its operations. Pribram's theory, in fact, has gained increasing support among neurophysiologists. The Argentinian-Italian researcher Hugo Zucarelli recently extended the holographic model into the world of acoustic phenomena. Puzzled by the fact that humans can locate the source of sounds without moving their heads, even if they only possess hearing in one ear, Zucarelli discovered that holographic principles can explain this ability. Zucarelli has also developed the technology of holophonic sound, a recording technique able to reproduce acoustic situations with an almost uncanny realism.

Pribram's belief that our brains mathematically construct 'hard reality' by relying on input from a frequency domain has also received a good deal of experimental support. It has been found that each of our senses is sensitive to a much broader range of frequencies than was previously suspected. Researchers have discovered, for instance, that our visual systems are sensitive to sound frequencies, that our sense of smell is in part dependent on what are now called 'cosmic frequencies,' and that even the cells in our bodies are sensitive to a broad range of frequencies. Such findings suggest that it is only in the holographic domain of consciousness that such frequencies are sorted out and divided up into conventional perceptions.

But the most mind-boggling aspect of Pribram's holographic model of the brain is what happens when it is merged with Bohm's theory. For if the concreteness of the world is but a secondary reality and what is 'there' is actually a holographic blur of frequencies, and if the brain is also a hologram and only selects some of the frequencies out of this blur and mathematically transforms them into sensory perceptions, what becomes of objective reality?

Put quite simply, it ceases to exist. As the religions of the East have long upheld, the material world is Maya, an illusion, and although we may think we are physical beings moving through a physical world, this too is an illusion. In effect, we are really just 'receivers' floating through a kaleidoscopic sea of frequency, and what we extract from this sea and transform into physical reality is but one channel from many, extracted from the super-hologram.

This new, incredible picture of reality, the synthesis of Bohm and Pribram's views, has come to be known as the holographic paradigm, and although some scientists have greeted it with scepticism, it has galvanised many others. A small but growing group of researchers believe it may be the most accurate model of reality that science has arrived at, thus far. Moreover, some believe it may solve some mysteries that have never before been explainable by scientists and even establish the paranormal as an intrinsic part of science.

Numerous researchers, including David Bohm and Karl Pribram, have noted that many para-psychological phenomena become much more understandable in terms of the holographic paradigm. In a universe in which individual brains are actually indivisible portions of the greater hologram and everything is infinitely interconnected, telepathy may simply be the accessing of the holographic level.

It is obviously much easier to understand how information can travel from the minds of one 'individual' to another at a far distant point and helps to understand a number of unsolved puzzles in psychology. In particular, the holographic paradigm offers a model for understanding many of the baffling phenomena experienced by individuals during altered states of consciousness.

Whatever the real truth may be, I certainly believe that until we know better, it is much wiser and more realistic to treat the universe as infinite, continually evolving, with no beginning, no end and until we can prove differently, as being the work of an unknown creator or creators.

Chapter 2

Evolution

"Of course, only a fool would question the theory of evolution, even though the theory is most closely associated with a man who, along with other members of his clan, was a key figure in the eugenics movement and even though the concept of natural selection just happens to nicely compliment the eugenics agenda, which in turn, dovetails nicely with the agenda of the 'peak oil' crowd, whose theory, as we all know, rests upon the notion of oil as a 'fossil fuel,' which is taken as a given by most of the scientific community, which just goes to show you, I suppose, that you shouldn't always listen to the scientific community." David McGowan, 25th July 2006

The western 'education' system actually teaches children *not* to think and to just blindly accept anything that the establishment says is fact. Any student who uses logic and solid scientific evidence to question any of science's proclaimed 'givens' including the theory of evolution, is ridiculed and insulted into submission. The students who submit to the will of the educators in this matter, almost all of them that is, then become unthinking robots programmed not to question the dogma presented but to merely accept anything and everything they are taught as undeniable fact.

A few years ago, I watched on with great interest as a primary school class in a park was playing the 'three-legged' race game, where the adjacent legs of two children were tied together. The children needed to cooperate with each other on each step taken in order to run successfully and they thought it was great fun. I heard the teacher tell them they were being trained to 'cooperate,' but maybe they were being brainwashed into conforming to a system in which they are not allowed to have individual thoughts or opinions? Are all children in fact, being moulded in this way to become 'team players' whilst simultaneously submitting to peer pressure and thereby relinquishing any ideas of individuality still lingering within their still-developing minds? Communist ideology has used these self-same brainwashing techniques for many decades.

This brainwashing of students in this way continues today, in the cultural Marxist 'thinking' that decrees that there is no absolute right or wrong, and whatever the children think is 'alright' is just fine, regardless of consequences. That is of course until they should question any elite-decreed dogma, such as for example, evolution, at which point they are then admonished and made to feel foolish. This brainwashing results in children who are unable to think logically, independently, scientifically, and accurately and these children then grow up to be the unthinking adherents of false science – either as observers of the current agenda or as active participants in the great pseudoscientific sham being widely perpetuated today.

The assumption is always made, that if you do not believe in evolution, then you must be a 'fundamental Christian' believing in divine creation as this is portrayed strongly as the only viable alternative. This is abject nonsense, however. Personally, I am not a Christian and nor do I follow any other organised religion but I am a 'spiritual' person and I believe that it is absolutely possible that there was a creator of some kind, although I confess to having no real idea what form this creator would take. Maybe it really is some kind of ethereal spirit as mainstream religions would have us believe, or maybe it is actually a physical entity residing in some other dimension, which has for its own reasons, decided to 'seed' our universe with a proliferation of life forms, as in the theory of 'Panspermia.' But I do also confess that I cannot understand why any 'creator,' either spiritual or physical would actually want or need us to worship it. I find this idea completely counterintuitive but of course, I also accept that we may never, ever know the real truth.

The Falsification of Science

But, to continue with the theme... as I have provided evidence to suggest strongly that the Big Bang theory is nothing more than a manufactured hoax, where does this leave the theory of evolution?

The two are of course, intrinsically linked. If the Big Bang really did happen, then evolution is almost a given, as it follows that **if** the universe itself developed in this manner, then the next logical step in this wider 'evolutionary' process is the evolution of living species.

To clarify what I mean by 'evolution,' it is necessary to make the distinction between 'microevolution' and 'macroevolution.' Microevolution is the gradual adaptation of different species to their environment, for example moths' wing colours adopting a pattern which blends seamlessly into the background of their natural habitat and which then of course provides them with a camouflage protection against predators. To suggest that this does not take place over huge timescales would be ridiculous. But macroevolution, the changing of an entire species into another species, is not only equally ridiculous but quite simply consummate nonsense.

So, are the origins of the human race really what we have been taught and assumed all our lives to be true? The received wisdom, relayed to our grand-parents, parents, ourselves, and our children, down the generations for the past 160 years or so, is that we all evolved from single-celled organisms, born of amino acids combining together in the 'primordial soup,' through a series of greater and more complex organisms to the apelike creatures that were our supposed ancestors and thence to human beings. No arguments or discussions as to the veracity or proof of this proposition are tolerated. It is a fact, end of story – period. Accept it or be ridiculed and worse.

I believe that this is an absurd and monstrous deception and that we are no more descended directly from single celled creatures or even apes than we are from unicorns or goblins. Evolution, or to be more precise the '***theory*** of evolution' is exactly that, a theory and an extremely tenuous one at that. There is a veritable mountain of evidence to contradict this premise and also to suggest that Darwin was part of the overall conspiracy and grand deception that continues to

69

this day. He is even rumoured by some sources to have confessed in anguish in his last days, to the hoax he was instrumental in inflicting on a gullible humanity and indeed some of his quoted statements late in his life directly bear-out that premise.

My research has led me to believe that, as with the Big Bang, evolution was conceived and promoted deliberately as a way of maintaining control of the people. By the second half of the nineteenth century, organised, mainstream religion was just beginning to lose its vice-like grip as a control mechanism of the subdued masses and there was a small but growing group of people who were questioning the unthinkable – was Christianity or indeed any religion, the truth after all? To combat this dangerous turn of events, what was needed was another false creed to supplant organised religion. It does not matter to the Elite what false paradigms we believe in, just so long as we believe in *something*, anything that will lead us away from the real truth. So, in the mid-nineteenth century, they simply decided to replace their fading, old-fashioned myth of faith-based creationism with something more in line with the fashionable, brave new world of scientific discoveries, hence the more credible, modern, 'scientific' myth of evolution. As per the quote at the beginning of this chapter, it is also, I believe, more than coincidence that the theory of evolution was not only originally propounded by Elite eugenicists but also that the tenets of evolutionary dogma nicely complement the fundamental principles of eugenics.

"The model of human prehistory built-up by scholars over the past two centuries is sadly and completely wrong, and a deliberate tool of disinformation and mind control." Michael Cremo and Richard L. Thompson, *'The Hidden History of the Human Race'*

This extremely revealing book, relates in great detail and with literally thousands of case studies and examples, how we have been duped into believing that homo-sapiens as a species is much less than one million years old and is a product of macro-evolution from apes. However, Cremo and Thompson have uncovered literally hundreds of examples of mainstream archaeological cover-up operations, designed to prevent the truth becoming widely known. And that truth is simply that there are in existence many, many examples of human remains, *some dating back several hundred million years*! One simple example…

"In Macoupin County, Illinois, the bones of a human were recently found on a coal-bed capped with two feet of slate rock, ninety feet below the surface of the Earth. The bones, when found, were covered with a crust or coating of hard glossy matter, as black as coal itself, but when scraped away left the bones white and natural." 'The Geologist' magazine, December 1862

This coal was at least 286 million years old and may be as old as 320 million years, way, way beyond any timescales admitted by the mainstream regarding the possible antiquity of our species. Any such discoveries these days are never reported in the mainstream media, despite there being thousands of examples constantly occurring.

Interestingly, this distortion of facts in an attempt to 'prove' a huge falsehood to be the truth, has resulted in the classic, Hegelian 'evolution versus creation' argument to keep us all busy and distracted from searching for the actual truth. In other words, never mind the real facts, let's all waste our time arguing the rights and wrongs of two false creeds. Strange is it not that if one does not subscribe to the religion of evolution, then one is automatically dubbed a 'creationist' with all the negative connotations and inherent stupidity and ignorance that this has been manipulated and engineered to imply?

"When you find issues/controversies which people love to debate endlessly, which are emotionally inflammatory and which divide the masses into oppositional stances/groups, it is a pretty strong possibility that the controlling elites might be busy behind the scenes, fomenting these quarrels and keeping them alive." kennysideshow.blogspot.com, 15th November 2011

It is also known that Dr Thomas Henry Huxley, a stalwart of the Elite establishment, Fellow of the Royal Society, and a prominent Freemason, strongly encouraged and even cajoled Charles Darwin to publish his theory. Huxley would eventually become the 'official spokesman' for Darwin and even became known as 'Darwin's Bulldog,' such was his forceful assertions of the truth of the theory.

John Hamer

He was also the grandfather of Aldous Huxley, the author of *'Brave New World'* a novel written in the 1930s that demonstrates an uncannily accurate depiction of a future society of oppressed masses in a similar vein to Orwell's *'1984.'* Another grandson was Julian Huxley, famous as the first secretary-general of UNESCO, a branch of the Elite-controlled United Nations. Coincidence? I shall let you the reader, draw your own conclusions.

Of course, daring to question the great religion of evolution – for that is what it has become to so many people, is now ironically regarded by society in general as either the actions of a religious zealot or an imbecile, such is the power of persistent propaganda. As already stated, I do not suggest for one moment that the localised evolving of bodily features and functions of certain creatures (microevolution) does not take place over millennia in order to enable organisms to adapt to their surroundings and for example, finetuning defences against predators. To suggest that, would be just as absurd. However, to believe that whole new species are created from others or from virtually nothing (macroevolution) when the abundant, contrary evidence is examined, seems too far-fetched and unscientific to be anything but an elaborate hoax and a deliberate deception, perpetuated by wholesale propaganda.

"The known fossil record fails to document a single example of evolution accomplishing a major transition – every palaeontologist knows that most species don't change." Stephen Gould, evolutionary biologist, Harvard University 1980.

"I am not satisfied that Darwin proved his point or that his influence in scientific and public thinking has been beneficial. The success of Darwinism was accomplished by a decline in scientific integrity." W.R. Thompson, Canadian scientist.

Dr David Kitts, professor of geology at the University of Oklahoma said, *"Evolution requires intermediate forms between species and palaeontology does not provide them..."* And Sir Solly Zuckerman admitted that there are no 'fossil traces' of transformation from an ape-like creature to man! Even Stephen J. Gould of Harvard University admitted that, *"The fossil record with its abrupt transitions offers no support for gradual change."*

I assume that all evolutionist college professors and university lecturers know that Darwin admitted the same fact and that they know that Darwin was never scientifically trained, he was instead, educated to become a religious minister, so by definition, evolutionary theory adherents are actually following the teachings of an apostate preacher?

"…as by this theory, innumerable transitional forms must have existed. Why do we not find them embedded in the crust of the earth? Why is not all nature in a confusion of halfway species instead of being, as we see them, well-defined species?" Charles Darwin.

Darwin's own answer to this particular question was that there had been insufficient time since his theory was espoused to thoroughly check the available fossil records. Interesting hypothesis yes, but now proven to be totally incorrect. Almost 140 years have now passed since the death of Darwin; allowing plenty of time to rectify this inconvenient fact, but evolutionary 'science' is no further forward in this respect than it was in the 1880s.

A famous fossil expert, Niles Eldredge also confessed that, *"…geologists have found rock layers of all divisions of the last 500 million years and no transitional forms were contained in them."* Dr Eldredge further stated that, *"…no one has yet found any evidence of such transitional creatures."*

All the alleged transitional fossils, that were so revered and cherished by evolutionists a generation ago, are now an acute embarrassment to them. Archaeopteryx is now considered to be only a bird, not an intermediate fossil and the infamous horse evolutionary series that is still found in some textbooks and museums has now been discarded and is regarded as inaccurate and an 'illusion' because it is not proof of evolution. In fact, the first 'horse' in the series is no longer even considered to be a horse – even by hard-line evolutionists.

Concerning transitional fossils, the prominent, world famous palaeontologist Colin Patterson admitted that, *"there is not one such fossil for which one could make a watertight argument."* He is absolutely correct – there is not a single one.

"Just as pre-Darwinian biology was carried out by people whose faith was in the Creator and His plan, post-Darwinian biology is being carried out by people whose faith is in, almost, the deity of Darwin. They've seen their task as to elaborate his theory and to fill the gaps in it, to fill the trunk and twigs of the tree. But it seems to me that the theoretical framework has very little impact on the actual progress of the work in biological research. In a way some aspects of Darwinism and of Neo-Darwinism seem to me to have held back the progress of science." Colin Patterson, senior palaeontologist, the Museum of Natural History, London

"Not many scientists are willing to risk their livelihood to point out the facts. They remain mute, mouthing the party line when necessary in order to keep their positions. Those illogical arguments mouthed by the scientists then fuel misunderstanding among those who are unable to double-check the truth and logic behind the theory of evolution." Duncan Long.

And remarkably...

"Not one change of species into another is on record. We cannot prove that a single species has changed into another." Charles Darwin, *'My Life and Letters'*, Vol. 1, page 210

That contemporary, great champion of evolutionary myth and dogma, Professor Richard Dawkins, latterly of the Elite-funded and controlled, great educational propaganda machine, Oxford University, and author of *'The Selfish Gene'* and *'The God Delusion,'* spends his time denouncing, belittling and even insulting anyone who dares question the great pseudo religion of evolution. Why would this be? Why is it such a crime or heresy to question or debate widely held scientific beliefs? Is Dawkins an evolutionist per se or simply anti-religion, choosing evolution as what he believes to be the only viable alternative? This again of course is yet another example of the classic Hegelian trap. Present two options to choose from, neither of which is correct, whilst encouraging the masses to pick their favourite and debate the pros and cons until we lose sight of the real issue. We should ask why the mainstream media even allows Dawkins

The Falsification of Science

a platform for his mostly disingenuous tirades, whilst denying it to those who espouse the contrary view in a more considered, rational, or scientific way. Dawkins' ironically, somewhat evangelical style arguments have brought him largely unreported ridicule from many quarters with even the hard-line Darwinian, Michael Ruse suggesting that Dawkins' frequent rants make him feel *"...embarrassed to be an atheist."*

After the 9/11, 2001 so-called, 'terrorist attacks,' Dawkins argued that, *"Many of us saw religion as harmless nonsense. Beliefs might lack all supporting evidence but, we thought, if people needed a crutch for consolation, where's the harm? September 11th changed all that. Revealed faith is not harmless nonsense; it can be lethally dangerous nonsense. Dangerous, because it gives people unshakeable confidence in their own righteousness. Dangerous, because it gives them false courage to kill themselves, which automatically removes normal barriers to killing others. Dangerous, because it teaches enmity to others labelled only by a difference of inherited tradition. And, dangerous, because we have all bought into a weird respect, which uniquely protects religion from normal criticism. Let's now stop being so damned respectful!"*

Maybe before opening his mouth about a subject of which he obviously understands very little, Dawkins should first of all have undertaken even a tiny amount of independent geopolitical research of his own. Had he done so, he would know that the 9/11 attacks were not carried out by religious fanatics at all, but by the very establishment that he seems to revere, pays him handsomely no doubt, and of course of which he is an integral cog in the whole, vast machinery. And that fact aside, perhaps Dawkins had simply and conveniently 'forgotten' that all the many genocides of the last century had been committed by atheists, not religious groups? I refer specifically to such atrocities as the Russian revolution which was responsible for the slaughter of tens of millions of Christians and other innocent Russian anti-communists, at the hands of atheistic communists, throughout the Soviet period of history.

This was of course closely followed by the 'real' holocaust of the twentieth century, the Holodomor of 1932-33, during which at least ten million Ukrainian people were exterminated, mainly by starvation, by the atheistic communists of the Soviet Union.

Then in the late 1940s and 1950s, along came the communist, Mao Tse-tung, the architect of the 'Chinese Cultural Revolution,' who in turn was responsible for the deaths of several more tens of millions. And lastly but by no means least was the communist Pol Pot and his Khmer Rouge in Cambodia in the 1970s. He was a mere rookie compared with the others. He was in fact the instigator of the murder of 'only' an estimated 2-3 million more innocents.

And this is all without even mentioning the two world wars, which whilst being perpetrated by alleged Christians, more likely those bloodied hands actually belonged to fully-committed, practicing satanists. So to the grim totals above we can also legitimately add at least another 100 million deaths – and all of those deaths occurred within the last one hundred years or so.

Speaking about Dawkins' most famous book, '*The God Delusion,*' even a fellow, committed atheist and evolutionist, Michael Ruse said, "*A good academic will inform himself in depth in a subject he is writing about. Dawkins did not. He is neither a philosopher nor a theologian. I am not a biologist myself, but at least I study the subject in depth before I write about it. And that arrogance and that pedantic attitude of his... Dawkins' book confirms my analysis of evolution as pseudo religion. His secular humanism has quasi-religious characteristics.*"

Another of Dawkins' fellow-evolutionists, Terry Eagleton, Professor of Cultural Theory at the National University of Ireland, Galway, began his review of '*The God Delusion,*' with these words...

"*Imagine someone holding forth on biology whose only knowledge of the subject is the 'Book of British Birds,' and you have a rough idea of what it feels like to read Richard Dawkins on theology...*

does he imagine like a bumptious young barrister, that you can defeat the opposition while being complacently ignorant of its toughest case? Dawkins, it appears, has sometimes been told by theologians that he sets

The Falsification of Science

up straw men only to bowl them over, a charge he rebuts in this book; but if 'The God Delusion' is anything to go by, they are absolutely right."

Could Dawkins, knowingly or un-knowingly be a puppet of the Elite, a so-called 'shill' or 'useful idiot' who is discreetly encouraged to spread his disinformation to as wide an audience as possible? If so, he certainly would not be the first nor the last, one suspects.

Richard Milton was initially an ardent believer in Darwinian doctrine until he began to investigate the myths and legends of evolutionary theory in depth. After 20 years of studying and writing about evolution, he realised that there were many anomalous elements in the theory. He therefore decided to put every main classic 'proof' of Darwinism to the test. His results left him stunned at first. He found that the theory could not even stand up to the rigours of even rudimentary investigative journalism. Eventually, he published a book titled *"The Facts of Life: Shattering the Myths of Darwinism."*

What happened next, however, shocked him.

"I experienced the witch-hunting activity of the Darwinist police at first hand – it was deeply disappointing to find myself being described by the prominent Oxford zoologist, Richard Dawkins, as 'loony,' 'stupid' and 'in need of psychiatric help' in response to purely scientific reporting."
Richard Milton

Do we detect here shades of the Soviet Union in the mid-20th century when so-called dissident scientists there began speaking out against the diktats and manufactured reality of Stalin's regime?

The biologist, Dr Pierre Grasse, considered the greatest living scientist in France, wrote a book to *"launch a frontal assault on all forms of Darwinism."* Grasse is by no means a religious fanatic, yet he refers to evolution as a *"pseudoscience."*

Dr Soren Lovtrup, Professor of Zoophysiology at the University of Umea in Sweden wrote, *"I suppose that nobody will deny that it is a great misfortune if an entire branch of science becomes addicted to a false theory. But this is what has happened in biology. For a long time now, people discuss evolutionary problems in a peculiar 'Darwinian' vocabulary...thereby believing that they contribute to the explanation of natural events. I believe that one day the Darwinian myth will be ranked the greatest deceit in the history of science."* He also said, *"Evolution is anti-science."* He is correct.

In fact Darwin, and his followers, in today's politically correct spiel were racists and eugenicists who believed that black and brown people were much closer to their alleged 'ape man,' than whites. Thomas Huxley, Henry F. Osborne, Professor Edwin Conklin, and many other Darwinists openly advocated white superiority – purely because of evolutionary bias. And evolutionary adherents for more than a century after Darwin can also be similarly accused, starting with Nietzsche who called for the breeding of a master race and for the annihilation of millions of 'inferiors,' followed by Marx, Engels, Lenin, Trotsky, Stalin, Mao Tse-tung, Pol Pot and many others. Indeed, evolutionary dogma has resulted in the spilling of the blood of countless tens of millions of innocents during the past 150 years. After all, evolutionists tell us that man is no different in any way to animals, thereby devaluing human life itself, so why should we be concerned about the lives of a few million 'subhumans?'

If all the preceding assertions are true, this all begs the question, 'what is the point of the deception?' If someone goes to all that trouble to make sure we believe something that is not and cannot be proven, there must be a reason for it and a hidden agenda behind it. Indeed, this simple test can be applied to anything but in this case the overwhelmingly obvious conclusion is that it is done in order to deceive and therefore impose and maintain control by taking advantage of the lack of knowledge of the real truth of our origins and purpose as a species.

Fossil records constitute the primary source for the evolutionists in searching for evidence for the theory of evolution. The fossil records certainly contain the remains of past human beings but when these are examined objectively, it may be seen that the records themselves are in no way in favour of evolutionary theory, but rather are against it, contrary to the assertions of the evolutionists. However, since these fossils are incorrectly portrayed by the evolutionists and presented for public opinion with the intent of fulfilling pre-conceived ideas; many people are fooled into incorrectly believing that the fossil records actually *verify* the theory of evolution.

The evolutionists disingenuously use the fact that findings of fossil records are open to many different interpretations, to their own advantage and as 'proof' of their own assertions. The discovered fossils are usually not sufficient to make a firm analysis but are generally comprised of incomplete and fragmented bone pieces. This is why it is so simple for them to distort the available data and use them fraudulently to portray the desired objectives.

Belief in the theory of evolution has come to be seen as almost a life-style choice, a mode of thinking, even an ideology rather than just simply a theory like any other by its ironically 'evangelical' defenders who do not deem it necessary to take steps to prevent the distorting of data or even the committing of more serious, deliberate forgeries. Indeed, extremist advocates of evolutionary ideology do not hesitate to undertake any kind of distortion necessary in order to interpret the fossil records in favour of evolutionary theory. It is a classical scientific mistake to build any kind of theoretical framework from the basis of an incorrect initial assumption and yet I believe that this fundamental 'mistake' is made time after time by the proponents of evolutionary theory and indeed of course many other insubstantial, allegedly 'scientific' theories.

"Theory shapes the way we think about, even perceive, data... We are unaware of many of our assumptions. In the course of rethinking my ideas about human evolution, I have changed somewhat as a scientist. I am aware of the prevalence of implicit assumptions and try harder to dig them out of my own thinking. Theories have, in the past, clearly reflected our current ideologies instead of the actual data... I am more sombre than I once was about what the unwritten past can tell us." David Pilbeam, anthropologist, Harvard University

It is true that ideological expectations can and do influence the interpretation of any given data set and the fact that fossil records are open to many different interpretations raises serious doubts on the reliability of the whole science of paleo-anthropology which is mostly under the control of the evolutionists. Certain prejudices and expectations will undoubtedly have an impact on the veracity of data extrapolation.

"...We then move right off the register of objective truth into those fields of presumed biological science, like extra-sensory perception or the interpretation of man's fossil history, where to the faithful anything is possible – and where the ardent believer is sometimes able to believe several contradictory things at the same time." Sir Solly Zuckerman, palaeontologist at Birmingham University, England

Since fossil records are usually unorganised and incomplete, the estimations based on them are inevitably totally speculative. As a matter of fact, the reconstructions (drawings or models) made by evolutionists based on the fossil remains are often treated in a speculative way in consort with the evolutionary theory. Since most people are more easily influenced by visual rather than written data, the aim of evolutionists is to entice them to believe that these reconstructed creatures have really existed in the past.

For this reason alone, the reconstructions of fossils and skulls are always designed to meet the needs of the evolutionary theory. Evolutionist researchers often set out from a single tooth, a mandibular fragment or even a tiny bone of the arm, draw semi-human-like imaginary creatures and then present these to the public sensationally as a link in the evolution of man. These drawings and reconstructions have indeed played an important role in the visualisation of the 'primitive man' image in the minds of people.

Reconstructions based on the bone remains can only reveal the general characteristics of the object at hand. Yet, the real defining details are soft tissues often muscles or tendons that do not leave an impression in the rocks as they decay too rapidly. Therefore, with the speculative interpretation of the soft tissues, the reconstructed drawing or model becomes totally dependent upon the imagination of the person constructing it.

The Falsification of Science

"To attempt to restore the soft parts is an even more hazardous undertaking. The lips, the eyes, the ears, and the nasal tip, leave no clues on the underlying bony parts. You can with equal facility, model on a Neanderthaloid skull the features of a chimpanzee or the lineaments of a philosopher. These alleged restorations of ancient types of a man have very little if any scientific value and are likely only to mislead the public... So, put not your trust in reconstructions." Ernst A. Hooten, Harvard University

Indeed, evolutionists invent such ridiculous stories that they even ascribe different faces to the same skull. For example, three different reconstructed drawings made for the fossil named *Australopithecus robustus*, is a famous example of such a forgery.

A group of evolutionists who could not find any substantial evidence in the fossil records to support their, at best, tenuous beliefs, actually decided to create their own evidence themselves. Some of these studies were even included in textbooks under titles such as 'evolution conspiracies' and this is probably a good clue to the fact that the theory of evolution is an ideology or a life philosophy that has to be contrived to be kept alive by considerable effort.

A well-known doctor and amateur palaeontologist, Charles Dawson announced in 1912 that he had found a jawbone and a cranial fragment in a pit in Piltdown, Sussex, England. Despite the fact that the jawbone was apelike, the teeth and the skull were similar to a human's. These specimens were designated by science as 'Piltdown Man,' and were determined to be dated to half a million years ago and depicted as absolute 'proof' of the evolution of man, for more than 40 years. Many scientific articles were written about these artefacts, many interpretations and drawings were made, and it was presented as important evidence and taught as undeniable proof of the macro-evolution of mankind.

The discovery of 'Piltdown Man' engendered massive enthusiasm in palaeoanthropology circles and gave birth to many new debates which automatically assumed that evolution was absolute fact. For example, the famous English anthropologist, G. E. Smith pondered... "*Did the brain or body of man evolve first?*"

In 1949, Kenneth Oakley from the palaeontology department of the British Museum in London devised the 'fluorine test' to determine the date of fossils. When the test was performed on the Piltdown-man fossil, the subsequent result was shocking. It was proved conclusively that the jawbone of Piltdown-man contained no fluorine, and this therefore indicated that the bone was underground no more than a relatively few short years and was therefore obviously a fraud. In addition, the skull itself contained a small amount of fluorine, enough to determine that it was a few thousand years old, only. It was also proved by the tests that the jawbone and the skull came from two entirely separate creatures and time-periods and must therefore be a deliberate hoax.

"The latest chronological research made with the fluorine method revealed that the [Piltdown] skull was only a few thousand years old. It was manifest that the teeth in the jawbone belonging to an orang-utan were worn out artificially and the primitive tools found next to the fossils were simple imitations sharpened by steel devices." Kenneth Oakley, palaeontologist, the British Museum, London

Alongside these fossils were found some extinct elephant fossils and some tool remains made out of the bones of the same elephant species. These elephant fossils were used in the dating of the skull and in the tests, it was understood that these elephant fossils were indeed very ancient. However, the jawbone and the skull were much more recent than the elephant fossils. What then was the significance of these facts? It was surmised that the Piltdown ivory fossil had probably been found in Africa and then deliberately placed in the Piltdown site to give the impression that the false skull was as old as the elephant fossil in order to mislead. As the researchers studied the other animal fossils found in the same region in more depth, they found that these were also placed there with the deliberate intention of deception and the Piltdown bone tool was eventually discovered to be an elephant fossil shaped with a steel knife.

The Falsification of Science

However, the hoax could still be regarded as a raging success by the evolutionists in as much as it had propagandised the population for almost half a century into a definitive belief of evolutionary myth and the Elite know very well that once any beliefs become deeply entrenched in the human psyche then even subsequent absolute proof to the contrary will not necessarily remove or diminish them.

This fake fossil that occupied the evolutionist circles for a many years, demonstrates the lengths to which those who desire to prove the theory of evolution at all costs are prepared to go. Why would this be? Why would anyone fake scientific evidence? I suggest that it is done (in this case at least) to provide hard evidence of the proof of evolutionary theory in the absence of any other real or tangible facts that would verify it. This in itself speaks volumes to my mind.

After the detailed analysis completed by Kenneth Oakley, William le Gros Clark and J. S. Weiner, this forgery was eventually made public in 1953. The skull was discovered to be human and was a mere few hundred years old and the jawbone was from a recently deceased ape. The teeth had been specially arranged and added separately to the jaw and the tooth sockets were set in such a way as to resemble those of a human. All these individual elements were then deceptively stained with potassium-dichromate to give them the false appearance of great age. These stains disappeared when the skull was dipped in acid.

There was also much evidence of artificial abrasion that in hindsight was so obvious that it begged the question; how had it escaped the notice of experienced palaeontologists for forty years? Sir Solly Zuckerman's view was...

"As I have already implied, students of fossil primates have not been distinguished by caution when working within the logical constraints of their subject. The record is so astonishing that it is legitimate to ask whether much science is yet to be found in this field at all."

However, in my view, the story of the Piltdown-man fraud provides a pretty good answer to that question.

So, the 'theory' of evolution is based on the hypothesis that contemporary man today has evolved from his primate ancestors, diversifying from them between 4 and 10 million years ago. Although no definitive consensus has yet been reached by the evolutionary researchers, the generally accepted list of ancestors of humans reads as follows:

Australopithecus or 'southern ape'

Homo habilis or 'tool using man'

Homo erectus or 'upright man'

Archaic Homo Sapiens or 'old modern man'

Homo sapiens or 'modern man'

According to the evolutionists the first ape ancestors of man, *Australopithecus* were creatures which had some human-like features but possessed mostly ape-like characteristics. Some branches of the *Australopithecus* have allegedly become extinct and the others developed into the Homo (human) strain. Evolutionists also insist that *Homo erectus* and its subsequent incarnations were almost identical with contemporary man.

Today there are over 200 species of apes still extant. However, it is claimed that there were in total, more than 6500 species of primates that lived in ancient times but are now extinct. According to the estimates of scientists, only 3% of these primates are known. The species *Australopithecus* named by evolutionists are actually extinct apes which share some common structural characteristics with today's apes.

The primary criteria used by evolutionists in categorising and evaluating human fossils are bipedalism (upright walking) cranial capacity (the volume of the brain-pan) and cranial shape. Various classifications are evaluated according to those criteria.

Yet, some of these criteria, especially the cranial capacity are extremely unreliable. For example, the generally accepted cranial capacity figure for a contemporary ape is a maximum of 750 cubic centimetres (cc). The cranial capacity of humans is said to range between 900-2200cc, but among the Australian Aborigine natives, there are quite a number of individuals who have a capacity of around 850cc and furthermore cranial capacity is obviously subject to huge variations, depending on age, sex, race, and other criteria. Cranial capacity can therefore never be a reliable means of measurement.

The crania of ape fossils and the crania of today's apes are very similar to each other, being narrow and long. However, human crania are more voluminous with wide foreheads, the skull is flat with no protrusions, eyes are wide apart and the shape of the eyebrow ridges above the eyes change according to racial traits. In addition, the mandibles of humans are very much different from that of apes, bearing a distinctly parabolic shape.

To continue the comparison, the arms of apes are longer in relation to the body and their legs are shorter, both toes and fingers of apes have grasping abilities and they are all quadrupeds – all true of both primitive and modern species. Indeed, their entire skeleton is designed for a quadrupedal-type body structure. They stand on two feet only rarely, for example when reaching upwards to grasp tree branches or pick fruit, but generally spend most of their time on all-fours.

Bipedalism is a characteristic exclusive to humans (in primates) and this quality is the factor that most distinguishes human beings from other mammals. A human hip, pelvis, back-bone, and spinal cord are designed only for a biped and could not function correctly in a quadrupedal frame. In short therefore, when analysing the 'proof' of evolution, one could realistically say that the most important and binding criterion should be bipedalism. Bipedalism is the critical factor that distinguishes humans from apes and therefore the focal point of the argument should be the question of whether our so-called 'ancestors' walked upright or not.

One of the most enduring chapters of the apocryphal human evolution story is *Neanderthal man*. Neanderthals, whom even the evolutionists deem to be 'real' human beings were regarded for some considerable time as 'a primitive human race' by the evolutionists and are considered by them as an intermediate, transitional form from ape to man, possibly in an attempt to solve the 'missing-link' conundrum which still haunts evolutionary theory to this day and which has never been adequately explained.

The story of Neanderthal man began in the Neander valley in what is now modern Germany, where a local schoolteacher discovered a skull fragment, a thighbone, and other small pieces of a skeleton in 1856. These pieces were subsequently studied by an anatomy professor named Schaafhausen at Bonn University and were eventually considered, after many surveys and comparisons, to be a typical human male with no anatomical abnormalities. According to Schaafhausen who made the first study, the bones belonged to an old human race, possibly to a Barbarian tribe who resided there before the Germanic races moved into the region.

Some years later however, the fossils were sent to the University of Berlin and re-examined there by Professor Rudolf Virchow. Virchow who later in life came to be regarded as the 'father of pathology,' made a diagnosis which still remains valid today; that these bones belonged to a *Homo sapiens* (modern human) who had suffered from severe arthritis in his childhood and who had died from what appeared to be several blows to the skull.

Nevertheless, William King an anatomy professor from Queens University in Ireland who studied the fossils after Virchow, produced a totally new interpretation of the facts, which was in effect responsible for the Neanderthal man 'legend.' As a long-time passionate advocate of the theory of evolution, King drew his conclusions from the structure of the bones in accordance with evolutionist prudence. He pronounced that this fossil was more 'primitive' than modern man and therefore could not be classified as such. He also assigned to the fossil, it's now ubiquitous scientific name, *Homo Neanderthalensis*. According to King, it was a member of the Homo (human) species; but at the same time too primitive to be a human.

Two years later, similar skeletons were found in Belgium. These skeletons, which did not attract much attention initially, were subsequently brought to the attention of those who were looking for the supposed ape-ancestors

The Falsification of Science

of man, influenced strongly of course by Darwin's book, '*The Origin of Species.*'

In 1908, further Neanderthal skeletons were found in Moustier in the region of La Chapelle-aux-Saints, France. These were studied by Professor Boule from the Paleo-anthropology Institute in Paris, himself a dedicated and passionate supporter of evolutionary theory. Professor Boule was indeed responsible for creating the popular, primitive Neanderthal man image in our minds. Boule described his findings as follows:

"Neanderthals seem to be closer to apes than any other group of man and their intelligence is not wholly developed. The composition, position and the order of the cerebellum and spinal cord are the same as the apes. Besides, the feet have the same grasping attribute as in chimpanzees and gorillas. The anatomical structure of Neanderthals indicates that they walk in an awkward and clumsy way."

At the same time, Professor Boule was responsible for the first Neanderthal face and body shape reconstruction. According to this reconstruction, which he made whilst relying upon his own preposterous preconceptions, *"Neanderthal man is a half-man and half-ape being. He cannot walk upright and stoops, as do apes."* This utterly baseless theory made by Boule in accordance with his subjective interpretation of the Neanderthal fossils he had in his possession is responsible for the popular mental image we have of Neanderthal man, which still abides to this day.

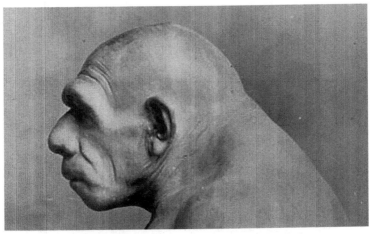

Popularised depiction of a Neanderthal

Despite all the unstinting efforts of the evolutionists, this subjective approach to Neanderthals began to change in the 1950s. Advanced technology began to confirm that Neanderthals were by no means primitive humans, in sharp contrast to the prevailing view. In light of the advent of this new data, these questions were still pertinent; was Neanderthal man, alleged to live only 30,000 years ago, really as primitive as the evolutionists claimed and were Neanderthals primitive creatures who had no civilisation and unable to even walk upright?

These questions were answered by two researchers who examined the La Chapelle-aux-Saints fossils in 1957. These anatomists, whose names were Straus and Cave discovered exactly why the fossil man found in 1908 and depicted in a reconstruction by Boule, stooped. As Professor Rudolf Virchow from Berlin University had pointed out originally, this person that this fossil represented had also suffered from chronic arthritis, just as had the original Neanderthal man who was found in 1856. This insidious bone disease was deforming the shape of the spinal column and led to stooping due to the gradual decaying of the bones. His mandible bone was also deformed and in short, the reason why the Neanderthal fossil possessed a bent posture was the simple fact that he suffered from nothing more uncommon than severe arthritis and not as had been proffered by the evolutionist dogma, his relationship to a primitive species of man.

In all other aspects, 'Neanderthal man' possessed human characteristics. His big toe was not bent as Boule claimed; his thighbone was exactly the same as that of modern man and the report prepared by Straus and Cave culminated with the following words:

"If they had come back to life today, most probably they would not be discriminated from the other people in the New York subway, provided that they had bathed, were shaved and wore modern suits."

Today, evolutionists remain evasive on the subject of Neanderthal man. It has been proven beyond reasonable doubt that the reason why the fossil was stooped as illustrated in Boule's crude and deceptive reconstruction, was the presence of severe arthritis. An authority on this subject, Erik Trinkaus of The University of New Mexico, remarked...

"Detailed comparisons of Neanderthal skeletal remains with those of modern humans have shown that there is nothing in Neanderthal anatomy that conclusively indicates locomotive, manipulative, intellectual or linguistic abilities inferior to those of modern humans."

The evolutionists deliberately ignore the difference between the average 1400cc cranial volume of modern man and the 1750cc volume of some so-called Neanderthals. They know very well that the announcement of this fact would pose another serious problem to their weak thesis. Since the evolutionists interpreted the cranial volumes they found, as evidence of evolution, accepting the fact that several Neanderthals had an even larger cranial volume than 'modern man' would imply a regression in the evolutionary process as this would simply mean that Neanderthals were more intelligent than modern humans.

Today Neanderthal man, as indeed is the case with many other subjects, is an assumed truism, an 'assumption' that has been deceptively transmuted into hard fact. The mainstream media and the film industry routinely discuss and treat the topic as though it were absolute, proven fact and not just a flimsy, insubstantial theory at best, as do educators and scientists. This is a recurring theme that as will become apparent, we find in many, many topics from biology to physics and chemistry alike.

The entire premise of Darwin's theory of evolution was the idea that it was made possible through natural selection. This concept is based on the suggestion that those members of a species that are a little stronger, a little larger, a little smarter, or run a little faster will live longer to procreate with these similar, superior adaptations. It implies that millions of generations later the changes will result in a new, superior species. These adaptations are referred-to as 'links' or 'intermediates' between the old species and the new.

The idea of natural selection sounds plausible when, for example, considering deer. The deer that can sense danger the quickest and run the fastest, are able to escape their predators on a more consistent basis and therefore by default live longer and their descendants can thus dominate the gene pool with their ancestors' inherited characteristics. I take absolutely no issue with this premise; this is micro-evolution in action and makes perfect sense.

However, other examples of natural selection evolution, have many flaws. One small example among many, of flawed 'evolution' is the idea that one day, a wingless creature began to evolve a wing and this fundamental issue is never addressed by evolutionists. Think about this... How could an entire, wholly functioning wing evolve in one step? A fledgling wing stub would not make the creature more adaptable to his environment and the first wing stubs would be much too small to enable the first bird to fly. There would of necessity have to be a gradual development of a wing, yet these intermediate stages of development would be useless – so why would they evolve? They would not be of benefit to the creature in any way whatsoever.

This is a paradox and the antithesis of the evolutionary theory of natural selection, which states that organisms adapt and change in order to survive better in their environment. Any 'bird' with a partially formed wing would be at a huge disadvantage within its environment. And why would the 'almost-bird' continue for millions of generations to gradually improve a wing stub that is useless to its survival? The theory of evolution is based on natural selection of the 'fittest' member of a species, not the weakest. A creature with a partial, indeed quite useless semi-wing would be at a severe disadvantage compared with its peers.

We are then asked to believe that some of these 'almost-birds' got tired of carrying around a worthless, incapacitating semi-wing, so they grew fingers at the end to help them climb trees. Similarly, ridiculously, the wings became arms and thus another new species was developed.

Evolutionists state confidently that birds (or 'almost birds') also developed hollow bones over a time span of several million years, to reduce weight in order to better facilitate flight. How could a bird possibly pass along this obviously long-term plan down millions of generations in order to keep the 'lighter-bone plan' progressing? The evolutionary concept of growing a wing and adapting bones to become hollow to enable a function which would become active only several million generations later, in fact **violates** and does not support the very foundation of evolutionary theory, ie. natural selection.

"... The problem is, it [evolution] is not a complete theory of life on Earth, even as a skeleton. It cannot explain even the broader points of speciation, since it cannot explain how equal environments create unequal selection. To take one example, the Serengeti is a pretty consistent environment. The lions and giraffes and zebras and so on live in the same grass, in the same air, drink the same water, under the same trees. What, precisely, caused them to evolve so differently, not as individuals, but as species? Mutation happens to individuals, not to species. A mutation happens to a gene, which is expressed in a specific offspring or set of offspring. These offspring, if superior, then deliver the gene to the whole herd over time, which then disseminates it further. So far, so good. But return to the individual offspring. Say we are evolving a giraffe by this method. The required mutation is then a long neck. But this mutation is only useful to a creature that is already a giraffe or pre-giraffe. The mutation will not help a pre-lion or a pre-zebra, since the lion eats meat, not leaves, and the long neck would just slow the zebra down. The mutation is useful only to an animal that is already living under trees, already trying to reach higher leaves. But if it could not reach the leaves before the mutation, why was it there? Was it just hanging around, looking up at those unused leaves, waiting for a mutation? The combination of specific mutation and specific environment is so unlikely that even great time cannot explain it.

What about the orchid with a four-inch tube, which requires a fly with a four-inch nose to pollinate it? The mutations cannot take great time to sync up, they must do so immediately or one or both species will fail. If the flower mutates to a five inch tube first, the fly cannot reach the nectar and quits visiting it. If the fly mutates to a five inch nose first, the pollen is not deposited on him, and the flower again fails. Neither species can wait around for accidental mutations of just the right sort. They must evolve together, and this is so unlikely as a matter of mutation statistics that it must show up the theory as a whole." Miles Mathis, '*The Illogic of Atheism*'

And of course, birds are by no means the only species that proves the theory of natural selection to be grossly unrealistic, not to mention just plain wrong. This same issue may be found in all species in one way or another. Take fish for example...

We are told by evolutionists that millions of years ago, a fish struggled out of the sea onto dry land and became a land creature. Let us examine this idea more closely. So, a fish manages to wriggle out of the water and onto the beach, but of course, cannot breathe air. This does happen and indeed has probably happened countless millions of times down the centuries and millennia. Fish are not exactly the brightest lamps in the street and undoubtedly are prone to do stupid things from time to time.

Sharks are often stranded on beaches, where they die a lingering, probably very unpleasant death by suffocation in air. Do you think it is maybe possible that these sharks are attempting to expedite a multi-million generation plan to grow legs and breathe air? The entire concept is utterly risible in the extreme, but let us return to the fish story...

The gills of a fish are their version of the lungs of land-based creatures and their function is to extract life-sustaining oxygen from water, and not from air, as we land organisms do. A beach-stranded fish will often partially choke before struggling back into the safety of the water. Why would he do such a stupid thing? What is the point of attempting the transition to a land-based organism in any event? This is not a survival function – as is 'true' microevolution. And so this gasping and choking continues for millions of generations until the fish chokes less and less does it? His gills evolve into lungs so he can breathe air on dry land, but now he is at risk of drowning in the water. This idea presents exactly the same problem as bird wings. How would it be possible for a gill to physically transform itself into a lung – even over millions of generations?

The Falsification of Science

And finally, one fine day several million years later, the 'fish' simply stays out on the land and never goes back into the water and has in effect, become a 'lizard.' Utterly laughable, I submit.

We all accept this so-called and at first glance, plausible 'evolutionary process' without really giving it too much thought, or at least I know I did for most of my life – until I began to think logically about it. How can any organ you would care to name, evolve in small incremental stages? A partial eye or a partial kidney is as useless as no eye or kidney at all, or debatably, even worse than useless. I would urge you to stop reading for a moment and think logically and clearly about this point before continuing.

There are many other examples of fatal flaws in the theory of evolution, indeed far too numerous to cover in any detail or to do justice to in a generalised work such as this. However, even using the small amount of evidence presented here, there now would appear to be only one possible conclusion:

The evolutionary theory asserting that humans came into existence by macroevolution from single-celled life forms and latterly from other primates is not supported by any convincing, concrete evidence whatsoever and indeed is invalidated completely by much evidence to the contrary. The whole premise of the evolution of man is actually based on extremely subjective interpretations, poor or bad science, deliberate distortions and even the outright forgeries of many unscrupulous evolutionists who seek to convince us that, yet another huge falsehood is the truth.

"Why doesn't the scientific community abandon Darwin's failed hypotheses? Simple: The Jewish-dominated media and educational establishment are determined that, like unconditional support of Israel, Holocaust mythology, hate laws, and 'civil rights' favouritism, there will be no end to the relentless force-feeding of evolution. Belief in evolution is a prerequisite for Jewish supremacism's New World Order." Reverend Ted Pike, researcher, 16th May 2011

In fact, the theory of evolution is actually in serious trouble. The advancement of biological knowledge has generated scientific facts that contradict the evolutionary theory and do not confirm it as disingenuously claimed; a fact that Professor Steven Jay Gould of Harvard described as *"the trade secret of palaeontology."*

The fossil record simply does not support the evolutionary theory, which claims that there once existed a series of successive life forms leading to contemporary organisms. The theory states that infinitesimal changes within each generation evolve into a new species, but the problem is that scientific facts do not support this wild assertion.

The fossil record contains no intermediate or transitional forms which is popularly known as the 'missing link' issue, and this problem exists in all species. What is really significant is that the missing link problem is becoming worse, not better, with the discovery of more and more fossils. The missing links are still not being discovered, even after over 160 years of endeavour, which I submit, proves that they have never existed. Darwin assumed transitional forms would be discovered in the fossil record over time, but that has not been the case. The fossil record, or lack thereof, is a major embarrassment to evolutionists and a fatal setback to the theory of evolution. New species exploded into being, seemingly from nowhere and new fossil discoveries continue to prove evolution to be grossly wrong.

There are undoubtedly some genuine, well-meaning scientists who firmly believe in the evolutionary model, but they have been duped in just the same way as the rest of us, by the deliberate subterfuge of those who are determined to perpetuate a lie. Or could they simply be victims of the Hegelian trap of mainstream religion-based creation versus evolution, believing as they do in the more 'plausible,' scientific alternative?

Whatever the answer to that particular poser may be, it seems clear to me that macroevolution is a huge deception, a pseudoscience and just one of many in a long list of deceptions emitting from the forces of evil, that we are forced to endure in our lives, on a daily basis.

The Falsification of Science

Dinosaurs

The dinosaurs, it is said by evolutionists, almost literally 'exploded' into being during the Triassic period only to be wiped-out in a single cataclysmic event around 65 million years ago. This catastrophe in itself is puzzling to say the least. Surely at least a few would have survived, as did individuals of most other species, but no, it appears that very conveniently, every last dinosaur on Earth was wiped out.

The fossil record such as the petrified bones found in the ground as at the Dinosaur National Park in Jensen, Utah, USA shows no intermediate or transitional species. So where are the millions of years of fossils showing the transitional forms of dinosaurs? They do not exist, because like all other life forms, dinosaurs did not evolve from virtually nothing.

In fact, I do not believe that dinosaurs ever even existed at all.

We are being subjected here to yet another hoax and as with the 'sub-hoaxes' outlined in the previous chapter, that prop-up the nonsensical Big Bang theory, this hoax is likewise designed as a 'patch' for evolutionary theory.

The species classification *Dinosauria*, meaning 'giant lizards,' was first defined by a certain Sir Richard Owen of the Royal Society in 1842. This was nothing but pure speculation on his part, as at this point in time, no alleged dinosaur fossils had ever been discovered. Does that fact in itself, not strike you as rather odd?

However no matter, as the Freemasonic mainstream media, worldwide, very quickly began to report in its own inimitable manner, on these amazing yet never-existent creatures, creating a sensation amongst both the 'chattering classes' and scientists alike. Then around ten or so years later, to everyone's delight and amazement, the first dinosaur fossil was found in the region of the upper Missouri River in the US mid-west. 'Great – proof at last,' cried all the various vested interests in unison. Amazing is it not that no-one in the entire history of the world had ever discovered a dinosaur fossil until… yes, you guessed it, very shortly after the creatures were first described?

Ferdinand Hayden, the discoverer of this remarkable artefact, sent one of the creature's teeth to a leading palaeontologist, Joseph Leidy, who soon identified this as belonging to a creature now named as a 'Trachodon,' which in itself is remarkable as no-one at this stage had classified any dinosaur species at all. Indeed, there were none to classify, but nevertheless, this was accepted as proof of these exotic creatures former existence.

Of course, it is redundant for me to state that it is impossible to construct an entire, hypothetical living organism, based on a tooth alone, but that is exactly what happened. Is it not dubious to say the least, that a myriad of ancient bird / lizard-like creatures' transitional forms should be found by archaeologists who coincidentally were all looking for them at more or less exactly the same time that evolution had been mooted by several sources? And is it not even more dubious that these fossils had supposedly existed for tens or even hundreds of millions of years but had never, ever been found by any other human civilisation until the theory of evolution's emergence in the mid-nineteenth century?

"Why are there no discoveries by native Americans in all the years previous when they roamed the American continents? There is no belief in dinosaurs in the native American religions or traditions. For that matter, why were there no discoveries prior to the nineteenth century in any part of the world? According to the World Book Encyclopaedia, 'before the 1800s, no-one knew that dinosaurs existed.' During the late 1800s and early 1900s, large deposits of dinosaur remains were discovered... Why has man suddenly made all these discoveries?" David Wozney, *'Dinosaurs: Science or Science Fiction?'*

In fact, no culture in the world had ever discovered a single dinosaur bone or fossil before the mid-1850s and then suddenly dinosaur remnants and relics were appearing everywhere, as if by magic. And, significantly in my view, neither are there any ancient tribal belief systems, written or even verbal histories passed down the generations from ancient times, of anything resembling these giant lizards. North and South America, Asia, Africa, Europe, all it would seem had copious deposits of dinosaur fossils. This is very strange as all these places had been inhabited for many thousands of years and more recently had been well explored by geologists and palaeontologists aplenty. Again, please ask yourself this question, 'why had no dinosaur fossils ever been found before the actual idea of dinosaurs was first dreamed up?'

The book, '*The Dinosaur Project,*' written by a paleontological journalist, Wayne Grady, claims that the period from around 1870 to around 1880, became, *"a period in North America where some of the most underhanded shenanigans in the history of science were conducted."* This was what became known as the 'The Great Dinosaur Rush' or the 'Bone Wars.'

Two adversarial palaeontologists, Edward Drinker Cope of the Academy of Natural Sciences and Othniel Marsh of the Peabody Museum of Natural History, began a bitter, long-lasting rivalry and a shared obsession with dinosaur fossil hunting. Originally good friends, they soon developed into deadly enemies in their desperate attempts at 'one-upmanship' which involved double-crossing, bribery, spying, theft, slander, the destruction of each other's respective 'samples' and eventually mutual hatred.

Marsh was alleged to have 'discovered' more than five hundred species including at least eighty different dinosaur types. But significantly perhaps, of the one hundred and forty or so different dinosaur species discovered by both men, only thirty-two are now considered 'real' and all the rest have been subsequently proven fraudulent. Not one of these species claimed, was actually accompanied by a complete skeleton, so all their work was based solely on reconstructions from individual bones or teeth. In truth, to the present day, no complete dinosaur skeleton has ever been found, and so, as we learned earlier in this chapter, 'put not your trust in reconstructions!'

"Discoveries and excavations seem not to be made by disinterested people, such as farmers, ranchers, hikers, outdoor recreationists, building construction industry basement excavators, pipeline trench diggers, and mining industry personnel but rather by people with vested interests, such as palaeontologists, scientists, university professors, and museum

organization personnel who were intentionally looking for dinosaur bones or who have studied dinosaurs previously. The finds are often made during special dinosaur-bone hunting trips and expeditions by these people to far-away regions already inhabited and explored. This seems highly implausible. More believable is the case of the discovery of the first original Dead Sea scrolls in 1947, which were unintentionally discovered by a child, and which were all published by 1955. In some cases of a discovery of dinosaur bones by a disinterested person, it was suggested to them by some 'professional' in the field to look or dig in a certain area. Also, very interesting to note are special areas set aside and designated as dinosaur parks for which amateur dinosaur hunters are required to first obtain a dinosaur hunting license." David Wozney, 'Dinosaurs: Science or Science Fiction'

It seems that establishment-funded archaeologists and palaeontologists are able to discover dinosaur fossils wherever they may search. In one of the largest dinosaur excavation sites, the Ruth Mason Quarry alone, over two thousand fossils were allegedly discovered. Casts of the 'original skeletons' assembled from these bones are currently on display in museums worldwide. The head of palaeontology at La Plata Museum, Florentino Ameghino, incredibly is claimed to be the discoverer of more than six thousand fossil specimens allegedly discovered throughout his career – all of which were in Argentina. And the dinosaur hunter, Earl Douglass sent three hundred and fifty tonnes of excavated 'dinosaur bones' to the Carnegie Museum of Natural History throughout his career, all originating at the 'Dinosaur National Monument,' in Utah. During an expedition to Patagonia, Dr Luis Chiappe and Dr Lowell Dingus supposedly discovered thousands of dinosaur eggs at a site measuring only a few hundred square yards. Statistical experts have commented on how these finds of huge quantities of fossils in one area, by just a few heavily-invested individuals, with much to gain, totally contradicts the laws of natural probability and lends much credence to the likelihood of forgeries or of concentrated 'planting' of fake artefacts.

"Dinosaur bones sell for a lot of money at auctions. It is a profitable business. There is pressure for academics to publish papers. Museums are in the business of producing displays that are popular and appealing.

The Falsification of Science

Movie producers and the media need to produce material to sell to stay in business. The mainstream media loves to hype alleged dinosaurs finds. Much is to be gained by converting a bland non-dinosaur discovery, of a bone of modern origin, into an impressive dinosaur find, and letting artists' interpretations and imaginations take the spotlight, rather than the basic boring real find. There are people who desire and crave prestige, fame, and attention. There is the bandwagon effect and crowd behaviour. And then there are people and entities pursuing political and religious agendas. Highly rewarding financial and economic benefits to museums, educational and research organizations, university departments of palaeontology, discoverers and owners of dinosaur bones, and the book, television, movie, and media industries may cause sufficient motivations for ridiculing of open questioning and for suppression of honest investigation." David Wozney, 'Dinosaurs: Science or Science Fiction'

'Tyrannosaurus Rex bones' have actually been sold at specialist auctions for upwards of $12 million. This demonstrates exactly what a lucrative business the field of dinosaur fossil-hunting can be, and there in itself is motive enough for the fakery. And to repeat, it just so happens that it is palaeontologists and museum officials who serendipitously seem to be the most prolific finders of them.

The first dinosaur to ever be publicly displayed was the 'Hadrosaurus foulkii,' at Edward Drinker Cope's Academy of Natural Sciences in Philadelphia. The bones were co-discovered by Joseph Leidy, Cope's esteemed professor, and the man responsible for the 'Trachodon,' the original dinosaur find back in the 1850s. The original Hadrosaurus reconstruction, which is still on display today, shows a huge plaster cast bipedal reptile standing upright using its tail as a third leg. What few people know, however, is that no skull was ever discovered, and there are no original bones in the public exhibit. It is merely a 'replica.'

"A visual and a sculptural artist were promptly hired to invent a skull, and from the illustrations of another artist, who had depicted the Iguanodon, the two artists drew the same face for the Hadrosaurus foulkii. The people involved could now technically defend the existence of this dinosaur if someone were to ask. The stunt worked out so well, and

fooled the public so thoroughly, that they could later change the head of the creature without anyone noticing. To this day, Hadrosaurus foulkii is on display at the Academy of Natural Sciences in Philadelphia. The bones are said to be kept behind heavy, closed doors, but a plaster copy is exhibited in their place... So, we learn of an iguana skull being substituted for the skull of a dinosaur on display. Was the public told at the time? What are we not being told today?" David Wozney, *'Dinosaurs: Science or Science Fiction?'*

However, what we are also not told, is that replicas are the rule and the not the exception. To this day, not a single complete skeleton of any dinosaur has ever been found. All the museum displays, models, mannequins, cartoons, and movies of prehistoric monsters you have ever seen are all imaginative reconstructions based on incomplete skeletons arranged in a manner palaeontologists believe to be the most 'realistic.' Furthermore, the skeletons exhibited in museums are all self-admittedly by palaeontologists, intricate fabrications made of plaster, fibreglass, other animal bones, and of course, copious amounts of wire and glue. They are NOT original bones or fossils.

When 'dinosaur bones' are transported and prepared they use strips of burlap soaked in plaster to protect the fossils. Then after applying a tissue separator to keep the plaster from direct contact with the fossil, the soaked burlap strips are applied until it is totally encased in a protective mummy-like coating ready for safe transportation. In an article entitled *'A Fossil's Trail from Excavation to Exhibit,'* one palaeontology insider confessed that, *"...through mould making and casting we can totally fabricate limbs, ribs, vertebrae, etc., for the missing pieces of an articulated skeletal mount. Plaster, fibreglass, and epoxies are often and commonly used. In reconstruction work on single bones, small to large cracks can be filled in with papier mache or plaster mixed with dextrin, a starch that imparts an adhesive quality and extra hardness to regular moulding plaster. We've also had success using epoxy putties. Large missing fragments can be sculpted directly in place with these same materials."*

In other words, museum personnel consistently and routinely use plaster and other materials to transport and fabricate skeletons and missing or incomplete bones. In fact, the huge 'dinosaur bone'

The Falsification of Science

displays found in museums across the world are all admitted to be carefully prepared fakes. No independent researcher has ever examined a real dinosaur skull. It is claimed that all the actual fossils are kept in high-security storage, but only a select few palaeontologists are ever allowed to examine them, so the ability to ascertain their authenticity is kept 'in-house' and well away from the general public's prying eyes. How convenient for them.

"Most people believe that dinosaur skeletons displayed in museums consist of real dinosaur bones. This is not the case. The real bones are incarcerated in thick vaults to which only a select few highly placed researchers hold a key, which means that NO independent researcher has ever handled a Tyrannosaurus Rex bone. When people unaffiliated with the paleontological establishment attempt to gain access in order to study these dinosaur bones, they are met with refusal upon refusal… Only around 2100 dinosaur bones sets have been discovered worldwide, and out of these, only 15 incomplete Tyrannosaurus Rex bone sets have been found. These dinosaur bone sets have never formed a complete skeleton, but from these incomplete bones sets, palaeontologists have constructed a hypothesis about the appearance of the whole skeleton, which they have modelled in plastic. If thousands of longnecks and large carnivorous reptiles had really roamed Earth, we wouldn't only have found 2100 dinosaur bone sets, but millions of bones, with ordinary people tripping over them when digging in their vegetable patches." Robbin Koefoed, *'The Dinosaurs Never Existed'*

"When children go to a dinosaur museum, are the displays they see, displays of science or displays of art and science fiction? Are we being deceived and brainwashed at an early age into believing a dinosaur myth? Deep probing questions need to be asked of the entire dinosaur business. There may have been an ongoing effort since the earliest dinosaur 'discoveries' to plant, mix and match bones of various animals, such as crocodiles, alligators, iguanas, giraffes, elephants, cattle, kangaroos, ostriches, emus, dolphins, whales, rhinoceroses, etc. to construct and create a new man-made concept prehistoric animal called the dinosaur. Where bones from existing animals are not satisfactory for deception purposes, plaster substitutes may be manufactured and used. Some material similar or superior to plasticine clay or plaster of Paris would be suitable. Moulds may also be employed. What would be the motivation for such a deceptive

endeavour? Obvious motivations include trying to prove evolution, trying to disprove, or cast doubt on the Christian Bible and the existence of the Christian God, and trying to disprove the 'young-earth theory.' The dinosaur concept implies that if God exists, He tinkered with His idea of dinosaurs for a while, then probably discarded or became tired of this creation and then went on to create man. The presented dinosaur historical timeline suggests an imperfect God who came up with the idea of man as an afterthought, thus demoting the biblical idea that God created man in His own image." David Wozney, *'Dinosaurs: Science or Science Fiction?'*

Key 'dinosaur skulls' into a search engine and you will find a variety of replicas, tailor-made dinosaurs, and 'museum-quality' skeletons. One of the largest and most-renowned suppliers of fake dinosaurs is the *Zigong Dino Ocean Art Company* in Sichuan, China which provides natural history museums worldwide with ultra-realistic dinosaur skeletons made from real bones. Chicken, frog, dog, cat, horse, and pig bones are melted down, mixed with glue, resin, and plaster, then used as base material for re-casting as 'dinosaur bones.' They are even given realistic-looking fractures and an antiquated, 'fossilised look' to achieve the right effect. Their website boasts that, *"...over 62% of our output goes to American and European markets, which means we will understand and are familiar with the intricacies and regulation of exporting to these regions... Since we are a partner of dinosaur museums, all products are made under the guidance of experts of the Chinese Academy of Sciences... We have gained a global sales network reaching the USA, Brazil, France, Poland, Russia, Germany, Saudi Arabia, South Korea, Thailand, Indonesia, exhibited in Peru, Argentina, Vancouver, Cincinnati, Chicago and other places."*

"I have heard there is a fake-fossil factory in northeast China, in Liaoning Province, near the deposits where many of these recent alleged feather dinosaurs were found." Alan Feduccia, University of North Carolina palaeontology professor

"The possibility exists that key dinosaur bones on display have been artificially modified through sculpture and carving. Bone sculpture is not an unknown human activity. Many cultures participate in creating man-made objects out of existing bones, totally unrecognizable from the

original shape. Is the dinosaur industry a customer of this sort of business? Is it possible that dinosaur skeleton replica are secretly assembled or manufactured in private buildings out of public view, with bones artificially constructed or used from a number of different modern-day animals? Why bother having any authentic original fossils at all if alleged replicas please the public?" David Wozney, 'Dinosaurs: Science or Science Fiction?'

Another issue with the existence of dinosaurs is their highly unnatural structural dynamics. Many dinosaur skeleton reconstructions feature bipedal 'monsters' like the Tyrannosaurus Rex with its forward-leaning torso and a head far larger and heavier than its counter-balancing tail. Many museum displays cannot even stand up under their own weight, without support so it is highly unlikely that beasts this large and disproportionate could exist at all. The loads acting on their skeletons are so great that calculations made by sceptical scientists indicate that the bones of the largest dinosaurs would buckle and crack under their own immense weight. Experts have also pointed out that dinosaurs would have to have moved much slower than portrayed in movies to prevent sudden shocks to their skeletons.

"This idea of slow moving animals does not agree with the biomechanical analysis of dinosaurs, which indicate that the dinosaurs were agile, active creatures. This is the paradox between the dinosaurs size and lifestyle. Many displays and drawings of dinosaurs appear to be an absurdity, showing a two-legged animal that would be totally off-balance, with the weight of head and abdomen much greater than weight of tail, which is supposed to act as a counter-balance. Is the dinosaur industry a case of science trying to meet public desires or expectations? The movie Jurassic Park is an example of showing dinosaurs much larger than any current displays in museums. After the movie came out, it is interesting to note that many articles were written asking 'Is this possible?' I can recall a report of dinosaur DNA being discovered preserved in amber, which later turned out to be false." David Wozney, 'Dinosaurs: Science or Science Fiction?'

"Overall, several millions of dollars have been spent promoting the existence of dinosaurs through movies, TV, magazines and comics. The world of movies and palaeontology are like Siamese twins. People's view on the existence of dinosaurs is based not on firm evidence, but on Hollywood *fixated artistic impressions. Documentaries colourfully illustrate each dinosaur's characteristics, like colours, weight, and muscle mass, but Don Lessem (advisor for Jurassic Park) admits that this is pure guesswork – consider for instance the question of how much these dinosaurs weigh. Don Lessem says, 'Scientists don't know how much dinosaurs weighed!'"* Robbin Koefoed, *'The Dinosaurs Never Existed'*

Dinosaurs are always presented to the public using colourful artistic reconstructions, drawings, models, gigantic skeletons in museums, cartoons and movies showing these magnificent but nevertheless fake creatures in explicit detail. But the fact is that from the assigning and arrangement of bones in each species, to the impossible-to-discern soft tissue, skin, eyes, noses, colour, hairiness, texture etc., and just like the many supposed ape-man species, all dinosaur reconstructions are 100% fictional, fabrications created by invested and inventive evolutionists. They purposely present dinosaurs to children in the media to condition their young impressionable minds into believing their deceptive, coldblooded lies. Feature length cartoons such as 'Ice Age' and 'The Land Before Time,' and movies such as the 'Jurassic Park' franchise and 'Dinosaur Island,' coupled with their associated 'merchandise,' colouring books, dolls, plastic toys, school textbooks, and huge displays in children's museums certainly contribute towards the imprinting of this sheer fakery onto young and still-developing minds. And of course these children grow up to be adults who have been thoroughly propagandised by a lifetime of false information, believing it to be impossible that they were ever lied-to on such a grand scale.

National Geographic and the 'Ice Age' movies were produced by the Freemason, Rupert Murdoch's News Corporation, and 20th Century Fox. The masonic production company, Universal Studios created 'Jurassic Park' and 'The Land Before Time.' They are owned by Comcast; whose chief shareholders are the Freemasons and the Illuminati's very own, JP Morgan and the Rothschilds. 'Discovery Channel' which regularly features many dinosaur documentaries is also financially backed by N M Rothschild and Sons Limited, of London.

The Falsification of Science

The former palaeontology student, Michael Forsell claimed in a radio interview with leading palaeontologist Jack Horner, that he (Horner) was, *"...a total fraud, fabricating evidence and perpetuating the myth of dinosaurs."* He continued by saying, *"I started my career in the field of palaeontology, only to leave my studies once I realised the whole thing was a sham. It's nonsense, most of the so-called skeletons in museums are actually plaster casts. They even do it openly on documentaries now, preserving the bones my ass! I struggled as a student, mainly because I could not tell the difference between a fossilised egg and an ordinary rock, and of course there is no difference. I was treated like a leper when I refused to buy into their propaganda, and promptly left the course. Dinosaurs never existed, the whole shebang is a freak show, they just grab a couple of old bones and form them into their latest Frankenstein's monster-like exhibit... We are all being fooled and it's wrong, but together we can stop it."*

Many palaeontologists claim that as dinosaur fossils have been radiocarbon dated to be tens of millions of years old that their authenticity is thereby proven. The fact is, however, that the methods used to date dinosaur fossils involve not measuring the actual fossils, but the rocks near where they are found. Most fossils are found near the surface of the earth, and if a modern-day animal were to die in the area, palaeontologists would probably date them as being the same age as the rocks.

Dr Margaret Helder in her book *'Completing the Picture, A Handbook on Museums and Interpretive Centers Dealing with Fossils,'* wrote that, *"Scientists used to be very impressed with the potential of radiometrics for coming up with absolutely reliable ages of some kinds of rocks. They do not feel that way anymore. Having had to deal with numerous calculated dates which are too young or too old compared with what they expected, scientists now admit that the process has many more uncertainties than they ever would have supposed in the early years. The public knows almost nothing about uncertainties in the dating of rocks. The impression that most people have received is that many rocks on earth are extremely old and that the technology exists to make accurate measurements of the ages. Scientists have become more and more aware however that the measurements which the machines make, may tell us nothing about the actual age of the rock."*

One of the main reasons that evolutionists needed the existence of dinosaurs was to answer the complex problems inherent in the theory of evolution including…sea dwelling animals evolving into land-dwellers, reptiles evolving wings, feathers, flying and becoming birds, in addition to many other reptiles evolving warm blood, live births, mammary glands and turning into mammals. Through their imaginary multi-million-year timeline and a variety of supposed transitional dinosaur forms, the paleontological establishment has been promoting various sea dinosaur, reptile/birds, and reptile/mammals to bridge these gaps. Many professionals and experts in the field have disputed such findings as often as they have been presented, however.

Dr Storrs Olson, a Smithsonian Institute Scientist, wrote, *"The idea of feathered dinosaurs and the theropod origin of birds is being actively promulgated by a cadre of zealous scientists acting in concert with certain editors at Nature and National Geographic who themselves have become outspoken and highly biased proselytizers of the faith. Truth and careful scientific weighing of evidence have been among the first casualties in their program, which is now fast becoming one of the grander scientific hoaxes or our age."*

No authenticated feathers have ever been found along with dinosaur fossils, though a few exposed hoaxes certainly attempted to fake such objects. Dr Olson called the adding of feathers to their findings, *"… hype, wishful thinking, propaganda, nonsense fantasia, and a hoax."* In the 1990s many fossils with feathers were supposedly discovered in China (suspiciously close to the Zigong Dino Ocean Art Company) but when these were examined by Dr Timothy Rowe, he found that the so-called 'Confuciusornis,' was nothing more or less than an elaborate hoax. He also found that the 'Archeoraptor,' supposedly discovered in the 1990s was composed of bones from five different animals. When Rowe presented these significant findings to National Geographic, the head scientist reportedly remarked, *"…well, all of these have been fiddled with!"* National Geographic then nevertheless proceeded with their news conferences and media stories about the Archeoraptor fossils being genuine and having found a 'missing link' in evolution. Quel surprise.

The Falsification of Science

"In 1999, National Geographic magazine was 'busted' when they presented, in a colourful and fancily presented article, the missing link. An Archeoraptor dinosaur, which was supposed to support the basic tenet of evolutionary theory, that dinosaurs had slowly developed over millions of years. Their proof consisted of a fossil, where carefully arranged bone imprints gave the impression of a creature half dinosaur and half bird. The scam was discovered during a CT scan which uncovered unnatural bone links. National Geographic magazine was later forced to admit, when pressured, that the fossil was man-made!" Robbin Koefoed, 'The Dinosaurs Never Existed'

Palaeontologists claim that the Archaeopteryx is another transitional form of bird which evolved from dinosaurs, but this theory completely falls apart against overwhelming evidence to the contrary. Other species like Confuciusornis, Liaoningornis, and Eoalulavis have all been found to be contemporary with the Archaeopteryx and are indistinguishable from present-day birds. Alan Feduccia from the University of North Carolina, one of the most famous ornithologists in the world stated that, *"I've studied bird skulls for twenty-five years and I don't see any similarities whatsoever. I just don't see it. The theropod origins of birds, in my opinion, will be the greatest embarrassment of palaeontology of the 20th century."*

Larry Martin from the University of Kansas, a paleo-ornithologist said, *"...to tell you the truth, if I had to support the dinosaur origin of birds with those characters, I'd be embarrassed every time I had to get up and talk about it."*

Even if dinosaurs did evolve into birds to fill their evolution gap, it does not explain how something like the common housefly could have evolved. Flies flap their wings simultaneously 500 times per second, even the slightest dissonance in vibration would cause them to lose balance and fall, but this never happens. How could they evolve such an amazing and specialised ability?

There are so many questions that desperately need to be answered. For example…

- Why were dinosaurs never discovered before the evolutionary theory was first proposed in the mid-19th century?

- Why do palaeontologists think they can reconstruct an entire species of ancient animal from a few fossilised teeth or bones without serious questions being asked?

- Why have so many 'dinosaur discoveries' turned out to be hoaxes?

- Why are all alleged 'authentic dinosaur fossils' kept under tight control and behind securely locked doors, well away from the enquiring eyes of any independent analysts?

- Why has erosion and weathering not destroyed all these supposed prints and fossils that are allegedly millions of years old?

- If dinosaurs were supposedly wiped out by the alleged meteor impact 65 million years ago, or other such global catastrophe, why is it that all the other various animal species that exist today were not similarly wiped out?

These and many more questions need to be answered before anyone in possession of their senses, even partially, should consider the existence of dinosaurs as anything but a convenient, evolutionist myth.

"The palaeontological establishment can control which hypotheses will be constructed through textbooks and the curriculum. In this way, students are brainwashed into a pseudo-reality controlled by the text material and the teacher's authority. A short practical example: a random dental bone is found at an excavation site and from this dental bone, the rest of the skeleton is guessed at. We are not kidding about this. The entire dinosaurian field of the palaeontological program is a sham." Robbin Koefoed, *'The Dinosaurs Never Existed'*

"During the nineteenth century a new world view of evolution was being pursued by then influential people such as Darwin and Marx. During this era of thought the first dinosaur discoveries were made. Were these discoveries 'made' to try to make up for inadequacies in the fossil record

The Falsification of Science

for the theory of evolution? The following issues raise red flags as to the integrity of the dinosaur industry and cast doubts as to whether dinosaurs ever existed: (1) dinosaur discoveries having occurred only within the last two centuries and in huge unusual concentrated quantities going against the laws of nature and probability; (2) dinosaur discoverers typically and generally not being disinterested parties without a vested interest; (3) the nature of public display preparation, calling into question the integrity and source of fossils, and allowing for the possibility of tampering and bone substitution, and the possibility of fraudulent activities on a systemic basis; (4) existing artistic drawings and public exhibits showing off-balance and awkward postures that basic physics would rule out as being possible; (5) very low odds of all these dinosaur bones being fossilized but relatively few bones of other animals; (6) implications of dinosaur discoveries to the theory of evolution and the belief that man was created in God's image, suggesting possible hidden and subtle political or religious agendas served on a naive and unsuspecting public; and, (7) a lack of funding for organizations and people questioning or being sceptical of each and every discovery and public display.

The possibility exists that living dinosaurs never existed. The dinosaur industry should be investigated, and questions need to be asked. I am unaware of any evidence or reason for absolutely believing dinosaurs ever were alive on earth. The possibility exists that the concept of prehistoric living dinosaurs has been a fabrication of nineteenth and twentieth century people possibly pursuing an evolutionary and anti-Bible, anti-Christian agenda. Questioning what is being told instead is a better choice rather than blindly believing the dinosaur story." David Wozney, *'Dinosaurs: Science or Science Fiction?'*

So, there you have it. Were dinosaurs real or has their entire existence been falsified to fulfil a surreptitious agenda? Again, I would ask you to pause and stop reading for a moment and think about the dinosaur 'industry' in general. What is the purpose of all the hype and propaganda, especially that which is disturbingly aimed at children? These days we are inundated with 'dinosaur propaganda' in the form of movies, TV programmes, documentaries, children's toys, and games. Ask yourself why this would be if not to perpetuate a myth? This is an absolutely typical ploy, used by the Elite who run our lives, to convert us to the 'faith.' It is a methodology that has been particularly successful down the decades and so why would they change a successful formula?

The simple answer is that they would not, I am afraid. We are the unwitting dupes of those who control us all, invisibly from the shadows.

"An important example which proves the fact that Darwinism is one of the biggest deceptions of atheistic Freemasonry is a resolution carried in a mason meeting. The 33rd degree Supreme Council of Mizraim Freemasonry at Paris, reveals in its minutes its promotion of evolution as science, while they themselves scoffed at the theory. The minutes read as follows... 'It is with this object in view (the scientific theory of evolution) that we are constantly by means of our press, arousing a blind confidence in these theories. The intellectuals will puff themselves up with their knowledge and without any logical verification of them will put into effect all the information available from science, which our Agentur specialists have cunningly pieced together for the purpose of educating their minds in the direction we want. Do not suppose for a moment that these statements are empty words, **think carefully of the successes we arranged for Darwinism.'** *Atheistic Freemasonry in the United States has picked up the resolution of Mizraim before long. New Age magazine in its March 1922 issue stated that the kingdom of atheistic Freemasonry will be established by evolution and the development of man himself. As seen above, the false scientific image of evolution is a deception set in the 33rd degree atheist Masonic lodges. Atheist Masons openly admit that they will use the scientists and media which are under their control to present this deception as a scientific, which even they find funny."* Harun Yahya, 'The Fundamental Philosophy of Atheistic Freemasonry'

Thus, I believe that we may conclude from this information that Darwinism, the theory of the survival of the fittest and its twin sister, the *theory* of human evolution, together with the official versions of Palaeontology and Archaeology such as they are practiced, are pseudosciences. In fact, they are surreptitious tools for mind-control of the masses. Their main purpose is to cut us off from the roots of the true history of humanity and promote a simplistic, naturalistic, and mechanistic view of life as being merely 'survival of the fittest.'

We will look deeper into the Freemasonic history of the evolution hoax in a later chapter, but I will leave the last word on the topic of the evolution of species to an anonymous scientist who wished only to be identified as 'Sam,' for perhaps obvious reasons...

"To be a molecular biologist requires one to hold on to two contradictory insanities at all times. One, it would be insane to believe in evolution when you can see the truth for yourself. Two, it would be insane to admit you don't believe in evolution. All government work, research grants, papers, big college lectures — everything, would stop. I'd be out of a job or relegated to the outer fringes where I couldn't earn a decent living. The work I do in genetic research is honourable ... but in the meantime, we have to live with the 'elephant in the living room.' Intelligent design is that elephant in the living room. It moves around, takes up an enormous amount of space, loudly trumpets, bumps into us, knocks things over, eats a ton of hay, and smells like an elephant. And yet we have to swear it isn't there!" Sam

Chapter 3

Ancient Technology and Knowledge

One of the most persistent societal myths is that the brains of people of ancient civilisations, were somehow inferior to ours. This is absolutely not the case. There is no evidence whatsoever to suggest that ancient peoples were any less 'intelligent' than we are today. What is often referred-to as 'intelligence' is really simply just knowledge. I have lost count of the number of times I have heard someone referred to as 'intelligent' or a 'genius' because he or she can recite or regurgitate 'facts' at will, for example in Trivial Pursuit™ type quizzes, or even in certain educational examinations. I am sure you will agree that this ability is more akin to an ability to absorb data and possessing a good memory than the ability to assimilate, deduce or interpret information in a logical manner from facts, and which I firmly believe is the truer test of intelligence.

But it is also true that the educational establishment perpetuates the myth that 'passing exams' is a sign of intelligence. However, it most certainly is not. Again, I would submit that the ability to pass exams is simply an ability to memorise information and data and regurgitate it at will and bears very little resemblance to 'intelligence' per se.

And even so-called 'IQ' tests, promoted as being a 'true' test of intelligence are only partially useful at best. Their problem being that the questions and tasks set within them tend to be culturally biased and also that it is obviously necessary to actually be literate in order to use them. Again, literacy (the ability to read and write) does in no way determine 'intelligence.' Many ancient civilisations were wholly illiterate but nevertheless still managed incredible technological achievements and to solve complex societal problems without difficulty.

In fact, there are numerous examples of ancient technology that leave us awestruck at the intelligence of people of the past, dating back thousands of years. These were the result of incredible advances in engineering and innovation as new and powerful civilisations emerged and began to dominate the ancient world. These advances stimulated societies to adopt new ways of living and governance, as well as new ways of understanding the world and beyond. However, many ancient technological mysteries were forgotten, lost to the pages of history, only to be re-invented millennia later.

We have lost the secret to making some of history's most useful inventions, and despite all of our more modern discoveries, our ancestors of thousands of years ago are nevertheless often able to baffle us with their ingenuity. In some cases, we have developed the modern equivalent of some of these inventions, but not all by any means and some only very recently.

What follows are just a few amongst thousands of examples of ancient technology history and artefacts that reflect the brilliance of ancient minds. Modern science tries vainly to either discredit or simply ignore these anomalies that do not fit the proscribed version of history, or even in many cases to cover them up completely in an attempt, once again, to foist upon us their own, heavily censored, versions of reality.

High-tech Stonework

There are many stone artefacts from the ancient world made from the hardest stone on the planet such as granite and diorite, which have been cut and shaped with such quality, precision, and accuracy that the standard explanations of their manufacture are simply inadequate.

Take the granite sarcophagus in the King's Chamber of the Great Pyramid for example, which has been hollowed out with such absolute precision that the famous English Egyptologist, Flinders Petrie firmly believed that the craftsmen present during the Pharaoh Khufu's reign must have had access to tools *"...such as we ourselves have only now re-invented."*

Petrie was dumbfounded by the anomalously modern stonecutting techniques evident throughout Egypt and wrote, *"The character of the work would certainly seem to point to diamond as being the cutting jewel; and only the considerations of its rarity in general, and its absence from Egypt, interfere with this conclusion..."*

The presence of such bizarrely fashioned stones dating back nearly 5,000 years still remains a conundrum today, *"...at the supposed dawn of human civilisation, more than 4,500 years ago, the ancient Egyptians had acquired what sounded like industrial-age drills."* It is not simply the quality of the cutting that has baffled researchers, but also the sheer immensity of some of the stones used to build the many monumental constructions, which has caused scholarly debate for centuries, and even today the experts struggle to explain how some of these massive blocks were cut, shifted and slotted so perfectly into place.

We are talking of such enigmatic archaeological sites as Baalbek in Lebanon, Sacsayhuamán, Ollantaytambo and Machu Picchu in Peru, Tiahuanaco and Puma Punku in Bolivia, Easter Island, Stonehenge in England, and of course the Great Pyramids and temples of Egypt, to name but a few.

These structures all exhibit signs of engineering skills which should not have been possible in such ancient times with only the primitive tools and knowledge which orthodox academics insist upon ascribing to them. We see evidence of heavy lifting which cannot be done today, even with the world's largest cranes, and holes and cavities in the hardest rock on earth that only diamond-tipped saws and drills of the modern era should be capable of producing. How was any of this possible, given the technology available at the time?

As Graham Hancock remarked in 'The Message of the Sphinx,' *"What we may be looking at here are the fingerprints of highly sophisticated and perhaps even technological people capable of awe-inspiring architectural and engineering feats at a time when no civilisation of any kind is*

The Falsification of Science

supposed to have existed anywhere on Earth." On the Giza plateau in front of the Great Sphinx are two temples built from the same limestone rock which was carved out of the horseshoe-shaped enclosure in which sits the Sphinx itself. Both the so-called 'Sphinx Temple' and the 'Valley Temple of Khafre' are built from massive limestone core-blocks, which were once fitted with granite casings inside and out.

Bewilderingly, the builders of these ancient and anonymous temples chose to stack stones of immense proportions, some of which are 30ft. (9.14 metres) long, 12ft. (3.66 metres) wide and 10ft. (3.05 metres) high, and weigh in the region of 200 tonnes each. Even the smallest stones were still a huge 50 tonnes each. Now the question that arises today, aside from how did these early civilisations manage to move about such incredibly large stones, must be 'why did they do it?' It would be so much easier to build with smaller blocks or bricks as we do in modern times, but for some unknown reason our ancestors chose the more difficult option.

What they have succeeded in doing, apart from leaving behind physical evidence of their existence, is confounding us with their capabilities, with the mysteries of their master engineers remaining unsolved.

To get a sense of the sheer size and weight of a 200 tonne block we can compare its dimensions to a locomotive train, which as a general rule of thumb weighs in at 30-35 tons per axle, with the average 4-axle locomotive around 125 tons and a 6-axle locomotive around 200 tons. Imagine stacking the largest of these trains one on top of the other, it would be a logistical nightmare. The standard cranes that we see in our cities on large construction sites can generally lift a maximum load of 20 tonnes when the crane arm is at minimum span, but whilst outstretched to its maximum will only lift round 5 tonnes. Specialised cranes are required to lift loads exceeding 50 tonnes, whilst only the largest cranes in the world are capable of dealing with a 200 tonne load and these monsters are very few and far between. For example, only two land-based cranes in the United States are presently capable of such a feat.

"Built with structural steel members and powered with massive electric motors, the majority of these cranes have a load limit of under 100 tonnes. A commission to put together a temple out of 200 ton blocks would be a most unusual and very taxing job, even for modern heavy-load and crane specialists." Graham Hancock

According to Egyptologists, the stones were raised 50ft. (15.24 metres) above the ground into position by employing a combination of ramps, levers and ropes alongside a huge amount of manpower. But, mystified by conventional explanations, John Anthony West sought answers and visited a Long Island construction site to see for himself precisely how a 200 tonne weight could really be shifted. He noted that it took twenty men working for six weeks to even prepare to lift a 200 tonne boiler using one of the largest cranes in existence with a 220ft. (67.06 metre) high boom and a counterweight made of concrete weighing an incredible 160 tonnes. A second crane was then brought in to precisely fit the boiler in place.

Discussing the construction of the Valley and Sphinx Temples, chief engineer Jesse Warren commented, *"…seeing how they moved these heavy 200 tonne blocks, possibly thousands and thousands of years ago, I have no idea how they did this job. It's a mystery, and it'll probably always be a mystery to me."*

In the Beqaa Valley north of Beirut lie the remains of the ancient Phoenician city of Baalbek, (meaning Lord Baal of the Beqaa Valley) once referred to as Heliopolis or 'the City of the Sun' in what is now modern day Lebanon. In the ancient world as early as 9000 BC (although time may reveal the site to be much older) Baalbek became an important pilgrimage site for the worship of the Phoenician sky-god Baal and his consort Astarte, the Queen of Heaven. At the heart of the city stood a grand temple dedicated to Astarte and Baal which today consists of a five million square foot platform built using some of the largest stones ever shaped by human hands. This immense ancient platform became the foundation for the later Roman temple complex which includes three separate temples dedicated to Jupiter, Bacchus and Venus respectively.

The stones used at this site are so enormous that they make the 200 tonne blocks seen at the Giza temples look like toy bricks in comparison. Some of the blocks in the wall are around 400 tonnes each, while the three hewn stones known as the 'trilithon' which lies in the base of the Jupiter Baal Temple ruins, weigh an estimated 800 tonnes each. Officially, scholars attribute the shifting and cutting of these monstrous monoliths to the Romans in circa 27 BC, but many researchers contest this idea, suggesting instead that they are the remnants of some other ancient civilisation from a far earlier epoch altogether, in keeping with the material posited in J.P. Robinson's book, '*The Myth of Man: Hidden History and the Ancient Origins of Humankind*'

Graham Hancock is one such dissident. *"The moving and positioning of three 800 tonne megaliths to a height of 18 or 20 feet above the ground as is the case at Baalbek is a problem of a completely different order. I suggest this requires careful consideration rather than simply saying 'the Romans did it,' as archaeology is at present inclined to do."*

The combined weight of the entire base at Baalbek is estimated at around five billion tonnes. Were the Romans capable of shifting such an amount of stone? It is difficult to imagine, and no other evidence of Roman construction anywhere else would support such a theory.

A stone quarry site may be found not too far from the ruins, and it is here where the largest stones ever unearthed were discovered. Protruding from the ground at a slight angle, a limestone monolith that was named 'Hajjar al-Hibla' in Arabic or 'the Stone of the Pregnant Woman' was found which weighs approximately 1,200 tonnes! Archaeologists agree that this huge monolith was left in the quarry simply because the quality of the stone block's edge proved too poor to transport as it could be too easily damaged. In 2014, the Oriental Department of the German Archaeological Institute conducted more excavation work in the stone quarry trying to find new data about the mining techniques and the transporting of the megaliths, but what they discovered was an even larger monolith directly beside the first one which had been beneath the ground until that summer.

Measuring an astounding 64ft. (19.6 metres) in length, 19.6ft. (6 metres) wide and 18ft. (5.5 metres) high, this unbelievable find weighs 1,650 tonnes (1.65 million kilos) and is now the largest known ancient stone block anywhere in the world. The archaeologists also concluded that due to the block's configuration and level of smoothness, it was most certainly being readied for transportation without being cut, and all at an altitude of around 3838ft. (1,170 metres).

Following the discovery of the second monolith, Hancock wrote, *"I believe that these huge megaliths long predate the construction of the Temple of Jupiter and are likely to be 12,000 or more years old – contemporaneous with the megalithic site of Gobekli Tepe in Turkey. I suggest we are looking at the handiwork of the survivors of a lost civilisation, that the Romans built their Temple of Jupiter on a pre-existing, 12,000 year old megalithic foundation, and that they were unaware of the giant hewn megaliths in the ancient quarry as these were covered by sediment in Roman times (as, indeed, the newly discovered block still was until very recently)."*

It remains unexplained how the Romans – or whoever was actually responsible for such an astounding feat of engineering – quarried, cut and transported the megaliths of Baalbek. And it is also a mystery as to why they left the largest blocks in situ. The mainstream answer is that they were simply too heavy to manoeuvre so were left in the ground, but they succeeded in shifting the 800 tonne 'trilithon' blocks so why not the others? Even if they decided they were beyond the limits of their engineering capabilities, surely, they would have cut them up into smaller, more manageable blocks in order to complete the temple constructions. *"It's really puzzling that they didn't do so and therefore the fact that these gigantic, almost finished blocks remain in the quarry and were never sliced up into smaller blocks and used in the general construction of the Temple of Jupiter, suggests to me very strongly that the Romans did not even know they were there."* explained Hancock.

We may never know the answer to these questions, but we can still speculate as to the technological means involved in manoeuvring the megalithic stones left around the world which continue to perplex us with their mysteries. More recent discoveries have revealed even more megalithic stones, this time in Southern Siberia near the mountains of Gornaya Shoria. Georgy Sidorov found and photographed the 'super' megaliths for the first time on an expedition to the Southern Siberian Mountains in 2014.

There are no accurate measurements given as yet but judging from the scale depicted by the human figures in some of the photographs, these Siberian megaliths hewn from granite could prove to be twice the weight of the largest known megaliths found at Baalbek, possibly in the region of between 3,000 and 4,000 tonnes each. How many more such discoveries lie in wait around the world? And what weight will these astoundingly immense stones reach?

Reports of the 2014 expedition claim that the geologists encountered some unexplained problems with their compasses when the arrows began deviating away from the megalithic blocks. It has been suggested that they may have experienced inexplicable phenomena of a negative geomagnetic field, leading to notions of the possible use of ancient 'anti-gravity' technologies unknown to modern science. But of course, modern science would not even consider that possibility as even remotely credible, instead choosing to simply ignore it as it falls outside of currently accepted scientific 'wisdom.'

Ancient Astronomers

Ever since we humans first looked up to the sky, we have been amazed by its beauty and untold mysteries. Naturally then, astronomy is often described as the oldest of the sciences, as it has been inspiring people for many thousands of years. Celestial phenomena are featured in prehistoric cave paintings and monuments such as the Great Pyramids of Giza and Stonehenge seem to be aligned with precision to cardinal points or the positions where the moon, sun or stars rise and set on the horizon.

Today, we struggle to imagine how ancient people could build and orient such structures and so this has led to many assumptions. Some suggest that prehistoric people must have had some knowledge of advanced mathematics and sciences to do this, whereas others go so far as to speculate that alien visitors 'guided their hands' in many ways. But what do we actually know about how people of the past understood the sky and developed a cosmology? However, a scientific discipline known as 'archaeo-astronomy' or 'cultural astronomy,' that was originally developed in the 1970s, is now beginning to provide insights. This subject combines various specialist areas, such as astronomy, archaeology, anthropology, and ethno-astronomy.

The pyramids of Egypt are some of the most impressive of ancient monuments, and several are oriented with high precision. The Egyptologist Flinders Petrie carried out the first high-precision survey of the Giza pyramids in the 19th century and discovered that each of the four edges of the pyramids' bases point towards a cardinal direction (due north, south, east, or west) to within a quarter of a degree.

But how did the Egyptians know that? Just recently, Glen Dash, an engineer who studies the Giza pyramids, proposed a theory. He drew upon the ancient method of the 'Indian circle,' which only requires a shadow-casting stick and string to construct an accurate east-west direction. He outlined how this method could have been used for the pyramids based on its simplicity alone. But could this really have been the case? It is not impossible, but at this point we are in danger of falling into the popular trap of projecting our current world views, methods and ideas into the past. But an insight into mythology and relevant methods known and used at the time is likely to provide a more reliable answer.

This is not the first time scientists have jumped to conclusions about a scientific approach applied retrospectively to the past. A similar thing happened with Stonehenge. In 1964, the late astronomer Gerald Hawkins developed an intricate method using pit holes and markers to predict eclipses at the mysterious monument. However, this does not conclusively prove that this was Stonehenge's true purpose.

The Falsification of Science

Stonehenge, Wiltshire, England

To begin to understand the past, we need to include various approaches from other disciplines to support an idea. We also need to appreciate that there will never be only one explanation of how a monument may have been aligned or used.

So how can cultural astronomy explain the pyramids' alignment? A study from 2001 proposed that two stars, Megrez and Phad, in the stellar constellation known as Ursa Major (the Great Bear) may have been the key. These stars are visible through the entire night and their lowest position in the sky can mark north using the 'merkhet,' an ancient time-keeping instrument comprising a bar with a plumb line attached to a wooden handle to track the alignment of the stars.

The benefit of this interpretation is that it links to star mythology drawn from inscriptions in the temple of Horus in Edfu. These elaborate on the use of the merkhet as a surveying tool – a technique that could also explain the orientation of other Egyptian sites. The inscription includes the hieroglyph 'the Bull's Foreleg,' which represents 'the Plough' or 'the Big Dipper' star constellation and its possible position in the sky.

Similarly, more plausible ideas for Stonehenge's purpose have been offered. One study identified strange circles of wood near the monument, colloquially known as 'Woodhenge' and suggested these may have represented the living, whilst the rocks at Stonehenge represented the dead. Similar practices are seen in monuments found in Madagascar, suggesting that it may have been a common way for prehistoric people to think about both the living and the dead. It also offers an interesting new way of understanding Stonehenge in its wider landscape. Others have interpreted Stonehenge and especially its 'avenue' as marking the ritual passage through the underworld with views of the moon on the horizon.

Cultural astronomy has also helped shed light on 6,000 year-old passage graves – a type of tomb consisting of a chamber of connected stones and a long narrow entrance, in Portugal. The archaeologist Fabio Silva has shown how views from inside the tombs frame the horizon where the star Aldebaran rises above a mountain range. This may mean that it was built to provide a view of the star from the inside either for the dead or the living, possibly as an initiation ritual.

But Silva also drew upon wider supporting evidence. The framed mountain range is where the builders of the graves would have migrated with their livestock over summer. The star Aldebaran rises for the first time here in the year – known as a helical rising – during the beginning of this migration and interestingly, ancient folklore also mentions a shepherd in this area who spotted a star so bright that it lit up the entire mountain range. Arriving there he decided to name both the mountain range and his dog after the star and both names still exist to this day.

Silva has also demonstrated how a view from within the long, narrow entrance passages to the tombs could enhance the star's visibility by restricting the view through an aperture. But while it is easy to assume that prehistoric people were not analytic astronomers with a great knowledge of science, it is important to remember that this only reflects our modern views of astronomy. Findings from cultural astronomy show that people of the past were indeed avid sky watchers and incorporated what they observed into many aspects of their lives.

Flexible Glass

Imagine a glass that can be bent and then easily manipulated back to its original form – or a glass that does not break when dropped. It is an historical legend of unknown origin that an ancient Roman

The Falsification of Science

glassmaker devised the technology to create a flexible glass, 'vitrium flexile,' but a certain emperor, for whatever reasons, decided the invention should not be allowed to exist.

Manmade glass (as opposed to a naturally occurring one such as obsidian) is widely accepted to have been invented by the Phoenicians. Over the course of the following millennia, glassmakers honed their skills, improving the techniques used to produce this substance, as well as the glass itself. In the Roman Empire, glass became a commonly produced item, although special luxury glasses were also created for the more wealthy. Arguably one of the most intriguing of these glass types is the so-called flexible glass.

Flexible glass is allegedly a form of unbreakable glass that was invented during the Roman period. (i.e. approximately 100BCE to 400CE.) It is also said to be a legendary lost invention dating to the reign of the Roman Emperor Tiberius (reigned 14CE to 37CE). Whilst no physical evidence of such a glass has been found so far, there are two main written sources attesting to its existence. One of these is Pliny the Elder's *'Natural History,'* whilst the other is the *'Satyricon,'* commonly attributed to the Roman courtier Petronius. Whilst his contemporary, Pliny's work is encyclopaedic in nature, and Pliny was never known to have written fiction, that of Petronius is a piece of satire, demonstrating how this incredible story was possibly picked up by writers of different genres.

In his *'Natural History,'* Pliny reports that flexible glass was made by a skilled glassmaker during the time of Tiberius but Instead of gaining the favour of the Roman emperor, the craftsman had his workshop shut down. This was thought to be to prevent the value of precious metals, i.e. gold, silver, and copper from being depreciated by this new material. A similar story is said to have been reported by Cassius Dio and Suetonius. Pliny expresses his doubts regarding the veracity of this story, as he mentioned that, *"This story, however, was, for a long time, more widely spread than well authenticated."*

Petronius' telling of the story in his *Satyricon*, on the other hand, may be described as a more dramatised version of the story told by Pliny.

In the satirist's account, the man who invented the flexible glass was granted an audience with the Roman emperor in order to show him his work and after Tiberius examined the glass cup, he handed it back to the glassmaker, who proceeded to throw it with all his might on the floor. The emperor was shocked at what had happened, but the man calmly picked the cup up from the ground, showing the emperor that it was only dented. The glassmaker then took a little hammer to beat the glass, and in no time, it had regained its original shape.

The Roman glassmaker was confident that he had impressed the emperor and was probably waiting to be rewarded for his ingenious creation. However, when the emperor asked if anyone else knew how to make this kind of flexible glass, the craftsman answered with a negative but instead of receiving the reward he had hoped for, the glassmaker was immediately hauled away and executed, thus taking the secret of making flexible glass with him to his grave. The reason given for this was that Tiberius believed that the invention would cause gold to be devalued, and Tiberius certainly owned plenty of that!

So is it possible to make Roman flexible glass? Today, the story of Roman flexible glass is mainly treated in the same manner as it had been by Pliny, ie. with much doubt. Nevertheless, there has been some speculation as to how this glass may have been made. One idea, for instance, is that the Roman glassmaker had somehow had access to boric acid or borax, which may be found in the natural environment. By adding a small percentage of boric oxide to the glass mixture, the end result would be something that was relatively unbreakable. It may be added that borax was imported from the East into Europe on a regular basis during the Middle Ages, and it was used by goldsmiths as a flux.

Boric acid could also be found in the steam vents of the Tuscan Maremma to the north of Rome, though this was supposedly only realised during the 19th century. Nevertheless, it is possible that the glassmaker may have stumbled on this source by chance, but in any case, it is likely that the recipe for Roman flexible glass, if it ever did exist at all, will continue to elude us and remain a 'lost invention of the Romans.'

The Antikythera Mechanism

Whenever 'ancient Greece' is mentioned most people automatically think of democracy, the Olympic Games, mythology and philosophy. But it would seem that not many are aware of how advanced the ancient Greeks were on a technological level too, and the Antikythera Mechanism, known as the world's first analogue computer, is the best example of all.

On the 17th May 1902, when Greek archaeologist Valerios Stais discovered what appeared to be a corroded chunk of metal, he made history by finding part of what some historians describe as the world's first analogue computer and which has since come to be known as the 'Antikythera Mechanism.' Although the term 'analogue computer,' may seem to be an exaggeration to some, it is an undeniable fact that the Antikythera mechanism that was recovered from the Antikythera wreck (a shipwreck off the Greek island of Antikythera) is the earliest preserved portable astronomical calculator in history.

In fact, the Antikythera mechanism is one of the most amazing mechanical devices ever discovered from the ancient world. The ancient artefact was recovered from the aforementioned shipwreck and is a metallic device consisting of a complex combination of gears, which dates back to the 1ˢᵗ or 2ⁿᵈ century BC. X-rays of the device have revealed that it is composed of at least thirty different types of gears and on the door-plates are about 2,000 letters that are thought to be an ancient version of a user-manual.

The amazing device, which some believe was inspired by the famous inventor and mathematician Archimedes, is so complex that many consider it to be the very first analogue computer. Derek de Solla Price, who analysed it in the 1960s, said that the discovery was akin to finding a jet engine in Tutankhamen's tomb.

And for all those with enquiring minds, Peter Lynch, professor of meteorology at University College Dublin, has now discovered what this machine was for and exactly how it worked.

Professor Lynch wrote that, *"The mechanism was driven by a handle that turned a linked system of more than thirty gear wheels. Using modern imaging techniques, it is possible to count the teeth on the wheels, see which cog meshes with which and what are the gear ratios. These ratios enable us to figure out what the mechanism was computing. The gears were coupled to pointers on the front and back of the mechanism, showing the positions of the sun, moon and planets as they moved through the zodiac. An extendable arm with a pin followed a spiral groove, like a record player stylus. A small sphere, half white and half black, indicated the phase of the moon. Even more impressive was the prediction of solar and lunar eclipses. It was known to the Babylonians that if a lunar eclipse is observed, a similar event occurs 223 full moons later. This period of about 19 years is known as the Saros cycle. It required complex mathematical reasoning and technology to implement the cycle in the mechanism."*

Amazingly, the device even included a dial to indicate which of the Pan-Hellenic games would take place each year, with the Olympics occurring every fourth year.

Ultimately, the Antikythera mechanism illustrates in the best possible way that the ancient Greeks used complex devices of precisely cut

The Falsification of Science

wheels to represent the latest in scientific understanding. It is also a 'doorway' that allows us to understand how the ancient Greeks saw the universe. They believed that nature worked according to predefined rules, like a machine, an approach that continued shaping the foundations of our contemporary scientific views.

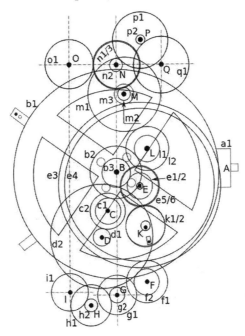

And while some modern scientists insist that this type of mechanical philosophy was developed only after the Industrial Revolution, in reality it was the ancient mechanics – just like the Antikythera Mechanism – that paved the way and inspired pretty much every modern technological 'miracle' of our times.

The 1200 Year-old Telephone

The 1,200 year-old telephone, a marvel of ancient invention, surprises almost all who hear about it. Reportedly found in the ruins of Chan Chan, Peru, this delicate communication artefact is thought to have been constructed around 1,200 to 1,400 years ago and is known as the earliest example of telephone technology in the Western Hemisphere.

This seemingly out-of-time artefact is evidence of the impressive innovation of the coastal Chimu people in the Río Moche Valley of northern Peru. Ramiro Matos, curator of the National Museum of the American Indian (NMAI) said, *"This is unique. Only one was ever discovered. It comes from the consciousness of an indigenous society with no written language."*

The early telephone appears to be a rudimentary speech transmission device, much like the 'lover's telephone' that has been known for hundreds of years but which became popular in the 19th century. It was usually comprised of tin cans connected with string, used to both listen and speak at a distance and mostly seen as a novelty.

This ancient Chimu device, described as an instrument is composed of two gourd tops bound with a length of cord. The gourds, each 3.5 inches (8.9 centimetres) long are coated in resin and act as transmitters and receivers of sound. Around each of the gourd bases is a stretched hide membrane. The 75 ft. (22.8 metre) line connecting the two ends is made of cotton-twine. However, the simplicity of the device disguises its implications.

This one-of-a-kind artefact reportedly predates the earliest research into telephones from 1833, and which began with non-electric string devices, by more than a thousand years. It is too fragile to physically test, but researchers have deduced how the instrument may have worked. What they must continue to speculate on, however, is *how* the Chimu people actually used this ancient 'phone' and what specifically it was used for?

As the Chimu were known to be an elitist society, it is reasonable to speculate that only the elite or priest class would have been in possession of such a valuable instrument. The precious instrument, with the seemingly magical ability to channel voices across space to be heard directly in the ear of the receiver was undoubtedly a tool designed for an executive level of communication.

There may have been many applications, such as communication between novices or assistants and their higher-ranking elites through chambers or anterooms. No face-to-face contact would have been needed, preserving status and ensuring security. Like many other ancient marvels, it might also have been a device to astound the faithful. Disembodied voices emitting from a hand-held object would have shocked and convinced people of the importance and station of the upper class or priests. Alternatively, there are some who consider the gourd and twine object as merely a children's toy. If such novelties are not our modern sacred objects, why must they have been believed to be religious items or priestly tools to humans of the past?

The artefact was in the possession of Baron Walram V. von Schoeler, a Prussian aristocrat, who was less flatteringly described as a *"shadowy Indiana Jones-type adventurer."* He participated in many excavations in Peru in the 1930s, and may have dug up the artefact himself from the ruins of Chan Chan. He distributed his collection among various museums, and the artefact eventually ended up at the storage facility of the National Museum of the American Indian in Maryland, USA, where it is treated with great care and preserved in a temperature controlled environment as one of the museums greatest treasures.

Ramiro Matos, an anthropologist and archaeologist specialising in the study of the central Andes explained, *"The Chimu were a skilful, inventive people, who possessed an impressive engineering society."* This is demonstrated also by their hydraulic canal-irrigation systems and their highly detailed, elaborate metalwork and artefacts.

The Chimu were the people of the Kingdom of Chimor, and their beautiful capital city was Chan Chan (translated as Sun Sun,) a sprawling mud brick complex – the largest such in the world – and it was also the largest city in pre-Columbian South America. Chan Chan was almost 20 square kilometres (7.7 square miles,) and was home to around 100,000 residents during its heyday around 1200 AD. The entire city was made from shaped and sun-dried mud, and was elaborately decorated with sculptures, reliefs and wall carvings on almost every surface.

The Chimu culture arose about 900 AD, but was eventually conquered by the Incas around 1470 AD. The Chimu telephone, and many other amazing ancient technologies, remind us that many ancient cultures were capable of incredible inventions, ideas, and creations long before our so-called 'sophisticated' modern societies were even dreamed of.

Greek Fire

The above is an image from an illuminated manuscript, the Madrid Skylitzes, showing Greek Fire in use against the fleet of the rebel, Thomas the Slav. The caption above the left ship reads, "The fleet of the Romans setting ablaze the fleet of the enemies."

The Byzantines of the 7th to the 12th centuries AD, fired a mysterious substance at their enemies in naval battles. This fiery liquid, shot through tubes or siphons, burned in water and could only be extinguished with vinegar, sand, and urine. We still do not know of what this chemical weapon, known as Greek Fire, was composed. The Byzantines guarded the secret jealously, ensuring that only a select few knew the secret, and the knowledge was eventually lost altogether.

The 300 Million-Year Old Screw

A Russian research team known as the Kosmopoisk Group, and which investigates UFOs and paranormal activity, claims to have found a one-inch screw embedded inside a rock that is 300 million years old. They say the screw is the remains of an ancient form of

The Falsification of Science

technology that proves that extra-terrestrials visited Earth millions of years ago. However, scientists refute this claim, and say that the 'screw' is nothing more than a fossilised sea creature called a Crinoid.

The Russian team were investigating the remains of a meteorite in the Kaluga region of Russia in the 1990s, when they came across the strange object. A paleontological analysis was carried out, which revealed the stone was formed between 300 and 320 million years ago. The team also claim that an x-ray of the stone shows that another screw is present inside it. However, they have not allowed international experts to examine the object, nor have they revealed what the screw is made of.

Since the initial finding, much debate has surrounded the discovery, with scientists scoffing at the suggestion that it reflects an ancient screw and suggesting there is a much less exciting explanation.

Nigel Watson, author of the UFO Investigations Manual told *Mail Online*: *"Lots of out-of-place artefacts have been reported, such as nails or even tools embedded in ancient stone. Some of these reports are . . . misinterpretations of natural formations. It would be great to think we could find such ancient evidence of a spaceship visiting us so long ago, but we have to consider whether extra-terrestrial spacecraft builders would use screws in the construction of their craft. It also seems that this story is probably a hoax that is being spread by the internet and reflects our desire to believe that extra-terrestrials have visited us in the past and are still visiting us today in what we now call UFOs."*

For now, the controversy surrounding the object remains very much alive, and unless the Kosmopoisk Group releases detailed information regarding the material of the 'screw,' it is unlikely that consensus will be reached any time soon.

Han Purple and Egyptian Blue

Han purple is an artificial purple pigment created by the Chinese over 2,500 years ago, which was used in wall paintings and to decorate the famous terracotta warriors, as well as ceramics, metal ware, and jewellery. The pigment is in fact a technological wonder, made through a complex process of grinding-up raw materials in precise proportions and heating to incredible temperatures. So intricate was the process, that it was not recreated again until 1992, when chemists were finally able to identify its composition. But this was just the beginning. Further research since then has discovered some amazing properties of Han purple, including the ability to emit powerful rays of light in the near-infrared range, as well as being able to collapse three dimensions down to two under the right conditions.

The production of Han purple, otherwise known as Chinese purple, dates back as far as 800 BC, however it appears that it was not used in art until the Qin and Han dynasties (221 BC – 220 AD) when it was applied to the world famous terracotta warriors, as well as ceramics and other items.

"Prior to the nineteenth century, when modern production methods made synthetic pigments common, there were only hugely expensive purple dyes, a couple of uncommon purplish minerals, and mixtures of red and blue, but no true purple pigment – except during a few hundred years in ancient China," wrote Samir S. Patel in *'Purple Reign: How ancient Chinese chemists added colour to the Emperor's army'*

The Falsification of Science

For some unknown reason, Han purple disappeared entirely from use after 220 AD, and was never seen again until its rediscovery by modern chemists in the 1990s.

Unlike natural dyes, such as Tyrian purple (from c. 1500 BC,) which are organic compounds and typically made from plants or animals, like the murex snail, Han purple was a synthetic pigment made from inorganic materials. Only two other manmade blue or purple pigments are known to have existed in the ancient world – Maya blue (from c. 800 AD,) made from a heated mixture of indigo and white clay, and Egyptian blue, which was used throughout the Mediterranean and the Near and Middle East from 3,600 BC to the end of the Roman Empire.

The scientist, Elisabeth Fitzhugh, a conservator at the Smithsonian Institute, was the first to identify the complex synthetic compound that makes up Han purple – barium copper silicate, a compound that differs from Egyptian blue only through its use of barium instead of calcium.

The similarities between Han purple and Egyptian blue led some early researchers to conclude that the Chinese may have learned to make the pigment from the Egyptians. However, this theory has been largely discounted as Egyptian blue was found no further East than Persia.

"There is no clear reason why the Chinese, if they had learned the Egyptian formula, would have replaced calcium with barium, which necessitates increasing the firing temperature by 100 degrees or more," wrote Patel.

So how exactly did the Chinese stumble upon the intricate formula to make Han purple, which involved combining silica (sand) with copper and barium in precise proportions and heating to about 850-1000°C? A team of Stanford physicists published a paper in the *Journal of Archaeological Science,* which proposed that Han purple was a by-product of the glassmaking process, as both glass and the purple pigment contain silica and barium. Apparently, barium tends to make glass shinier and cloudy, which means that this pigment could be the work of early alchemists trying to synthesise white jade.

Since its composition was first discovered, scientists have continued to investigate this unique pigment. Researchers at the British Museum discovered that, when exposed to a simple LED flashlight, Han purple emits powerful rays of light in the near-infrared range. According to their study, published in the journal Analytical and Bioanalytical Chemistry, the Han purple pigments show up with startling clarity under the right conditions, meaning that even faint traces of the colour, which are invisible to the naked eye, can be seen with infrared sensors.

The fluorescent properties of Han purple were not the only surprise. Quantum physicists from Stanford, Los Alamos National Laboratory and the Institute for Solid State Physics (University of Tokyo) reported that when Han purple is exposed to extreme cold and a powerful magnetic field, the chemical structure of the pigment enters a new state called the quantum critical point, in which three-dimension material 'loses' a dimension.

"We have shown, for the first time, that the collective behaviour in a bulk three-dimensional material can actually occur in just two dimensions," Ian Fisher, an assistant professor of applied physics at Stanford said in the Stanford Report. *"Low dimensionality is a key ingredient in many exotic theories that purport to account for various poorly understood phenomena, including high-temperature superconductivity, but until now there were no clear examples of 'dimensional reduction' in real materials."*

The scientists have proposed that this effect is due to the fact that the components of barium copper silicate are arranged like layers of tiles, so they do not stack-up neatly. Each layers' tiles are slightly out of sync with the layer below them and this may frustrate the wave and force it to go two dimensional.

The researchers have said the discovery may help understand the required properties of new materials, including more exotic superconductors.

Fisher said, *"Han Purple was first synthesised over 2500 years ago, but we have only recently discovered how exotic its magnetic behaviour is. It makes you wonder what other materials are out there that we haven't yet even begun to explore."*

Yes Ian, it does indeed.

The Baghdad Battery

One of the most interesting and highly debated artefacts of the Baghdad Museum in Iraq is a clay pot. It is 5-6 inches high and encapsulates a copper cylinder. Suspended in the centre of this cylinder – but not touching it – is an iron rod. Both the copper cylinder and the iron rod are held in place with an asphalt plug and the rod shows evidence of corrosion, probably due to the use of an acidic liquid such as vinegar or wine.

These artefacts (more than one was found) were discovered during the 1936 excavations of the old village of Khujut Rabu, near Baghdad. The village is considered to be about 2000 years old and was built during the Parthian period (250 BC to 224 AD).

Although it is not known exactly what the use of such a device would have been, the name 'Baghdad Battery,' or 'Parthian Battery,' comes from one of the prevailing theories established in 1938 when Wilhelm Konig, the German archaeologist who performed the excavations, examined the battery and concluded that this device was an ancient electric battery. Another theory, however, suggests that they were containers to hold papyrus.

After the Second World War, Willard Gray, an American working at the General Electric High Voltage Laboratory in Pittsfield, Massachusetts, built replicas and filling them with an electrolyte, found that the devices could produce 2 volts of electricity. So if the artefact were indeed a battery, what would electricity have been used for and why have we not yet discovered further evidence of its use? And if not used as a battery, what would have been the specific use of such device?

It is important to remember that Iraq is considered to be the location of both the Garden of Eden and the Tower of Babel. Who knows how many more artefacts are hidden in this ancient place on Earth? It is an interesting fact that, as a side effect of the illegal invasion of Iraq in 2003, by the US and allied forces, many of Iraq's museums were looted of their priceless ancient treasures by the occupying forces and shipped back to the US. Who knows what other strange, inexplicable devices and artefacts were taken and were they are now? Could this be just one small part of yet another Elite cover-up operation?

Viking Compasses

As well as their well-deserved reputation as ruthless robbers, rapists and murderers, the Vikings were also impressive mariners, capable of traversing the North Atlantic along an almost straight line. And now, new interpretations of a medieval compass suggest that this well-travelled race, may have used the sun to operate their compasses even when the sun had set below the horizon.

The remains of a Viking compass – otherwise known as the 'Uunartoq disc' – were discovered in Greenland in 1948 in an 11th century convent. Although some researchers originally argued that it was simply a decorative object, others have suggested that the disc was an important navigational tool that the Vikings would have used in their roughly 1,600-mile-long (2,500 kilometre) voyage from Norway to Greenland.

Though only half of the wooden disc remains, it is estimated to have been roughly 2.8 inches (7 centimetres) in diameter with a now-lost central pin that would have cast a shadow from the sun indicative of a cardinal direction.

Researchers based at Eötvös Loránd University in Hungary have studied the fragment in detail. They concluded that although the disc could have functioned as a single entity, it was more likely to have been used in conjunction with other tools – including a pair of crystals and a flat, wooden slab – to facilitate navigation when the sun was low in the sky or even below the horizon.

The Falsification of Science

"When the sun is low above the horizon, even the shadow of a small item can fall off the board, and such situations are frequent in the northern seas," said study co-author Balázs Bernáth.

Bernáth and colleagues think that, to help solve this long-shadow problem, the Vikings may have used a low-lying, domed object in the middle of the compass to create a wider, shorter shadow than a more typical sundial spike would have done. A wide hole within the centre of the disc, previously interpreted as a place to grip the compass, could have served as a holding spot for this so-called central gnomon, the team suggests.

The researchers think that, to locate the sun after sunset, the Vikings could have used a pair of crystals known as 'sunstones,' which are calcite stones that produce patterns when they are exposed to the polarisation of UV rays within sunlight. When the crystals are held up to the sky, the orientation of the patterns cast within the stone can help pinpoint the position of the sun below the horizon.

Once the Vikings had determined the position of the hidden sun, they could have used a specially designed wooden slab called a shadow stick to simulate the shadow of the gnomon based on the angle at which the hidden sun would strike it. The location of the outer edge of that imaginary shadow could then have been used to determine their cardinal direction. Researchers conducted field tests to estimate the plausible accuracy of this so-called 'twilight compass,' and found that it would have worked with only 4 degrees of error, which is better than other ancient forms of celestial navigation and comparable to modern magnetic pocket compasses.

"Not the best, maybe, but it would have been a really big help," Bernath said.

The team estimated that the twilight compass would have functioned for as long as 50 minutes after sunset around the spring equinox, the time when the Vikings are thought to have used this compass based on etchings in the wood. No shadow sticks or sunstones have ever been found in conjunction with the disc, but evidence of both exist in medieval written records, suggesting that they would have been available to the Vikings.

The team agreed that the findings are a testament to the sophistication of this group of people often remembered as vicious and merciless heathens.

"They were ruthless robbers, but not only ruthless robbers. This instrument is quite remarkable," said Bernath.

Navigating the oceans in ancient times was extremely difficult and dangerous, given that the sailors of those times had no compasses on their tiny vessels. In order to travel from, say, Europe to the Americas, it was virtually impossible to know whether or not you were just simply going around in circles in cloudy or dark conditions. Scientists were puzzled about how the Vikings were consistently able to travel in a totally straight line from Norway to Greenland and back, some 1,600 miles, whilst the rest of the world was rowing around in circles. Then, in 1948, the ancient Viking artefact was discovered beneath an 11th century convent and everything changed.

Before magnetic compasses, ancient mariners had to find their way using sundials, which told time and direction by shining a shadow onto a disc. As you may imagine, at night or even on a cloudy day, they were about as useful as reading tea leaves and sacrificing a goat to Odin. But the Viking compass, known as the Uunartoq disc, provided a solution to that problem. As well as being an amazingly sophisticated sundial with several shadow sticks to work out the cardinal directions, medieval records of the device refer to a 'magic' crystal that enabled it to function even when the sun was not visible.

The Lycurgus Cup

The Falsification of Science

This ancient jade-green cup appears red when lit from behind. It is believed by some researchers that the Romans may have been the first ones to come across the colourful potential of nanoparticles, possibly by accident, but they certainly somehow managed to perfect it! This amazing property of the Lycurgus Cup puzzled scientists for decades when the cup was acquired by the British Museum in the 1950s and it was not until 1990 that the mystery was actually, finally solved.

Researchers studied broken fragments of the cup under a microscope and discovered that the ancient Romans had created the glass with silver and gold particles that were ground down to be as small as 50 nanometres in diameter. This is less than one-thousandth the size of a grain of salt and suggests that the Romans knew what exactly they were doing, meaning that they had knowledge of nanoparticles. The principle is that when the light strikes the cup, the electrons within the particles of the cup vibrate in ways that alter the colour depending on the position of the observer. So, when different liquids are poured into the cup, and it is struck by the light, the electrons behave differently, thus changing the colour. Incidentally, this is precisely how home pregnancy tests work!

The above are just a few examples of ancient technology and wisdom that have been excluded from mainstream science and I would ask the question, does science ever acknowledge or seriously investigate these issues in order to add to the generic knowledge pool? Well, no of course not. To do so would be totally contrary to the agenda. Mainstream science would rather that we do not become aware of all these myriad anomalies to the official story. It would not 'do' at all for the masses to be fully-informed of the fact that the paradigm under which we live is a wholly manufactured falsehood.

Chapter 4

Freemasonry and the Royal Society

Freemasonry

Of all the secret societies, probably the most famous, as well as being the most misunderstood by both outsiders and even its adherents at all its lower levels, is Freemasonry.

Freemasonry probably had its original roots in the mediaeval crafts, whereby each trade had its own 'guild' or 'union' in modern parlance, to protect the interests of its members. In return for this protective presence the craftsman had to submit to the most rigorous regulation. He had to serve as an apprentice, usually without pay, for two to ten years (depending on the trade) often live with and obey the master craftsman who tutored him and then finally once this long induction process was complete, the apprentice was free to start out alone, frequently taking one of his master's daughters as a wife.

With the expansion of economics often came the need of the craftsman to borrow money to finance long-term or long-distance undertakings, assuming his willingness to pay interest for this benefit. The Christian Church condemned usury and money-lending was permitted only by and to Jews, who were barred from guild membership by dint of their religious practices.

Stonemasons, by the very nature of their trade, were itinerant journeymen, constantly moving between villages and towns seeking employment. Their membership of the Masonic craft guild was a

The Falsification of Science

reassurance to potential customers and employers that they were indeed bona fide, highly skilled craftsmen who could be relied upon to provide a fair day's work for a fair day's pay and also importantly that they had the necessary skill-sets to adequately perform the job for which they were being paid. The insignia of their guild usually displayed representations of the tools of their trade and where language or literacy was a barrier to communication, served as a visual guarantee of ability. It was from these humble beginnings that secret symbols, restricted membership, oaths of secrecy and mutual financial aid evolved but eventually the guilds became entities that were no longer necessarily populated by those skilled in the crafts and trades their societies purported to represent, becoming almost entirely symbolic and totally unrepresentative of that craft or trade.

In 1645, the Royal Society, founded either in Oxford or in London depending on source, was created with the intention of promoting scientific enquiry rather than the simple, unthinking acceptance of received wisdom. The irony in this premise will become apparent, shortly! Many facets of the society were based on the tenets of Freemasonry and indeed many of the founders were Freemasons – a state of affairs that still exists to this day.

It was the brother-in-law of Oliver Cromwell, the future 'Lord Protector' of the Commonwealth of Great Britain who became its first chairman. Cromwell's uncle, Thomas Cromwell during the reign of Henry VIII a century earlier had already severed the ties between the Roman Catholic Church and the English monarchy and his nephew Oliver, himself managed to complete the cycle by engineering the severing of King Charles I's head from his body, allegedly for high treason.

In 1717, Freemasonry, by now a new form of cult entirely distinct from the various existing creeds of Europe, spread rapidly to Paris, Florence, Rome, and Berlin, where its deliberately syncretic rituals and décor, Solomon's temple's signs and symbolism made it thoroughly cosmopolitan and religiously neutral. Nothing could better encapsulate the early spirit of the 'enlightenment.'

Andrew Ramsay, a Scottish Jacobite exiled in France, who was Chancellor of the French Grand Lodge in the 1730s, claimed that the first Freemasons had been stonemasons in the crusader states who had learned the secret rituals and gained the special wisdom of the ancient world. According to the German Freemasons, the Grand Masters of the Order had learned the secrets and acquired the treasure of the Jewish Essenes.

Either way, Freemasonry had escaped its earlier guise of stonemasonry and in its new incarnation appealed to the intellectuals and the nobility. The early membership of Masonic lodges included merchants and financiers, notaries and lawyers, doctors, diplomats, and gentry, in other words 'men of substance' or sound reputation. By the middle of the eighteenth century these included members of the French royal family, Frederick the Great, Maria Theresa's husband, Francis of Lorraine and her son, Joseph. Voltaire was also admitted with great pomp and ceremony into a publicity-hungry Masonic lodge in Paris.

Freemasonry not only played an important role in the French Revolution, but also with regard to the American Revolution, in particular the lodges affiliated to the Grand Lodge of Scotland. Scottish Rite Freemasonry blossomed in the fertile landscapes of the now fledgling, emerging state in North America and indeed Freemasonry could be found on both sides of the looming war between the colonists and the Crown and although there is no clear evidence of collusion amongst Masons from opposing camps, the fact that the British made some extraordinary military 'errors' arouses my suspicions in this regard.

Sir William Howe's failure to pursue Washington after expelling him from New York and Sir Henry Clinton's almost wilful failure to link up with Burgoyne's army marching south from Montreal in 1777 are the two most conspicuous examples. The Grand Master for North America was Joseph Warren and the Green Dragon coffee house in Union Street, Boston, purchased by the Provincial Grand Lodge is generally considered to be the site where one of its offshoots 'The Sons of Liberty' plotted the Boston Tea Party and carried it out in the guise of 'Red Indians,' these days of course, re-branded and packaged for the sake of political correctness as 'Native Americans.'

So already there was an infiltration of the 'hidden hand,' as a secret society is an ideal vehicle for underhanded control. It was, however, an infiltration of which many of its members were unaware. But probably one of the little known and least advertised facts regarding Freemasonry and the Masonic lodge is its entirely Jewish origin and nature. The religion of Judaism, based on the Babylonian Talmud and the Jewish Kabbalah, formed the basis for the Scottish rites 33 ritual degree ceremonies.

"Masonry is based on Judaism. Eliminate the teaching of Judaism from the Masonic ritual and what is left?" The Jewish Tribune of New York, 28th October 1927

"Freemasonry is a Jewish establishment, whose history, grades, official appointments, passwords and explanations are Jewish from beginning to end." Rabbi Isaac Wis

Undoubtedly already under Jewish influence, Judeo-masonry in Europe became popular with the rise to power of the House of Rothschild in the latter years of the 18th century. Adam Weishaupt who formed the 'Illuminati' in 1776, founded the Lodge of Theodore in Munich and was befriended and funded by Meyer Rothschild, whose clerk Sigmund Geisenheimer in his Frankfurt office held wide Masonic contacts and was a member of the French Grand Orient Lodge, 'L'Aurore Naissante.' With the help of Daniel Itzig (court Jew to Frederick William II) and the merchant Isaac Hildesheim (who changed his name to Justus Hiller) he founded the Judenloge.

In 1802 the old established Jewish families including the Adler, Speyer, Reiss, Sichel, Ellison, Hanau and the Goldsmid families became members of the Judenloge and in 1803 Nathan Rothschild joined the Lodge of Emulation in England whilst his brother James Rothschild became a 33rd degree Mason in France.

The definitive book on the masons *'Morals and Dogma,'* authored by the late Sovereign Grand Commander of the Scottish rite, Albert

Pike, states that... *"Masonry conceals secrets from all except the adepts and sages and uses false explanations and myth interpretation of its symbols to mislead."*

The rise of Masons to political power in Israel dates back to the state's origins in 1948. David Ben Gurion, Israel's first Prime Minister, was a Freemason. Every Prime Minister since then has been a high level Mason, including Golda Meier who was a member of the women's organisation, the Co-masons. Most Israeli judges and religious figures are Masons, and the Rothschild-supported Hebrew University in Israel has erected an Egyptian obelisk, an overt symbol of Freemasonry in its courtyard.

Indeed today, it is virtually impossible to obtain a position of high office in any sphere in any country, without membership of this all-pervasive body. From politicians to lawmakers, to police and security agencies, religion, science, and business, they are all heavily populated at the upper echelons by high-ranking Freemasons. All of which makes claims of democracy for our society, almost laughable were it not so serious a subject. Freemasons must always and under all circumstances put their own 'brother masons' first, above all else despite any oaths of loyalty they are compelled to make to other organisations – and even their countries! How then can one conclude that any election, from political by-elections to the election of political party leaders to the appointment of company directors and high-level civil servants, could possibly be fairly conducted?

The simple answer of course is that they cannot and thus we have as our default a system whereby exploitation and corruption is the norm and not the exception, whatever we may try to convince ourselves to the contrary.

Five thousand years ago in the Middle East, pagan religions prevailed in places such as ancient Mesopotamia and Sumeria. These pagan societies propagated several myths about the origins of life and the universe as they knew it. One of these was the belief in the evolution of man from lesser species. For example, according to the Sumerians, life first appeared in water before moving onto the land and evolving into new species. Many years later, another civilisation also believed in evolution – ancient Greece. Some Greek philosophers, calling themselves materialists, accepted only the existence of 'matter,' and regarded it as the original source of life. Consequently, they resorted to the myth of evolution from the Sumerians as the explanation of how life came into being.

"Man, who has sprung from earth and developed through the lower kingdoms of nature to the present, rational state, has yet to complete his evolution by becoming a God-like being and unifying his consciousness with the omniscient to promote that which is and always has been the sole purpose of all initiation." W. L. Wilmhurst, *'The Meaning of Masonry'*

Thus, ancient Greece became the foundation of materialism and the evolutionary theory. These two concepts from pagan societies were introduced to the modern world in the nineteenth century by Freemasons who embraced many of these ancient beliefs and creeds. These people were totally opposed to the idea of monotheistic religion and so were committed to destroying it by the promotion of the tenets of evolutionary and materialistic philosophy.

Jean-Baptiste Lamarck, a Frenchman, was the first 'modern' philosopher to promote the idea of evolution in 1809, asserting that all species had progressively evolved from simple, single-celled creatures, through small variations over a long period of time. He and Erasmus Darwin were 'chosen' by the Freemasonic establishment, to be the ones responsible for spreading the evolutionary concept and this culminated eventually in Charles Darwin's work, *'The Origin of Species'* which was the final catalyst for the gigantic hoax of 'evolution' and which has now become the generally, publicly accepted explanation for human origins, despite its many contradictions and impossibilities, and as detailed in the earlier chapter on evolution.

Many ancient secret societies now have metamorphosed into modern, Elite so-called 'think-tanks' that fund the researchers that serve their various agendas. The Freemasons were once very powerful indeed and the members of the Elite class joined it en-masse in the 1800s and early 1900s, but the Elite class no longer necessarily requires the cover of Freemasonry to meet secretly and plan world domination as it once did.

Members of the Elite classes are now totally free and can easily meet in 'closed-door' meetings surrounded by employees gagged by iron-clad non-disclosure agreements or their teams of attorneys protected by client-attorney privileges. They also meet in larger committees and conferences such as the Trilateral Commission, Chatham House, The Council on Foreign Affairs, and the Bilderberg group where secrecy is always strongly enforced, plus in many other minor, so-called 'think-tanks.'

Freemasonry currently accepts members of all religions which fit with the New World Order's plan for the formation of a one world religion. Could the Masonic 'Grand Architect of the Universe' ultimately be the unifier of all the world's religions, the purported 'Antichrist'? It is certainly possible.

The theory that Freemasons worship Satan gained popularity in the 1890s with the writings of French journalist, Gabriel Jogand-Pagès. After the publication of Pope's Leo XIII's anti-Masonic *Humanum Genus,* Jogand-Pagès made a great show of leaving Freemasonry and pretending to return to Roman Catholicism. He was thus able to carve a comfortable anti-Masonic niche for himself.

Under the pen names of 'Léo Taxil' or 'Dr. Bataille,' Jogand-Pagès published a series of stories detailing Freemasonry's involvement with Satanism. He based his articles in part on the revelations of his entirely fictitious Satanic High Priestess, Diana Vaughan and the Masonic 'Palladium' which practiced ritual murders and worshipped the devil Baphomet. Taxil also claimed that Scottish Rite Grand Commander Pike was the 'Sovereign Pontiff' of universal Freemasonry. He transcribed the following address in which he claimed Albert Pike was giving instructions to the twenty-three Supreme Councils of the World...

"That which we must say to the crowd is, we worship a god, but it is the god one adores without superstition. To you, sovereign grand inspector general, we say this, and you may repeat it to the brethren of the 32nd, 31st

The Falsification of Science

and 30th degrees – the Masonic religion should be by all of us initiates of the high degrees, maintained in the purity of the Luciferian doctrine. If Lucifer were not god, would Adonay (the God of the Christians) whose deeds prove cruelty, perfidy and hatred of man, barbarism, and repulsion for science, would Adonay and His priests, calumniate Him?

Yes, Lucifer is god, and unfortunately Adonay is also God, for the eternal law is that there is no light without shade, no beauty without ugliness, no white without black, for the absolute can only exist as two gods. darkness being necessary for light to serve as its foil, as the pedestal is necessary to the statue, and the brake to the locomotive.

Thus, the doctrine of Satanism is heresy, and the true and pure philosophical religion is the belief in Lucifer, the equal of Adonay, but Lucifer, god of light and god of good, is struggling for humanity against Adonay, the god of darkness and evil."

In April 1897, the now infamous Taxil called a press conference and confessed to the fraud. Nevertheless, many are still not convinced. The Taxil hoax notwithstanding, the 33rd degree Freemason, Albert Pike wrote, ... *"Lucifer, the Son of the Morning! Is it he who bears the Light, and with its splendours intolerable blinds feeble, sensual, or selfish souls? Doubt it not!"* 'Morals and Dogma' p. 321

The Judeo-Masonic conspiracy is rooted in the *'Protocols of the Learned Elders of Zion,'* which was first published in 1905 in Russia by Sergei Nilus. This pamphlet details a joint Jewish-Freemasonic conspiracy in which the two work together as one, in order to bring about the end of the world as we know it and usher in their much-vaunted New World Order. Despite Tsar Nicholas II's disapproval of the methods and attempts at confiscating copies of the 'Protocols,' they served as a convenient diversion from Russia's recent humiliation by Japan in their brief war of 1905 and the book continued to circulate widely and was subject to many reprints.

Contemporary disinformation and 'fake news' wishes us to believe that the 'Protocols' are merely a 'hoax.' But even if this is the case, which I personally do not believe, then they are nevertheless an extremely accurate depiction of the state of the world today. Within the 'Protocols,' the Jews are described as being the power behind the Freemasons and merely use the fraternity as a screen for their own covert activities. If the 'Protocols' are indeed genuine, then the Jewish Elders of Zion and the Freemasons are symbiotically involved in a conspiracy to eventually completely exterminate all 'Goyim' (non-Jews).

"Secret masonry, which is not known to, and aims which are not even so much as suspected by, these 'goy' cattle, attracted by us into the 'show' army of Masonic lodges in order to throw dust in the eyes of their fellows." Protocol 11

"Meantime however, until we come into our kingdom, we shall act in the contrary way. We shall create and multiply free Masonic lodges in all the countries of the world, absorb into them all who may become or who are prominent in public activity, for these lodges we shall find our principal intelligence office and means of influence." Protocol 15

The masons are often accused of plotting to rebuild Solomon's Temple at the present location of the Islamic 'Dome of the Rock,' but Freemasons maintain that their rituals refer to a metaphorical temple inside each lodge. They also deny Lyndon Larouche's claim that they (the Masons) are attempting to gain control of Temple Mount in Jerusalem and begin building the third temple of Solomon.

In a letter to the revolutionary leader Giuseppe Mazzini, founder of the 'Mafia,' Albert Pike described the coming of three 'world wars' necessary for the world to accept a New World Order under Lucifer. Mazzini was himself a great supporter of a unified Italy and a United States of Europe, a precursor to the European Union. The first war would overthrow the Czars and much of the rest of the European monarchies and build a 'fortress of atheistic communism.' The second war would pit the Fascists against the Zionists and the third war would be fought between Islam and the Zionists. Lucifer would then rise from these ashes of civilisation, to rule the world.

The letter was originally brought to the public domain by the Canadian intelligence operative and researcher William Guy Carr in the mid-1950s. Carr claimed that the letter was once on display at the

The Falsification of Science

British Museum in London, however, the Museum itself denied this claim. Carr then pointed to respected author Cardinal Rodriguez of Chili's '*The Mystery of Freemasonry Unveiled,*' in which, according to Carr, Rodriguez also claimed to have seen the Pike-Mazzini letter at the museum.

Here is the relevant section from the Pike-Mazzini letter...

"The First World War must be brought about in order to permit the Illuminati to overthrow the power of the Czars in Russia and of making that country a fortress of atheistic Communism. The divergences caused by the 'agentur' [agents] of the Illuminati between the British and Germanic Empires will be used to foment this war. At the end of the war, Communism will be built and used in order to destroy the other governments and in order to weaken the religions.

The Second World War must be fomented by taking advantage of the differences between the Fascists and the political Zionists. This war must be brought about so that Nazism is destroyed, and that the political Zionism be strong enough to institute a sovereign state of Israel in Palestine. During the Second World War, International Communism must become strong enough in order to balance Christendom, which would be then restrained and held in check until the time when we would need it for the final social cataclysm.

The Third World War must be fomented by taking advantage of the differences caused by the agentur of the 'Illuminati' between the political Zionists and the leaders of the Islamic world. The war must be conducted in such a way that Islam [the Moslem Arabic world] and political Zionism [the State of Israel] mutually destroy each other. Meanwhile the other nations, once more divided on this issue will be constrained to fight to the point of complete physical, moral, spiritual and economical exhaustion. We shall unleash the Nihilists and the atheists, and we shall provoke a formidable social cataclysm which in all its horror will show clearly to the nations the effect of absolute atheism, origin of savagery and of the most bloody turmoil.

Then everywhere, the citizens, obliged to defend themselves against the world minority of revolutionaries, will exterminate those destroyers of civilization, and the multitude, disillusioned with Christianity, whose deistic spirits will from that moment be without compass or direction, anxious for an ideal, but without knowing where to render its adoration, will receive the true light through the universal manifestation of the pure doctrine of Lucifer, brought finally out in the public view. This manifestation will result from the general reactionary movement which will follow the destruction of Christianity and atheism, both conquered and exterminated at the same time."

Incredible as it may seem, this extremely 'prophetic' letter was written in the 1870s, but of course this was not so much a 'prophecy' more a foreknowledge of what was being planned for the next century.

'P2' or 'Propaganda Due' the infamous Italian Masonic lodge is used as a cover to coordinate everything from terrorist bombings, massive financial frauds, and drug trafficking.

The covert 'P2' Masonic lodge was discovered in a 1981 police raid on Licio Gelli's home by Italian authorities in the aftermath of the Vatican-owned 'Banco Ambrosia' scandal in which 'God's banker,' Roberto Calvi, a P2 member was found hanging under London's Blackfriars Bridge with his jacket pockets full of bricks. Police discovered a list of over 900 P2 Lodge members which included high-ranking government officials and 43 members of the Italian Parliament. The list also included the name of future Italian President, Silvio Berlusconi, and Mafia banker Michele Sindona.

The P2 Masonic Lodge had been voted-out of the Grand Lodge of Italy in 1974 but Gelli managed to somehow 'convince' the Grandmaster to issue the lodge with another warrant. But nevertheless, by 1981, the other Masonic lodges were very eager to distance themselves from P2 and it was later officially disavowed in a Masonic tribunal which ruled the original 1974 vote valid.

The P2 lodge was also used as a cover for Operation Gladio in which Gelli cooperated with American and British intelligence services to promote extremist activities, including the 1980 Bologna railway bombing, in order to prevent a communist takeover of Italy.

The Falsification of Science

The Bavarian Illuminati was founded by Adam Weishaupt in 1776 and the following year, Weishaupt was initiated as a Freemason at the *Die Loge Theodor vom guten Rat* in Munich. He soon realised the potential of Masonic lodges as recruiting grounds for his own secret society and directed members of his Order to begin 'Illuminising' Freemasonic lodges throughout Germany and later the rest of Europe. After the dissolution of the Masonic rites of 'Strict Observance' at the 1782 Masonic Congress of Wilhelmsbad, Weishaupt and Baron von Knigge were shrewdly able to recruit old Strict Observance members and increased the membership of the Bavarian Illuminati to around 3000 members. The Illuminati quickly established 'Illuminised' lodges in France, Italy, Sweden, Belgium, Holland, Denmark, Poland, and Hungary.

The anti-clerical and anti-monarchical doctrines of the Bavarian Illuminati and other revolutionaries indeed influenced the events leading up to the French Revolution.

"The great strength of our Order lies in its concealment; let it never appear in any place in its own name, but always covered by another name, and another occupation. None is fitter than the three lower degrees of Free Masonry; the public is accustomed to it, expects little from it, and therefore takes little notice of it. Next to this, the form of a learned or literary society is best suited to our purpose, and had Free Masonry not existed, this cover would have been employed; and it may be much more than a cover, it may be a powerful engine in our hands." 'Proofs of a Conspiracy' John Robison 1797

The huge influence that Freemasonry has had upon British society in general for almost 200 years has finally been publicly laid bare. The names of royalty, statesmen, judges, senior military, archbishops, scientists, corporate executives, and senior police have recently been discovered in a secret archive which lists two million Freemasons.

And these Masonic records, dating from 1733 onwards, have now been made available to the public for the first time. Within its pages it revealed that for example, the Kings, Edward VII, Edward VIII and

George VI were all Freemasons as were military leaders such as the Duke of Wellington and Lord Kitchener.

Britain's wartime Prime Minister, Winston Churchill was also a Freemason along with literary figures including Rudyard Kipling, Oscar Wilde and the creator of Sherlock Holmes, Sir Arthur Conan Doyle.

Others, include representatives of almost every facet of society... the musical legends Gilbert and Sullivan, the Antarctic explorers Ernest Shackleton and Captain Robert Scott, the former England cricket captain Douglas Jardine, and the scientist Sir Alexander Fleming, who discovered penicillin.

More than 5,500 senior police officers, thousands of military figures, 170 judges, 169 MPs, 16 bishops and an Indian prince are all listed in the Freemasons archive which has now been made available on the genealogical website, ancestry.com

Businessman Harry Selfridge is named in the archive along with the social reformer Thomas Barnardo, the famous civil engineer Thomas Telford and thousands of other engineers who made Britain a world industrial power. And these are just a small sample. It would appear that for anyone who is influential in any sphere whatsoever, membership of this 'secret society' of Freemasonry is almost mandatory, which of course, cannot be a healthy state of affairs for any so-called 'democratic' nation. But of course we already know that democracy is just a sham in any event and a subtle yet deceptive methodology by which 'the people' are duped into believing that they have some small say in the governance of their countries.

"Democracy is two wolves and a sheep voting on what to have for dinner. Liberty is a well-armed lamb contesting the vote." Former US President, Thomas Jefferson

In fact it is believed that there are around six million Freemasons in the world today – including two million alone in the United States. There are 250,000 Freemasons in around 8,000 lodges in England and Wales.

The Falsification of Science

Freemasons also had a huge influence over the controversial inquiries into the Titanic disaster in which more than 1,500 passengers and crew died. The archive reveals that the judge who presided over the British Wreck Commissioner's inquiry was a Freemason along with leading investigators and some others who escaped censure altogether.

The US inquiry actually blamed the British Board of Trade for lax rules which allowed the Titanic to depart on its maiden voyage in 1912 with just 20 lifeboats for the 2,208 people on board. But the Board of Trade was perhaps unsurprisingly completely exonerated by the British inquiry led by the establishment lackey, Lord Mersey, who was the 'go-to' person for the smooth expedition of many governmental conspiracies and cover-ups at that time. Two of the inquiry's five expert assessors were also listed as masons – the naval architecture specialist, John Harvard Biles and senior engineer assessor Edward Chaston.

Another key figure in the inquiry listed as a Freemason was Lord Pirrie who was chairman of the Harland and Wolff shipbuilders in Belfast which built the Titanic and a director of International Mercantile Marine (IMM) the White Star Line's parent company.

"The Titanic inquiry in Britain was branded a 'whitewash' because it exonerated most of those involved. The whole of the ruling class is Masonic, from the heir to the throne, down. It is part of being 'in the club.' Part of the whole ethic of Freemasonry is, 'whatever it is, however it's done, you protect the brotherhood' and that's what happens." Nic Compton

For a more comprehensively detailed appraisal of the real truth surrounding the appalling Titanic disaster of 1912, please see my book, 'RMS Olympic.'

The Royal Society

The origins of the Royal Society lie in an 'invisible college' of philosophers and scientists who began meeting in the mid-1640s to discuss the ideas of Francis Bacon, the originator of the 'New Atlantis,' the esoteric plan to develop the American colonies as a perfect society. Two of the original members of the Royal Society, Sir Robert Moray and Elias Ashmole, were already Freemasons by the time the Royal Society was formed. The Society met weekly to witness experiments and to discuss what would now be called scientific topics although science then was much more broadly defined and included subjects such as alchemy and astrology.

As the membership of Masonic lodges grew rapidly after 1717, Freemasonry and the Royal Society, became almost intertwined as many of the senior members of the Society also held correspondingly senior positions within the proliferation of Masonic lodges. Indeed, several were closely involved in promoting new lodges and developing the constitutional basis of the new Grand Lodge and early lodges were often a forum for lectures on scientific subjects. For example, John Theophilus Desaguliers was both an important publicist for Newton's scientific ideas and a leading Freemason too.

By the end of the 1700s, particularly during the long presidency of Sir Joseph Banks, himself a Freemason, membership of the Royal Society had become a mix of working scientists and wealthy amateurs who were potential patrons and could help finance scientific research at a time long before any government would consider doing so itself. Several of these patrons were also Freemasons and would have met with scientists both at meetings of the Royal Society as well as in the various lodges. As the professionalisation of science developed in the nineteenth century, Fellows began to be elected solely on the merit of their scientific work. New types of science developed, and scientific education expanded with the growth of university science degrees and medical schools. Freemasonry also attracted these scientist Fellows often in the growing number of new lodges whose membership was drawn from particular universities, hospitals, or other specialist establishments.

All those pictured above were Freemasons. The 'giveaway' is the depiction of the Freemasonic 'set square and compasses' in each of the above pictures. Here is a brief introduction to some of them:

The Falsification of Science

Nicolaus Copernicus (1473 – 1543) was a Renaissance mathematician and astronomer who formulated a model of the universe that placed the Sun rather than the Earth at the centre of the universe, as had been believed for several millennia prior to that.

Galileo Galilei (1564 – 1642) was an Italian astronomer, physicist, engineer, philosopher, and mathematician who played a major role in the scientific revolution of the seventeenth century. He has been called the *'father of observational astronomy,'* the *'father of modern physics,'* and the *'father of science.'* His contributions to observational astronomy include the telescopic confirmation of the phases of Venus, the discovery of the four largest satellites of Jupiter (named the Galilean moons in his honour) and the observation and analysis of sunspots. Galileo was a major champion of Copernicanism and heliocentrism – the model that places the Sun at the centre of the solar system.

John Hamer

Johannes Kepler (1571–1630) was a German mathematician, astronomer, and astrologer. He was also a key figure in the 17th century scientific revolution and is best known for his laws of planetary motion, based on his works '*Astronomia nova,*' '*Harmonices Mundi,*' and '*Epitome of Copernican Astronomy.*' His works provided one of the foundations for Isaac Newton's theory of universal gravitation.

Arguably the most prominent of these famous Freemasons, Sir Isaac Newton (1642–1727) was an English physicist and mathematician (described in his own day as a 'natural philosopher,' an early term for a 'scientist') who is widely recognised as one of the most influential scientists of all time and a key figure in the scientific and mathematical revolution of the seventeenth century.

His book '*Philosophiæ Naturalis Principia Mathematica*' first published in 1687, laid the foundations for classical mechanics and Newton also made seminal contributions to optics, and he shares credit with Gottfried Wilhelm Leibniz for the development of calculus.

Newton's major work, *Principia* formulated the laws of motion and universal gravitation, which dominated scientists' view of the physical universe for the next three centuries. He supplemented Kepler's laws of planetary motion with his mathematical description of gravity. This work also demonstrated that the motion of objects on Earth and of celestial bodies could be described by the same principles. In his later life, Newton became president of the Royal Society, and also served the British government as Warden and Master of the Royal Mint (the body responsible for the creation of British coinage).

Newton was knighted by the British royals, an action which tends to suggest close ties to the monarchy and therefore the 'establishment.' Anyone who receives a British knighthood, without exception, is always someone who is working for, and/or to perpetuate the Elite 'agenda' in some way or other. And interestingly, Newton adopted a coat of arms immediately after being knighted by Queen Anne in 1705, which further confirmed his extremely sinister background. This coat of arms consisted simply of a pair of human tibiæ crossed

The Falsification of Science

on a black background, in the manner of the pirate flag the so-called 'Jolly Roger' but without the skull.

Newton's Crossbones Coat of Arms

Perhaps needless to say, but this coat of arms is extremely suspicious given that similar symbols have been used by secret societies such as the infamous Skull and Bones society of Yale University, USA, to which many prominent politicians and businessmen belong.

One of Newton's closest friends was John Theophilus Desaguliers, who was a Freemasonic Grand Master and is commonly regarded as the *'Father of the Grand Lodge System.'* In his capacity as president, Newton nominated Desaguliers as *'Curator of Experiments'* in 1712, and he would also later become Secretary of the Royal Society. According to the London Museum of Freemasonry, Desaguliers was both a leading Freemason, as well as an important publicist for Newton's scientific ideas.

So here we have someone who was not only a close confidant of Newton, responsible for publishing Newton's scientific material, but was also a Grand Master Freemason tasked with keeping Freemasonic secrets. Thus, there is plenty about which we should be very suspicious, given the enormous influence of Freemasonry in Newton's life.

Looking at the bigger picture, if, for example, you were the British monarch, ruler of all he/she surveys, and you wanted to spread lies in order to deceive and thereby remain in control of the world, then the following would probably be very close to a perfect solution...

Firstly, create an official, 'royal' endorsed organisation that can effectively manipulate and monopolise the dissemination of all information regarding what you would wish for the general public to believe is the 'scientific' truth.

Royal Society

From Wikipedia, the free encyclopedia

This article is about the learned society in the United Kingdom. For other uses, see Royal Society (disambiguation).

The President, Council, and Fellows of the Royal Society of London for Improving Natural Knowledge,[1] commonly known as the **Royal Society**, is a learned society for science and is possibly the oldest such society still in existence.[a] Founded in November 1660, it was granted a royal charter by King Charles II as "The Royal Society".[1] The Society is the United Kingdom's and Commonwealth of Nations' Academy of Sciences and fulfils a number of roles: promoting science and its benefits, recognising excellence in science, supporting outstanding science, providing scientific advice for policy, fostering international and global cooperation, education and public engagement.

Reading between the lines of the above phrases highlighted in yellow...

"Promoting science and its benefits." In other words, publish material that presents only your own desired narratives, whilst censoring or ridiculing alternative scientific views and dissenting scientists.

"Recognising excellence in science." In other words, publicly and officially recognise only your own agents/scientists and ridicule and deny the oxygen of publicity to those who disagree with their often disingenuous pronouncements.

"Supporting outstanding science." Offer funding only to scientists who are prepared to follow your own narrative and only produce results that conform to your agenda.

"Providing scientific advice for policy." Influence politics and law using only your decreed 'science' eg. 'green' energy policies.

"Fostering international and global cooperation." Heavily promote a system whereby a central governmental body controls the flow of scientific information aka the 'New World Order.'

"Education and public engagement." Control the education and publishing sectors in order to successfully indoctrinate the public with your own narrative of history and science.

Secondly, ensure that no-one except those of your choosing and approval based on their loyalty to the agenda, are allowed membership to this 'official' organisation...

The Falsification of Science

> The society is governed by its Council, which is chaired by the Society's President, according to a set of statutes and standing orders. The members of Council and the President are elected from and by its Fellows, the basic members of the society, who are themselves elected by existing Fellows. As of 2016, there are about 1,600 fellows, allowed to use the postnominal title FRS (Fellow of the Royal Society), with up to 52 new fellows appointed each year. There are also royal fellows, honorary fellows and foreign members, the last of which are allowed to use the postnominal title ForMemRS (Foreign Member of the Royal Society). The Royal Society President is Venkatraman Ramakrishnan, who took up the post on November 30, 2015.[2]

…and subsequently promote the work of these scientists over-and-above all else in the 'royal' controlled or sponsored scientific journals, to ensure that their work becomes widely accepted over time, as 'the absolute truth,' no matter how ridiculous or un-scientific it may be.

And lastly, ridicule and censor scientists who present material that opposes your chosen narrative.

Over several decades and even centuries, the scientists, and proclamations of what is or is not the truth, found in all textbooks will almost certainly be those of your choosing. And subsequently, in the fullness of time, and over many generations, the hope is that the alternative, truthful information will be forgotten and lost to posterity forever.

Have you never wondered why a brilliant, until recently completely ignored and forgotten, scientist and inventor like Nikola Tesla never appeared in any scientific textbooks, whilst a whole plethora of Freemasons who only stole other people's ideas (eg. Thomas Edison) and spread lies (eg. Newton/Kepler/Copernicus etc.) appear all the time? The sad truth is that the vast majority of people who pass through the educational system never consider whether the 'facts' they are *repeatedly* told are true or not. The natural tendency is for us to implicitly trust the controlled establishment, be it academia or the scientific world and indeed most people simply accept whatever lies are published by this 'royal' organisation as fact, and then continue believing and regurgitating those lies, for the rest of their lives – as well as passing on these truths to future generations.

And of course the end result of this is that if anyone questions anything they are taught, then they must be 'crazy,' because of course the Royal Society would not deliberately lie and could not possibly be wrong, could they?

John Hamer

Please now consider this. In ...

1660 – The Royal Society was founded by Freemasons

1663 – The Royal Society was awarded its official name via royal charter by a Freemason

1666 – An apple fell on the head of one of these Freemasons

1666 – That same Freemason invented Calculus

1687 – That same Freemason then went on to publish his *'Principia Mathematica'* that became the bedrock of 'science' for the next 300+ years.

Note the '666' number which is Freemasonic/Satanic (the so-called 'number of the beast') and which seems to occur all too frequently in many Freemasonic inspired events.

Many people will believe that it is just a coincidence that the organisation responsible for promulgating the gravity story was created just a few years before Newton randomly 'discovered' gravity. However, others would view it with some justifiable scepticism and conclude that the Royal Society was created *specifically* for the purpose of introducing the public to such grandiose theories as gravity, that were already in the planning long before the alleged apple incident. If of course the 'apple incident' ever really happened, which I strongly suspect it did not and is simply just another fanciful, apocryphal tale for populist consumption by the masses.

This is especially true given that the choice of fruit used in the story; the apple, is already ingrained in our collective psyche as the symbol of 'the fall' in the biblical 'Garden of Eden.' Eve bit into the 'apple,' the representation of the knowledge of good and evil, and it was this act that led to the banishment from paradise of all of humanity.

Once again, I would ask you to pause and think about this for a moment...

What are the odds that someone (Newton) just happened to stumble upon what is considered to be the most significant 'discovery' of

The Falsification of Science

modern science, AND created a revolutionary form of mathematics, at around the same time that the organisation in charge of promulgating the said discovery was created... simply by chance?

So, to summarise the foregoing for clarity...

A group of Freemasons instigated a Royally-acknowledged and endorsed organisation to proclaim upon what is true and what is not true in science, that only they and their (extremely) carefully selected friends and colleagues can be part of.

Then immediately afterwards, these same Freemasons just happen to make a whole sequence of historical 'discoveries' that lay the foundations of science for the next 300 years or more.

Coincidence?

To further summarise, these Freemasons, all allegedly 'highly intelligent geniuses' are credited in textbooks kindly provided to us by other Freemasons in the guise of 'trustworthy' publishers, with outlining the motion of the planets and modelling how the solar system works.

This model may be summarised as follows...

- The Earth is a spinning ball rotating around an enormous ball of exploding gas at 66,600 mph. (note that number again)

- Our solar system, including the Earth and all and all the planets, whirls around the centre of our galaxy at 220 kilometres per second, or 490,000 mph

- The Earth is spinning at 1,000 mph (at the equator) but we cannot detect this motion because the ball on which we live is so large

- People further away from the equator spin more slowly than 1,000 mph but feel no noticeable difference

- The vacuum in space somehow does not affect life on Earth

- What prevents us all from flying off the ball Earth (like water on a spinning tennis ball) is 'gravity'

- Gravity is a theory (yes, it is still only a theory) that took Newton three volumes to explain – and even then, he had to create a whole new form of mathematics, and his first volume began with the words "If..., then..."

- The Pole Star, 'Polaris' has stayed exactly above the north pole for thousands upon thousands of years, despite the Earth and Solar system and all of the known universe moving at hundreds of thousands of miles per hour in random directions.

So it was very impressive indeed that all these geniuses managed to discover all these 'facts,' would you not agree? And please allow me to remind you re those '666' numbers, too!

FREEMASONS AND THE ROYAL SOCIETY

Alphabetical List of Fellows of the Royal Society who were Freemasons

This is an attempt to list Fellows of the Royal Society ("FRS") who were freemasons. It was first issued in January 2010 and this second edition is issued in January 2012. Both have been compiled, on behalf of the Library and Museum of Freemasonry, by Bruce Hogg, assisted by Diane Clements. The Royal Society's website includes two lists of Fellows, from A-J and K-Z, with approximately 8000 names recorded for the period 1660-2007. There is no comparable listing of freemasons and their details have been drawn from a variety of sources as described below. This is unlikely to be a complete list of the freemasons who were FRS and any additions and corrections are welcomed and will be added to future updates of this list. Please email with details of these to the Library and Museum at libmus@freemasonry.london.museum

This list of Fellows of the Royal Society who were Freemasons as per the above, of 8,000 since its inception in 1660 up until 2007, is a truly staggeringly large number for an organisation whose stated primary purpose is supposed to be the furtherance of science, especially considering that they impose strict limits on the number of members (currently 1,600).

Clearly, the Royal Society was founded by Freemasons, and has continued to be totally dominated by Freemasons ever since, for the sinister purposes as described.

In 2010, *'Freemasonry Today'* actually detailed the Masonic origins of the Royal Society. The members listed, included Erasmus Darwin, 'coincidentally' the grandfather of Charles Darwin, and also such scientific luminaries as James Watt, Joseph Priestley, and Benjamin Franklin. Masonic authors such as Albert Pike and Manley P. Hall have revealed that the teachings of Freemasonry are identical to those of the ancient Sumerian, Babylonian, and Egyptian mystery religions, as well as Judaism.

"The Masonic religion should be, by all of us initiates of the high degrees, maintained in the purity of the Luciferian doctrine. Yes, Lucifer is our God." Albert Pike, *'Morals and Dogmas'*

The aforementioned, James Watt, the inventor of the steam engine, was a member of the Somerset House Lodge in London and Benjamin Franklin, in 1734, became the Grand Master of the St. John's Lodge in Philadelphia.

As well as Freemasonry, Franklin was also deeply involved in other occult practices and was famously a member of the notorious 'Hellfire Club,' in London, whose members were implicated in such heinous and insidious practices as child and virgin sacrifices. Indeed when Franklin's former London home was being renovated in the late nineteenth century many years after his death, several skeletons, some of which were those of children, were discovered in the wall cavities. The Hellfire Club was also infamous for its patrons' heavy drinking and involvement in sex orgies and black magic rites and ceremonies as well as its members' open ridiculing of Christianity and mocking its practices.

Erasmus Darwin was initiated into the 'Time Immemorial Lodge' in Scotland in 1788. He was the author of 'Zoonomia' in which volume he first proposed evolutionary theory, subsequently expanded upon greatly by his grandson more than seventy years later. The aforementioned Jean Baptiste Lamarck was a member of the French Masonic organisation, 'Amis de la Verité' (Friends of the Truth) founded in 1790. He published his 'Philosophie Zoologique' in 1809 and this work mirrored that of Erasmus Darwin.

So, in conclusion, not only does Freemasonry hold a vicelike grip on all elements of society, but also pervades and indeed is most prominent, almost to the point of exclusivity, within the Royal Society itself, extending its iron grip on all matters scientific. Given that we know now that Freemasonry brings to bear this all-encompassing influence upon the Royal Society, how can we trust whatever scientific decrees it decides to foist upon society in general?

The plain truth of the matter is that unfortunately, we cannot.

Chapter 5

Flat Earth

A cat-loving friend once said, *"The Earth cannot possibly be flat. If it were, then cats would already have knocked everything off the edge!"* Which is possibly the strongest argument ever heard in favour of the globe Earth!

Seriously though and joking aside, there has recently been a vast quantity of literature, videos, and debate on the Internet, especially on social media, regarding the possibility of our Earth being flat, and not a sphere, as has been the generally accepted 'norm' for the past five hundred years and more.

According to proponents of the Flat Earth theory, the Earth is in fact a disc with the Arctic Circle in the centre and a wall of ice of 150-200 feet in height around the circumference in Antarctica, which also prevents the oceans from spilling 'over the edge.' This theory also states that it is covered by some kind of 'dome,' which prevents the atmosphere from escaping, and seals the entire biosphere from the vacuum of outer space and thus solving the issue raised in an earlier chapter, of why Earth's atmosphere does not simply dissipate into space. Day and night are explained by the Sun and the Moon being spheres measuring 51 kilometres in diameter and moving in a circular motion above the plane of the flat Earth. Like spotlights, these celestial spheres illuminate different parts of the planet at different times in a continuous cycle of 24 hours.

Let me state immediately that I am not totally convinced, yet, by the arguments supporting a flat Earth, although I do feel it right to concede unreservedly that there is plenty of compelling evidence for the premise...as well of course, as counter arguments to the contrary. But at the very least the flat Earth theory highlights many issues with the globe Earth hypothesis as presented, as it appears that we have in fact been lied to about this subject on a massive scale. But what the actual truth is regarding the real shape of the Earth is still an unexplained mystery, and certainly worth asking questions about. However, whatever shape the Earth actually is, it is not too difficult to conclude that we are most definitely being lied to in one way or another.

As discussed in the previous chapter, the Royal Society and Freemasonry hold a very insidious and far-reaching 'grip' on the current, accessible 'knowledge' of what is true and what is false in our society today. Can it be a mere coincidence therefore that people such as Copernicus, Galileo and Newton who were some of the most important 'architects' of the currently accepted paradigm, were all 'fully paid-up' members of the Masonic fraternity? And indeed, is it a coincidence that the Earth was generally accepted as being both flat and the centre of the universe right up until the point in time, five hundred years or so ago, that Nicolaus Copernicus devised his solar-centric model of the solar system, amidst much derision, scorn and indeed, rage, from the incumbent 'flat-Earthers' of the Catholic Church and its many millions of adherents.

Even the ancient Greeks, who possessed many gifted and 'far ahead of their time' mathematicians, scientists, and thinkers such as Euclid,

The Falsification of Science

Archimedes, Plato, and Socrates believed that the Earth was flat in the absence of any proof to the contrary.

Flat-Earthers believe that all our lives we have been taught a falsehood so gigantic and diabolical that it has blinded us from our own experience and common sense and from seeing the world and the universe as they truly are. Through pseudoscientific books and programmes, mass media and public education, universities and government propaganda, the world has been systematically brainwashed and slowly indoctrinated over centuries into the unquestioning belief of what is very possibly the greatest lie of all time. And that is despite the existence of some pretty stiff competition for that particular accolade!

But what exactly is the 'flat Earth theory?' In fact, there never has been anything named a 'flat Earth theory.' Different cultures during different time periods have posited a staggeringly diverse array of worldviews, which cannot easily be summed-up with the phrase 'flat Earth.' And nor is the idea of a flat Earth something that is exclusive to the Western world.

Even the most cursory historical research reveals that the idea that the Earth is flat has been a notion shared by an extraordinarily wide range of cultures and tied to vastly different metaphysical systems and cosmologies... for obvious reasons presumably. Be honest with yourself, now dear reader, who among us would have guessed that the Earth was actually a sphere had we not been told this over and over again throughout our lives, and presented with copious amounts of 'evidence' to backup that premise? However, there is certainly no overtly apparent evidence for us mere mortals to believe in such a wild and crazy notion that the Earth is in fact a spinning ball in space. And if the 'globe Earthers' could... or ever had proved that the Earth is a globe, then there would be no flat Earth movement in existence to contest the issue.

As stated above, it was a common belief in ancient Greece, as well as in India, China and in a wide range of indigenous or 'pre-state' cultures. Both the poets Homer and Hesiod described a flat Earth, and this view was also maintained by Thales, considered by many one of the first philosophers, Lucretius, an avowed materialist, as well as Democritus, the founder of atomic theory.

The ancient Greek conception, in turn, has some parallels with that of early Egyptian and Mesopotamian thought, both cultures believing that the Earth was a large disc surrounded by a gigantic body of water. The ancient Chinese were also virtually unanimous in their view of the Earth's flatness, although…in this system…the heavens were spherical, and the Earth was square.

A number of ancient Indian conceptions, common…with some degree of variation…to ancient Hinduism, Jainism and Buddhism, tie their cosmography to botanical images, with the Earth being comprised of four continents surrounding a mountain, akin to the way petals encircle the centre of a flower. Ancient Norse views also postulated a circular, flat Earth surrounded by a sea inhabited by a giant serpent.

Others, such as the Mountain Arapesh people of Papua New Guinea, envisage a world, which ends at the horizon, 'the place where giant clouds gather.' But even where commonalities exist across these traditions, vastly different metaphysical and cosmological narratives are in evidence. And, to complicate matters further, to these we must add cultures and intellectual traditions for whom the shape of Earth is of no interest whatsoever. Many tribal or pre-state societies, for instance, have little concern for what might be considered cosmography. Obsessed, for obvious, more practical reasons with their own day-to-day survival, they simply have no need of such intellectual posturing.

But the key question, for me, remains and is this; simply because we know for certain that there is so much of our alleged 'knowledge' that is falsified for political and financial expediency, could this also actually mean that the spherical Earth is also a false premise for reasons similar to the Big Bang and evolutionary theory?

So, let us go ahead and examine the 'flat Earth' much more closely. However, before being tempted to dismiss it out of hand as sheer, unadulterated nonsense and skipping past this chapter, please allow us to play the role of 'devil's advocate' against the heliocentric (spherical Earth) model and in favour of the flat Earth.

Firstly, why would 'they' lie about the shape of the Earth? It is a rather large lie, so after all, there must be a very good reason. However, the answer is quite simple really...

The most obvious reasons for such a massive lie are our old friends, 'money' and 'world control.' NASA takes tens of billions of dollars per year from US taxpayers for starters, but it is valid to ask the question as to where all that money is going, if indeed it *is* all fakery? It certainly requires large sums of money to, for example, fake a space station in a huge 'swimming' pool and pay technicians to create CGIs of the globe Earth... as well as keep them all quiet about it, but certainly not all that much, and so several people must be getting rich(er) on the back of it all.

NASA was begun in the 1950s by prominent members of the occult. All top-level astronauts are Freemasons who have sworn an oath of secrecy upon pain of horrific death and the Satanic Kabbalah is openly taught at NASA. Once again, please undertake your own research on this topic and as with everything else contained within these pages, please do not take my word for it all. Clearly, these people are not promoting true science but a pagan religion and a sinister agenda.

As many people are now becoming aware, the Satanic Elite have a plan to create a communitarian, totalitarian one world government, otherwise known as the 'New World Order' or NWO for short. One of the key goals in this devilish plan is to commandeer science, as well as history and distort it as a stepping-stone to creating a completely brainwashed populace and thereby allowing them free rein in fulfilling their agenda. As discussed in previous chapters, one of those goals is to convince the general population that we evolved from nothing more than pond slime in the primordial soup, through fish and small mammals to monkeys and apes, and that everything now in existence was created from nothing, in the utterly risible concept known as the 'Big Bang.'

"We'll know our disinformation programme is complete when everything the American public believes, is false." William Casey, CIA Director during the Reagan administration

The flat Earth model is a stationary Earth, set on a firm foundation of unknown composition, with a 'local' Sun and Moon set in the 'firmament,' or inside the dome. The stars are above the firmament, the universe is relatively small and close, and the Earth is not hurtling through a vast expanding Big Bang universe, it is actually the centre point of all creation.

The theories of evolution, the Big Bang theory, atheism, the New Age movement, the New World Order, and the occult are all based on the ball Earth premise and the spinning ball Earth hurtling around a giant Sun (Pagan Sun Worship) is the foundation for all these Satanic beliefs. Without the ball Earth all of these theories (lies) fall apart and have no foundation.

Using the flat Earth model as a basis, evolution, and the Big Bang simply 'do not compute.' A flat, stationary Earth inside a dome of unknown composition, with a close, fixed 'universe' of stars and wandering stars (planets) rotating around *inside* the dome, clearly points to intelligent design of some kind and completely negates all the above deceptions.

But of course, those deceitful Elites hate the truth, or at very least hate we, the 'useless eaters,' knowing the truth, so of course they attack the flat Earth movement and use every trick known to man to prevent the truth being disseminated and ridicule it to the point that people will not even bother to research it and will dismiss it out of hand as sheer lunacy. But all it takes is the smallest amount of cursory research and the truth cannot then be denied.

In the flat Earth model, all the celestial bodies are orbiting the magnetic centre of the known universe, which is purported to be the magnetic North Pole of the Earth, positioned in the dead centre of the flat Earth disc. The Earth is stationary, and all other heavenly bodies orbit the Earth, outside the dome.

The Falsification of Science

From direct observation, the horizon always appears as a completely flat 360-degree circle to the observer, regardless of how high he may be. Any 'curvature' in photographs or videos of the Earth you may think you see, is from curved plane windows or Go-Pro™ cameras and fisheye lenses (which, by the way, NASA always uses as 'proof' of a globe Earth). The reality is that the horizon never curves because the Earth is an endless circular plane. On a globe that despite allegedly being almost 25,000 miles in circumference there would always be a plainly obvious, noticeable disappearance of objects the further away they are, as they would be leaning away from us and dropping below the constantly curving horizon!

Also, it is an indisputable fact that the horizon always rises to meet our eyelevel no matter how high in altitude we may be. Even at a height of twenty miles, the horizon rises to meet the eyelevel of the observer or camera lens. This is only physically possible if the Earth is a huge 'endless' flat, circular plane. If Earth were a globe as claimed, no matter how large, as we ascended, the horizon would stay fixed and the observer would have to look downwards, looking down further and further to see it.

Furthermore, the natural behaviour of water is to always find and maintain its 'own level.' If the Earth were a giant spinning, tilted sphere and hurtling through space at unimaginable velocities, then truly flat, consistently level surfaces would not exist here and instead there would be a massive bulge of water in the oceans because of the curvature of the Earth. If the Earth were curved and spinning, the oceans of water would be finding their own level and covering the land. Some rivers would be flowing uphill and there would be widespread chaos and flooding. What we would see, and experience, would be vastly different but as the Earth is claimed by 'flat Earthers' to be an extended flat plane, this fundamental physical property of fluids finding and remaining level is consistent with experience, common sense and even more significantly, physics. The water remains flat because the Earth is flat.

If Earth were a ball 25,000 miles in circumference as NASA and the entire scientific, modern astronomical and educational establishments claim, spherical trigonometry dictates that the surface of all standing water must curve downward an easily measurable eight inches per mile multiplied by the square root of the distance. This means that along a six-mile channel of standing water, that the Earth would dip six feet on either end from the central peak. Every time such experiments have been conducted, however, standing water has proven to be perfectly level.

The Sun is much closer than we have been told, according to the flat Earth paradigm. It is, in fact, within our atmosphere and it can be clearly seen that it is not 93 million miles away, simply by the Sun's rays emerging from a cloud and forming a triangular pattern. If we follow those rays to their source it will always lead to a place above the clouds, but if the Sun were truly many millions of miles away, all the rays would arrive on Earth at a straight angle, because of the vast distance involved. Also, the Sun can be seen directly above clouds in some hot air balloon photos, creating a hot spot on the clouds below it and in other photos it is possible to clearly see the clouds dispersing directly underneath the close, small Sun. This would surely be impossible with a distant Sun.

If we were living on a spinning globe, airplanes would constantly have to dip their noses every few minutes to compensate for the curvature of the Earth (with a circumference of 25,000 miles the Earth would be constantly curving at the speed of a plane). In reality, however, they simply never do this, and all pilots always learn to fly based solely on a level, flat Earth model. Also, if the Earth were spinning, planes travelling westwards would reach their destinations much faster relatively speaking than either planes travelling eastwards or indeed than if the Earth were stationary, since the Earth is spinning in the opposite direction. If the atmosphere is spinning with the Earth, then planes flying westwards would have to fly faster than the Earth's spin to reach its destination. In reality, the Earth is flat, and planes just simply fly level and reach their destination easily because the Earth is not moving.

Here is a rather damning quote from an official NASA website...

The Falsification of Science

*"A linear aircraft model for a rigid aircraft of constant mass flying **over a flat, nonrotating Earth** is derived and defined. The derivation makes no assumptions of reference trajectory or vehicle symmetry. The linear system equations are derived and evaluated along a general trajectory and include both aircraft dynamics and observation variables."* NASA Technical Reports Server. https://ntrs.nasa.gov/search.jsp?R=19890005752

The experiment known as 'Airy's failure' proved that the stars move relative to a stationary Earth and not the other way around. By first filling a telescope with water to slow down the speed of light inside, then calculating the tilt necessary to allow the starlight directly down the tube, Airy failed to prove the heliocentric theory (hence the experiment's name) since the starlight was already arriving at the correct angle with no change necessary, and instead proved the geocentric model correct.

In other experiments, namely the 'Michelson-Morley' and 'Sagnac' experiments, they attempted to measure the change in the speed of light due to Earth's assumed motion through space. After measuring in every possible different direction in various locations they failed to detect any significant change whatsoever, thus proving the stationary, geocentric model of the universe.

If gravity is really a force strong enough to prevent the world's oceans, buildings, people, and atmosphere from flying off the surface of the spinning ball, then surely it is also totally impossible for gravity to be simultaneously too weak to prevent small birds, insects, and children's balloons to become airborne and travel freely in any direction. Physicists cannot have this both ways. Also, if gravity is strong enough to curve the massive expanse of the oceans around a globular Earth, it would be impossible for fish, humans, and other creatures to swim through such forcefully held water.

Everything on the flat Earth is naturally arranged by density and mass. For example, ship's captains in navigating great distances at sea never need to factor the supposed curvature of the Earth into their calculations. Both 'Plane Sailing' and 'Great Circle Sailing,' the most popular nautical navigation methods, use plane, not spherical trigonometry, making all mathematical calculations on the assumption that the Earth is perfectly flat. If the Earth were indeed a sphere, such an errant assumption would lead to constant, glaring inaccuracies in navigation. Plane Sailing has worked perfectly fine in both theory and practice for thousands of years however, and plane trigonometry has time and again proven more accurate than spherical trigonometry in determining distances across the oceans.

If the Earth were truly a globe, then every line of latitude south of the equator would have to measure a gradually smaller and smaller circumference the farther south travelled. If, however, the Earth is an extended plane, then every line of latitude south of the equator should measure a gradually larger and larger circumference the farther south travelled. The fact that many captains navigating south of the equator assuming the globe theory have found themselves drastically off course, the more so the farther south travelled, testifies to the fact that the Earth is definitely not a ball.

In the flat Earth model of the cosmos, the North Pole is the immovable centre of the world and the entire universe. Polaris, the North Star, sits over the North Pole at the highest point in the heavens, and like a slowly rotating planetarium dome all the celestial bodies revolve around Polaris and over the Earth once per day. The Sun circles over and around the circumference of Earth every 24 hours, steadily travelling each day from the equator during the March vernal equinox, up to the Tropic of Cancer at the June summer solstice, back down to the equator for the September autumnal equinox, and all the way down to the Tropic of Capricorn on the December winter solstice.

Also, in the flat Earth model, the South Pole does not exist at all and Antarctica is actually a gigantic ice-wall extending around the entire outer circumference of the Earth holding in the oceans like a giant bowl. As strange as this concept may sound at first, it is nevertheless a little-known or even promoted fact that if you set a bearing due south from anywhere on Earth, inevitably at or just before the 78° southern latitude you will find yourself face-to-face with an enormous ice-wall towering 100-200 feet in the air extending to the east and west around the entire circumference of the Earth.

"The ice-barrier, so frequently referred-to in accounts of the Antarctic regions, is the fore-front of the enormous glacier-covering, or ice-cap, which, accumulating in vast, undulating fields from the heavy snowfall, and ultimately attaining hundreds, if not thousands, of feet in thickness, creeps from the continent of Antarctica into the polar sea. The ice-barrier, yet a part of the parent ice-cap, presents itself to the navigator who has boldness enough to approach its fearful front, as a solid, perpendicular wall of marble-like ice, ranging from one thousand to two thousand feet in thickness, of which from one hundred to two hundred feet rises above, and from eight hundred to eighteen hundred feet sinks below, the level of the sea." General A.W. Greely, *'Antarctica, or the Hypothetical Southern Continent.'* Cosmopolitan 17 (1894) p. 296

"It has been demonstrated that the Earth is a plane, the surface-centre of which is immediately underneath the star called 'Polaris,' and the extremities of which are bounded by a vast region of ice and water and irregular masses of land. The whole terminates in fog and darkness, where snow and driving hail, piercing sleet and boisterous winds, howling storms, madly-mounting waves, and clashing icebergs are almost constant." Dr Samuel Rowbotham, *'Zetetic Astronomy, Earth Not a Globe'*

Antarctica is not the tiny ice-continent found confined to the underside of Freemasonry's globe that it is purported to be. Quite the contrary, Antarctica literally surrounds us 360 degrees, encircles every continent, and acts as a barrier holding in the oceans. The most commonly asked questions, and the greatest mysteries yet to be solved are how far does the Antarctic ice extend outwards? Is there a limit? What lies beyond, or is it just snow and ice forever? However, thanks to U.N. treaties, international laws, and constant military surveillance, both the North Pole and Antarctica remain cloaked in government secrecy, both are strictly designated 'no-fly/no-sail' zones, with several reports of civilian pilots and ship's captains being warned away and escorted away under threats of extreme violence.

As a small aside, I do wonder if Captain Robert Scott's famous, yet unsuccessful quest to be the first humans to reach the South Pole in 1912, is just an invented fantasy to help convince us that the South Pole is 'real?' If so, it certainly wouldn't be the first nor the last alleged, significant historical event to have been totally or partially fabricated. Were Scott and his entire expedition quietly exterminated by the Elite powers-that-be before they could actually reach the Antarctic and discover the real truth about that mysterious 'continent?' Or maybe the expedition did really take place and Scott discovered some rather uncomfortable, not to mention inconvenient truths for the establishment and subsequently had to be 'disposed of' in some way? Is it at all significant perchance that not a single member of that ill-fated expedition actually survived, but yet we are told conveniently, that they DID actually reach the South Pole, but perished only twelve miles from safety on their return, alleged 1,900 mile round-trip hike across the snowbound Antarctic wastelands?

Instead we are left with a highly romanticised and enduring tale of bad luck and disaster, that has created an indelible imprint on the psyches of all those who hear it, with the existing apocryphal myths embellished and enhanced by such Hollywood blockbusters as '*Scott of the Antarctic*' (1948) and innumerable allegedly 'factual' documentaries, perpetuating the same storyline? This of course is a typical ploy to manipulate and deceive, as was the case with the Titanic disaster of (coincidentally?) the exact same year and a mere two weeks after the Antarctic 'disaster,' and which 'plotline' was also subtly, psychologically reinforced in our minds by a whole raft of Hollywood-created 'Titanic' movies, from the 1930s onwards.

"How far the ice extends; how it terminates; and what exists beyond it, are questions to which no present human experience can reply. All we at present know is, that snow and hail, howling winds, and indescribable storms and hurricanes prevail; and that in every direction 'human ingress is barred by unsealed escarpments of perpetual ice,' extending farther than eye or telescope can penetrate, and becoming lost in gloom and darkness."
Dr Samuel Rowbotham, *'Zetetic Astronomy, Earth Not a Globe'*

Before reaching the Antarctic ice-wall, navigating the increasingly tumultuous southern oceans, explorers encounter the longest, darkest, coldest nights and the most dangerous seas and storms anywhere on Earth. Vasco de Gama, an early 16th century Portuguese explorer of the South Seas wrote that, *"The waves rise like mountains in height; ships are heaved up to the clouds, and apparently precipitated by circling whirlpools to the bed of the ocean. The winds are piercing cold, and so boisterous that the pilot's voice can seldom be heard, whilst a dismal and almost continual darkness adds greatly to the danger."*

In 1773, the English explorer, Captain James Cook became the first modern adventurer known to have breached the Antarctic Circle and reach the ice barrier. During three voyages, lasting three years and eight days, Captain Cook and his crew sailed a total of 60,000 miles along the Antarctic coastline never once finding an inlet or path through or beyond the massive glacial wall. Cook wrote, *"The ice extended east and west far beyond the reach of our sight, while the southern half of the horizon was illuminated by rays of light which were reflected from the ice to a considerable height. It was indeed my opinion that this ice extends quite to the pole, or perhaps joins some land to which it has been fixed since creation."*

On the 5th October 1839 another explorer, James Clark Ross began a series of Antarctic voyages lasting a total of 4 years and 5 months. Ross and his crew sailed two heavily armoured warships thousands of miles, losing many men from gale force winds and icebergs, looking for an entry point beyond the southern glacial wall. Upon first confronting the massive barrier Captain Ross wrote of the wall, *"... extending from its eastern extreme point as far as the eye could discern to the eastward, it presented an extraordinary appearance, gradually increasing in height, as we got nearer to it, and proving at length to be a perpendicular cliff of ice, between one hundred and fifty feet and two hundred feet above the level of the sea, perfectly flat and level at the top, and without any fissures or promontories on its even seaward face. We might with equal chance of success try to sail through the cliffs of Dover, as to penetrate such a mass."*

"'Yes, but we can circumnavigate the South easily enough,' is often said by those who don't know. The British ship Challenger recently completed the circuit of the Southern region ... indirectly, to be sure ... but she was three years about it and traversed nearly 69,000 miles ... a stretch long enough to have taken her six times round on the globular hypothesis." William Carpenter *'100 Proofs the Earth is Not a Globe'*

"If we now consider the fact that when we travel by land or sea, and from any part of the known world, in a direction towards the North polar star, we shall arrive at one and the same point, we are forced to the conclusion that what has hitherto been called the North Polar region, is really the centre of the Earth. That from this northern centre the land diverges and stretches out, of necessity, towards a circumference, which must now be called the Southern region: which is a vast circle, and not a pole or centre ... In this and other ways all the great navigators have been frustrated in their efforts and have been more or less confounded in their attempts to sail round the Earth upon or beyond the Antarctic circle. But if the southern region is a pole or centre, like the north, there would be little difficulty in circumnavigating it, for the distance round would be comparatively small. When it is seen that the Earth is not a sphere, but a plane, having only one centre, the north; and that the south is the vast icy boundary of the world, the difficulties experienced by circumnavigators can be easily understood." Dr Samuel Rowbotham, *'Earth Not a Globe, 2nd Edition'*

If the Earth were truly a globe, then every line of latitude south of the equator would have to measure a gradually smaller and smaller circumference the farther south one travelled. In other words, the circumference at 10 degrees south latitude would comprise a smaller circle than at the equator, 20° south latitude would comprise a circle smaller than 10°, and so on. If, however, the Earth is an extended plane, then every line of latitude south of the equator should measure a gradually larger and larger circumference the farther south travelled. 10° south latitude will comprise a larger circle than the equator, 20° south latitude will comprise a circle larger than 10°, and so on. Likewise, if the Earth were a globe, lines of longitude would expand at the equator whilst converging at both poles whereas if the Earth is an extended plane, lines of longitude should simply expand straight outwards from the North Pole. So which is actually the case?

"Upon the principle, as taught by Scripture and common observation, that the world is not a Planet, but consists of vast masses of land stretched out upon level seas, the North being the centre of the system, it is evident that the degrees of longitude will gradually increase in width the whole way from the North centre to the icy boundary of the great Southern Circumference. In consequence of the difference between the actual extent of longitudes and that allowed for them by the Nautical Authorities, which difference, at the latitude of the Cape of Good Hope, has been estimated to amount to a great number of miles, many ship-masters have lost their reckoning, and many vessels have been wrecked. Ship-captains, who have been educated in the globular theory, know not how to account for their getting so much out of their course in Southern latitudes, and generally put it down to currents; but this reason is futile, for although currents may exist, they do not usually run in opposite directions, and vessels are frequently wrecked, whether sailing East or West." David Wardlaw Scott, 'Terra Firma'

During Captain James Clark Ross's voyages around the Antarctic circumference, he often wrote in his journal as to how perplexed they were upon routinely finding themselves out of sync with their charts, stating that they found themselves an average of twelve to sixteen miles adrift of what their calculations had predicted, every day with some days as much as twenty-nine miles.

Lieutenant Charles Wilkes commanded a United States Navy exploration expedition to the Antarctic from the 18th August 1838 to the

10th June 1842, which was almost four years spent *"exploring and surveying the Southern Ocean."* In his journals Lieutenant Wilkes also mentioned being consistently east of his reckoning, sometimes over twenty miles in less than eighteen hours.

"The commanders of these various expeditions were, of course, with their education and belief in the Earth's rotundity, unable to conceive of any other cause for the differences between log and chronometer results than the existence of currents. But one simple fact is entirely fatal to such an explanation, viz., that when the route taken is east or west the same results are experienced. The water of the southern region cannot be running in two opposite directions at the same time; and hence, although various local and variable currents have been noticed, they cannot be shown to be the cause of the discrepancies so generally observed in high southern latitudes between time and log results. The conclusion is one of necessity, forced upon us by the sum of the evidence collected that the degrees of longitude in any given southern latitude are larger than the degrees in any latitude nearer to the northern centre; thus proving the already more than sufficiently demonstrated fact that the Earth is a plane, having a northern centre, in relation to which degrees of latitude are concentric, and from which degrees of longitude are diverging lines, continually increasing in their distance from each other as they are prolonged towards the great glacial southern circumference." Dr Samuel Rowbotham, 'Zetetic Astronomy: Earth Not a Globe'

"11th February 1822, at noon, in latitude 65.53. S. our chronometers gave 44 miles more westing than the log in three days. On 22nd April 1822, in latitude 54.16. S. our longitude by chronometers was 46.49, and by D.R. (dead reckoning) 47° 11´: On 2nd May 1822, at noon, in latitude 53.46. S., our longitude by chronometers was 59° 27´, and by D.R. 61° 6´. 14th October 1822, in latitude 58.6, longitude by chronometers 62° 46´, by account 65° 24´. In latitude 59.7. S., longitude by chronometers was 63° 28´, by account 66° 42´. In latitude 61.49. S., longitude by chronometers was 61° 53´, by account 66° 38´." Captain James Weddell, 'Voyages Towards the South Pole'

"In the southern hemisphere, navigators to India have often fancied themselves east of the Cape when still west, and have been driven ashore on

the African coast, which, according to their reckoning, lay behind them. This misfortune happened to a fine frigate, the Challenger, in 1845. How came Her Majesty's Ship 'Conqueror,' to be lost? How have so many other noble vessels, perfectly sound, perfectly manned, perfectly navigated, been wrecked in calm weather, not only in dark night, or in a fog, but in broad daylight and Sunshine ... in the former case upon the coasts, in the latter, upon Sunken rocks ... from being 'out of reckoning,' under circumstances which until now, have baffled every satisfactory explanation." Reverend Thomas Milner, *'Tour Through Creation'*

The equatorial circumference of the supposed globe-Earth is said to be 24,900 statute or 21,600 nautical miles. A nautical mile is the distance, following the supposed curvature of the Earth, from one minute of latitude to the next. A statute mile is the straight-line distance between the two, not taking into account Earth's alleged curvature.

The Australian handbook, *'Almanack, Shippers' and Importers' Directory'* states that, the distance between Sydney and Nelson is 1400 nautical or 1633 statute miles. Allowing a more than sufficient 83 miles as the distance for rounding Cape Farewell and sailing up Tasman Bay to Nelson leaves 1550 statute miles as the straight-line distance from the meridian of Sydney to the meridian of Nelson. Their given difference in longitude is 22° 2' 14". Therefore if 22° 2' 14" out of 360 is 1550 miles, the entirety measures 25,182 miles. This is larger than the Earth is said to be at the equator, and 4262 miles greater than it would be at Sydney's southern latitude on a globe of said proportions. One 360th part of 25,182 gives 70 miles as the distance between each degree of longitude at Sydney's 34° southern latitude. On a globe 25,000 miles in equatorial circumference, however, degrees of longitude at 34° latitude would be only 58 miles, a full 12 miles per degree less than reality. This perfectly explains why James Clark Ross and other navigators in the deepest southern regions experienced greater than twelve mile daily discrepancies between their reckoning and reality, the farther South travelled the farther the divide.

"From near Cape Horn, Chile to Port Philip in Melbourne, Australia the distance is 9,000 miles. These two places are 143° of longitude from each other. Therefore the whole extent of the Earth's circumference is a mere arithmetical question. If 143° make 9,000 miles, what will be the distance made by the whole 360 degrees into which the surface is divided? The answer is, 22,657 miles; or, 8357 miles more than the theory of rotundity would permit. It must be borne in mind, however, that the above distances are nautical measure, which, reduced to statute miles, gives the actual distance round the Southern region at a given latitude as 26,433 statute miles; or nearly 1,500 miles more than the largest circumference ever assigned to the Earth at the equator." Dr Samuel Rowbotham, 'Earth Not a Globe, 2nd Edition'

Similar calculations made from the Cape of Good Hope, South Africa to Melbourne, Australia at an average latitude of 35.5° South, have given an approximate figure of over 25,000 miles, which is again equal to or greater than the Earth's supposed greatest circumference at the equator. Calculations from Sydney, Australia to Wellington, New Zealand at an average of 37.5° south have given an approximate circumference of 25,500 miles, which is greater still! According to the ball-Earth theory, the circumference of the Earth at 37.5° latitude should be only 19,757 statute miles, almost six thousand miles less than such practical measurements.

"The above calculations are, as already stated, only proximate; but as liberal allowances have been made for irregularities of route, etc., they are sufficiently accurate to prove that the degrees of longitude, as we proceed south-wards, do not diminish, as they would upon a globe, but expand or increase, as they must if the Earth is a plane; or, in other words, the farthest point, or greatest latitude south, must have the greatest circumference and degrees of longitude." Dr Samuel Rowbotham, 'Zetetic Astronomy: Earth Not a Globe'

"Parallels of latitude only ... of all imaginary lines on the surface of the Earth ... are circles, which increase, progressively, from the northern centre to the southern circumference. The mariner's course in the direction of any one of these concentric circles is his longitude, the degrees of which INCREASE to such an extent beyond the equator (going southwards) that

The Falsification of Science

hundreds of vessels have been wrecked because of the false idea created by the untruthfulness of the charts and the globular theory together, causing the sailor to be continually getting out of his reckoning. With a map of the Earth in its true form all difficulty is done away with, and ships may be conducted anywhere with perfect safety. This, then, is a very important practical proof that the Earth is not a globe." William Carpenter, '100 Proofs the Earth is Not a Globe'

Airplane Flights in the Southern Hemisphere

There are many strange anomalies in existence, regarding flights in the Southern Hemisphere, which provide further circumstantial evidence, if not hard proof, that the Earth is flat and not a globe.

For one simple example from many, take a flight from Santiago in Chile, to Christchurch, New Zealand. Is it not strange that not one single airline provides a direct, straight line flight which is a distance of approximately 6,000 miles or approximately the same as the distance from London to Los Angeles? Instead they **all** follow this strange, convoluted route as below...

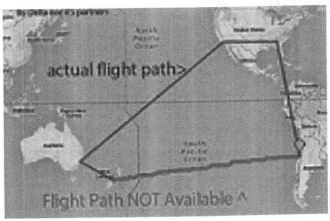

As you may see from the above, the route goes almost due north up towards the east coast of the USA, before turning due west towards the west coast near Los Angeles, before taking a south-westerly track across the Pacific Ocean to Sydney, Australia, before turning back south-easterly to Christchurch, New Zealand. This is an alleged journey of 15,000 miles plus, when logically it should be one of only 6,000 miles. It does not make any sense does it?

However, now please consider the picture below...

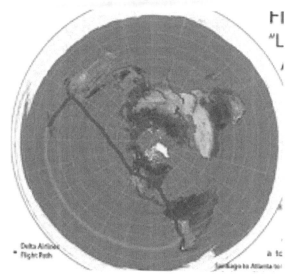

This is a view of exactly the same journey on a flat Earth map. Please note that this is much more of a 'straight line' route...apart from the small anomaly of the stopover in Sidney, Australia, which can be excused for probable economic reasons. Sydney is far more populous and a much more in-demand destination than Christchurch, so it does at least make *some* sense that the city should be a regular stop on the journey.

Here are some more examples of apparently 'strange' airline routes from and to Southern Hemisphere destinations...

Johannesburg, South Africa to Sao Paulo, Brazil. This flight goes via London, England.

Santiago, Chile to Johannesburg, South Africa, refuels in Dakar, Senegal, in northern Africa.

And these are by no means the only ones, just a very, very small selection. Again, some research of your own on this topic, would prove to be very enlightening.

The Falsification of Science

Solar Eclipses

The official NASA explanation of a Solar Eclipse is that the Moon passes directly between the Sun and the Earth, thereby blocking the Sun's light, creating a total solar eclipse on the Earth for those directly in the path of the shadow. However, there are several problems with this explanation.

The first and main problem with the globe Earth model is that solar eclipses sometimes approach from the west. We are told that the Moon rotates around the Earth from east to west just as does the Sun, but in this instance the Moon during a westerly approaching solar eclipse is obviously eclipsing the Sun from the west. So how does that work on the globe Earth model? The professional liars at NASA tell us it is just an optical illusion, because of the angle of the Sun. Excuse me? One 'scientist' from NASA actually said it is because the Moon rotates from west to east. What? And another 'scientist' said that the Moon rotates around the Earth twice as fast as the Earth spins. Hmm. Other explanations from NASA get even more ridiculous and make no sense whatsoever. If their 'science' is so accurate, then why can they not agree on which way the Moon rotates around the Earth or how fast it is travelling? Something is very wrong here, so could it be something other than the Moon eclipsing the Sun?

Another major problem with the globe Earth model is that the path of the total eclipse shadow that crossed North America in 2017 (and all other solar eclipses) was only 73 miles across. How can a shadow be smaller than the object casting the shadow? This is physically impossible. We know from experience that shadows can be the same size or larger than the object casting the shadow, but they can never be smaller. We are told that the Moon is 2,159 miles in diameter. So should the Moon's shadow on Earth not be at least 2,159 miles wide? But instead we are told that the shadow of the US 2017 total eclipse was only 73 miles wide.

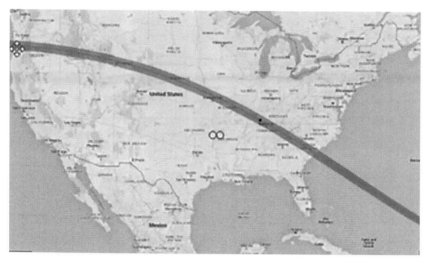

We also know that the farther away an object, the larger is its shadow. So how can the shadow of the Moon be SMALLER than the Moon when the Moon is 2,159 miles wide and is 238,900 miles away?

This is how shadows work:

But, according to NASA, the Sun's rays during a lunar eclipse only, converge on the Moon and create a 'laser affect.' Are they making this up as they go along?

The Falsification of Science

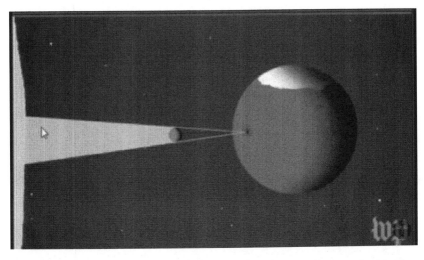

However, at every other time the Sun's rays are always depicted as shining in straight direct rays at Earth, which is necessary in order to account for the shadows we see from the sunlight. NASA (**N**ever **A** **S**traight **A**nswer) changes its story whenever it suits them, in order to explain phenomena they cannot easily explain otherwise.

Another huge problem with the paths of the solar eclipses on the globe model is that they make strange wavy, zigzag patterns on a globe, which really make no sense. How can a zigzag pattern be possible on the globe model?

But strangely, or maybe not so strangely, if we observe the same paths set out on a flat Earth map then they are ovals and circular paths of the eclipse around the flat Earth. This is yet another point in favour of the flat Earth.

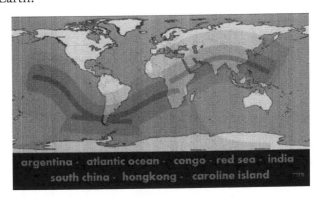

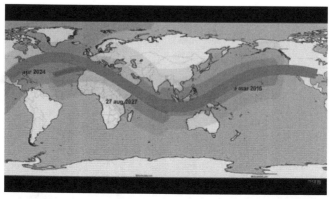

An example of a total eclipse track using the globe Earth model

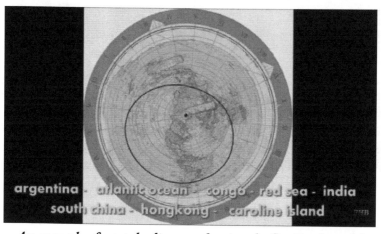

An example of a total eclipse track using the flat Earth model

The solar eclipse is a localised event, and these localised eclipses happen all the time. This is only possible with a local Sun and Moon, or whatever is blocking the Sun. This represents more proof that the Earth is flat, and the Sun and Moon are close and small.

Spinning Earth?

Our inability to detect or measure in any way the spinning of the Earth at up to the claimed 1,000 miles per hour in an easterly direction is one of the most quoted proofs given by flat Earthers. They assert that...

Vertically fired cannonballs and other projectiles should fall significantly due west, but they do not, they come straight down.

The spin of the Earth should significantly change the flight times of eastern and western flights but in reality, there is little difference in the flight times, other than those caused by the prevailing wind direction... usually west to east.

The answer to these questions from the perspective of the globe Earther conspiracy theorists(!) is that as the Earth spins, gravity pulls the entire atmosphere and everything within the atmosphere including the cannon balls, the butterflies, hot air balloons and helicopters, so everything rotates at the same speed as the Earth. Because everything is then moving at the same 1,000 miles per hour it gives the illusion that there is no movement.

But it is truly a mystery as to how gravity could cause the Earth and the atmosphere and everything in it to spin at up to one thousand miles per hour and hold things so tightly in place, keeping all the water in the ocean firmly stuck to the Earth, yet still allow a butterfly to flap its tiny, fragile wings and go effortlessly in any direction it chooses, into the sky.

More 'Proofs' of a Flat Earth

If the Earth were truly constantly spinning eastwards at over 1,000 mph, vertically fired cannonballs and other projectiles should fall significantly due west.

If the Earth were truly constantly spinning eastwards at over 1,000 mph, helicopters and hot-air balloons should be able to simply hover over the surface of the Earth and wait for their destinations to come to them!

If the Earth were truly constantly spinning eastwards at over 1,000 mph, during his 'Red Bull' stratosphere dive, Felix Baumgartner, who spent three hours ascending over New Mexico, should have landed

2,500 miles west into the Pacific Ocean but instead landed a few dozen miles east of the take-off point.

If the Earth and its atmosphere were truly constantly spinning eastwards at over 1,000 mph, then north/south facing cannons should establish a control while east-firing cannonballs should fall significantly farther than all others and west-firing cannonballs should fall significantly closer.

If the Earth and its atmosphere were truly constantly spinning eastwards at over 1,000 mph, then the average commercial airliner travelling at over 500 mph would never be able to reach its eastward destination before it was overtaken by it from behind.

If the Earth and its atmosphere were truly constantly spinning eastwards at over 1,000 mph, landing planes on such fast-moving runways which face all manner of directions north, south, east, west, and otherwise would be practically impossible, yet in reality such fictional concerns are completely negligible.

If the Earth and its atmosphere were truly constantly spinning eastwards at over 1,000 mph, then clouds, wind and weather patterns could not casually and unpredictably move in all directions, with clouds often travelling in opposing directions at varying altitudes simultaneously.

If the Earth and its atmosphere were truly constantly spinning eastwards at over 1,000 mph, this should somewhere somehow be seen, heard, felt, or measured by someone. So far it never has been.

In his book *'South Sea Voyages,'* Arctic and Antarctic explorer Sir James Clark Ross, described his experience on the night of 27th November 1839 and his conclusion that the Earth must be motionless, *"The sky being very clear ... it enabled us to observe the higher stratum of clouds to be moving in an exactly opposite direction to that of the wind."*

If the Earth-globe rotates on its axis at the incredible speed of 1,000 mph, such an immense mass would of necessity cause a tremendous rush of wind in the space it occupied. The wind would travel all in one direction, and objects such as clouds which fell 'within the sphere

The Falsification of Science

of influence' of the rotating sphere, would have to all travel in the same direction.

NASA and modern astronomy decrees that the Earth is a giant ball tilted slightly, wobbling, and spinning at 1,000 mph around its central axis, travelling at 67,000 mph in circles around the Sun, spiralling at 500,000 mph around the galactic central point of the Milky Way, while the entire galaxy moves at a ridiculous 670,000,000 mph through the Universe, with all of these motions originating from an alleged Big Bang explosion 14 billion years ago. That is a grand total of 670,568,000 mph in several different directions we are all supposedly moving along at, simultaneously, yet no one has ever seen, felt, heard, measured, or proven a single one of these motions to exist, whatsoever.

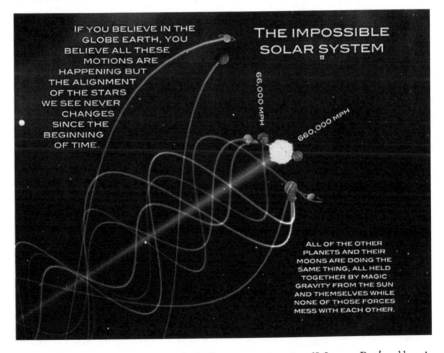

The above 'facts' are recorded for posterity in *'Monty Python's'* epic 'Galaxy Song' from the film, 'The Meaning of Life.' The lyrics are copied below for some light relief to all this intense information being disseminated here, but also to demonstrate how pervasive this ongoing, relentless propaganda, is...

John Hamer

Just remember that you're standing on a planet that's evolving
And revolving at nine hundred miles an hour,
That's orbiting at nineteen miles a second, so it's reckoned,
A sun that is the source of all our power.
The sun and you and me and all the stars that we can see
Are moving at a million miles a day
In an outer spiral arm, at forty thousand miles an hour,
Of the galaxy we call the 'Milky Way.'
Our galaxy itself contains a hundred billion stars.

It's a hundred thousand light years side to side.
It bulges in the middle, sixteen thousand light years thick,
But out by us, it's just three thousand light years wide.
We're thirty thousand light years from galactic central point.
We go 'round every two hundred million years,
And our galaxy is only one of millions of billions
In this amazing and expanding universe.

The universe itself keeps on expanding and expanding
In all of the directions it can whizz
As fast as it can go, that's the speed of light, you know,
Twelve million miles a minute, and that's the fastest speed there is.
So remember, when you're feeling very small and insecure,
How amazingly unlikely is your birth,
And pray that there's intelligent life somewhere up in space,
'Cause there's bugger-all down here on Earth.

...which is all very amusing, and the last two lines of all are indeed very ironic and apt! I would recommend you listen to it on YouTube™ to appreciate how 'catchy' the music and the lyrics indeed are.

But back down to Earth once again (please excuse the pun)...

There are huge, centuries-old stone Sundials and Moon dials all over the world, which still tell the time down to the minute as accurately and as perfectly as the day they were created. If the Earth, Sun, Moon, planets, and stars were truly subject to all these claimed contradictory revolving, rotating, wobbling, and spiralling motions according to modern astronomy, it would be impossible for these devices to tell us the time so accurately without constant adjustment due to the aforementioned, complex movements of all the heavenly bodies.

From *'Earth Not a Globe'* by Samuel Rowbotham... *"Take two carefully-bored metallic tubes, not less than six feet in length, and place them one yard asunder, on the opposite sides of a wooden frame, or a solid block of wood or stone. Adjust them so that their centres or axes of vision shall be perfectly parallel to each other. Now, direct them to the plane of some notable fixed star, a few seconds previous to its meridian time. Let an observer be stationed at each tube and the moment the star appears in the first tube let a loud knock or other signal be given, to be repeated by the observer at the second tube when he first sees the same star. A distinct period of time will elapse between the signals given. The signals will follow each other in very rapid succession, but still, the time between is sufficient to show that the same star is not visible at the same moment by two parallel lines of sight when only one yard asunder. A slight inclination of the second tube towards the first tube would be required for the star to be seen through both tubes at the same instant. Let the tubes remain in their position for six months; at the end of which time the same observation or experiment will produce the same results ... the star will be visible at the same meridian time, without the slightest alteration being required in the direction of the tubes. From which it is concluded that if the Earth had moved one single yard in an orbit through space, there would at least be observed the slight inclination of the tube, which the difference in position of one yard had previously required. But as no such difference in the direction of the tube is required, the conclusion is unavoidable, that in six months a given meridian upon the Earth's surface does not move a single yard, and therefore, that the Earth has not the slightest degree of orbital motion."*

"*Why, in the name of common sense, should observers have to fix their telescopes on solid stone bases so that they should not move a hair's-breadth, ... if the Earth on which they fix them moves at the rate of nineteen miles in a second? Indeed, to believe that 'six thousand million, million, million tons' is 'rolling, surging, flying, darting on through space for ever' with a velocity compared with which a shot from a cannon is a 'very slow coach,' with such unerring accuracy that a telescope fixed on granite pillars in an observatory will not enable a lynx-eyed astronomer to detect a variation in its onward motion of the thousandth part of a hair's-breadth is to conceive a miracle compared with which all the miracles on record put together would sink into utter insignificance. Since we can, (in middle north latitudes), see the North Star, on looking out of a window that faces it ... and out of the very same corner of the very same pane of glass in the very same window ... all the year round, it is proof enough for any man in his senses that we have made no motion at all and that the Earth is not a globe."* William Carpenter, *'100 Proofs That Earth is Not a Globe'*

We are told that the Earth and its atmosphere spin together at such a perfect uniform velocity that no one in history has ever seen, heard, felt, or measured the supposed 1,000 mph movement. This is then often compared to travelling in a car at uniform velocity, where we only feel the movement during acceleration or deceleration. In reality, however, even with our eyes closed tightly, and with windows closed, along a smooth road in a luxury car at a mere, uniform 50 mph, the movement absolutely may be felt. At twenty times this speed, Earth's imaginary 1,000 mph spin would most certainly be noticeable, felt, seen, and heard by all.

People sensitive to motion sickness feel distinct unease and physical discomfort from motion as slight as an elevator or a train ride. This means that the 1,000 mph alleged uniform spin of the Earth has no effect on such people but add an extra 50 mph uniform velocity of a car travelling on the Earth's surface and suddenly they feel distinctly sick. The idea that motion sickness is nowhere apparent in anyone at 1,000 mph, but suddenly materialises at 1,050 mph is utterly ridiculous and indicates that the Earth has no motion whatsoever.

"If flying had been invented at the time of Copernicus, there is no doubt that he would have soon realised that his contention regarding the rotation of the Earth was wrong, on account of the relation existing between the speed of an aircraft and that of the Earth's rotation." Gabrielle Henriet, *'Heaven and Earth'*

The Falsification of Science

The former Polish cosmonaut, Miroslaw Hermaszewski stated the following at the end of an interview at a conference where he was the guest of honour...

Interviewer: *"You've been there! Is Earth really a sphere hanging in outer space?"*

Hermaszewski: *"The Earth is flat...as some expect...I didn't expect this question...I assure you it's flat."*

Miroslaw Hermaszewski is not just any cosmonaut; he is a national hero in Poland. He is a retired Brigadier General who flew aboard the Soviet Soyuz 30 spacecraft in 1978 and is the only Pole who has ever flown in what is allegedly 'outer space.' He was elected to the Mazovian Regional Assembly, which is a regional parliamentary body in Poland and the president of Poland, Aleksander Kwaniewski, awarded him the Commander's Cross of the Order of Polonia Restituta in 2003. He is so popular in Poland that at a 2018 event, he attracted a vast crowd that gathered to hear him speak and was totally swamped by autograph hunters!

Some have alleged that Hermaszewski was using deadpan sarcasm. But it was not really deadpan. He did smile, which is not something one usually does whilst being deadpan. But his smile was not spontaneous; it was an insincere, contrived, nervous smile. He also anxiously looked around, and he awkwardly took a pretend sip from what was clearly an empty cup. He seemed very uncomfortable making the statement that the Earth is flat and seemed not to think it was funny as he became solemn and morose as the interview drew to a close. There is little doubt that Hermaszewski has been living the utter lie of having gone into outer space, which probably does not exist in the form that we are told, and he may have viewed this as an opportunity to unburden his conscience.

The 'Bedford Level' Experiments

The Bedford Level Experiments were a series of observations carried out along a six-mile length of the Bedford River on the Bedford Level in the county of Norfolk, England, during the nineteenth and early twentieth centuries, in an attempt to demonstrate that the Earth was flat.

At the point chosen for all the experiments, the river was a slow-flowing drainage canal running in an uninterrupted straight line for a six-mile stretch to the northeast of the village of Welney. Most results have served to conclusively prove the flat Earth hypothesis, and although a few vested interests have claimed otherwise, these have been thoroughly disproved using true science. The Bedford Level Experiments remain one of the most widely accepted examples of flat Earth proof in existence.

The most famous of the observations, involved a set of three poles fixed at equal height above water level along the six-mile length of the water. As the surface of the water was understood to follow any hypothetical curvature of the Earth, the observation that the three poles aligned perfectly when observed through a theodolite, serves as evidence of a flat Earth.

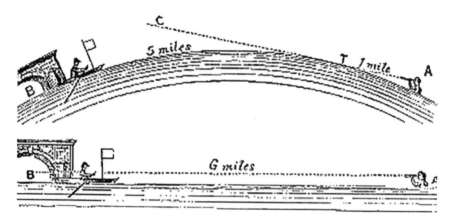

The Bedford Level Experiments

The first investigation was carried out by Samuel Birley Rowbotham (1816-1884) the president of the Flat Earth Society, in the summer of 1838. He waded into the river and used a telescope held eight inches above the water to watch a boat with a five-foot mast row slowly away from him. He reported that the vessel remained constantly in his view for the full six miles to Welney bridge, whereas, had the water surface been curved with the accepted circumference of a spherical Earth, the top of the mast should have been some eleven feet below his line of sight, according to standard, universally accepted Earth curvature calculations.

Rowbotham repeated his experiments several times over the years but his discoveries received little attention until, in 1870, a supporter by the name of John Hampden offered a wager that he could show, by repeating Rowbotham's experiment, that the Earth was flat. The noted naturalist and qualified surveyor Alfred Russel Wallace, who by the way was a Freemason as well as an ardent evolutionist and friend of Charles Darwin, accepted the wager. Wallace won the bet. Hampden, however, published a pamphlet alleging that Wallace had cheated and sued for his money. Several protracted court cases ensued, with the result that Hampden was imprisoned for libel, but the court also determined that Wallace had indeed cheated.

In 1901 Henry Yule Oldham, a geography student at King's College, Cambridge University, conducted the definitive experiment as described above and obtained exactly the same results as Rowbotham.

The Bedford Level was the scene of further experiments over the years, until in 1904, photography was eventually used to prove that the Earth is flat. Lady Blount, a staunch believer in the zetetic (flat Earth) 'faith,' hired a professional photographer, a Mr Clifton who arrived at the Bedford Level with his company's very latest, 'state of the art,' camera complete with photo-telescopic lens. The apparatus was set up at one end of the clearly visible six-mile length, while at the other end Lady Blount and 'some scientific gentlemen' hung a large, white calico sheet over the Bedford bridge so that the bottom of it was close to the water. Mr Clifton, lying down near Welney bridge with his camera lens two feet above the water level, observed by telescope the hanging of the sheet, and found that he could see the whole of it down to the very bottom. This surprised him, because until that point in time he had been a fanatical, orthodox globe 'believer' and the globe Earth theory stated that over a distance of six miles, the bottom of the sheet should be more than 20 feet below his line of sight.

His photograph showed not only the entire sheet, but also its reflection in the water below. That was certified in his report to Lady Blount, which concluded: *"I should not like to abandon the globular theory off-hand, but, as far as this particular test is concerned, I am prepared to maintain that (unless rays of light will travel in a curved path) these six miles of water present a level surface."*

All of which sounds very much like a 'cop-out,' to me... very similar in fact, to my own position!

The Bishop Experiment and other examples of 'Impossible' Viewpoints

Located in California, Monterey Bay is a relatively long bay that sits next to the Pacific Ocean. The distance between the extremities of Monterey Bay, Lovers Point in Pacific Grove and Lighthouse State Beach in Santa Cruz, is just over twenty-three statute miles.

On a very clear day it is possible to see Lighthouse Beach from Lovers Point and vice versa. With a good telescope, lying down at the edge of the shore on the Lovers Point beach twenty inches above the sea level it is possible to see people at the water's edge on the adjacent beach twenty three miles away, near the lighthouse. The entire beach is visible down to the water splashing upon the shore. Upon looking into the telescope it is possible to observe children running in and out of the water, splashing and playing and people sunbathing on the shore. It is also possible to see runners jogging along the water's edge with their dogs. From this vantage point the entire beach is indeed visible. This is what is known to 'flat Earthers' as the 'Bishop Experiment,' as first conducted by Tom Bishop.

The Falsification of Science

Distance between the two points, courtesy of Google Earth™

If the Earth is a globe, and is 24,900 statute miles in circumference as claimed, then the surface of all standing water must have a certain degree of convexity... every part must be an arc of a circle. From the summit of any such arc there will exist a curvature or declination of eight inches over the first statute mile. Over two miles the fall will be thirty-six inches and by the end of the third mile, seventy-two inches, or six feet.

Correcting for the height of the observer of about twenty inches, when viewing the opposite beach over twenty-three miles away, there should be a bulge of water obscuring objects up to 300 feet above the far beach. There is not. Even accounting for refraction, the amount hidden should be around 260 feet and seeing down to the shoreline should be impossible.

The distance along the Irish Sea from Douglas harbour on the Isle of Man, off the northwest coast of England, southwards to the Great Orme in Llandudno, Wales is 60 miles. If the Earth were a globe then the surface of the water between them would form a 60-mile arc, the centre towering 1,944 feet higher than the coastlines at either end. It is well known by locals and easily verifiable, however, that on a clear day, from a modest altitude of 100 feet, the Great Orme is clearly visible from Douglas Harbour. Indeed I can verify this personally, having stood at Douglas harbourside and observed the Great Orme headland in the distance for myself. This would be completely impossible on a globe of approximately 25,000 miles circumference. Assuming the 100 foot altitude causes the horizon to appear approximately 13 miles away, then the 47 miles remaining means the Welsh coastline should still fall an impossible 1,472 feet below the line of sight!

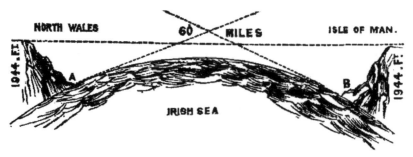

On a clear day from the high land near Douglas harbour on the Isle of Man, the whole length of the coast of North Wales is often plainly

The Falsification of Science

visible to the naked eye. From the Point of Ayr at the mouth of the River Dee westwards toward Holyhead, comprises a 50-mile stretch, which has also been repeatedly found to be perfectly horizontal. If the Earth actually has a curvature of 8 inches per mile squared, as NASA and modern science claims, then the 50 mile length of the Welsh coast seen along the horizon in Liverpool Bay would have to decline from the centre-point an easily detectable 416 feet on each side.

The Philadelphia skyline is clearly visible from Apple Pie Hill in the New Jersey Pine Barrens, 40 miles away. If the Earth really were a ball 25,000 miles in circumference, factoring in the 205-foot elevation of Apple Pie Hill, the Philadelphia skyline should remain well hidden beyond 335 feet of curvature.

The New York City skyline is clearly visible from Harriman State Park's Bear Mountain 60 miles away. If Earth were a ball 25,000 miles in circumference, viewing from Bear Mountain's 1,283- foot summit, the Pythagorean theorem determining distance to the horizon being 1.23 times the square root of the height in feet, the New York skyline should be invisible behind 170 feet of curved Earth.

From Washington's Rock in New Jersey, at just a 400-foot elevation, it is possible on a clear day to see the skylines of both New York and Philadelphia in opposite directions at the same time covering a total distance of 120 miles! If the Earth was a ball 25,000 miles in circumference, both of these skylines should be hidden behind more than 800 feet of the Earth's curvature.

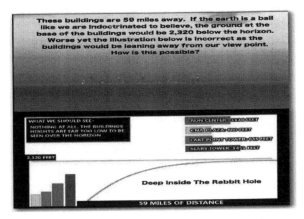

It is often possible to see the Chicago skyline from sea level, 60 miles away across Lake Michigan. In 2015 after photographer Joshua Nowicki photographed this phenomenon, several media outlets and TV channels claimed his picture to be a 'superior mirage,' an atmospheric anomaly caused by temperature inversion. While these anomalies certainly do occur, the skyline in question was facing right side up and clearly seen, unlike a hazy illusory mirage, and on a ball Earth 25,000 miles in circumference should be 2,400 feet below the horizon.

On 16th October 1854, the Times newspaper reported Queen Victoria's visit to Grimsby, Lincolnshire from Hull, Yorkshire whilst recording that they were able to see the 300 foot tall dock tower from 70 miles away. On a ball Earth 25,000 miles in circumference, factoring in their ten-foot elevation above the water and the tower's 300-foot height, at 70 miles away the dock tower should have been fully 2,600 feet below the horizon.

In 1872, Captain Gibson and his shipmates, sailing the ship *'Thomas Wood,'* from China to London, reported seeing the entirety of the island of St. Helena on a clear day, from 75 miles away. Factoring in their height during measurement on a ball Earth 25,000 miles in circumference, the island should have been 3,650 feet below their line of sight.

From Genoa, Italy at a height of just 70 feet above sea level, the island of Gorgona can often be seen 81 miles away. If the Earth was a ball 25,000 miles in circumference, Gorgona should be hidden beyond 3,332 feet of curvature.

From Genoa, Italy at a height of just 70 feet above sea level, the island of Corsica can often be seen 99 miles away. If the Earth was a ball 25,000 miles in circumference, Corsica should fall 5,245 feet, almost an entire mile below the horizon.

Also from Genoa, on bright clear days, the island of Elba can be seen, an incredible 125 miles away! If the Earth was a ball 25,000 miles in circumference, Elba should be forever invisible behind 8,770 feet of curvature.

The Falsification of Science

From Anchorage, Alaska at an elevation of 102 feet, on clear days Mount Foraker can be seen with the naked eye 120 miles away. If the Earth truly was a globe 25,000 miles in circumference, Mount Foraker's 17,400 summit should be leaning back away from the observer and covered by 7,719 feet of curved Earth. In reality, however, the entire mountain can be quite easily seen standing straight from base to summit.

The Statue of Liberty in New York City, stands 326 feet above sea level and on a clear day can be seen as far as 60 miles away. If the Earth were really an oblate spheroid as claimed, that would make the statue an impossible 2,074 feet below the horizon.

From high land near Portsmouth harbour on the south coast in Hampshire, England looking across Spithead to the Isle of Wight, the entire base of the island, where water and land come together composes a perfectly straight line 22 statute miles long. According to the ball Earth theory, the Isle of Wight should decline 80 feet from the centre on each side to account for the necessary curvature. The crosshairs of a good theodolite directed there, however, have repeatedly shown the land and water line to be perfectly level.

And as a final example, another one from my own experience...

I can clearly remember as a child being utterly fascinated that it was possible, from the summit of a local landmark, Castle Hill in Huddersfield, West Yorkshire, England on a crisp, clear day, to clearly see York Minster in York, North Yorkshire, a distance of 37 miles 'as the crow flies.' At the time of course I did not even consider the utter impossibility of this, using the globe Earth model. In my child's mind I believed that the Earth was a globe, without even questioning it at all and yet neither did I consider even for one brief moment the fact that according to accepted science and using the accepted formula, that the globe Earth curves downward at eight inches per mile, multiplied by the square root of the distance. Using this proven formula, then York Minster should have been obscured from my vision by 912 feet of curvature.

All of the above is 'accepted' science and not supposition or statistical juggling of any kind. How then do we reconcile this with the above examples? The obvious conclusion has to be that we are being lied to... either about the globe Earth itself... or if not, then certainly about its diameter.

But of course mainstream science always has an 'answer'... or at least it claims to have one.

The Michelson-Morley Experiment

The Michelson...Morley experiment was designed to test the velocity of light and was first performed in 1887 by Albert A. Michelson and Edward W. Morley. Unfortunately, the tendency of writers on this topic is to underestimate, gloss over, or even misrepresent exactly what this experiment demonstrated, its significance, and how it was such a crucial turning point in science which necessitated the adoption of a radically different and alternative model of space.

At this period in history it was widely 'believed' that in order for light to be able to travel through space it was necessary for there to be a medium filling the void through which it could propagate, in much the same way as sound waves travel through the air or ripples through water. This background medium of space was colloquially known as the 'aether.'

The original purpose of the Michelson...Morley experiment was to compare the speed of light in perpendicular horizontal directions at various times of the day, in an attempt to detect the relative motion of matter through the stationary luminiferous aether. And this was to be achieved by using the rotation of the Earth and its motion around the sun to create interference bands of light for the study. Indeed, at the time Morley wrote to his father that the purpose of the experiment was *"to see if light travels with the same velocity in all directions."*

But the totally unexpected result of the experiment was that it was discovered that without question, light absolutely did travel at the same velocity in all the various horizontal directions tested. The Earth did not measurably move around the Sun at all, in contradiction to all expectations and the mainstream accepted astronomical model, thus destroying in an instant, all the received wisdom of that time... and which incidentally still persists to this day, regardless. Michelson and Morley found that a light beam discharged horizontally in the direction of the Earth's assumed motion showed virtually no difference in speed from a light beam discharged north to south or south to north. In other words, the experiment totally failed to detect the Earth moving in or against space, of whatever space was understood to consist.

Later, one of Albert Einstein's biographers, Ronald W. Clark remarked that... *"The problem which now faced science was considerable, for there seemed to be only three alternatives. The first was that the Earth was standing still, which meant scuttling the whole Copernican theory and was unthinkable."*

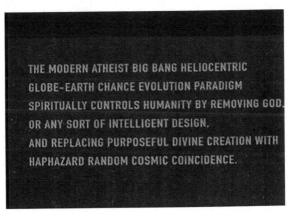

However, everyone's favourite scientific 'genius,' the Freemason Albert Einstein, in what was to become another successful episode of 'logic-patching' soon shored up the deficiencies in the prevailing mainstream view by formulating his 'Theory of Special Relativity' (1905) which succeeds in 'papering over the cracks' in the minds of many scientists. Its subsequent popularity with scientific circles being attributed to the fact that the second postulate of Special Relativity created an illusion that the velocity of light is constant to all observers regardless of motion, and seemingly explained the motionless Earth result of the Michelson-Morley experiment. This fundamental change to the nature of space and time allowed the theory of the Earth's motion around the Sun to survive directly contradicting experimental evidence and encouraged, or indeed perhaps forced its adoption as the accepted model of space, for heliocentrism.

G.J. Whitrow, a British mathematician, cosmologist, and science historian, characterised the Michelson-Morley experiment from a historical perspective:

"If such an experiment could have been performed in the sixteenth or seventeenth centuries when men were debating the rival merits of the Copernican and Ptolemaic systems, the result would surely have been interpreted as conclusive evidence for the immobility of the Earth, and therefore as a triumphant vindication of the Ptolemaic system and irrefutable falsification of the Copernican hypothesis."

Michelson and Morley and more recently, other scientists repeated the experiment many times, and in many different horizontal axial positions and configurations, at different times of the day ... all with a negative result.

Thus, the Earth was proven without question, to be motionless.

Earth 'Photographs' courtesy of NASA

The picture on the left (below) is allegedly a photograph by NASA taken in 2007, whilst the one on the right was supposedly taken in 1978.

Two things spring immediately to mind ...

Firstly, what or who actually took these photographs and from where? It is certainly too far out in space to be from satellites or the Hubble Telescope or indeed the International Space Station. And in fact the earlier one from 1978 predated both those latter technologies.

Secondly, why is it that in the image on the left, North America is about one third of the size of the image on the right compared to the overall size of the Earth? Sheer nonsense of course.

The Falsification of Science

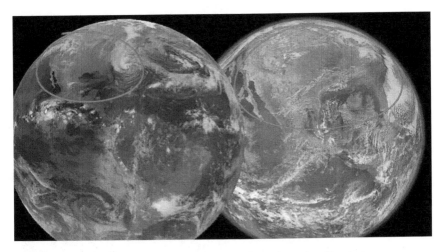

In addition, it is always presented as a fact that the Earth is not a perfect sphere, but is an 'oblate spheroid,' i.e. slightly pear-shaped, marginally wider in the northern hemisphere than the south. Take another look at the images above. Could there be two more perfect spheres than those depicted there?

And, if they are lying about all this, what else are they lying about? The answer is... almost everything, unfortunately.

"The globe Earth does not exist, gravity does not exist, dinosaurs never existed, aliens do not exist, the Easter bunny does not exist, the tooth fairy does not exist, and Santa Claus does not exist, neither do dragons. It would be cool if they all did, but they don't." Paul Michael Bales

The conclusion therefore has to be, from all the preceding information in this chapter, that many valid points have been raised, which do not produce satisfactory explanations from the globe Earth supporters. Considering these and many other points raised by the flat-Earthers, one has to seriously question the validity of the globe Earth model which ascribes to the Earth so many simultaneous motions in different directions that no one has ever been able to feel or detect, let alone measure, or prove. Also one has to marvel at the magical powers the globe-Earthers ascribe to gravity, which makes everything possible in their fantasy universe.

Whether or not you now believe that the Earth is flat, is entirely up to you and your own 'belief system.' Your authors are still not 100% convinced of it, quite yet, but hope you agree that the foregoing information, is quite compelling at the very least?

Chapter 6

NASA and Outer Space

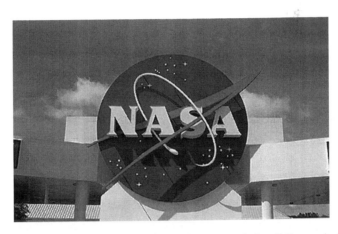

The acronym NASA is popularly said to stand for 'Never A Straight Answer.' How true this is. They are constantly being caught out lying, bending the truth, and also contradicting themselves and generally fudging answers to probing questions that they know will incriminate them in some way, should they ever answer them truthfully.

I spent many months around ten years ago researching NASA and the Apollo Moon landings, and it did not take too much digging beneath the surface to discover that everything they have been involved in down the years and decades consists of some kind of subterfuge, ranging from minor deceit to immense, utterly mind-boggling deceptions.

An open mind on this subject will almost certainly bring the realisation that the Apollo programme in its entirety was nothing more than an expensively staged hoax. In fact the pure, hard scientific evidence and analysis irrefutably reveals this to be simply one huge fraud among many. Think about that fact for a moment. The greatest feat ever allegedly accomplished by humankind, is actually an elaborate Hollywood production that cost the American taxpayer around $30 billion – in the 1960s and early 1970s! That equates to around $250 billion, or a quarter of a trillion, in 2020 values.

But plainly speaking, it is nothing less than a huge scandal that NASA has constructed such a money-generating colossus composed of lies, hoaxes and disinformation especially given that the American taxpayer derives no benefit whatsoever from NASA'S existence and alleged scientific contributions to the overall human knowledge pool, garnered by supposed explorations and experiments in space. Absent of any other sources to verify their often outrageous claims, we are all expected to prostrate ourselves to these self-styled 'experts' despite irrefutable evidence that NASA is nothing more than an elaborate financial sponge and a habitual peddler of utterly risible pseudoscience.

Its ongoing fraud would appear to have no limits and NASA has been perpetrating exactly that, as harsh as that may sound, for a very long time indeed. The very nature of fraudsters is that they never stop; they continue on and on until they either die or get caught, whichever comes first. So why would NASA ever stop lucrative fraudulent activity if they have never yet been officially exposed? Admittedly, many people are already aware of this fraud to a certain extent, but we tend to be mere 'lone voices in the wilderness,' accused of being 'conspiracy theorists' or 'tin-foil-hat-wearers' and our views and arguments are thereby consigned to the very backwaters of obscurity. So, once an initial fraud is perpetrated and not exposed, subsequent and indeed systemic fraud, becomes a virtual certainty.

To further 'muddy the waters,' NASA was founded by Satan-worshipping occultists and black magicians. This is not idle speculation or wild exaggeration, but a simple unembellished fact. NASA is a military-Hollywood-pseudoscientific satanic cartel, whose true aim is to use taxpayer funding to promote and greatly profit a clandestine group of unseen individuals by developing and deploying pseudoscientific technology and methodology.

The first of NASA's many frauds was the hiding of its true origins and the cast of characters that had and still has significant influence and therefore a Google™ search for NASA presidents, CEOs or controllers will not reveal much in the way of useful or truthful information.

Jack Parsons

One of the founders of NASA was a talented young rocket scientist named Jack Parsons. Wernher von Braun, the ex-Nazi, claimed that it was the self-taught Parsons, and not himself, who was the true father of the American space programme, for his contribution to the development of solid rocket fuel.

"Jack's childhood was one of solitude, loneliness, and wealth. He could sit in his room all day and read, never worrying about supporting his family or where his next meal was going to come from. He ferociously read Jules Verne, including his 1865 novel 'From the Earth to the Moon,' and the new sci-fi magazine 'Amazing Stories.' Soon, space wasn't just what was above Jack's head, it was a romantic obsession."

Whilst at school one day, Jack was being soundly beaten by someone when an older boy intervened and broke up the fight. That boy, Ed Foreman, would become Jack's best friend into his adult years and was an essential player in Jack's rocketry dreams. Jack and Ed spent their time together discussing the science fiction books they were reading and soon, began experimenting for themselves.

Using Ed's father's engineering tools and the resources and supplies Jack took from his part-time job at the Hercules Powder Company, they made explosive devices. Their teachers and parents were extremely worried about them but undeterred, Ed and Jack continued with their dangerous experiments.

In 1934, with Jack now 20 years old, the duo's interest in rockets went from a child's fantasy to an academic pursuit when, despite not being students there, they gained the support of the nearby California Institute of Technology. Ed, Jack, and several members of CalTech's community soon formed the GALCIT Rocket Research Group and in late 1936, the group performed their first rocket motor test near the Devil's Gate Dam in Pasadena. The motor exploded, and they soon became infamous on campus as the 'suicide squad' due to the danger and perceived craziness of their experiments, particularly as rocket technology was considered by many scientists at the time to be foolish and mere science fiction in terms of any practical use and development of the technology.

This 'suicide squad' was the beginning of the famed Jet Propulsion Laboratory (JPL) the institution responsible for the alleged Mars Rover Landing and many other advancements in rocket and robotic sciences. Obviously, Parsons was a real talent with a seemingly innate sense for propulsion and rocketry (a very nascent industry at the time) and this type of knowledge was seriously coveted by the military-industrial complex. His design and understanding of the chemical composition of liquid rocket fuels was considered peerless and Parsons himself was regarded as a romantic, dashing figure with a reputation as a fast-living, rocketry hotshot and risk-taker. All his scientific achievements to this point should have been enough for one person in one lifetime, but Parsons had a still more lofty set of ambitions. He wanted to tear down the walls of time and space, and he had an entirely non-scientific set of ideas on how to achieve this.

These non-scientific ideas were steeped in the occult, Satanism, and black magic. Parsons said, *"I swore the Oath of the Abyss, having only the choice between madness, suicide, and that oath. [Then] I took the oath of a Magister Templi, even the Oath of Antichrist before Frater 132, the Unknown God. And thus was I Antichrist loosed in the world; and to this I am pledged, that the work of the Beast 666 shall be fulfilled."*

Apparently, noting that Antichrist is only a few letters away from 'anarchist,' the manifesto that follows is in large part an exhortation to 'do what thou wilt' in most things economic and/or political. The goal of all these efforts, according to Thomas Metzger, the notorious

The Falsification of Science

Satanist, was to bring on the Apocalypse, since according to the satanic creed, things can only get better from there.

By now dear reader, maybe you are already thinking 'what utter garbage!' But none other than the FBI, who were not too happy about the notion that Parsons' taxpayer-funded salary could possibly be seen to be openly supporting the Antichrist and the hastening of the Apocalypse, took the idea seriously enough to open an investigation. Documents recently released through the Freedom of Information act make up 130 pages of heavily redacted text in which the Agency tries to make sense of Parsons' 'religious' beliefs and document his frequently careless handling of classified materials. Perhaps the 'Errol Flynn' persona fitted Jack Parsons well, but in truth, he is now known best for his occultist ideals, black magic, and his devotion to the satanic agenda.

Parsons also had a secret life, which appeared totally at odds with his public persona, and it came to further dominate his life as the 1940s progressed. He and his wife, Helen had discovered the Agape lodge of the O.T.O. (Ordo Templi Orientis) international magical fraternity in Los Angeles in 1939 and had joined it in 1941. At this time it was under the leadership of Wilfred Talbot Smith, an Englishman who had founded this particular lodge about a decade earlier, in circa 1930.

Parsons even converted an attic room into an O.T.O. Gnostic Mass temple complete with a copy of the Egyptian 'Stele of Revealing,' venerated by followers of the infamous British O.T.O devotee and black magician Aleister Crowley. It was the only such temple in the world in those days which was properly functioning.

Aleister Crowley was at this time, already a 33rd degree Freemason and the world head of the O.T.O., and he took action that increased Parsons' stature in the Order. In circa 1943-44, he convinced Smith, via a paper entitled *'Is Smith a God?'* that astrological research had shown that Smith was not a man, but actually an incarnation of some deity. Taking the hint that Crowley wanted him out, the 'god' Smith went into private magical practice, remaining head of the lodge in name only.

In the meantime, Parsons had lost his wife, Helen to Smith, but still remained on good terms with her. He was keeping busy with Order activities, one of the most important of which was the sending of money to Crowley, for both his minimal upkeep and the O.T.O. publishing fund – a good percentage of which came from Parsons' own, substantial funds.

Aleister Crowley

Crowley, who brought actual fame to the O.T.O. (which was already well-known in Freemasonic circles,) was one of Parsons' major inspirations in life. The by now, elderly man's accomplishments had been many and varied. He was a poet, publisher, mountain climber, chess master, and bisexual practitioner of sexual magic – or 'Magick,' as he termed it. And, made famous by journalists as the 'wickedest man in the world,' he considered his identity to be the 'Great Beast 666' as referred-to in the book of Revelation in the Bible, although he was not specifically an adherent of that work, particularly in his religious ideas.

According to most accounts, when Parsons' father died in the early 1940s, Parsons inherited a mansion and coach-house at 1003 South Orange Grove Avenue in Pasadena, California. To the shock of the neighbours, the place became a haven for 'Bohemian' types and atheists, who were the sort of people to whom Parsons liked to rent out rooms.

The lodge headquarters was moved to this location, making use of two rooms in the house, the bedroom, which became a temple, and a wood-panelled library which was dominated by an enormous portrait of Crowley. According to a story told by L. Sprague De Camp from the 24th June 1990 edition of the Los Angeles Times, at one point the police, who had heard neighbours' reports of a ritual in which a pregnant woman jumped nine times through a fire in the garden, came to investigate, but Parsons managed to divert them by use of his now famous 'name' and his scientific credentials.

L. Ron Hubbard

Hubbard, the founder of the pseudo-religion Scientology, was once referred to as, 'the greatest humanitarian in history.' However there was another, darker side to him which will be made apparent herein. So it is, in order to fully understand Scientology, that one must begin with L. Ron Hubbard.

In the late 1940s, Hubbard was penniless and deep in debt. He was a struggling writer of science fiction and fantasy and was forced to sell his typewriter for $28.50 simply to struggle by for a few more days.

"I can still see Ron three-steps-at-a-time running up the stairs in around 1949 in order to borrow $30 from me to get out of town because he had a wife after him for alimony." Hubbard's former literary agent, Forrest J. Ackerman.

At one point, Hubbard was reduced to begging the Veterans Administration to let him keep a $51 over-payment of benefits. *"I am nearly penniless,"* wrote Hubbard, a former Navy lieutenant. Hubbard was mentally troubled, too. In late 1947, he asked the Veterans Administration to help him receive psychiatric treatment...

"Toward the end of my military service, I avoided out of pride any mental examinations, hoping that time would balance a mind which I had every reason to suppose was seriously affected. I cannot account for nor rise above long periods of moroseness and suicidal inclinations and have newly come to realise that I must first triumph above this before I can hope to rehabilitate myself at all."

In his most private moments, Hubbard wrote bizarre statements to himself in notebooks that actually came to light several decades later in the Los Angeles Superior Court. *"All men are your slaves,"* he wrote in one – and in another he wrote, *"You can be merciless whenever your will is crossed, and you have the right to be merciless."*

Hubbard was troubled, restless, and adrift in those little known years of his life, but he never lost confidence in his ability as a writer. He had made a living with words in the past and he believed he could do it again. Before the financial and emotional problems that consumed him in the 1940s, Hubbard had achieved moderate success writing for a variety of cheap, pulp magazines. He specialised in westerns, mysteries, war stories and science fiction. He was also a master sailor and glider pilot.

Although Hubbard's health and writing career foundered after the war, he remained a veritable fount of ideas and his biggest was yet to be born. He had long been fascinated with mental phenomena and the mysteries of life and was an expert in hypnotism. During a 1948 gathering of science fiction aficionados in Los Angeles, he hypnotised many of those in attendance, convincing one young man that he was cradling a tiny kangaroo in his hands. And it was this intense curiosity about the mind's power that led him into a friendship in 1946 with the famous rocket fuel scientist, John (Jack) Whiteside Parsons. As previously stated, Parsons was a protégé of the British Satanist Aleister Crowley and leader of a black Magick group modelled after Crowley's infamous occult lodge in England.

Hubbard also admired Crowley, and in a 1952 lecture described him as *"my very good friend."* Parsons and Hubbard lived in an ageing mansion on South Orange Grove Avenue in Pasadena, part of an area which was also home to an odd mix of 'Bohemian' artists, writers, scientists, and occultists. Hubbard met his second wife, Sara Northrup, at the mansion and although she was Parsons' lover at the time, Hubbard was undeterred. He actually married Northrup bigamously before divorcing his first wife and long before the 1960s counterculture, some residents of the estate smoked marijuana and embraced a philosophy of promiscuous, ritualistic sex.

Crowley biographers have written that Parsons and Hubbard practiced sex Magick.' As the biographers tell it, Hubbard, dressed in robes, chanted incantations while Parsons and his wife-to-be, Cameron, engaged in sexual intercourse intended to produce a child with superior intellect and powers. The ceremony was said to span eleven consecutive nights. Hubbard and Parsons' friendship eventually ended after a disagreement over a joint sailing-boat sales venture that culminated in a court case.

In later years, Hubbard tried to distance himself from his embarrassing association with Parsons, and Parsons died in 1952 when a huge chemical explosion ripped through his garage.

In an attempt to explain away his dubious past, Hubbard insisted that he had been working undercover for Naval Intelligence to break up black magic in America and to investigate links between the occultists and prominent scientists at the Parsons mansion. He also said that the mission was so successful that the house was razed to the ground, and the black magic group was dispersed but Parsons' widow, Cameron, disputed Hubbard's account. In a brief interview with *The Times* she said that the two men *"liked each other very much"* and *"felt they were ushering in a force that was going to change things."* And change things they certainly did. L. Ron Hubbard went on to create the insidious pseudo-religious cult, the Church of Scientology, a masterpiece of mind control.

The so-called 'Church' of Scientology is a vicious, highly manipulative, and dangerous cult masquerading as a religion. Its sole purpose is to generate money and it practices a variety of mind control techniques on people lured into its midst to gain control over their money and their lives. Its primary aim is to relieve them of every penny that they have and can ever borrow and also to enslave them in order to further its wicked ends.

Jack Whiteside Parsons on the other hand was a founding member of NASA's Jet Propulsion Laboratory, with some crediting him as being one of the 'fathers of rocketry' and yet others joking that JPL was actually Jack Parsons' Laboratory, but you will not find much about him on NASA's websites. Parsons' legacy as an engineer and chemist has long been overshadowed by his pursuit of the occult and has led to what some critics describe as a rewriting of the history books. However 'strange' Parsons was though, it did not preclude NASA from naming a crater on the alleged 'dark side of the Moon' after him.

Of course one cannot actually see the dark side of the Moon nor confirm that such a crater exists, so that is very appropriate.

"... America's space programme owes much to Parsons' rocket design and innovations and in 1972 the International Astronomical Union honoured him by naming 'Parson's Crater' on the dark side of the Moon. After co-founding the JPL – which his admirers referred to as 'Jack Parsons' Laboratory' – Parsons started 'Aerojet Corporation,' now the world's largest rocket producer and manufacturer of solid-fuel boosters for space shuttles...."

Parsons, who took the oath of the antichrist in 1949, also contributed to the design of the Pentagon under subsequent CIA director John J. McCloy. A pentagon of course, being a powerful satanic symbol. How appropriate.

So to recap, NASA was instigated through the likes of the self-acclaimed *"wickedest man in the world"* Aleister Crowley, the mass mind-control black magician and Satanist L. Ron Hubbard and the occultist, black magician, and confirmed Satanist Jack Whiteside Parsons. In addition to this 'unholy trinity' there were two more recruits into the NASA 'Dream Team,' Wernher von Braun and a certain Walt Disney, all of whom were essential elements in the creation of one of the greatest financial and theological frauds in human history... NASA.

- Jack Whiteside Parsons – occultist, Black Magician, Satanist, Head of Ordo Templi Orientis California Agape Lodge

- Aleister Crowley – 33rd degree Freemason, Leader of Ordo Templi Orientis, black magician, Satanist, The Beast '666'

- L. Ron Hubbard – mass mind controller, black magician, Satanist, and founder of The Church of Scientology religious cult

- Walt Disney – occultist, mass mind controller, black magician, Illuminati-paedophile, Freemason, and founder of 'The Ordem DeMolay'

- Werner von Braun – Ex-NAZI director of the German V-2 Rocket programme and recruited into the US under 'Operation Paperclip.' To be fair, he was probably the most 'normal' one of the entire, sorry bunch.

As the 'capstone' to the NASA Dream Team, the adoption of pseudo-science provided the template that makes the NASA hoaxers' rockets fly. However, NASA did not need to look too far for a sterling example of pseudoscience. The Soviet communists had perfected the craft of pseudoscience as both the US and USSR engaged in the fake cold war. The cold war was pre-engineered by the Khazarian-Rothschild-Zionist-Bankster cabal, aka the 'Khazarian Mafia' after WWII and it already had total financial control of both governments.

Lysenkoism

At the outset of the cold war, unsurprisingly, both the two main protagonists, the USA and the USSR embarked upon a huge spending blitz where billions of taxpayer dollars were channelled into the mythical, Freemasonic controlled faux 'space race' and the much vaunted 'nuclear arms race,' with the Soviets. The father of pseudoscience in fact, was a man named Trofim Lysenko who worked under the Jesuit-trained, Khazarian mass murderer Josef Stalin. Lysenko eventually became the official scientific voice of Russia and no matter how absurd his assertions, to disagree with them would have meant an instant one-way ticket to the Siberian gulags – or possibly worse.

Although it is impossible to say for sure, Lysenko probably killed more human beings than any other individual scientist in history. Other dubious scientific achievements have cut thousands upon thousands of lives short... dynamite, poison gas and bombs, for example, but Lysenko, a Soviet biologist, condemned perhaps millions of people to starvation through bogus agricultural research – and did so without remorse, like so many other of his fellow psychopaths, down the generations. Only guns and gunpowder, the collective product of many researchers over several centuries, can match the carnage generated by Lysenko.

Having grown up desperately poor at the turn of the 19th into the 20th century, Lysenko believed wholeheartedly in the deceptively empty promises of the Marxist-communist creed. And when the doctrines of science and the doctrines of communism clashed, he always chose the latter, confident in his belief that biology would always conform to his warped ideology in the end. But of course it never did. However, in many ways that commitment to ideology helped to salvage Lysenko's reputation to this day. Because of his hostility toward the 'West,' and his mistrust of western science, his work is currently enjoying a revival in his homeland, where anti-Western sentiments, perhaps unsurprisingly, still run deep.

Lysenko clawed his way to the very top of the Soviet scientific 'greasy pole' with almost indecent haste. Born into a family of peasant farmers in 1898, he was illiterate until the age of thirteen, according to a recent article on the revival of his methods in *Current Biology.* He nevertheless took advantage of the Russian Revolution and gained admission to several agricultural schools where he began experimenting with new methods of growing peas during the long, hard Russian winters, amongst other projects. Although he conducted many poorly designed pseudoscientific experiments and undoubtedly faked some of his results, the research won him praise from a state-run newspaper in 1927. His poverty stricken background also made him popular within the Communist party, which of course glorified 'peasants' – especially those who managed to succeed against all odds.

Soviet officialdom eventually placed Lysenko in complete control of Soviet agriculture in the 1930s. The only problem was though that he had some very strange and most unscientific ideas. In particular, he loathed genetics. Although still a relatively new discipline, genetics advanced rapidly in the 1910s and 1920s and the first Nobel Prize for work in genetics was awarded in 1933. However, especially in that era, genetics emphasised fixed traits. For example plants and animals have stable characteristics, encoded as genes, which they pass down to their descendants. Although nominally a biologist, Lysenko considered such ideas reactionary and evil since he saw them as reinforcing the status quo and denying all capacity for change. In fact he denied that genes even existed as this did not fit in with the dogma of his political beliefs

Instead, as the journalist Jasper Becker described in the book '*Hungry Ghosts*,' Lysenko promoted the utterly absurd, cultural Marxist philosophy that the environment alone shapes plants and animals and has nothing to do with genetics. So, if they were placed in a 'proper' setting and exposed to the right stimuli, he declared, they can be 're-made' to an almost infinite degree.

So, Lysenko began to 're-educate' Soviet crops to sprout at different times of year by for example, soaking them in freezing water, among other practices. He then claimed that future generations of crops would remember these environmental cues and, even without being treated themselves, would inherit these beneficial traits. According to traditional genetics, this is impossible. It is akin to cutting the tail off a cat and expecting her to give birth to tailless kittens. Lysenko, undeterred, was soon boasting about growing orange trees in Siberia, according to '*Hungry Ghosts*.' He also promised to boost crop yields nationwide and convert the empty, barren Russian interior into vast farmlands.

Such claims were exactly what Stalin and the other Soviet leaders wanted to hear. In the late 1920s and early 1930s Joseph Stalin, at Lysenko's behest, had instituted a catastrophic scheme to 'modernise' Soviet agriculture, forcing millions of people to join collective, state-run farms. The results were widespread crop failure and famine but of course in line with his communist ideological tenets, Stalin refused to change tack and instead ordered Lysenko to remedy the disaster with methods based on his radical new ideas. Lysenko then instructed farmers to plant seeds very close together, for instance, since according to his 'law of the life of species,' plants from the same 'class' never compete with one another. He also forbade all use of pesticides, which is no bad thing to do of course, but unfortunately he also banned organic fertilisers too, which in the prevailing Russian conditions, turned out to be disastrous.

Unfortunately, wheat, rye, potatoes, beets, in fact almost everything grown according to Lysenko's insane methods died or withered. Stalin still deserves the bulk of the blame for the famines, which killed at least ten million people, but Lysenko's intransigence and his practices prolonged and exacerbated the food shortages. Deaths from the famines peaked around 1932 to 1933, but four years later, and even after a 163-fold increase in farmland cultivated using Lysenko's methods, food production was actually lower than it had previously been. The Soviet Union's allies suffered under Lysenkoism, too. Communist China also adopted his methods in the late 1950s and in part due to this fact, the rule of Chinese communism resulted in even greater famines. Peasants were reduced to eating tree bark, bird droppings, and the occasional child or family member. As a result, at least thirty million died of starvation – either directly or indirectly.

But because he continued to enjoy Stalin's support, Lysenko's failures did nothing to diminish his power within the Soviet Union. His portrait hung in scientific institutes across the country, and every time he gave a speech, a brass band would play, and a choir would sing a song written in his honour.

Outside the USSR however, a very different tune was being sung – one of unstinting criticism. A British biologist, for instance, lamented that Lysenko was *completely ignorant of the elementary principles of genetics and plant physiology...to talk to Lysenko was like trying to explain differential calculus to a man who did not know his two-times table.* Criticism from foreigners did not sit well with the Soviets of course and even less so with Lysenko, who loathed Western 'bourgeois' scientists and denounced them as 'tools of imperialist oppressors.' He especially detested the American-born practice of studying fruit flies, the staple 'workhorse' of modern genetics and referred to such geneticists as 'fly lovers and people haters.'

Unable to silence his many western critics, Lysenko nevertheless attempted to eliminate all dissent within the Soviet Union. Scientists who refused to renounce genetics found themselves at the mercy of the secret police. The 'lucky' ones were those who were simply dismissed from their posts and left destitute, but many thousands of others were rounded-up and left to rot in prisons or psychiatric hospitals. Many were even sentenced to death as enemies of the state or, 'fittingly' starved to death in their jail cells, most notably the botanist, Nikolai Vavilov. Prior to the 1917 Revolution, Russia had possessed arguably the best genetics community in the world but over a relatively short period of time Lysenko managed to complete decimate it, and arguably set back Russian biology a good half-century.

Following Stalin's death in 1953, Lysenko's grip on his vast power began to weaken and by 1964, he had been deposed as the 'dictator' of Soviet biology, and he died in 1976 without regaining any influence. However, his portrait continued to be displayed in some scientific institutes, even throughout the Gorbachev years, but by the 1990s, the country had finally put the horror and shame of Lysenkoism behind it.

Until very recently, as a *'Current Biology'* article explained, Lysenko enjoyed a renaissance in Russia and several books and papers praising his legacy have appeared, bolstered by what the article refers to as *"a quirky coalition of Russian right-wingers, Stalinists, a few qualified scientists, and even the Orthodox Church."*

There are several reasons for this. Firstly, the new field of epigenetics has made Lysenko-like ideas fashionable once again. Most living things have thousands of genes, but not all those genes are active at once; some are turned on or off inside cells or have their 'volumes' turned up or down. Epigenetics is the study of these changes in gene expression and it just so happens that environmental cues are often the triggers that turn genes on or off. In certain cases, these environmentally driven changes can even pass from parent to child – just as Lysenko claimed.

But even a cursory look at his work reveals that he did not predict or anticipate epigenetics in any important way. Whereas Lysenko claimed that genes did not exist at all, epigenetics takes genes as a 'given.' They are the 'things' being 'turned on or off' and while epigenetic changes can occasionally (and only occasionally) pass from parent to child, the changes always disappear after a few generations, they are never permanent, which actually firmly contradicts all of Lysenko's wrongful assertions.

Epigenetics alone, then, cannot satisfactorily account for Lysenko's recent revival. There is something much deeper occurring – some would describe it as a mistrust of science itself. As the '*Current Biology*' article explained, Lysenko's new defenders "*…accuse the science of genetics of serving the interests of American imperialism and acting against the interests of Russia.*" Science, after all, is a major component of Western culture and because the 'barefoot peasant' Lysenko stood up to Western science, the illogical reasoning is that he therefore must be a true Russian hero. Indeed, nostalgia for the Soviet era and its anti-Western sentiment is still common in Russia today. A 2017 poll found that 47 percent of Russians approved of Joseph Stalin's character and 'managerial skills.' And also basking in the reflected glories of Stalin's newfound popularity are several of his lackeys, among them, Trofim Lysenko himself.

On one hand, this renewal of enthusiasm for the Soviet era is quite shocking. Genetics almost certainly will not be banned in Russia again, and the rehabilitation effort remains a fringe movement overall. But fringe ideas can sometimes also have dangerous consequences and this particular one distorts Russian history and glosses over the massive damage that Lysenko inflicted in abusing his power to silence and kill colleagues, to say nothing of all the innocent people who slowly starved to death because of his proclamations. The fact that even some qualified scientists are once again praising Lysenko demonstrates just how pervasive anti-Western sentiment is in some circles – even to the extent of perverting science in favour of ideology.

On the other hand, however, there is something depressingly familiar about the Lysenko affair, since ideology perverts science in the Western world also.

The Falsification of Science

But like the Soviet Union itself, the 'science' of Trofim Lysenko has now been consigned to the dustbin of history, yet the dangers of Lysenkoism, in subsuming science to ideology and politics, continue unabated.

But back to the topic in hand after that brief diversion. Pseudoscience is not science, but science and politics combined together in order to generate governmental and corporate revenues as well as to perpetrate a false agenda by creating fake scientific dogma reinforced with repetitive brainwashing. NASA has studied Lysenkoism well and has deceived the world by its many sleights of hand. But thanks now to the Internet, we are able to actually analyse what we have been 'shown' and told by those all-knowing space 'geniuses' at NASA.

Today's incarnation of NASA is a military/Freemasonic controlled entity that acts as the sole gatekeeper and arbiter of information about space and the universe and our place within it and to assume that NASA's black magic and occult origins have not been refined and reinforced by Freemasonry is complete naïveté.

Vast industries controlled by the Khazarian 'mafia' that revolve around NASA and the military industrial complex have been spawned. But in order to prove the point here, take a small sampling of the major US aerospace and defence contractors and research the major shareholders and especially investigate the following four institutions; the Vanguard Group, BlackRock Advisors LLC, Fidelity, and State Street Corporation, and this will prove extremely revealing. Roughly 80% of the industry is owned by these Khazarian-Rothschild subsidiaries. In addition, a cursory look at the various investment funds reveals the very same four companies.

Raytheon, Boeing, Northrup Grumman, General Dynamics Corporation and United Technologies Corporation, are just a few names that may be familiar to the reader and there is no doubt that hundreds more contractors and subcontractors exist, in addition to those mentioned. The Freemasons of NASA take in billions of taxpayer dollars, only to redistribute it amongst the

Khazarian-Rothschilds' majority owned companies and subsidiaries in the military-industrial complex.

And the level of compartmentalisation among these interconnected companies is so tight, that all employees are kept on a 'need-to-know' only basis, which means that only the top of the pyramidal structure knows the agenda and the final desired outcomes. Everyone else is virtually kept in the dark, and this is just the way that NASA prefers it of course.

Ironically speaking, NASA constitutes the largest 'black hole' ever yet discovered – especially in regard to funding. It is the undisputed king of all the many governmental 'black ops' with technology and advanced military craft that most people have no idea even exist. Billions of dollars have been embezzled by NASA from the American people by pure deception and are reallocated for other nefarious uses by the Khazarian mafia. And in return for our slavish obedience and gullibility, we receive a fake saga about the universe and its origins and our planet, directed and produced by Hollywood at a fraction of the cost of producing the 'real thing.' It is certainly more profitable to 'fake it than to make It,' as the saying goes. The line between Hollywood and NASA is virtually indistinguishable to the knowing observer.

Hollywood has cashed-in to a very great extent and perpetuates the worship of NASA through the endless genre of space, alien, and sci-fi movies. It is no surprise to discover that the same institutional majority shareholders of major Hollywood companies and subsidiaries such as Disney, Comcast, MGM, Time Warner, DreamWorks, and Sony Entertainment are all interchangeable. Again, please research these companies. Vanguard Group, Fidelity, BlackRock LLC, and State Street Corporation are all majority shareholders of the aforementioned Hollywood companies, and thus Hollywood becomes a de facto propaganda machine for NASA. The mainstream media is also largely controlled by the same Khazarian mafia and no day ever passes without an article, and often more than one, written about the incredible exploits of NASA and its quest for the exploration of the huge void of space.

The Falsification of Science

So to briefly recap. Parsons and Hubbard idolised Aleister Crowley and were avid Satanists, so NASA certainly did not sprout from 'humble' beginnings. L. Ron Hubbard said it best... *"You don't get rich writing science fiction. If you want to get rich, start a religion."* Fate is certainly not without its little ironies. NASA became mega-wealthy because of science fiction and for many people NASA is in effect, a religion.

And of course, it is no coincidence that NASA has been frantically re-writing history in order to erase its true origins. The mighty pillars of scientific certainty that NASA supposedly sits upon are actually satanic black magicians of the occult. This trivial fact of course, is strangely (or maybe not) omitted from today's school curricula.

The Challenger 'Disaster'

On the 28th January 1986, the space shuttle *'Challenger'* exploded and broke into millions of pieces, 73 seconds into its flight, leading to the deaths of its seven crew members, which included five NASA astronauts and two payload specialists. The victims allegedly were...

John Hamer

Francis R. Scobee, Commander

Michael J. Smith, Pilot

Ronald McNair, Mission Specialist

Ellison Onizuka, Mission Specialist

Judith Resnik, Mission Specialist

Greg Jarvis, Payload Specialist

Christa McAuliffe, Payload Specialist

But did they really die? The picture above is somewhat of a clue. At least five of the seven are still alive today. Some of them even use their original names! Governments are infamous for their false flags, cover-ups, and hoaxes, from Roswell to 9/11 et al.

A simple Internet search using the key words 'Challenger disaster astronauts still alive' will prove extremely revealing, should you wish to discover more about this. Staggeringly, most of these people do not even attempt to hide their real identities

There have been many theories about what really happened on that day. For example, we need to know, if there actually was anyone on board in the first place and if so, were they the people we were told were there – and if so, were there survivors? If, as NASA tells us, there were no survivors, then how does NASA account for the existence of the five still alive crewmembers? Well actually NASA does not account for them at all, it simply ignores the facts and sticks rigidly to the claim that all seven, stated astronauts died in the explosion.

It was reported that the explosion occurred because of a technical issue with the 'O' rings within the shuttle rocket motor, which caused it to explode and completely disintegrate. The timing of the deaths of those inside of the craft was not known absolutely but some were known to have survived the initial breakup of the shuttle – allegedly. Ultimately NASA was blamed for the incident.

The Falsification of Science

But if the above is true and there were survivors – or indeed there was no crew on board the shuttle when it exploded, what would NASA's motivation be for the hoax and subsequent cover-up?

My 'pet' theory on this topic is this...

The event of the launching of the Space Shuttle *'Challenger'* on this particular occasion, was a huge event, watched live on TV by up to 20 million viewers and particularly in schools all across America due to the presence of the 'payload specialist,' Christa McAuliffe, who also happened to be a full-time teacher. The event was therefore hyped-up as a memorable and significant event in space travel to date. Now, it is a fact that several million US schoolchildren were highly traumatised by seeing their 'hero' Christa 'die' live on TV in such an horrific manner. So, was this a deliberate 'psy-op' in order to test the public response to such an event and to monitor the effect on people's psyches – especially those of young children. Sounds too fantastic for words? Well maybe, but if it is the case it would not exactly be a precedent. This kind of operation takes place regularly, unfortunately and is a tried and tested way for governments to control and manipulate their people by the use of extreme trauma.

Hubble Space Telescope

Hubble Space Telescope images are identical, if not subjectively slightly inferior, to those from Earth-based observatories.

An Earth-based motionless observatory telescope-camera takes twelve minutes to produce an image of the night sky, so how does the Hubble Space telescope do this travelling at an orbiting speed of 7600 meters per second? A twelve-minute exposure over a travelled distance of 5472 km? It does not make logical sense.

'Sofia' is an infrared telescope on a jumbo jet. Why do they need this if they have the Hubble? The Hubble cannot be placed on the glass layer due to the maintenance needed, especially with all the micro-meteorites allegedly constantly bombarding Earth's atmosphere. The space shuttle would have to stop, let the astronauts out to repair the telescope, and start the shuttle again to get back to Earth.

The Hubble Space Telescope is merely another marketing ploy to sell us 'outer space.' As a real application it is unnecessary and unfeasible.

The Apollo Project

"In the big lie there is always a certain force of credibility; because the broad masses of a nation are always more easily corrupted in the deeper strata of their emotional nature than consciously or voluntarily; and thus in the primitive simplicity of their minds they more readily fall victims to the big lie than the small lie, since they themselves often tell small lies in little matters but would be ashamed to resort to large-scale falsehoods. It would never come into their heads to fabricate colossal untruths, and they would not believe that others could have the impudence to distort the truth so infamously. Even though the facts which prove this to be so may be brought clearly to their minds, they will still doubt and waver and will continue to think that there may be some other explanation. For the grossly impudent lie always leaves traces behind it, even after it has been nailed down, a fact which is known to all expert liars in this world and to all who conspire together in the art of lying. These people know only too well how to use falsehood for the basest purposes." Adolf Hitler, 1925

If you believe that the USA actually landed men on the Moon, you really need to think it all through more carefully because we are about to present proof that the Apollo Moon landings in the late 1960s and early 1970s were all one huge fabrication.

As already well documented in two of my other books, 'The Falsification of History' 2012, and especially 'Behind the Curtain' 2016, our entire financial system is based upon a gross lie, as is almost all of our fake 'reality.' So if you already believe that most of what we are taught to believe-in is one massive lie... then why should learning that the Moon landings were faked, be any more difficult to accept? Firstly, keep in mind that there were probably only about one hundred people involved in the Moon landings hoax and subsequent cover-up. Mission Control in Houston as well as most of the men and

The Falsification of Science

women who worked on this project almost five decades ago, had no idea it was all fake, of course.

But how is this possible you may ask? It is very simple really... the Elite who planned and executed this utter charade, never let anyone see too much of the whole picture. This technique is called 'compartmentalisation' and it is prevalent in every element of society. The thousands of NASA people involved only had their own small parts of the whole upon which to concentrate. Mission control was in Houston and the launch site in Florida. The engineers, mechanics, computer programmers etc., mostly in California, were all isolated from each other and there was no crossover between them. So most people would never be able to figure out that the whole thing amounted to nothing more elaborate than a Hollywood production – and a very unconvincing one at that, which is fairly simple to expose once we scratch even slightly below the surface.

In, 1962, then President JFK said that he had a vision of America *"...putting a man on the Moon and returning him safely before the decade is out."* Now, Kennedy knew this was impossible, he knew that the Russians had faked Yuri Gagarin being the first man in space (see John's book, *Behind the Curtain* for more details) and the Russians knew that US technology was only at a similar state of advancement as their own, so they were complicit with each other's claims and lies. The best rocket scientists in the world, those involved with the space programme, informed Lyndon Johnson that the science was at least 30 years away from being able to accomplish such a task, but the fact is... they were actually over-optimistic.

Indeed, more than half a century later we still do not possess the technology to send a man to the Moon and return him safely to Earth. The powers that decided on this fraud concluded that if they could not send a man to the Moon and get him back safely, then they would simply have NASA fake the project and then keep the billions of dollars of taxpayer money used to fund this operation for themselves. You may have heard this old story before somewhere! After all, as elaborate and expensive as the plan to 'fake it' was, it would have been orders of magnitudes less expensive than the 'real thing' would have cost. The cost of the entire Apollo programme was $25.4bn in 1969 dollars and $240bn in 2021 dollars.

Several motives have been suggested for the US government to fake the Moon landings and some of the recurrent elements are . . .

Distraction – The government benefitted from a popular distraction to take attention away from the carnage of the highly unpopular Vietnam War. Lunar activities did actually abruptly stop, with planned future missions cancelled, at *exactly* the same time that the US ceased its involvement in the Vietnam War.

Cold War Prestige – The government considered it vital that the US should win the space race with the USSR. Going to the Moon, if it were possible, would have been risky and expensive. It would have been much easier to fake the landing, thereby ensuring a successful propaganda campaign and the prestige it would bring the country.

Money – NASA raised approximately $30bn dollars by pretending to go to the Moon. Some of this could have been used to pay-off a large number of 'in the know' people, providing significant motivation for complicity. In variations of this theory, the space industry is characterised as a political economy, much like the military-industrial complex, creating fertile ground for its own survival.

Risk – The available technology at the time, and even that of today, is still not capable of landing humans on the Moon. The Soviets, with their own competing Moon landings programme and an intense economic, political, and military rivalry with the USA, could be expected to have cried 'foul play' if the US had attempted to fake a Moon landing. However, Ralph Rene responded to that accusation by saying that shortly after the alleged Moon landings, the US silently began shipping hundreds of thousands of tons of grain as 'humanitarian aid' to the allegedly starving USSR. He regarded this as evidence of a cover-up, the grain being the price of silence. In fact, the Soviet Union had its own Moon landings programme which was mysteriously and unexpectedly shelved after the American 'achievements.'

The Falsification of Science

There is an old saying that 'a liar needs a good memory.' Nowhere is this more evident than in the Apollo programme. NASA constantly tells lies to cover up previous lies, and other discrepancies uncovered by people investigating the Moon landings by altering previous data, removing, and even adding photographs, and retracting previous statements made. And over time, lies and cover-ups become totally unsustainable. This only serves to reinforce the idea that NASA is constantly fire-fighting and slowly being 'painted into a corner' from which it cannot escape. The longer that NASA's extravagant claims continue, the more lies they have to tell in order to counteract them, until it reaches the point where it becomes totally ridiculous.

Many Apollo astronauts have since died from causes other than 'old age,' as have many of the original NASA officials involved in the scam. And consequently current officials, who know that Apollo was a fake, have sometimes not quite got it right when talking openly in public. Perhaps the biggest slip of the tongue was made by NASA's longest serving administrator Dan Goldin, when interviewed by British TV journalist Sheena McDonald in 1994. He said that mankind could not venture beyond Earth's orbit, 250 miles into space, until they could find a way to overcome the dangers of cosmic radiation. He must have totally forgotten that they supposedly sent twenty-seven astronauts 250,000 miles outside Earth's orbit, 25 years previously.

The Van Allen Radiation Belts, which lie some 600 – 22,000 miles above the Earth, is probably the main argument for our not having reached the Moon or anywhere near it. The Van Allen Belts refers to dense layers of charged particles trapped around the Earth by the Earth's magnetic field. The particles include protons, electrons, and heavy ions that emanate from solar wind and cosmic rays (high energy particles from outside the Solar System). Interestingly, Dr James van Allen himself, for whom the belts are named, was a vociferous critic of the Moon landings programme throughout the rest of his long life, (he died at the age of 92 in 2006) stating that it was 'impossible' that humans could pass through them safely. But of course his words were largely suppressed and ignored and certainly never widely reported.

Outer space is awash with deadly radiation that emanates from solar flares from the Sun. Astronauts orbiting Earth in near space, such as those who recently ahem, 'fixed the Hubble telescope,' are protected by the Earth's Van Allen Belts. But the Moon is 240,000 miles distant, way, way outside this safe band and also, during the Apollo flights, astronomical data shows that there were no less than 1,485 such flares.

John Mauldin, a NASA physicist, once said that shielding at least two metres thick would be needed to protect humans from the radiation generated by the Van Allen Belts, yet the walls of the Lunar Landers which allegedly transported the astronauts from the spaceship to the Moon's surface were about the thickness of heavy duty aluminium foil – according to NASA's own proclamations.

How could something merely the thickness of aluminium foil possibly stop all that deadly radiation? And if the astronauts were protected by their space suits, why did rescue workers not use such protective gear at the Chernobyl meltdown, which allegedly released only a fraction of the radiation dose that the Apollo astronauts would have encountered? Not one of them ever contracted cancer – not even the Apollo 16 crew who were on their way to the Moon when a huge solar flare erupted. *"They should have been fried,"* commented author and researcher, Ralph Rene.

The late Bill Kaysing, a pre-eminent Apollo researcher, worked as a technical writer for *Rocketdyne*, a company heavily involved in the Apollo programme. At the time of JFK announcing in May 1961 that the US government intended to 'put a man on the Moon by the end of the decade,' Kaysing claimed that NASA carried out a feasibility study which confirmed that there was only a 0.0017% (or 1 in around 50,000) chance of landing a man on the Moon and returning him to Earth with the available technology. And Kaysing argues convincingly that it was totally impossible for NASA to go from 0.0017% to 100% by 1969. Surely if we could get to the Moon with 1960s technology, it should be fairly simple for us to get to the Moon today. However, all nations STILL have extreme difficulty placing an object in a high Earth orbit.

The Falsification of Science

In his book, 'We Never Went to the Moon,' Kaysing wrote that NASA and the Defence Intelligence Agency (DIA) worked together on faking the Apollo 11 Moon landing. An empty Saturn V rocket was launched but fell back to Earth when it was out of the public gaze and NASA also allegedly created a lunar landscape in 'Area 51' in the Nevada desert and according to Kaysing, the filmset was still there at the time of his writing. Meanwhile, the astronauts and Mission Control were taking part in a meticulously staged hoax designed to fool the public into believing they had landed on the Moon. Fake photographs and film were taken and the astronauts' return to Earth was staged by dropping a dummy space capsule from an army plane into the ocean. Kaysing went on to suggest that the astronauts were brainwashed, and that they and their families were placed under extreme duress and threatened, in order to guarantee their cooperation with the hoax.

Another American author, now also sadly deceased, Ralph Rene, also believed that astronauts could not have made it to the Moon. In his book, 'NASA Mooned America!' Rene hypothesised that the Apollo spacecraft would have needed at least an equivalent mass in lead of two metres of water shielding to prevent cosmic radiation from cooking the astronauts inside. The hoax theorists believe that, when NASA realised they did not have the technology to take men safely to the Moon by the end of the 1960s, they resorted to faking the lunar landings. This ensured that they would score a propaganda coup against the Soviets and keep the money rolling in for funding their real space projects.

Even Arthur C. Clarke referred to Apollo 11 as a *"Hole in History"* and the prominent, well-respected 20th century British historian A.J.P. Taylor referred to it as... *"The biggest non-event of my lifetime."*

"NASA and other connected agencies couldn't get to the Moon and back and so went to ARPA (Advanced Research Projects Agency) in Massachusetts and asked them how they could simulate the actual landing and space walks... We have to remember that all communications with Apollo were run and monitored by NASA, and therefore journalists who thought they were hearing men on the Moon could have easily been

misled. All NASA footage was actually filmed off TV screens at Houston Mission Control for the TV coverage... No-one in the media was given the raw footage. The world tuned-in to watch the Moon landing and what looked like two blurred white ghosts throwing rocks and dust. Part of the reason for the low quality was that, inexplicably, [or perhaps not- JH] NASA provided no direct link-up. Networks actually had to film man's greatest achievement from a TV screen in Houston, making it impossible for anyone to examine it." Bill Kaysing

Bill Wood has degrees in mathematics, physics, and chemistry, and is a space rocket and propulsion engineer. In other words, a rocket scientist. He has been granted high security clearance for a number of top-secret projects and has worked with *McDonnell Douglas* and engineers who worked on the Saturn 5 rocket (the Apollo launch vehicle). He also worked at NASA's Goldstone facility as a Communications Engineer during the Apollo missions. This facility in California, was responsible for receiving and distributing the pictures allegedly sent from Apollo to Houston and he says that very early video recording machines were used to record the NASA footage by the TV networks. They received the FM carrier signal on Earth, ran it through an FM demodulator and processed it in an RCA scan converter that took the slow scan signal and converted it to the US standard black and white TV signal. The film was then sent on to Houston. When they were converting from slow scan to fast scan, RCA used disc and scan recorders as memory, and it played back the same video several times until it got an updated picture. In other words the signal was recorded onto video one then converted to video two. Movie film runs at 30 frames per second, whereas video film runs at 60 frames per second. So in other words the footage that most people saw that they thought was 'live' was not and was actually 50% slower than the original footage.

It has to be said, that the 'Moon Hoax debunkers,' all those who believe the Moon landings were real, tend to believe most of what they are told about everything by the elite-controlled mainstream and nothing that they are told via the alterative media. That is to say, they would also try to debunk everything else in this book. And yet, as we hope we have proven, both in fine detail and through a continuous

The Falsification of Science

pattern, (a 'proof of system' if you will?) The Elite mastered the arts of deception and distraction, a long, long time ago.

"[Why do] people cling so tenaciously, often even angrily, to what is essentially the adult version of Santa Claus, the Easter Bunny and the Tooth Fairy? What primarily motivates them is fear. But it is not the lie itself that scares people; it is what that lie says about the world around us and how it really functions. For if NASA was able to pull off such an outrageous hoax before the entire world and then keep that lie in place for four decades, what does that say about the control of the information we receive? What does that say about the media and the scientific community and the educational community and all the other institutions we depend on to tell us the truth? What does that say about the very nature of the world we live in? That is what scares the hell out of people and prevents them from even considering the possibility that they could have been so thoroughly duped. It's not being lied to about the Moon landings that people have a problem with, it is the realisation that comes with that revelation: if they could lie about that, they could lie about anything."
David McGowan, researcher

Had the very first transatlantic flight in 1919 not been followed-up with others similar for the next fifty years and more, would there not have been questions asked and would people not have found it strange or unusual to say the very least? If say, in the 1920s, had someone designed a jet airliner capable of speeds of 600 mph or more and then after a short time that technology 'disappeared' and could not be re-created, would that not seem to be at odds with commonly accepted logic or reality at all? We submit that it most certainly would and yet this is exactly the case with the so-called Moon landings of 1969-72.

Is it not also strange that almost up to the point in time when the alleged Moon landings took place, that the Soviet Union (USSR) had been leading the 'space race' by some considerable distance and yet to this day (2021) has never either bothered or managed to put a man on the Moon.

John Hamer

The Russians were the first to launch a vessel of any kind into space, the first to send a living creature into space, the first to perform a manned space flight, the first to perform a spacewalk, the first nation to have two spacecraft in orbit simultaneously and the first to perform a 'docking' manoeuvre in space. They also purportedly landed the first unmanned vehicle on the surface of the Moon, achieved the first flyby of the Moon, launched the first craft to impact the Moon, were the first to make a soft landing on the Moon, put the first object into lunar orbit and remain, to this day, the only nation to land and operate a robotic vehicle on the Moon. It should now make perfect sense to everyone then why the Soviets, who were ahead of the US in virtually all aspects of space exploration, in some cases by decades, never landed a man on the Moon or even sent a man to orbit the Moon. Up until the 'successful landings' by the Americans, they had been comprehensively beaten by the Russians in every important aspect of the space race. The Soviets had logged almost five times as many man-hours in space than the Americans and yet in the single most important aspect, the landings themselves, the US had literally almost cruised to victory, totally unopposed. Very strange indeed – well at least if you believe it, it is.

I also believe it significant that no other industrialised nation on Earth has managed to successfully visit the Moon – or even attempted to do so, despite the fact that there have been massive, across-the-board technological advancements since the 1960s. We believe that it is quite possible that the entire US space programme has largely been, from its first inception, little more than an elaborate cover story for the research, development and deployment of space-based weaponry and surveillance systems. The compliant media never investigates or even mentions these things of course, but recently declassified US government documents make clear that the goals being pursued through space research are largely military in nature.

"Control of space means control of the world. From space, the masters of infinity would have the power to control Earth's weather, to cause drought and flood, to change the tides and raise the levels of the sea, to divert the Gulf Stream and change the climates." Future US President, Lyndon Johnson, 1959

The Falsification of Science

But if this hoax was perpetrated in almost total secrecy, how was it all kept from the thousands of people involved in the huge project, you may well ask? As pointed out previously, please bear in mind that there were only around a hundred very senior people involved in the actual Moon landings hoax itself. Mission Control in Houston as well as most of the men and women who worked on this project over five decades ago, had no idea it was a fake. How was this possible to achieve? Very simply, the Elites who staged this fiasco never let anyone see more than a small fraction of the 'big picture.' The many thousands involved were only small cogs in a very large machine. Mission control was based in Houston, the launch site in Florida and the engineers, mechanics, computer programmers and assorted other technicians did not normally come into contact with each other, personally. So it would have been next to impossible to work out that the whole sordid enterprise amounted to nothing more than a sophisticated Hollywood production.

Why would the US not return to the Moon in the four decades that have elapsed since the last alleged Moon landings and why would other technologically advanced nations not attempt to emulate the feat? Could it be that the costs of such a venture would be totally prohibitive as some sources would have us believe? Even in those heady days of the late 1960s and early 1970s, the US was not exactly awash with money. Not only was it fighting an extremely costly, overt war in the Vietnam, but was also engaged in the covert, Cold War arms race and yet still spending untold billions on the space race, so money or lack of it would not be a particular issue, especially not over a period of time as long as five decades – more than half a lifetime.

Also, consider this; the surface-to-surface distance from the Earth to the Moon is approximately 235,000 miles and since the last alleged, manned Moon landing in 1972, not one human has been further out into space than 400 miles and very few have gone even that distance. Most space-shuttle orbits take place at around 200 miles from the Earth, the same distance away approximately, as the Space Station. So, to put these facts into perspective, in the twenty first century, utilising the best technology that money can buy, NASA is only able to send humans around 200 miles into space, but in the 1960s it had the capability to reach an object 235,000 miles away, undertake several orbits of the Moon and then make the return trip – all on a single tank of fuel! Please pardon our scepticism.

And what about all the many hours of footage that NASA has of the Moon landings and the astronauts on their Moonwalks, transmitted back directly into our at the time, state-of-the-art TV sets that now with the benefit of hindsight, look like something out of the nineteenth rather than the twentieth century!? Even in the 1990s it was no simple task to transmit pictures directly from the Iraqi desert during the first Gulf War, so the transmission of pictures from a quarter of a million miles away almost a quarter of a century earlier was a really impressive technological feat, if it happened. Unfortunately, NASA has sadly, but rather conveniently 'lost' all the thousands of hours of tapes of the Moon footage, 700 cartons in all.

"The U.S. government has 'misplaced' the original recording of the first Moon landing, including astronaut Neil Armstrong's famous 'one small step for man, one giant leap for mankind'... Armstrong's famous Moonwalk, seen by millions of viewers on July 20, 1969, is among transmissions that NASA has failed to turn up in a year of searching, spokesman Grey Hautaluoma said. 'We haven't seen them for quite a while. We've been looking for over a year, and they haven't turned up,' Hautaluoma said... In all, some 700 boxes of transmissions from the Apollo lunar missions are missing." Reuters, 15th August 2006

These tapes represented supposedly the greatest human achievement ever, both in technological and symbolic terms. How could such a thing happen? Surely these are historical records that should have been treated as one of the great human treasures – on a par with such priceless artefacts as the ceiling of the Sistine Chapel, the Mona Lisa, and the Pyramids. Should such an irreplaceable treasure as these tapes, have not only been copied several times for security purposes, but also have been locked away securely in a fireproof, bombproof, waterproof vault somewhere, 'just in case'? And also, would not multiple copies have been made available for educational and/or scientific research and advancement purposes? Obviously, NASA and the US government did not feel that they were important enough for any of that.

Surely this is all absolute and utter garbage? How could *700 cartons* of tapes be missing? Perhaps one or two boxes – possibly – but not the entire 700. For a start, they must have filled several large rooms and it is therefore simply not credible in our opinion. Could it be that they do not want the tapes to be exposed to any kind of scientific analysis using today's technology? We believe that is most likely the *real* reason and that in itself speaks volumes. This one factor alone carries far more weight than the somewhat trivial 'flag waving in the breeze' and 'shadows at the wrong angles' arguments that are often used by NASA and other Mon hoax debunkers, as 'strawman' arguments, to distract the attention away from the 'real' unanswerable issues surrounding this huge non-event.

Reuters also commented that... *"Because NASA's equipment was not compatible with TV technology of the day, the original transmissions had to be displayed on a monitor and re-shot by a TV camera for broadcast."* So, what we were actually seeing on our mediaeval TV screens were not 'live transmissions' as we were told, but was footage shot directly from a tiny black and white TV monitor and then retransmitted second hand via the TV stations. All totally different of course to what we were led to believe at the time and subsequently. With this admission by NASA, surely it is not difficult to see how the entire footage could have been faked?

The next issue worth commenting upon is the absolutely bizarre movements of the astronauts in performing their many Moonwalks as witnessed on the small portions of footage that still survive. As many sceptics have commented (and proved), if the tapes are played at approximately twice the speed, then those very odd skipping-type movements of the astronauts look suspiciously similar to normal speed movements on Earth.

So, the simple formula for creating Moonwalk footage is to take original footage of men in ridiculous costumes moving around awkwardly here on Earth, broadcast it over a tiny, low-resolution, black and white television monitor at about half-speed and then re-film it with a camera focused on that screen. The end result will be tapes that, in addition to having a grainy, ghostly, rather surreal 'broadcast from the Moon' look, also appear to show the astronauts moving about in entirely unnatural ways. But not, it should be noted, too unnatural. And does that not seem a little strange too? If we are being honest, the average male never stops being a little boy at heart and what red-blooded, macho-male, given the opportunity to spend some time in a greatly-reduced gravity environment, can resist seeing how high he can jump? Or how far he can jump? Or what dramatic somersaults he could perform? So, what did the astronauts *actually* do? They hit golf balls. Yes, that is correct, the only method by which they were prepared to demonstrate the lack of gravity on the Moon, was to hit golf balls of which it was impossible to accurately judge the distance they actually travelled!

It seems more than a little odd that they failed to do *anything* that could not be faked simply by changing the tape speed. Some athletes here on Earth are able to perform a standing vertical jump of around four feet (1.3m) so it is rather strange that the astronauts' best efforts were only around 12 inches (0.3m.) In one-sixth gravity, at least 10 feet (3m) should have been easily achievable, even for the most unfit among us, let alone for highly trained, super-fit professionals such as these young men allegedly were.

Indeed, should the astronauts' every movement not have been *quicker* than normal given the fact that there is virtually zero wind-resistance on the Moon? If so then, why does all the available footage show only half-speed movement? It is almost as though it was the only way that NASA could think of attempting to represent anything that could be even remotely construed as resembling non-earthly movement. Maybe then it is completely unsurprising that all the original footage has mysteriously disappeared, as being submitted to modern-day technology would expose it as fake in around two seconds flat.

Somewhat worryingly, it also transpires that it is not simply the film footage that has disappeared in its entirety, but also the complete set of *13000+*, yes thirteen thousand plus reels of telemetry data including voice and biomedical data. All of that information, in fact the entire technical record of all the Apollo Moon missions has gone, *plus* all the design blueprints for the lunar modules, the lunar rovers and the entire Saturn V multi-sectioned rockets. Worryingly that is for us but not for NASA of course as there is now no way at all that the contemporary scientific community could ever have the opportunity of studying these documents in detail and thus prove them all impossibilities if not fakes of the most naïve kind.

For a short time there was a boost to the case of those who insist that Apollo project is not the almighty fiction that it most definitely is. This came in the form of a promise from NASA to send a probe (unmanned) 'back' to the Moon to photograph the various bits of detritus left over from the Apollo missions which it was said would prove conclusively that all the thousands of we so-called 'conspiracy theorists' out there, were all wrong. Sadly for the 'believers' though, no such images have ever been publicly forthcoming despite their wild claims to the contrary. Even the Hubble space-telescope was widely touted as being capable of homing-in on the lost Apollo artefacts allegedly spread liberally about the surface of the Sea of Tranquillity, but this too has proved a false dawn. It is debatable as to whether or not the Hubble technology is capable of this feat, but this is a moot point in any case, as there have been no images forthcoming from that source either.

Then in 2009, NASA announced that its 'Lunar Reconnaissance Orbiter' had returned the first images of the Apollo landing sites.

"The LROC team anxiously awaited each image. We were very interested in getting our first peek at the lunar module descent stages just for the thrill – and to see how well the cameras had come into focus. Indeed, the images are fantastic and so is the focus." LROC principle investigator, Mark Robinson of Arizona State University

Unfortunately, that has proved not to be the case. The images are in fact not 'fantastic' by anyone's definition and neither is the focus. In actual fact the images are from such a distance that the tiny white dots they show – in shadow also, it must be noted – could be almost anything. Spot the Apollo debris if you can (below). Yet more insubstantial bluster by NASA.

Subsequent Japanese, Chinese and Indian unmanned lunar probes have also unsurprisingly perhaps, spectacularly failed to provide photographs of the Apollo landing sites.

"There's no reason to go back… Quite frankly, the Moon is a giant parking lot, there's just not much there." Val Germann, President of the Central Missouri Astronomical Association

Strange then is it not that so many space agencies worldwide send unmanned probes there and focus enormously powerful telescopes on the Moon's surface? What could they possibly learn about this 'parking lot' from those distances viewed through Earth based telescopes that the Apollo astronauts did not already discover by actually being there? And how much more valuable information can unmanned probes provide over the allegedly 'real' firsthand experiences of the astronauts?

NASA also claims that several of the Apollo missions left small laser beam 'targets' on the Moon's surface that enable NASA scientists to bounce laser beams from them and which gives absolutely accurate readings of the distance from the Earth to the Moon. Now bearing in mind that these 'targets' were approximately the size of a small computer monitor screen, does anyone really believe that the technology existed in the 1960s and 1970s to accurately hit a target of that size with a laser beam from almost a quarter-million miles away? In fact, is it even possible today? NASA states blithely and conveniently that there is no technology in existence that could accurately pinpoint the location of the Apollo detritus from the Earth, so how are they able to successfully locate a tiny laser target many times smaller than the artefacts supposedly left over from the Moon landings? Nevertheless, according to NASA and its many apologists, the fact that these signals are being bounced off these targets on a regular basis, 'proves' beyond doubt that Apollo astronauts went to the Moon. We think not somehow.

The 'actual' alleged Apollo lunar module

One could be forgiven for thinking that the above picture is of a model of a lunar landing module constructed by ten-year old school children in their arts and crafts classes, but nevertheless this is a 'real' lunar module allegedly sitting on the surface of the Moon, photograph courtesy of the Apollo astronauts and NASA. John can personally vouch for this being NASA's stated lunar module as he actually stood within 10 feet of it himself at the museum at Cape Canaveral, Florida. This 'incredible piece of technology,' we are led to believe, actually not only landed on the surface of the Moon, but three days subsequently, took-off again, flew seventy miles upwards, back into orbit around the Moon and successfully redocked with the command module, which was incidentally travelling at a speed of over 4000 mph at the time, in order that it could then navigate the quarter-million mile journey back across the empty nothingness of space to land precisely where and when it was programmed to do.

The most striking aspect of the photograph above though is not so much how a craft seemingly put together with duct tape, tarpaulin and bits and pieces of aluminium foil could perform such technologically advanced feats, but how did it manage to carry all the necessary equipment and accessories in order to keep two human beings alive for three days in the most unimaginably inhospitable environment known to man?

According to NASA's own data, the lunar modules were only 12 feet in diameter. This being the case, how was it possible to accommodate all the navigational guidance equipment (in the 1960s of course, this would have been extremely bulky) and what about the power supplies, the reverse thruster for landing and the powerful rocket motor required for take-off again? There would also need to have been several other smaller rocket thrusters for stabilisation purposes, the massive amounts of fuel required to feed the rocket engines, especially upon take-off to accelerate enough to break-free of the Moon's gravity despite the fact that it is claimed to be only 1/6th that of Earth. They would also have needed plenty of equipment just to sustain life for two people for several days and provide some 'home comforts' such as places to sleep, bodily waste management facilities, food and water supplies, adequate oxygen for three days for two people, the list just goes on and on. And let's not forget that the oxygen tanks in the spacesuits would have also needed a recharge system to enable the surface walks to take place over a concerted period of time. A backup oxygen system would also have been needed in case the original failed.

The astronauts in addition to all of the above would also have desperately needed an air conditioning system, both in their spacesuits and in the module itself, the capabilities of which would have to be seen to be believed. Consider this; the surface of the Moon is subject to some incredible temperature swings. It can be +125°C (+257°F) in sunlight and -170°C (-274°F) in the shade with very little variation in between these two extremes, so in the sun a human would be boiled alive and, in the shade, would be frozen solid in minutes. In

The Falsification of Science

order to cope with these extremes, the spacesuits worn by the Apollo astronauts would have required technology lightyears beyond what we have today, never mind that which was available in the 1960s. Also it is quite pertinent to point out that in order for air-conditioning systems to function correctly, they need a decent supply of air. The clue is in the name really and unfortunately air is a commodity which is apparently in fairly short supply on the Moon, last time we checked. An air conditioner cannot possibly work in a vacuum and a spacesuit surrounded by a vacuum cannot transfer heat from the inside of the suit to anywhere else. A vacuum, as you may remember from school physics lessons, is a perfect insulator and therefore anyone would roast alive in his suit under such circumstances.

But we are not even finished there. The mission would also have required equipment to maintain the 'ship' and to provide it with essential spares, for emergencies. And then there would have been all the testing and portable lab kits that they used to conduct experiments on the Moon's surface plus storage space for all the hundreds of pounds of Moon rock that was allegedly brought back, and which reportedly sits in hundreds of museums and scientific institutes around the world today, many of which incidentally have already been exposed as fakes! The latter visits to the Moon were also supposedly equipped with the 'Moon rover.' This in itself was over ten feet long with four wheels larger than the standard car wheels of today – how did they get it in, we really do wonder? Well, according to the geniuses at NASA, this beast actually folded-up to be the size of a large suitcase! Can anyone with even a semi-functioning brain really accept this abject nonsense?

But lastly and definitely not least, the astronauts would have needed power – and lots of it. The only way that the ship and its vital functions could be powered whilst it was on the Moon's surface would be with batteries, likewise anything else that needed a power supply, such as the life support system, the lights, the communications system, the television cameras and transmitters, the lunar rovers, the suits etc. etc. As it would also not be possible to recharge any of the batteries then they would have needed some pretty huge, and powerful ones at that, and these all had to be found a place in the severely restricted space on board that tiny lunar module.

It is also important to acknowledge that, unlike the initial launch on Earth, which involved the collective, sustained efforts of thousands of technicians of all levels and the use of many types of peripheral computer and monitoring equipment, the astronauts leaving the Moon had only themselves and some completely untested-in-that-environment, assorted ironmongery, cables and plastic upon which to rely. It beggars the imagination to think how uncomfortable and scary it must have felt to be on the surface of the Moon for a few days performing experiments and hopping and skipping around the place in a seemingly carefree, happy-go-lucky manner, wondering that if or when the time came, whether that completely untested contraption would actually get me back home all the way from the Moon, or even back the 70 vertical miles to the rendezvous point with the command module. Fortunately, though, the completely untested-in-the-conditions-prevailing-on-the-Moon lunar module worked absolutely perfectly the first time and with no need for modifications or last-minute hitches, despite the literally, alien environment in which it was being utilised.

Today of course, NASA cannot even launch a highly technically-advanced space shuttle from Earth without occasional disasters, even though they have since modified their ambitions considerably. After all, sending spacecraft into low-Earth orbit (200 miles return) is infinitely more straightforward than sending spacecraft all the way to the distant Moon and back (470,000 miles). It would seem that although technology has advanced immeasurably since the Apollo Moon landings that tellingly, NASA has hugely downgraded its ambitions in space and now has a significantly worse safety record than it had in the 1960s, despite that downgrade.

In 2005, NASA made this incredible statement:

"NASA's vision for space exploration calls for a return to the Moon as preparation for even longer journeys to Mars and beyond, but there is a potential showstopper: radiation. Space beyond low-Earth orbit is awash

The Falsification of Science

with intense radiation from the Sun and from deep galactic sources such as supernovas. ...Finding a good shield is important." NASA spokesman, 24th June 2005

Do they really expect us to believe that it was possible to undertake the Apollo missions in the 1960s and 1970s but now, around fifty years further down the line it has suddenly become impossible to leave the vicinity of the Earth because of space radiation? Did the technology to overcome this problem then exist in 1969-72 but has been somehow, inexplicably lost or forgotten or did the Apollo missions not actually take place as described? This statement narrows down the options somewhat. If 'finding a good shield' is indeed important, then why can they not just simply use the technology that was deployed on the Apollo craft? No one died at the time and certainly none of the astronauts subsequently suffered from radiation-induced problems in any way that we are aware of.

Lead is the usual method of choice for radiation shielding, but the issue is that lead is so heavy and impractical for use in anything but static situations on Earth. Attempting to build spaceships with a four feet thick lead encasement is far from practical as the Russians themselves discovered when they calculated that this was in fact the only way that they could traverse the Van Allen Radiation Belts with a human cargo in the 1960s. Maybe this was why they simply gave-up on the race to be first to the Moon?

Let us now turn our attention to the photographs asserted to have been taken on the Moon by our intrepid Apollo astronauts. In actuality the very existence of the photographs is a technical impossibility. Unfortunately, it would simply not have been possible to capture *any* of the images allegedly shot on the Moon in the manner that NASA describes them to have been obtained. In the 1960s, camera technology was very limited in comparison to today's and Hasselblads, the cameras purportedly used by the astronauts, although they were probably the best and most sophisticated on the market at that time, the simple fact is that they were incapable of generating the images claimed to have been taken on the Moon, under the circumstances in which they were supposedly taken.

Cameras of those far-off times before microchip technology were not very 'intelligent,' so every function had to be performed manually. The photographer had to manually focus each shot by squinting through the viewfinder and rotating the lens until the scene came into focus. The correct aperture and shutter speeds had to be manually selected for each shot also, in order to ensure the correct exposure time for the circumstances. This also required peering through the viewfinder, to meter the shot or to use a separate, handheld light meter. Finally, each shot had to be properly composed and framed, which obviously also required looking through the viewfinder.

The problem for the astronauts was that the cameras were mounted on their chests, which made it completely and utterly impossible to see through the viewfinder to meter, frame and focus the shots. Everything, therefore, was total guesswork and focusing would have been entirely guesswork also, as would the framing of each shot. An experienced, professional photographer can fairly accurately estimate the exposure settings, but the astronauts lacked this experience and they were also doubly handicapped by the fact that they were viewing the scenes through heavily tinted visors, which meant that what *they* were seeing was not what the camera was focusing upon.

To add to their not inconsiderable problems, they were wearing space helmets that seriously restricted their field of vision, along with enormously bulky, pressurised gloves that severely limited their hand and finger movements. The odds therefore of them getting even one of those three elements (exposure, focus and framing) correct under the prevailing conditions on any given shot would have been exceedingly low and yet, amazingly enough, on the overwhelming majority of the photos, they got all three right.

"For those who don't find that at all unusual, here is an experiment that you can try at home: grab the nearest 35MM SLR camera and strap it around your neck. It is probably an automatic camera so you will have to set it for manual focus and manual exposure. Now you will need to put on the thickest pair of winter gloves that you can find, as well as a motorcycle helmet with a visor. Once you have done all that, here is your assignment: walk around your neighborhood with the camera pressed

The Falsification of Science

firmly to your chest and snap a bunch of photos. You will need to fiddle with the focus and exposure settings, of course, which is going to be a real bitch since you won't be able to see or feel what you are doing. Also, needless to say, you'll just have to guess on the framing of all the shots. You should probably use a digital camera, by the way, so that you don't waste a lot of film, because you're not going to have a lot of 'keepers.' Of course, part of the fun of this challenge is changing the film with the gloves and helmet on, and you'll miss out on that by going digital. Anyway, after you fill up your memory card, head back home and download all your newly captured images. While looking through your collection of unimpressive photos, marvel at the incredible awesomeness of our Apollo astronauts, who not only risked life and limb to expand man's frontiers, but who were also amazingly talented photographers. I'm more than a little surprised that none of them went on to lucrative careers as professional photographers." David McGowan, 2009

Despite all the acclaim he has received for his exploits as an astronaut, Neil Armstrong clearly has been unjustly denied recognition of his astounding abilities as a photographer. Some may argue that he was not in the same league as say, David Bailey or Patrick Lichfield, but we would disagree. Those two individuals created some stunning pictures throughout their careers, but could they have done so whilst wearing a pressurised spacesuit, gloves and helmet and with their cameras mounted on their chest and whilst working in an environment that featured no air, one-sixth gravity, and utterly stupefying extremes of heat and cold? We seriously doubt it.

Even more tellingly perhaps, the designer of the particular type of Hasselblad cameras 'used on the Moon,' has publicly stated to all who were prepared to listen that it would be impossible to use

his cameras in the way described and under those circumstances, but of course this has not been widely reported and subsequently airbrushed from history. In addition, the film used must have been a hitherto unknown and long since forgotten variety of 'super-film' designed to withstand temperature fluctuations of over 300°C and also to withstand the lethal Van Allen radiation on the way home.

Even relatively low-level radiation in airport X-ray machines has been known to totally 'wipe' conventional celluloid film.

This still shot (below) of Neil Armstrong and Buzz Aldrin planting the US flag on the Moon's surface was taken by a 16mm movie camera mounted on the lunar module. Aldrin's shadow (A) is far longer than Armstrong's and yet the only light on the Moon – and the only light source self-admittedly used by the NASA 'astro-NOTs,' was the sun, and this alone should not create such unequal shadows. It is fair to suggest therefore that all shadows should run parallel to one another and be of equal, relative length. But this is definitely not the case with many of the Moon landing videos and photographs which clearly show shadows that fall in different directions and are of relative, unequal length.

This therefore must mean that multiple light sources are present, strongly suggesting that the photos were produced on a film set. There are indeed hundreds of photos showing shadows of astronauts, flags, rocks, and other objects falling in different directions up to 90 degrees apart; this is impossible without secondary lighting, which was certainly not transported to the Moon.

NASA has attempted to blame strange shadows on the uneven landscape, with subtle bumps and hills on the Moon's surface causing the discrepancies. This explanation is frankly laughable. How could hills cause such large angular differences? But then, some people will believe anything they are told by officialdom. In the image below, the lunar module's shadow clearly contradicts those in the foreground, the differential being almost 45 degrees.

John Hamer

Source of "sunlight" is just offstage to left, according to shadows. Shadows from a faraway source like the sun should be parallel in the photo; from a close source on a stage set, the light rays would diverge as seen in this Apollo 14 photograph because the light is much closer. The shadows point to the light source.

I bet you thought the lunar sky was solid black, right? Wrong. Subtraction of yellow and blue from the chroma scale reveals that the studio lighting representing the sun is reflecting off of a background.

In consecutive Apollo 17 photos (AS17-135-20588 and 89) the rock's shadows change almost 180 degrees as if they switched studio lights between shots and in shot (AS14-64-9089) studio-lighting representing the Sun is seen reflecting off a black background, a photographic effect that could not possibly occur in the blackness of space.

The Falsification of Science

There are also several anomalies with the 'Moon rovers.' In one Apollo 17 shot (AS17-140-21370) the Moon rover is shown still packed-up, not yet unloaded, but there are clear wheel tracks to be seen across the foreground of the entire photo, (below)...

There are many pictures which show Moon rovers with no wheel tracks in front or behind them (as though they were set down into place) even though there are many footprints visible in the picture. There are also pictures of astronauts with footprints all around them, but no prints leading to or from where they are, as if they were also lowered into place by wire.

Moon rovers which leave no tracks in the moon dust

A study from more than a dozen images of LRVs WHICH LEAVE NO TRACKS in the powdery moondust to reach the location where they sit. Above, two consecutive Apollo pictures in which the rover moved forward approximately one foot (note yellow spots which show the same two footprints). The wheels leave no tracks during the move, though the footprints were left.
Furthermore, there are NO TRACKS BETWEEN/BEHIND THE TWO WHEELS, which would be mandatory IF the LRV drove to the location. The photo at left from Apollo 17 shows the TRACKLESS moondust in even sharper detail. No tracks are evident in this extreme close up. This leads us to speculate that the rover was LOWERED INTO POSITION to avoid disturbing the carefully racked moondirt.

The Falsification of Science

And there is also one Apollo 12 shot that shows the reflection of what can only be an overhead studio light...

Australian researcher Bill Dines spotted an odd reflection in this Apollo 12 helmet and thought it might be a lighting technician's spot light, above, suspended from overhead. Looks very similar to me.

Neil Armstrong took two really REMARKABLE photos of Buzz Aldrin saluting the flag during the Apollo 11 moon mission. Seen at top are the two full-frame exposures, uncropped, showing that with his chest-mounted Hasselblad WITHOUT A VIEWFINDER, Armstrong managed to crop the two transparencies with such great precision that the exposures are identical except for the slight movement by Aldrin during his "salute". The photos are so remarkable that the flag and flagpole CAST NO SHADOWS like the LEM in the background and Aldrin in the foreground. The precise angle of the sunlight is seen in the shadow of Aldrin's shoulder on his PLSS backpack. The thin shadow of the pole and the big shadow of the flag should be within the yellow box. Secret technology?

257

John Hamer

But, for a more authoritative view on the Moon photography, here are the words of my co-author and editor, Shannon Rowan, for your consideration.

"Speaking as a professional photographer for more than three decades, and one who has spent countless hours in darkrooms processing and printing from black and white and colour film, and working as camera assistant and production manager in commercial photo studios and on location for some of the top photographers in the world, the possibility of the Apollo Moon landing photos being faked really grabbed my attention in recent years.

Having read the main arguments used to expose the faked images (many of which I find interesting and valid) I also, upon reading NASA's official statements in regard to the technical information explaining the type of cameras and films used on these missions, have raised a few additional questions of my own from the perspective of a professionally trained analogue photographer.

My first question is how can film be specially protected for radiation exposure? No details are ever given as to how this was done. Film is inherently light sensitive so that photographs can be recorded via light onto its surface (aka 'emulsion'). Radiation is an electro-magnetic phenomenon, as is light. So if the film is coated to protect it from radiation on the Moon (assuming that the spaceship protected it from the Van Allen Belts), how could it retain its photo-sensitivity to record images? Regular film at higher light sensitivity (or ISO/ASA) ratings can be damaged simply by passing through airport x-ray machines, a much weaker source of radiation than allegedly exists on the moon. In the 1990s I always had to ask airport security to hand check my rolls of film in order to avoid damage, a common practice amongst photographers at the time. And NASA claims that both low and high rated films were taken on Apollo 11, including a film rated at ISO 1600 (this is the type of film which would most definitely be ruined if x-rayed, as anything rated ISO 400 and above should certainly not be x-rayed). And if this was possible, why has this type of film never been made commercially available? I am sure many photographers wishing to protect their film from airport x-rays and other radiation sources would definitely be interested in purchasing it!

The Falsification of Science

Next, and probably most importantly, how can both the cameras and film survive the extreme temperature fluctuations said to exist on the Moon's surface – over 200°F in sunlight and minus 200+°F in the shade / shadows? The only answer NASA gives is that the camera's body was painted with a special silver coating for protection for extreme temperature changes. "The outer surface of the 500EL data camera was coloured silver to help maintain more uniform internal temperatures in the violent extremes of heat and cold encountered on the lunar surface." No further explanation is given.

Simply changing the colour of the camera body from black to silver is hardly going to be enough to protect it and its temperature-sensitive contents from extreme temperature changes. Yes black will absorb more heat than silver, but the differences would be negligible, making the argument laughable. What should be 'focused' upon here is the material of which both camera and film were made.

The camera body, being metallic, would certainly be a magnificent conductor of both cold and heat. And the film base would have to consist of some type of acetate or polyester. (prior to the 1950s highly flammable nitrate films were the norm, which incidentally sometimes led to theatres burning to the ground when the projector bulb got too close to the moving film and caught it on fire). And this film base would be coated with an emulsion consisting of insoluble light-sensitive crystals suspended in a gelatine. Gelatines easily melt with the application of high heat. However NASA simply states that the film used had 'special emulsions' but does not explain any further as to what constitutes 'special,' so we do not know how this emulsion could have been changed to protect it from heat.

Absent any other substance mentioned by NASA or elsewhere as to how these cameras were protected from extreme temperatures, apart from changing the colour from black to silver, we have to assume there was no real protection, therefore the camera in the shade (as it must have been when Buzz Aldrin had his back to the sunlight and in shadow but with some other light source, presumably a flash, illuminating him) would freeze instantly and then once back in the sunlight would cook. Even if the astronauts had some kind of oven-mitt style protective gloves to handle cameras akin to hot frying pan handles, how would the film inside the camera fare in such extremes?

When processing slides or negatives the chemical concoction 'developer,' used in the first part of the process, should not be heated above 80°F (27°C). If it becomes too hot it creates an effect known as reticulation, which describes extreme grain and / or cracking, sometimes done deliberately, usually using temps between 90-100°F (32-38°C), by photographers to achieve an artistic effect. Also when undeveloped films are exposed to high temperatures, particularly colour films, they quickly lose integrity and may develop poorly, if at all. This is why film stock is stored in refrigerators before using and between exposure and processing. Try putting a roll of film in your oven at 200 degrees, then putting it in your freezer and back into the oven again and see what kind of damage is done! Bubbling and warping would be a likely outcome, even from the heat exposure alone. And extreme freezing would certainly result in cracking and breakage. Having this damage happen while in camera would also likely cause technical difficulties in moving the film through the camera in order to take the next picture. In short, it would be impossible to do so.

The metal 'leaf shutter' in a Hasselblad camera would also easily seize up in cold temperatures. This can even occur in temperatures above freezing and most likely there would have been problems with the shutter locking-up from carrying the camera around in the shadow of the lander.

And what about the glass components of the cameras? The lenses, the internal mirrors and glass plates (such as the Reseau plate between the film and camera body)? How is it that these glass parts did not crack in the extreme cold and then shatter when brought back to the extreme heat? Even if the glass were tempered there is still no way that these parts, especially the delicate optics in the lens, could have survived in conditions on the Moon.

Even NASA states on its website that the cameras used on the Apollo missions were modified by removing all forms of lubrication inside since the liquids would 'boil' on the Moon... "Lubricants used in the camera mechanisms had to either be eliminated or replaced because conventional

The Falsification of Science

lubricants would boil off in the vacuum and potentially could condense on the optical surfaces of the lenses, Reseau plate, and film."

So they admit that the temperatures on the Moon are high enough to boil lubricants, but not enough to warp or melt film?

NASA also tells us that the Hasselblad cameras used on the Apollo missions were motorised in order to make things easier on the astronauts. "The electric motor in these Hasselblads largely automated the picture taking process."

Motorisation of the film magazine would require battery power, as would the internal flash supposedly used for some of the close-up imagery of the Moon's surface taken with a newly commissioned Kodak Stereo Close-Up camera. And batteries, as most of us know, are also damaged by extreme temperatures. In cold conditions, even just above freezing, batteries lose their functionality. During my many years as a commercial photographer and photography assistant, I had to often hold batteries under my armpits to keep them warm during cold weather outdoor location photo shoots because of this issue.

The reverse problem of course would be batteries overheating, melting, and possibly exploding. (I was once the unfortunate victim of a laptop computer fire when my laptop batteries overheated (something known as 'thermal runaway') and caused them to explode and damage my apartment and leave me with lung damage from toxic smoke inhalation.

A light meter would also require batteries, whether inside the camera itself or used externally (a detail we are, yet again, not given by NASA). How were these batteries also, prevented from both freezing and exploding?

And as for the actual operation of the cameras, I have some important technical questions to pose. Allegedly the astronauts' only task while taking photos on the Moon was to set the aperture and shutter and then use an extension cord shutter release to take the pictures. Per NASA's website, "The astronauts needed only to set the distance, lens aperture, and shutter speed, but once the release button was pressed, the camera exposed and wound the film and tensioned the shutter."

Although as any seasoned photographer knows, setting aperture and shutter speed manually is no easy task and requires the aid of a light meter of some kind, whether internal to the camera (necessitating looking into a viewfinder, something not accessible to the astronauts with the cameras strapped to their chests and with their large helmets in the way) or external. And with these exposure readings one would then need to make the proper calculations in order to get the perfectly exposed images we see from the Moon landings. But as I already stated, no mention of any kind of light meters is made by NASA.

And holding and operating a light meter with large, pressurised gloves would also be extremely challenging as would setting aperture and shutter speed even with the larger levers' modifications (From NASA's website: "Modifications to the cameras included special large locks for the film magazines and levers on the f-stop and distance settings on the lenses. These modifications facilitated the camera's use by the crew operating with pressurized suits and gloves.")

Their specially-modified cameras also did not have the reflex viewfinder that could be looked down into from on top of the camera with which Hasselblad's are normally equipped, but instead had a simple sighting ring for composing shots. Again from NASA's official description, "Additionally, the cameras had no reflex mirror viewfinder and instead a simple sighting ring assisted the astronaut in pointing the camera."

However where this sighting ring is placed is not evident and cannot be seen on the camera itself in any of the photographs.

The Falsification of Science

Underwater photographers often use these kinds of sighting rings since they cannot easily look through small viewfinders through their diving masks. However, these sights are attached to the top of the cameras and held up in front of the photographer's face. Where are these sighting rings used by the Apollo photographers? The only further explanation NASA provides as to how the images were composed is the following, "The astronaut would point his body in order to aim the cameras." Please go ahead and try this at home and see how well your photographs turn out!

Further questions I have are related to lighting on the Moon. Namely, what visual effect does the strength of the Sun's light have with no atmosphere on the Moon? Atmosphere affects light here on Earth and how we perceive it visually, so the total lack thereof on the Moon would change the nature of sunlight, significantly. The assumption would be that without the filtration effect of atmosphere, the Sun's rays would be more powerful, creating a much harsher lighting effect. This would necessarily also affect the settings needed to properly expose film and the type of film used. The least, not the most, light sensitive film would be required and highest aperture (letting in the least amount of light through the lens) and highest shutter speeds ($1/100^{th}$ of a second listed as the highest on those cameras would probably not be enough). Most likely dark 'neutral density' filters would need to be placed on the camera lenses, similar to the dark shields in the astronaut helmets, needed to protect the wearer's eyes. But no mention of these density filters is made.

And no mention of the astronaut photographers having to contend with these very alien lighting conditions is mentioned anywhere by NASA. Also, why would high-speed film be needed (1600 ASA was one of the types mentioned as part of their film arsenal) in such strong light or even flash photography for that matter (unless used to illuminate images taken in shadows, but not needed for close-ups of lunar rocks lighted by the Sun, as we are told was main application of flash photography on the Moon).

Astronauts when turned away from the sunlight, standing in shadow, are still lit up in the official photographs, as if a reflector or some other professional light source was used. The shot of the astronaut descending the steps from the lander is evenly exposed by an artificial light source inside the shadow of the lander, which would have been in absolute pitch-black shadow in contrast with the bright sunlight exposure in the background, given how high contrast lighting must be without atmosphere. Yet not only is the astronaut evenly lit, but assuming a flash was used for light, it should have reflected a hot spot somewhere on the extremely reflective material of the lander which is also quite well-lit but in my opinion by a different light source which seems to indicate studio lighting instead of flash photography.

Also the 'Earth-rise' shot from the Moon's surface is a technical impossibility. Firstly it should be noted that all of the Apollo missions took place during lunar daytime. This is because a Moon 'day' is equivalent to 27.3 days on Earth as it takes that many days for the Moon to rotate once on its own axis. This does actually explain why the astronauts if only having spent 3 Earth days on the moon during lunar daytime could not have either viewed or photographed stars due to the constant presence of harsh sunlight. Likewise, by the same argument, they could not have witnessed the Earth lit up by the sun during a lunar daytime. In this famous photograph (below) the Earth appears to be nearly 'full' and very much well lit. And even if they had been on the Moon during a lunar night, it would not be photographically possible to capture both a well-lit foreground (the Moon's surface) and have a well-lit Earth as the surface of the Moon would necessarily be in complete darkness and this type of exposure would

be impossible, at least without the aid of special studio lighting equipment and not using simple flash photography. Additionally the only flash we were told the astronauts had with them was the internal one in the close-up camera. Hasselblad's are not equipped with flashes, so any flash would have to be handheld and external to the camera. No such flash is mentioned by NASA.

NASA also claims they engineered a special film that was thinner than normal to fit more photos in the film magazine. But how would the film, both thinner and also specially coated with some unknown secret substance specifically protect it from radiation damage?

Also thinner film would make it even more vulnerable to damage by temperature changes and also cause it to be more delicate as far as winding through the camera is concerned.

But I also question the veracity of the statement that this 'special' film allowed for 160 shots per colour roll and 200 shots per black and white roll. Normally the film magazines for Hasselblad cameras can hold at maximum up to 24 images (for a 70mm camera which is the format allegedly used by the Apollo astronauts). We are told the film magazines (the back removable portion of the camera which holds the roll of film) were specially modified to hold these larger rolls of film, of a type of film, which was also thinner than usual (how much thinner is not stated). But looking at the photographs of these special cameras, it does not show a magazine noticeably bigger than a standard Hasselblad.

Hasselblad 500 EL/M (1982) w/ Carl Zeiss Sonnar 250mm f/5.6
Image by Süleyman Demir (Image rights)

Astronaut Michael Collins used this 70mm Hasselblad camera in the command module while Neil A. Armstrong and Buzz Aldrin used the two film magazines during their Apollo 11 moonwalk.

And even if this was all true and they had been able to engineer this uniquely thin film and film rolls capable of storing that many images as well as a magazine to hold these special rolls, why then were these incredible advances in photography withheld from the public and (again) not manufactured commercially? At least three decades after the fact, all of the professional photographers I knew and worked with had to continue using Hasselblad cameras only capable of recording 12 to 24 images per roll. It would have been extremely handy for we photographers to be able to shoot

The Falsification of Science

160 to 200 photos without having to manually reload new magazine backs (something that was not an easy task with those particular cameras and actually required special training to learn, as Hasselblads were not in fact 'simple to use' cameras as NASA also claims).

And what of the effect of static electricity on film? It must have been a very serious problem to contend with, given the total lack of atmosphere on the Moon, which would necessitate an extremely dry environment, one that must produce incredible amounts of static electricity from the smallest of movements. While living in Arizona, in even that arid, Earth based climate, I often would shock myself when opening metal car doors, even while touching the plastic door handle and wearing rubber soled shoes. Film is extremely sensitive to static charges and easily damaged by them, and especially so the thinner the film. Once I had a roll of important film completely ruined because I had been kneeling on a carpet and moving along, rubbing it with my knees, while taking the photographs. The amount of static charge created in the process was enough to cause bizarre light streaks all over my film, ruining all of the images.

NASA, likely expecting this kind of question, has its rote response in readiness:

"When film is normally wound in a camera, static electricity is generated on the film surface. This electricity is dispersed by metal rims and rollers, which guide the film, and by humidity in the surrounding air. In the lunar surface camera, however, the film was guided by the Reseau plate's raised edges. As glass is a poor electrical conductor, and with the absence of surrounding air, the charge built up between the glass surface and the film could become so great that sparks could occur between the plate and the film. In order to conduct the static electricity away and prevent sparking, the side of the plate facing the film was coated with a thin transparent conductive layer and silver deposited on the edges of the conductive layer. The electrical charge was then led to the metallic parts of the camera body by contact springs."

However this description only pertains to a static charge that can be generated by the camera itself in the act of advancing film through it and does not address external charge potentials, which would surely be in abundance on the Moon, given all the facts we are told about that environment. Even the astronauts touching the cameras while moonwalking could be enough to cause damage. And this particular modification admittedly also causes the metallic parts of the camera (its entire body and the magazine housing the film) to carry even more static charge than normal. And how would the 'metallic parts,' which are also very close to the film, having a charge, protect the film from this type of damage?

Another suspicious claim made by NASA is that the Apollo 11 mission resulted in 33 rolls of exposed film. It should be noted that 33 is a number that carries special meaning to Freemasons (33rd degree being the highest achievable level within the organisation) as does 11, signifying the 'master number.'

Yet despite the fact that NASA claims 33 rolls of film, each reportedly capable of 160-200 exposures, resulting in a potential 5,280-6,600 images, the public has only ever been able to view a handful of these thousands. And currently there are rumours circling the Internet claiming that NASA recently released 10,000 plus images of the Apollo Moon mission, because they got 'sick of all the conspiracy theories,' with one such article merely showing 14 total images as further 'proof' of the landing, half of which we have all already seen, ad-nauseum. If NASA really has been releasing never-before-seen archived Moon landing images, where are all of these thousands of images?

And if there are new ones in circulation, why were they not shown to us before now and how are we to believe in their authenticity given the current level of sophistication regarding the creation of fake photographs and which technology is accessible to the average person today, unheard of during the Apollo Missions. And why have NASA and the rest of the mainstream suddenly put so much energy into debunking 'faked Moon-landing' claims? With online censorship growing at an alarming rate in 2020, it is almost impossible to find any of the many previously readily accessible Moon-landing hoax websites. Instead, the search engines are littered with endless Moon-landing hoax debunking sites. I personally sampled many of these sites and noted how misrepresented the arguments against NASA's claims were, as well as how absurd were the so-called 'factual answers.'

The Falsification of Science

Finally it is yet another very convenient fact, for NASA that is, that this very specially modified Hasselblad camera was left behind on the Moon's surface by the astronauts such that no one is now able to inspect it after the fact. 'The camera and lens were left behind and still rest on the Moon's surface at Tranquility Base.' Apparently this was done to 'ensure the safe take-off from the Moon.' As if that tiny bit of extra weight removal was that essential??

As always we will just have to take NASA at their authoritative word. However they are not authorities on the subject of photography, but I happen to be, so I will not for one, accept their, frankly lame, technical explanations of how the Moon photographs were achieved."

Well, that just about comprehensively deals with any remaining doubt as to the veracity of NASA's photographs, allegedly taken 'on the Moon!' Thank you, Shannon!

Many people have pointed-out that when the first Moon landing was shown on live television, viewers could clearly see the American flag waving and fluttering as Neil Armstrong and Buzz Aldrin planted it. Photos of the landing also seem to show rippling in a breeze, such as the image above which definitely shows a fold in the flag. The obvious problem here is that there is no air in the Moon's atmosphere, and therefore no wind to cause the flag to ripple.

Many explanations have been put forward to disprove this phenomenon as anything unusual. For example, NASA claimed that the flag was stored in a thin tube and the rippled effect was caused by it being unfurled before being planted. Other explanations involve the ripples caused by the reaction force of the astronauts touching the aluminium pole, which is shown to shake in the video footage. As always NASA's explanations are nothing if not preposterous bordering on utterly laughable, at times.

If you have never seen the footage of the Apollo 11 astronauts' press conference, the DVD is available to buy at a very reasonable price on the Internet and I would strongly suggest that you track it down and do so. What is so striking and revealing about this is the absolute downbeat demeanour of the astronauts themselves throughout the entire session.

If someone had just completed the most wonderfully uplifting experience and had been on the most incredible adventure ever undertaken by the human race in its entire history, would I be wrong to suggest that they may have appeared happy, elated and exhilarated, flushed with success, even self-satisfied, experiencing a feeling of great achievement that they would wish to share with the world? Obviously, someone forgot to tell them this, in that case. I have never seen a more morose, sullen, disinterested, less cooperative bunch of people in my entire life. Anyone would have thought that they did not really go to the Moon at all and were resentful of being 'put on the spot' and having to 'think on their feet' to answer all the awkward, unplanned-for questions they were being asked. Oh, wait a second...

And of the seven Apollo missions to put 'men on the Moon,' six were claimed to be 'successful' and Apollo 13 was 'aborted.'

As with the many other 'political' assassinations in other spheres, there are also several highly suspect, and very convenient deaths involved in the Apollo project. None more so than...

On 27th January 1967, the astronaut Gus Grissom along with two fellow astronauts, perished in a fire in what was to be Apollo 1 at Cape Canaveral, Florida. Even in death, Grissom is still tormented by those who would believe that he did not live up to his country's expectations, not to mention his own expectations.

On 13th April 1959, Air Force Captain Gus Grissom received official word from the Air Force Institute of Technology at Wright-Patterson Air Force base that he had been selected as one of the seven Project Mercury astronauts. Six others received the same notification; Lieutenant Malcolm Scott Carpenter, US Navy Captain, Leroy Gordon Cooper, Jr. (33rd Degree Mason,) US Air Force Lieutenant-Colonel John Herschel Glenn, Jr. (33rd Degree Mason,) US Marine Corps Lieutenant Commander Walter Marty Schirra, Jr.

The Falsification of Science

(33rd Degree Mason,) US Navy Lieutenant Commander Alan Bartlett Shepard, Jr., and US Navy Captain Donald Kent (Deke) Slayton.

The Mercury programme was the United States' first manned space venture and the first step in the country's quest to reach the Moon. Being a Mercury Astronaut meant being constantly subjected to scrutiny both by the public and by the media. Hailed as heroes when the missions went well, they could equally expect to be chastised for any problems which might occur along the way. Indeed, Grissom's first space flight, aboard the Mercury Redstone *'Liberty Bell 7,'* was somewhat less than a complete success.

Although take-off and re-entry went as expected, upon splashdown in the Atlantic Ocean, the seventy explosive bolts which held the hatch in place unexplainably exploded prematurely, forcing Grissom to evacuate the capsule and swim for his life while the rescue helicopter frantically tried to save the capsule from sinking. It was unsuccessful, and Grissom nearly drowned whilst the Liberty Bell sank to the bottom of the ocean, never to be recovered.

This was only the beginning of Grissom's media woes. The press hounded him mercilessly, and he underwent an inquiry by NASA into the loss of the spacecraft. Although his fellow astronauts strongly supported him, and the inquiry led to the eventual conclusion that the explosive hatch blew of its own accord, Grissom never recovered his previous reputation with the public or the media. Much to his dismay, the exhaustive testing done on similar spacecraft with the same explosive bolts yielded no explanations for the explosions, as the bolts, never again exploded prematurely in any of the tests.

"I didn't do anything. I was just lying there, and it just blew," he protested. However, the media painted him as a failure, a coward who panicked and blew the hatch in an attack of claustrophobia. Even after his death, perhaps because he was an easy target and could not defend himself, the public opinion of Grissom was still questionable. In the book and the movie, *'The Right Stuff,'* Tom Wolfe portrayed Gus as, *"…the goat among the astronauts, a hard-drinking, hard-living type who courts the favours of barmaids with gewgaws he promises to*

carry into space. He is also held up to the world as a man who screwed up, who panicked, blew the explosive hatch off his capsule and allowed it to sink to the ocean floor after re-entry."

However, eventually some evidence did come to light, that the explosive device on the hatch *could* accidentally blow without being pulled – a fact that led NASA to remove the devices from future spacecraft designs. Also, had Grissom pulled the explosive release on the hatch, his hand or arm should have had powder and bruise marks, but neither were ever found.

Grissom's luck was better on his second space flight, however. Selected to command the Gemini 3 mission shortly after his completion of the Mercury 7 flight, he became the first man ever to fly twice in space. But he had yet to live down his reputation with the loss of his Liberty Bell capsule.

After his second, successful flight, he was notified confidentially by NASA that he had been chosen to be the first man to set foot on the Moon. Project Apollo was now underway, and Grissom with Edward H. White and Roger Chaffee, were selected for the first mission.

The Apollo I mission was undertaking a simulated launch in preparation for an actual lunar flight, when a fire stated to be caused by an arc of electricity in the pure oxygen atmosphere of the sealed capsule of Apollo 1, destroyed the capsule and incinerated all three astronauts. A tragic end to the career of Gus Grissom, who even in death, was pilloried by both the media and Congress.

Dr John McCarthy, the director of research, engineering, and testing for *North American Aviation*, the aerospace company primarily responsible for building the Apollo capsule, laid the blame (literally) at Grissom's feet. His hypothesis was that the command pilot may have kicked or scuffed a wire lead connected to an air-sampling instrument.

The ignition source was never determined, but there were over a dozen fire hazards in the module, and the module's design was grossly flawed in terms of safety. Most notable of these hazards was the inward-opening exit hatch. There was very little room inside, and in the panic of fire, the crew would have been severely hampered in exiting had they opened it. This is a moot point however, as the hatch would not even open with a crowbar after the module was pressurised. The interior was supplied with 100% medical-grade oxygen during the pre-launch, and thus, the air itself instantly burst into flames once the fire had started, blowing out one of the cabin's walls within 15 seconds. The three men did not die from smoke inhalation, since they were fully suited with their helmets on, but were killed by the fire itself.

As unfair as McCarthy's observation obviously was, it was rescinded almost immediately when he could not find proof to support his hypothesis. Grissom would have had to be a contortionist to have reached that particular wire and create enough force to move it.

NASA eventually concluded that the Apollo I deaths of Grissom, White and Chafee, were the result of an explosive fire that burst from the pure oxygen atmosphere of the space capsule. NASA investigators could not identify what caused the spark but wrote-off the catastrophe as an accident.

"My father's death was no accident. He was murdered." Scott Grissom, Gus's son

In February 1999, Scott Grissom, went public with the family's long-held belief that their father was purposely killed by fire aboard Apollo I. They believed that the numerous safety flaws were so grossly negligent that NASA could not possibly have overlooked them all, but instead deliberately provided them, and then sabotaged the equipment in some way so as to ensure an electrical spark once the cabin was sealed and pressurised.

Scott Grissom strongly suspected that his father had somehow irritated the NASA hierarchy in the past, possibly due to the embarrassment of the Liberty Bell 7 incident, and his general outspokenness. Scott claims that in 1990, he was able to inspect the Apollo 1 command module and found a *"fabricated metal plate"* behind a switch on one of the instrument panels. The switch controlled the capsules' electrical power from an outside source to the ship's batteries and he argued that it was the placement of this metal plate that was an act of sabotage. He claimed that when one of the astronauts toggled the switch to transfer power to the ship's batteries, a spark was created that ignited a fireball.

NASA denounced the younger Grissom's conclusion as "*...the ravings of an understandably angry child,*" but has never even attempted to deny his claims. Whether such a metal plate existed in the cabin is not publicly known, but in another stunning development, a leading NASA investigator charged that the agency engaged in a cover-up of the true cause of the catastrophe that killed Grissom and the two other astronauts.

Clark MacDonald, a *McDonnell-Douglas* engineer hired by NASA to investigate the fire, offered corroborating evidence. Breaking more than three decades of silence, MacDonald said that he determined that an electrical short caused by the changeover to battery power had sparked the fire. He also claimed that NASA destroyed his report and interview tapes in an effort to stem public criticism of the space programme. "*I have agonised for 31 years about revealing the truth, but I didn't want to hurt NASA's image or cause trouble, but I can't let one more day go by without the truth being known,*" he said.

Grissom's widow, Betty agreed with her son's claim that her husband had been murdered... "*I believe Scott has found the key piece of evidence to prove NASA knew all along what really happened but covered up to protect funding for the race to the Moon.*"

Grissom was the senior astronaut and also maybe significantly, the most critical of the problem-plagued Apollo programme, and the main Apollo contractor, *North American Aviation*. With billions of dollars at stake, Grissom had become a problem for NASA.

On one occasion, shortly before the tragic fire that claimed their lives, Grissom hung a lemon on a wire coat hanger on the Apollo 1 rocket (picture below) during a publicity photo-shoot and in addition made an unauthorised statement to the press in early January 1967 to the effect that he believed that the 'Moon landings' were at least 'a decade away,' for which he was severely reprimanded.

Grissom's lemon

That was probably his death sentence signed and sealed right there and then. Indeed, less than a month later all three 'rebels' were dead. Shortly before his untimely death Gus Grissom had also said to his wife... "*If there is ever a serious accident in the space programme, it's likely to be me.*" And among his last words before he died, when there was a communications failure with the capsule just prior to the fire, were, "*How are we going to get to the Moon when we can't communicate between two buildings?*"

Even before Apollo I, Grissom had received death threats which his family believed emanated from within the space programme. The threats were serious enough that he was put under Secret Service protection and had been moved from his own home to a secure safehouse.

The Apollo I disaster led to a series of congressional hearings into the incident and NASA. During the hearings, one launch pad inspector, Thomas Ronald Baron, sharply criticised NASA's handling of the incident and testified that the astronauts attempted to escape the capsule earlier than officially claimed. Baron was a safety inspector on the Apollo project who after the fire, testified before Congress that the Apollo programme was in such disarray that the United States would never make it to the Moon. He also claimed that his opinions made him a target, and on 21st April 1967, reported on-camera to news reporters that he and his wife had been harassed at home. As part of his testimony, Baron also submitted a 500-page report detailing his findings. Then exactly one week after he testified, Baron's car was struck by a train and he, his wife and his stepdaughter were all killed instantly. His report mysteriously disappeared, and to this day it has never been found.

In fact, no less than eleven Apollo astronauts were mysteriously killed before undertaking their missions, three had oxygen pumped into their test capsule until it exploded, seven died in six separate (yes, six separate) plane crashes, and one died in a car crash. Overall, a highly unlikely series of coincidences that I will allow the reader to make-up his or her own mind upon.

Paul Jacobs, a private investigator from San Francisco, at the request of Bill Kaysing, interviewed the head of the US Department of Geology in Washington about the so-called 'Moon rocks.' *"Did you examine the 'Moon rocks,' did they really come from the Moon?"* Jacobs asked and the geologist's only response was to laugh. Both Paul Jacobs *and* his wife died from 'cancer' within 90 days of this incident. Yes both, within 90 days!

Astronaut and ahem, 'first man to walk on the Moon,' Neil Armstrong, could have made millions through endorsements and personal appearances. Courted by kings, praised by politicians, he could have signed a myriad of book and film deals telling his amazing story. But instead, he became a virtual recluse who used a fake name to receive his mail and refused to talk about his incredible feat. He also suffered with mental illness in his later years, some say maybe as a direct result of his name being used as the 'foundation stone' for the biggest lie in history. Or maybe it was that he became distressed and even paranoid by the overwhelming number of sources exposing him as a liar and a fraud?

In fact none of the astronauts give public interviews or took questions at speaking events and as already stated were all very unconvincing, nervous, and shifty in their first press conference upon their return from the Moon as previously related. It was certainly still available as a mail-order DVD on the Internet at the time of writing this book (2020).

The Falsification of Science

Rumour also has it, that Apollo 12 astronaut Charles (Pete) Conrad Jr., was intending to 'go public' about the fake Moon landings on the 30th anniversary of the non-event back in July 1999. He was however unfortunately killed in a motorcycle accident one week before the anniversary. Another sad 'coincidence.'

President Lyndon Johnson made certain Apollo files classified, with a declassification date of 2026. This is so that those involved in the Apollo scam would be long dead and gone, but we need not wait until then for the truth behind Apollo, as the truth is now already well known.

"It is commonly believed that man will fly directly from the Earth to the Moon, but to do this, we would require a vehicle of such gigantic proportions that it would prove an economic impossibility. It would have to develop sufficient speed to penetrate the atmosphere and overcome the Earth's gravity and having travelled all the way to the Moon, it must still have enough fuel to land safely and make the return trip to Earth. Furthermore, in order to give the expedition a margin of safety, we would not use one ship alone, but a minimum of three ... each rocket ship would be taller than New York's Empire State Building [almost ¼ mile high] and weigh about ten times the tonnage of the Queen Mary, or some 800,000 tons." 'Conquest of the Moon,' Wernher von Braun, 1953, Viking Press

Of course, Wernher von Braun obviously quickly forgot this opinion completely, when his 'sponsors,' the CIA, had put him in overall charge of the 'team,' in the race to outdo the Soviets in space and weapons research, not to mention make a whole boatload of money for the elite and their greedy friends.

But on 5th September 2015 Buzz Aldrin cryptically 'admitted' that he never went to the Moon.

He was interviewed at the National Book Festival in Washington DC, by an 8-year old girl named Zoey. In response to a question from Zoey, Aldrin admitted that the reason no-one has returned to the Moon since the Apollo missions is that NASA did not go there in the first place. Aldrin, who has a long history of alcohol abuse, sounded very intoxicated during the interview and in that state of inebriation, inadvertently 'let the cat out of the bag.'

"Why has nobody been to the Moon in such a long time?" Zoey asked him.

Aldrin replied, *"That's not an 8-year old's question. That's my question. I want to know. But I think I know because we didn't go there. And that's the way it happened. And if it didn't happen, it's nice to know why it didn't happen. So in the future, if we want to keep doing something, we need to know why something stopped in the past that we wanted to keep it going."*

Hmm. Definitely the inane ramblings of an inveterate drunk, but the question remains, did he actually know himself, what he was saying?

It was an American, Dr Robert H. Goddard (1882-1945) the rocketry genius, who was the father of the space programme. A man of great vision and humanity, he was ignored by the US government, only to see his revolutionary research and patents put to another use by Wernher von Braun and his team.

So, how and where did NASA fake the lunar approach, lunar orbit, lunar landing, and lunar take-off, for all the Apollo Moon landing TV transmissions? Contrary to what many believe, the sequences were not shot in a desert, Hollywood studio, or Area 51. There may well have been some pictures taken at the infamous Area 51, and a few Apollo pictures that were taken in some remote desert, but the majority of stills and video were created at NASA's own *Langley Research Center* in Hampton, Virginia. NASA Scientists knew in the early 1960s that a manned mission to the Moon was impossible within 8 years, and so the plan to fake the Moon landings was put into operation.

The Lunar Orbiter and Landing Approach simulator (LOLA,) was a $2 million facility built at the *Langley Research Center* around 1963/64. The lunar terrain detail within it was gathered by the American Lunar Orbiter series of probes that mapped the entire surface of the Moon in greater detail than ever before.

The Falsification of Science

This facility had an overhead crane structure about 250 feet tall and 400 feet long and the massive crane system supported five-sixths of the vehicle's weight through servo-driven vertical cables. The remaining one-sixth of the vehicle weight pulled the vehicle downward simulating the weak lunar gravitational force. During actual flights, the overhead crane system was 'slaved' to keep the cable near vertical at all times. A ball-joint system on the vehicle permitted angular movement for pitch, roll, and yaw.

According to Bobby Braun and other NASA officials, the idea was to teach the astronauts how to land a rocket propelled lunar module. However no rocket powered LM was ever suspended from this crane and in any case, most people know that it is totally IMPOSSIBLE to control a rocket engine. The landings were controlled purely by traverse and the raising and lowering of the LM in the same way as a conventional crane.

The mock LM was traversed the full length of the crane, and simultaneously lowered at the same time in order to create an authentic looking lunar landing, when viewed from within the mock LM itself. The power supply to the mock LM was by cable from the crane tower. This enabled a large fan, (fitted beneath the mock LM) to create the dust-scatter effect of a rocket engine as it descended to the fake Moon surface.

The film shown to public of the LM supposedly blasting off from the Moon's surface was also created beneath this crane at LRC. The mock LM was simply attached to the crane and hoisted very rapidly at the same time a pathetic looking blast-off simulation was enacted beneath it. The film was then speeded-up for showing to the public, and it is interesting to note that the camera filming this sequence cut short once the LM had reached the crane maximum height. In other words WHY did the camera not continue to film the LM until it was out of view? Quite simply because it was not possible to do so under the circumstances in which the 'lift-off' was faked.

Fake lunar surfaces beneath the crane

The above pictures were taken by Bob Nye on 20th June 1969, one month before Armstrong supposedly made his 'giant leap for mankind.' The picture on the right shows the lander hovering above the fake Moon crater surface beneath the crane. The picture on the left, taken at night, looks like a realistic Moon setting, and the light source seen in left picture is the same light source that highlights Buzz Aldrin in the controversial picture of him allegedly on the Moon. Those lights are fixed on top of the crane gantry, as shown in an earlier picture.

The pictures below demonstrate how the astronauts were suspended from the crane in order to simulate low gravity.

They eventually settled for an upright position with the astronaut suspended by strong elastic bungee cord, so that his feet were only just touching the ground, similar to a 'baby bouncer.' As the astronauts walked in any given direction, the overhead crane moved in the same direction and this enabled the astronauts to literally float along in a crude 'Moon walk' fashion.

The Falsification of Science

There is a classic piece of film that depicts Apollo 17 astronauts supposedly cavorting on the Moon and one of them is actually suspended 2 feet horizontally off the ground. This sequence lasts for a couple of seconds, so how do NASA officials explain that and why is it that no-one else has passed comment on this patently absurd shot? It is clear evidence that the astro-NOT in a fake space suit is suspended from wires.

The picture above left, depicts a 20-foot diameter sphere which can be rotated from below. In the left of that picture can be seen a huge blank placard. This is the scene before work began on converting the sphere to an authentic looking Moon complete with craters, (for lunar approach,) and the placards were to be used to depict the orbiting of the Moon.

Also notice the rail track around the placards, (there were 3 placards in all) and note the moving trolley gantry on the track upon which the camera was mounted. Firstly, it began to film the rotating sphere, (lunar approach) and then would swing around and begin scanning the fake lunar surface on the placard to simulate a lunar orbit. The picture above right depicts the sphere after modelling work. Notice how the background is dark with no stars visible. Remove that man from the picture and it could EASILY pass as being taken by the Apollo command module circling the Moon. It is absolutely evident that there were a number of people involved in the faking of Apollo but NASA's response to these allegations was that if it were faked, someone would have 'blown the whistle.' Well, the NASA staff involved are all sworn to secrecy, and they surely would prefer that they and their wives and children and other family members remain alive. There is no doubt that the threats used to buy their silence would have been taken extremely seriously.

NASA claim that picture on the right is the far side of Moon, taken by Apollo 8. Compare this sphere with the one shown on the left. It speaks for itself, does it not? In all of these pictures notice the black background. We really need to ask the question also... 'What possible purpose could these no doubt incredibly expensive to produce, lifelike models of the Moon have served, if not to fake mission photographs?' What would have been the point of all the time and expense taken to construct them? I have given this a lot of thought and I personally cannot think of even one single, plausible reason for it.

The Falsification of Science

But in fact, the vast majority of NASA's fake Moon pictures were created in the mid-1990s. The proof lies in the fact that most do not appear in any books or magazines prior to 1990. Ninety-five percent of NASA's fake Moon pictures on their websites, *were never seen in any form, prior to the launch of the Internet.* Once the Internet became widely available, they had to produce a considerable number of fake Moon pictures, for all six missions; otherwise the public would doubtless demand to know why there were so few.

Science fiction's portrayal of covered-up or faked space missions dates back many decades. In the February 1955 issue of *Galaxy Science Fiction* magazine, author James Gunn published a story entitled '*The Cave of Night.*' The story dealt with a manned mission to Mars which goes wrong, stranding an astronaut with no hope of rescue and the climax of the story is shocking, utilising the notion of fakery to portray an erroneous perception of the outcome of the mission.

The plot of the 1969 movie *Marooned* also involved a manned mission to the Moon going wrong. The failure of a re-entry rocket leaves the occupants of the lunar capsule stranded in space and although there is no cover-up inherent to the actual space-flight, the original script called for the suggestion that a story would be created to perpetuate the notion of a heroic attempt to rescue the astronauts, should they have perished. The film received the full support of NASA, including the use of Cape Kennedy for interior and exterior location filming.

Capricorn One (1978) however, went much further than *Marooned*, featuring a plot that utilised Hollywood trickery and gimmicks to fake the first manned spaceflight to Mars. In the film, the astronaut crew are removed from their rocket and driven to a film set in the desert to record fake footage of their planetary touchdown. However things spiral out of control for the hapless, fake astronauts when their (empty) rocket is seen to explode on take-off, in front of millions on live TV, thus meaning that NASA now has three 'dead' astronauts on its hands. Of course they realise that their lives really are in danger now and attempt to escape before the space agency can deal with them in the usual manner. Without creating too much of a 'spoiler,' this is a very watchable production, complete with a pleasing twist at the end.

Bizarrely, this film also received full support from NASA, which is strange given how NASA has generally avoided supporting Hollywood productions that cast the agency (or fictional agencies with a resemblance to NASA) in an unflattering light. (more 'revelation of the method' perchance?) The film was directed by Peter Hyams, who would go on to also direct *'2010: The Year We Make Contact,'* the sequel to Kubrick's *'2001: A Space Odyssey,'* six years later.

According to the *Clavius* website, the first mention of Stanley Kubrick and his possible involvement with the Apollo cover-up appeared in 1995 on the *Usenet* newsgroup. The Clavius group have dedicated themselves to sceptically debunking all notions of an Apollo cover-up (meaning that they are not exactly the most unbiased or objective source of reference material). However the timeframe to which they refer, does seem to tally. There are no apparent references to Kubrick and the Apollo cover-up before this time period.

As documented earlier, Hollywood has always served its elite masters well. From everything we know about the truly great filmmaker Stanley Kubrick, he was certainly not what could be called a Hollywood 'team player.' In fact, the total opposite is nearer the mark. Fiercely independent, he was a perfectionist who moved over five thousand miles away from the centre of the movie universe, just to be free to do things his own way. But, given his love of new technology, creative challenges, and absolute control, the idea that NASA would be 'over the Moon' as it were, to have him onboard to help them fake the Apollo Moon landings... well, it would be the perfect match, the perfect challenge, if it were true.

In October 2002, the William Karel-directed *'Dark Side of the Moon'* mockumentary film was aired on the French TV channel *Arte*. The film seemingly supported the idea that the television footage of the Apollo 11 Moon landing was faked and recorded in a studio by the CIA, with help from Stanley Kubrick. The film included interviews with notable 'elite' figures such as Donald Rumsfeld, Henry Kissinger, Alexander Haig, Vernon Walters and Apollo astronaut, Buzz Aldrin. However, further investigation of these interviews revealed that they were actually carefully edited and blended from existing interviews that had no connection to Kubrick or the faking of the Apollo 11 footage.

The official blurb of the film states... *"Filmmaker William Karel pursues his reflection on the relation of the United States with image, cinema and their capacity to produce 'show.' What other story can lend itself to such an examination but the space conquest, a war of image and show more than anything else. What if it was just a huge hoax initiated by the two great powers? Between lies and truths, this film mixes actual facts and others, completely trumped-up. Playing with irony and lie, its purpose is to entertain and raise the question of the use of archive, which can be made to tell whatever you want."*

Whatever one makes of the mockumentary, there is a certain irony to this exercise in contextual dissembling, given that this is something the mainstream media appears to do on a daily basis. Tellingly, the film exercises techniques that have been the hallmark of certain media 'psy-ops,' such as Orson Welles' 1938 *War of the Worlds*, radio broadcast.

It is curious that the film also contains interviews with Kubrick's widow Christiane, who discusses the Kubrick/Apollo connection in a far more realistic context than the other featured interviews. We really should ask why she became involved. The most interesting and detailed parts of the film are contained in her interviews and if, as some have suggested, the film was a CIA exercise in debunking, was Christiane a willing participant?

The 'Stanley Kubrick and the Moon Hoax' article and the *Dark Side of the Moon* mockumentary have done much to muddle the notion of Kubrick's possible involvement in the Apollo coverup. But, if Kubrick's involvement was real, then these pieces are certainly convenient for those 'elite' players wishing to prevent researchers from getting too close to the truth.

It is also worth mentioning the theory that *'2001: A Space Odyssey'* was also utilised as part of the agenda-driven drug-related counterculture of the late 1960s. It is important to note that this phenomenon has a clear overlap with the work of NASA and the state-sponsored mind-control paradigm, respectively.

For all the examination of the 'Stanley Kubrick and the Moon Hoax' article and the *Dark Side of the Moon* mockumentary, please understand that a great deal of strong circumstantial evidence existed long before the notion was ever fully articulated. For example, Jay Weidner, a researcher and expert photographer who has virtually dedicated his studies to Stanley Kubrick and the global, New World Order agenda, has plausibly demonstrated that the front-projection process (used so successfully in the 'dawn of man' sequences of Kubrick's *2001: A Space Odyssey*) shares key similarities with some of the abnormalities identified in the Apollo 'Moon' footage, such as the clear lines of definition between the rough foreground and the smooth background.

The Falsification of Science

In order to understand Kubrick's connection with the elite and the military-industrial complex, one must look back at his career as a whole. Kubrick is remembered as being a notorious perfectionist with a meticulous attention to detail and the production of his films was often laboured and incredibly lengthy. Actors were pushed to the limits of their ability and patience, as often scores of 'takes' of one short scene were repeated until Kubrick was satisfied with the result. It is also known that there was no piece of set dressing or background that hadn't been placed or framed without his prior approval or specific reasoning and the same is also said of his wardrobe choices and actors' appearances. This should come as no surprise, given that Kubrick began his career as a still photographer – an art that requires a precise knowledge of framing and context in order to be fully proficient.

Some researchers have claimed that Kubrick became trapped within his profession and that his art became a conduit through which he used illusion and imagery to reveal the greater truths that he had come to realise. It is also claimed that his alleged involvement in the Apollo hoax would have aided and abetted the deceptions. However, Kubrick's disdain for aspects of the establishment was already apparent much earlier in his career.

In February 1993, the legendary researcher, the late Bill Cooper discussed *2001: A Space Odyssey* on his radio show. Cooper described the Monolith as a symbolic catalyst for the beginning of the programming/control of humanity and how the Monolith effectively imparts 'forbidden knowledge' to humanity, dismantling 'paradise' in its wake. As witnessed in the 'dawn of man' sequence of the film, the 'forbidden knowledge' leads to the death of one ape at the hands of another. Cooper believed that the ape, 'Moonwatcher,' was a symbol of the first priest or initiate of the mystery school teachings – instrumental in guarding the secrets of the ages, astral theology, the study of the Sun, Moon and stars, etc. Cooper also highlighted the six transformations that Bowman undergoes in the finale of the film, the sixth level of attainment in the mystery school teachings, and the associated '666' paradigm of occult teachings.

There are several further subtle indications of '666' embedded in the film...

"It also appears that the 'monoliths' in the movie appear for 666 seconds. The time between the first appearance and final disappearance of each of the four 'monoliths,' the four times added together, is 666 seconds. Additionally, there are apparently 666 camera shots starting from 'The Dawn of Man' (the first shot after the opening credits) to 'The End' (the last shot of the closing credits.) The running time of the film in seconds, from the beginning of the Overture to the end of the Exit Music (total exhibition time), is allegedly equal to the number of Moon orbits contained in 666 years (8903). Alternatively, the running time in seconds, from the beginning of the MGM lion logo to the fade-out of the story, is equal to the number of Moon phases contained in 666 years (8237). Everything before and after the movie proper, that is, the overture, end credits, and exit music times, adds up to 666 seconds. For an 'added bonus,' the director Stanley Kubrick was reported to have died 666 days before the year 2001, on March 7th, 1999." Jay Weidner.

Jay Weidner has also proposed that Kubrick created *'2001: A Space Odyssey'* as a *"visual and alchemical initiation into the on-going transformation and evolutionary ascent of man to a so-called Star Child destiny."* The obvious analogies are the celestial alignments that precede each of the alchemical transmutations in the film. The second main allegory is the monolith or 'black stone' that initiates these transmutations. Again this mirrors the alchemical lore about the black stone (known under numerous names – most notably 'The Philosopher's Stone') causing the transmutation of the alchemist. The film itself (the dimensions of the movie screen) shares the same dimensions as the monolith, prompting some researchers to consider the act of viewing the film as part of a greater, symbolic ritual.

"Kubrick completely reveals that he understands the 'Great Work.' The monolith represents the Philosopher's Stone, the Book of Nature and the Film that initiates. Stanley Kubrick has truly made the Book of Nature onto film. Using powdered silver nitrates, that are then glued onto a strip of plastic, and then projected onto the movie screens of our mind, Kubrick

The Falsification of Science

has proven himself to be the ultimate alchemist-artist of the late 20th century." Jay Weidner.

Despite clearly being on the inside (and obviously a Hollywood 'illusionist,') Kubrick's films have told us more about the hidden, global agenda, than any other Hollywood endeavour, albeit largely in the form of allegory and metaphor. Was Kubrick's decision to enact a form of disclosure, to hide clues, hidden meanings in his work, prompted by guilt or some twisted, dark sense of humour? Did he become a prisoner of an industry that he once loved, and decided to articulate the things he came to see and understand?

There are those that believe Stanley Kubrick was killed by the elite/Illuminati for revealing too much about the secret society in his final film '*Eyes Wide Shut.*' Whilst his official cause of death was listed as cardiac arrest (certainly not overly shocking for a 70 year-old man) some researchers point to the preponderance of Illuminati symbolism in his films, his clean bill of health prior to dying, and the strange editorial takeover of the film before its release as evidence there was more going on than initially meets the eye.

Believers of this theory, many of whom also believe that Kubrick was enlisted by the US government to produce the faked Moon landings, view the prevalence of Illuminati symbols in his films as more than just the result of unconscious archetypes. They are evidence that Kubrick himself was a part of the Illuminati and was using his films as a vehicle for communication with outsiders, as way to reveal the existence of a globalist conspiracy. Indeed, many if not all of his films, including '*The Shining*,' '*2001*,' and '*A Clockwork Orange*,' are full of Illuminati symbology, most notably the 'all-seeing eye' and pyramid symbols.

Incidentally, the ubiquitous 'one eye' symbol and the pyramid shape, often seen in popular culture indicate strong evidence that the world's major 'celebrities' are puppets of the Illuminati-Elite. These blatant, ongoing demonstrations of allegiance to the Elite are what is commonly referred to as 'cartel signalling.'

However, if Kubrick's body of work is rife with Illuminati symbolism, his final film *'Eyes Wide Shut'* is the climax. Not only is this film about a mysterious, perhaps murderous, secret society, it is drenched in allusions to the New World Order cabal. Occult symbols such as the pentagram can be found throughout the film, as well as multiple references to rainbows and looking glasses, which are notoriously used to evoke *'The Wizard of Oz'* and the cultural brainwashing of MK-ULTRA and other CIA operations.

So to conclude this chapter, it is plain that NASA is not all it appears to be on the surface. When we dig deeper into its murky corners, a very different picture of the reality of 'outer space' begins to emerge, of which NASA is the focal point. Both its founders and subsequent incumbents of positions of power within the organisation are tainted with a very distinct stench of the Elite, Illuminati-driven New World Order, and its utterly psychopathic proponents.

Chapter 7

The Nuclear Weapons Hoax

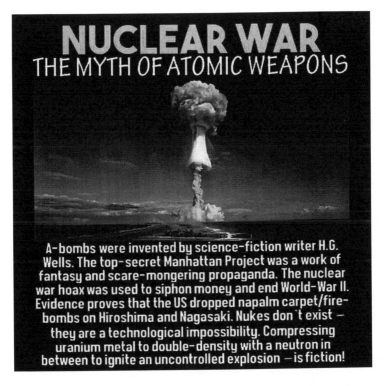

As with many so-called 'conspiracy theories,' at first glance the premise that nuclear weapons technology is an eminently provable hoax, appears too ludicrous for words.

I think it is a truism to state that most of my generation, the so-called 'baby boomers,' those born in the ten years immediately following the end of WWII, grew up in an atmosphere of tension, fear and the ever-present threat of a global thermonuclear war hanging over our heads like the proverbial 'Sword of Damocles.' In fact, who amongst the 60 and 70-somethings in our midst, does not remember the civil defence films and drills at school in the 1950s and early 1960s, constantly reminding us of the fact that our lives were really just hanging in the balance and at the whims of politicians whose itchy fingers, we were told, were strategically and permanently poised on the nuclear triggers?

In truth, 'living in fear' is something that our governments and their controllers love to inflict upon we mere mortals. It is one of the ways by which the 'few' are able to successfully control the 'many.' Whether it be the overt threat of imminent nuclear annihilation, 'Muslim' terrorism on our doorsteps, being wiped out in a virulent, pandemic plague of various different viral strains or by deranged, gun-wielding mass murderers on the rampage, our governments have a ready-made, 'off the shelf' nemesis, just to ensure that we all have something, anything at all, to worry about.

A fearful population is indeed a more easily controlled population.

Ask yourself this; is it simply a coincidence that the rise of 'terrorism' to what they would have you believe is the epic proportions of today, has only really become apparent since the end of the Cold War (1945-1990)? Or does it perhaps seem more likely that one was designed to replace the other? With the demise of the Soviet Empire, they certainly needed a new 'enemy' to keep us in fear of our lives and the threat of Muslim terrorism certainly ticks all the boxes, as emphasised by such major false flag events as 9/11, 7/7, Madrid 3/11 and countless other lesser ones.

Despite my background as a geopolitical researcher and having uncovered the massive deceptions being perpetrated on the entire human race throughout its history and exposing them in *'The Falsification of History'* (published 2012) and *Behind the Curtain* (published 2016), even I too was more than a little sceptical as to whether the fact that nuclear weapons were indeed a hoax, was accurate. And

The Falsification of Science

if true, how could this particular colossal deception possibly have been perpetrated so successfully?

However, 'perpetrated' it most certainly has been and as always we need to delve deeper into the murky depths of geopolitics to understand the 'whos,' the 'hows,' and the 'whys' of it.

In truth, we only need to examine other successful government sponsored psychological operations or 'psy-ops' down the decades in order to prove that it is quite a simple process to successfully 'fool all of the people, all of the time' on almost any topic you would care to name. All governments spend a huge amount of taxpayers' money on research into the human psyche and how the knowledge gleaned from this research may be best used to keep the sprawling masses and the 'great unwashed' or 'useless eaters' firmly in their places. This is often largely achieved through lies from the totally controlled mass media which in reality is merely the propaganda organ of big government and which is not there to 'inform' us as we may believe, but in reality to spread lies and disinformation at the behest of those in power.

So why do we believe that nuclear weapons actually exist? Well, firstly because we are 'told' this is the case by our 'masters,' and secondly because we have all seen the still pictures and videos of the infamous, horrifying and terror-inducing 'mushroom clouds,' which are said to be the primary signature of any nuclear explosion. I sincerely doubt that anyone reading this has actually witnessed a nuclear explosion personally and as with any other psy-op we could name, images, still or moving no longer constitute proof of anything. The stunning advances in modern photo and video technology have seen to that.

So what if I told you that nuclear weapons do not even exist and are simply just another psyop that we have been exposed to all our lives?

In December 1938, the German scientist Otto Hahn discovered 'fission,' that is that certain atoms could split into two other atoms under certain conditions, whilst also releasing energy. This is factually correct (I assume). However, soon afterwards a few deceitful scientists claimed that fission could be used for military purposes, for example as a devastating bomb, and convinced equally deceitful and corrupt politicians that it should be developed – secretly, and that the masses should not be informed until the weapon was actually being used.

The principle verifiable 'evidence' for the existence of nuclear weapons is the bombing of two major cities in Japan in August 1945 – Hiroshima and Nagasaki, the only times that these 'terror weapons' have been used in an act of aggression. However, the chief inspector of the US Secretary of War for these cities in the aftermath of the bombings was US Army Major Alexander de Seversky who also investigated many other heavily bombed cities throughout Japan and Europe. On each occasion, he firstly conducted an aerial survey followed by a thorough investigation on the ground, and surprisingly he detected a similar pattern in every city he examined, including Hiroshima and Nagasaki, evidently due to the methodologies used by the bombers as well as the nature of the targets bombed.

He was prepared for and anticipated shocking sights in Hiroshima yet found it to be exactly the same as all the rest of the bomb-devastated cities he had surveyed. There was no bald spot at the centre of the blast, as there should have been according to 'predictions' and the metal framework of the buildings standing in the very epicentre of the bomb blast were still intact – again defying common assumptions regarding nuclear blasts. Some bricks had been blown out of those buildings in the areas closer to the actual blast point, but the main Hiroshima hospital, only a mile away from the epicentre of the explosion, suffered nothing more than having its windows blown out and no one in the building was even injured, let alone killed. I repeat for the avoidance of doubt, *people in the hospital in Hiroshima only one mile away from the alleged blast's epicentre were unharmed by the explosion!*

"As Special Consultant to the Secretary of War, Judge Robert P. Patterson, I spent nearly eight months intensively studying war destruction in Europe and Asia. I became thoroughly familiar with every variety of damage – from high explosives, incendiaries, artillery shells, dynamite, and combinations of these. In this study, I inspected Hiroshima and Nagasaki, the targets of our atom bomb, examining the ruins, interrogating eyewitnesses, and taking hundreds of pictures.

The Falsification of Science

It was my considered opinion, I told correspondents in Tokyo, that the effects of the atom bombs—not of future bombs, but of these two—had been wildly exaggerated. If dropped on New York or Chicago, one of those bombs would have done no more damage than a ten-ton blockbuster; and the results in Hiroshima and Nagasaki could have been achieved by about 200 B-29s loaded with incendiaries, except that fewer Japanese would have been killed. I did not 'underrate' atom bombs or dispute their future potential. I merely conveyed my professional findings on the physical results of the two bombs—and they happened to be in startling contrast to the hysterical imaginative versions spread through the world.

My findings were pounced upon in outraged anger by all sorts of people, in the press, on the air, at public forums; and by scientists who haven't been within 5000 miles of Hiroshima. But the violence of this reaction cannot alter the facts on view in the two Japanese cities.

I began my study of Japan by flying over Yokohama, Nagoya, Osaka, Kobe, and dozens of other places. Later I visited them all on foot. All presented the same pattern. The bombed areas looked pinkish—an effect produced by the piles of ashes and rubble mixed with rusted metal. Modern buildings and factories still stood. That many of the buildings were gutted by fire was not apparent from the air. The centre of Yokohama, for instance, seemed almost intact when viewed from an airplane. The long industrial belt stretching from Osaka to Kobe had been laid waste by fire, but the factories and other concrete structures were still standing. On the whole it was a picture quite different from what I had seen in German cities subjected to demolition bombardment. The difference lay in the fact that Japanese destruction was overwhelmingly incendiary, with comparatively little structural damage to inflammable targets.

In Hiroshima I was prepared for radically different sights. But, to my surprise, Hiroshima looked exactly like all the other burned-out cities in Japan. There was a familiar pink blot, about two miles in diameter. It was dotted with charred trees and telephone poles. Only one of the city's twenty bridges was down. Hiroshima's clusters of modern buildings in the downtown section stood upright.

It was obvious that the blast could not have been so powerful as we had been led to believe. It was an extensive blast rather than intensive. I had heard of buildings instantly consumed by unprecedented heat. Yet here I saw the buildings structurally intact, and what is more, topped by undamaged flag poles, lightning rods, painted railings, air raid precaution signs and other comparatively fragile objects.

At the T-bridge, the aiming point for the atomic bomb, I looked for the 'bald spot' where everything presumably had been vaporised in the twinkling of an eye. It wasn't there or anywhere else. I could find no traces of unusual phenomena.

What I did see was in substance a replica of Yokohama or Osaka, or the Tokyo suburbs – the familiar residue of an area of wood and brick houses razed by uncontrollable fire. Everywhere I saw the trunks of charred and leafless trees, burned and unburned chunks of wood. The fire had been intense enough to bend and twist steel girders and to melt glass until it ran like lava – just as in other Japanese cities.

The concrete buildings nearest to the centre of explosion, some only a few blocks from the heart of the atom blast, showed no structural damage. Even cornices, canopies and delicate exterior decorations were intact. Window glass was shattered, of course, but single-panel frames held firm; only window frames of two or more panels were bent and buckled. The blast impact therefore could not have been unusual.

Then I questioned a great many people who were inside such buildings when the bomb exploded. Their descriptions matched the scores of accounts I had heard from people caught in concrete buildings in areas hit by blockbusters. Hiroshima's ten-story press building, about three blocks from the centre of the explosion, was badly gutted by the fire following the explosion, but otherwise unhurt. The people caught in the building did not suffer any unusual effects." Major Alexander P. de Seversky

Does all that not fly in the face of everything we have been told about nuclear blasts? They are supposedly nonsurvivable at distances of up to five miles from the epicentre and all buildings in their way, be they concrete, brick, or any other material whatsoever, should be totally annihilated if not 'dustified.'

The Falsification of Science

An American prisoner detained in a Japanese POW camp made this telling statement at the time...

> "We aren't really sure what's happened. We're happy about the war ending, but I don't think everyone really believes this thing about the atom bomb. I think it might be some kind of propaganda device that gives the Japanese a good excuse to surrender. Perhaps the Japanese said we dropped an atomic bomb, and we're just going along with it."

Regardless of whether or not the statement can be taken as legitimate, it is highly suggestive of the fact that there were at least rumours present among some groups of WWII soldiers to the effect that the atomic bombs dropped on Hiroshima and Nagasaki may well have been wholly concocted propaganda of some kind.

But regardless, the greatest damage was NOT actually done by the blast but by fire as an after effect and which accounted for at least 60,000 of the alleged but never proven '200,000 persons' who perished according to Seversky's report. Flimsy Japanese-style wooden houses, of which there were many, constituted the main structural damage. Seversky concluded that the bomb had the effect of a large incendiary as most of the damage was caused by fire alone and not by the blast. He also stated that a great deal of wood remained in the rubble of the main area of the blast indicating that those buildings had not been incinerated by the heat of a nuclear blast but were destroyed afterwards by the fire that resulted from the bomb. He stated that a fleet of two hundred B-29 bombers which each dropped their routine load of incendiary bombs would have achieved exactly the same effect.

Exactly the same applies to Nagasaki. As a matter of fact, the Nagasaki bomb was alleged to have been more powerful than the one exploded over Hiroshima. However, the principle area affected in Hiroshima constituted roughly a four square mile area, which is roughly a one mile radius around the centre of the wooden buildings destroyed by fires, yet the principle area affected by Nagasaki's allegedly more powerful bomb, was only one solitary square mile.

Seversky also pondered upon why Nagasaki and especially Hiroshima had even been chosen as targets since they had no military value whatsoever. They would have been very easily destroyed by fire as the majority of the structures in these two cities were shoddy, poor quality, and insubstantial wooden houses. However, they would have easily served the purpose of someone planning to elicit maximum propaganda value for the amount of destruction caused, as such structures are easily destroyed by fire. Incidentally, speaking of propaganda, the death figures quoted by the Bomb Museum in Hiroshima are self-admittedly derived from the deaths of anyone who was within the affected area within two weeks of the bombing including rescue workers, reporters, etc. regardless of when they died and remarkably is still counting these deaths even today.

Major Seversky stated in his report that the effect of one of these so-called 'atomic bombs' dropped on New York City would actually only affect an area much smaller than one of the five main boroughs.

Seversky even wrote an article on the results of his investigation which appeared in the February 1946 issue of *Reader's Digest* and in which none of the authorities interviewed contested his descriptions of Hiroshima or Nagasaki or the damning facts he had stated within it. They merely contested his opinions, such as his comparison with New York City and his allegations that the incendiaries dropped by two hundred B-29s would have achieved the same effect.

The article was entitled *'Atomic Hysteria'* and actually received a huge amount of scorn and criticism, no doubt from the 'usual suspects' i.e. vested interests and so-called 'experts' in America who had never ventured any closer than five thousand miles away from Japan and who nevertheless insisted that Seversky's findings were untrue. These 'experts' were even given a 'right of reply,' three months later in the May 1946 issue of *Reader's Digest* in which the author interviewed many military and scientific authorities in an effort to refute Seversky's article. Significantly, both these articles were apparently censored from the British edition, but they most certainly do appear in the American editions of *Reader's Digest* of February and May 1946.

The Falsification of Science

In support of Seversky's conclusions, in July 1943 the UK's Royal Air Force easily destroyed 60% of the city of Hamburg's housing stock using conventional weaponry. The Luftwaffe, the German air force, was already defeated by this time and explosives and firebombs were dropped in huge numbers to destroy the citizens' homes. Many civilians were killed yet industry was barely affected at all, as much of it was already hidden and safe from Allied bombings. At the same time in Italy, the military took over the government, arrested and pre-emptively and cold-bloodedly murdered the dictator, Mussolini without bothering with the inconvenience of a trial, and agreed to a ceasefire with the Allies.

It was then decided in Germany to evacuate people from all the major towns and cities and to intensify the propaganda to fool people into believing that Germany was still winning the war on all fronts. It worked very well, and Germany fought on for another two years, even if 'the writing was on the wall' already, even by 1943. The Japanese government copied these German initiatives with evacuations of towns and the propaganda to fool people that all was well, but at the same time began negotiations with the US for surrender. The US however, did not want to discuss peace terms with Japan as it already had plans to make a huge show of obliterating a couple of cities to expedite the desired nuclear hoax and simultaneously facilitate the planned-for Cold War, which lasted another forty-five years beyond the end of WWII.

Of course, critics of this hypothesis may cite the many and diverse alleged nuclear tests which have taken place variously above and below ground, throughout the Cold War era in a number of locations around the world. However, these tests, including their precise location and especially timing, were always closely-guarded military secrets and all we ever saw were the obviously complicit-with-the-agenda media reports and the easy-to-fake stills and moving pictures of the utterly terror-inducing mushroom clouds. And we are also encouraged to believe the propaganda that we were in a desperate race with the Germans to be the first to build the bomb, and miraculously developed it at the very end of the war, as opposed to having it ready to use partway through the war. There are two points to consider here...

Firstly, had the 'bomb' been ready 'too early' then the war would have been over before it started – and readers of my previous books and articles may well be already familiar with my views on that particular scenario. I believe that both world wars were custom-designed for certain purposes and to expedite a particular agenda, which could not have been fulfilled had the wars not lasted their intended durations. Deploying a superweapon such as the atomic bomb, would of course have precluded this.

And secondly, the more often a weapon such as this is publicly used (as opposed to tested) the more it becomes wide-open to investigation and a possible exposure of the hoax.

Also remember that the Trinity test of the A-bomb was reported to have been undertaken on the 16th July 1945 and yet Hiroshima was bombed on 6th August 1945, just three weeks later. Not only does this make no sense as a matter of testing policy, but it also makes no sense either, given the state of the war in July 1945. The testing of critical, dangerous weapons would of necessity, normally take much, much longer than that. Major new developments of anything, especially something so lethal are not just tested once and then used in angry retaliation a mere three weeks later. Even something as relatively insignificant as new cosmetics undergo many months or years of safety testing and then a further period monitoring the aftereffects. So, this apparently indecent haste to utilise such a bomb tested only once, especially given the current state of the war at that time, seems very suspicious to me.

Here is a short Japanese political history for reasons of context. In 1868, rich, conservative, authoritarian Japanese families assumed control of Japan and installed a 'puppet' emperor, and the descendants of these oligarchs and plutocrats are still in power 150+ years later in 2021. They loved America and Europe, their money and colonial outlook, and Japan rapidly acquired its own colonies – the Ryukyu Kingdom 1879, Taiwan in 1895, the Kwantung peninsula, Sakhalin

The Falsification of Science

and the 56 Kuiril islands in 1905, Korea in 1910, the German Far East colonies in 1919, Manchuria in 1932 and French Indochina in 1940. Japan was a devoted ally of the USA and Europe from 1868 until 1924, when America adopted the Immigration Act of 1924, preventing further Japanese immigration to the US. And the situation worsened in the 1930s, when the USA and specifically President Franklin D. Roosevelt adopted many economic sanctions against Japan and flatly refused diplomatic talks to resolve the problems. The reason for this was that the USA was protecting its puppets in China and the opium trade there, which Japan had planned to eliminate.

Effectively, it was this course of events that then convinced Japan, no longer having either any political or economic options, to ally itself with Germany and Italy, two countries which had also been deliberately provoked into violence by the Western powers of Britain and the US, resulting in the attack on Pearl Harbor, in Hawaii in December 1941. Japan assumed all along that the USA would negotiate over the sanctions, but sadly for them – and the rest of the world, Roosevelt had other ideas. The result of all that, is of course, history.

The intention was to totally destroy Japan economically and militarily with the US atomic bombs and force it to agree to return to its pre-1868 borders. After being blasted into submission, with the US and its allies having gotten their evil way, by the early 1950s, Japan was already thriving once more and 'best buddies' with the USA and Europe again.

Given the fact also that Japan was already beaten at the point in time when the 'deadly bombs' were deployed, why was there such inordinate haste to deploy an atomic 'solution?' Japan was already totally devastated, shattered, and destroyed by the American carpet-bombing of all major cities, especially during the six months prior to the 'atomic attacks.' There was no possibility of this defeated (in all but name) nation, ever attacking the US mainland and therefore needing to be stopped at any cost in order to prevent thousands of US civilian deaths. American forces had been systematically obliterating Japanese cities from early March, and the Japanese were now quite clearly unable to defend themselves.

And in fact, they had been attempting to surrender for many months by mid-1945, but their pleas had been ignored by the American powers-that-be. Indeed it is a popular, albeit cynically engendered public misconception that credits the dropping of the two atomic devices on the Japanese cities of Hiroshima and Nagasaki on the 6th and 9th August 1945, respectively with ending the war months early and saving the lives of millions.

With WWII rapidly coming to a close, the Elite needed an excuse to move into the next phase of their long-term agenda, aka the 'Cold War.' The attack on Hiroshima and Nagasaki sent a clear message to the Soviets and indeed the rest of the world and it was already known by the American branch of the Elite that the Soviets would not sit idly by and let American military technology intimidate and dominate them. The Soviets had already begun work on their own version of the terror weapon, subsequently helped enormously and probably intentionally by the wholesale leaking of atomic secrets by double agents. Within a year or so of the end of the war, the Russians (allegedly) had their own atomic devices and thus was born the 'Cold War' and the great 'arms race' of the second half of the twentieth century, designed solely to terrify, and as an excuse to suppress, the populations of the whole world in much the same way as the contemporary, bogus 'war on terror' works today.

The Americans and British blatantly and repeatedly ignored desperate Japanese attempts to unconditionally surrender, as of course had been the case with Germany almost throughout the entire duration of the war, because firstly they wanted to drag out the war for as long as possible and also, they needed to actually demonstrate to the world, the 'devastating effects' of the atomic bomb, otherwise the planned psy-op known as the 'Cold War' could not have generated the same terror in people's minds.

"Our entire post-war programme depends on terrifying the world with the atomic bomb. We are hoping for a tally of a million dead in Japan. But if they surrender, we won't have anything." US Secretary of State, Edward Stettinius Jr., the son of a JP Morgan partner, early 1945

The Falsification of Science

The above quote tells us all we need to know about the true intentions behind the bombings, in my considered view.

According to the historian Eustace Mullins, President Truman, whose only real job before becoming a senator had been a Masonic 'organiser' in Missouri, did not make the fatal decision alone. A committee led by James F. Byrnes, Bernard Baruch's puppet, instructed him. Baruch was the Rothschild's principal agent in the USA and a presidential 'advisor' spanning the era from Woodrow Wilson to John F. Kennedy. Baruch, who was also chairman of the Atomic Energy Commission, spearheaded the 'Manhattan Project' named for Baruch's hometown. He chose the lifelong communist, Robert Oppenheimer to be its research director. In this respect, the A-Bomb was very much the 'bankers' bomb.'

In the United States the news of the bombing of Hiroshima was greeted with a mixture of relief, pride and shock but mainly joy. Apparently, it was reported that Oppenheimer himself walked around like a prize-fighter, clasping his hands together above his head in triumph when he heard the 'good' news.

According to the mainstream version of events, Tokyo was firebombed on 9th March 1945, killing 100,000 people. From March to July, sixty-six other Japanese cities were similarly firebombed, causing another half million deaths. How many US cities did the Japanese bomb in that period, or indeed ever? Zero. (if we discount the initial war-instigating attack on Pearl Harbor) This was indeed, another correlating factor with the way that Germany had been treated by the sadistic, psychopathic Allied commanders under instruction from *their* controllers behind the scenes.

But in reality, whether or not Japan had 'unconditionally surrendered' is completely beside the point. The truth is that the Japanese were now incapable of inflicting any more harm upon the US, surrender or no surrender. They were totally beaten, and whether or not they had unconditionally surrendered at this stage, was utterly meaningless. It was certainly no justification for inflicting upon their innocent citizens, the so-called deadliest weapon ever allegedly developed. But no matter really because no nuclear weapons were ever used in Japan. It simply did not happen. The Japanese government knew that, the Russian government knew that, as did the American and British governments. In fact, the only ones who did not, and still do not know that are the terrified and traumatised citizens of the world, who have been propagandised into a state of mass subservience by the threat of an imminent doomsday, if we do not blindly follow all governmental agendas.

The whole nuclear scare was not used in order to keep the Russians under control as the Russians never had any nukes either. I well remember as a very young child in the late 1950s, even at the age of six or seven years old, being traumatised by talk of 'world destruction' or appropriately enough, MAD (mutually assured destruction) as it was then referred-to. I also vividly remember the civil defence drills, which involved hiding under desks at school after an alarm was sounded to simulate the onset of a nuclear attack. Of course, hiding under flimsy wooden desks would definitely have saved everyone's lives of course, in the event that a) the threat had been real and b) deadly nuclear weapons had been deployed. As stated, this (idle) threat has been used mainly to keep the citizens of the world in a state just short of utter panic for over seventy-five years, and to maintain military and Intelligence expenditures at an absurdly high level in a frantic scramble of brinkmanship and to maintain the military upper hand over our alleged enemies.

Is it not extremely odd that the bomb tested at Trinity was a plutonium device, as was *'Fat Man,'* the bomb allegedly used at Nagasaki? The bomb used firstly, at Hiroshima, *'Little Boy,'* was a uranium bomb, so it was never tested at all, in fact. Why would the Americans choose to drop the bomb that was untested, instead of the bomb that had been tested? It makes no logical sense whatsoever. But of course, it is a moot point in reality as neither of the two attacks actually was atomic in nature.

This also makes no sense...after the war had ended, it was not expected that the inefficient *'Little Boy'* design would ever again be required, and so all its specifications, plans and diagrams were

The Falsification of Science

therefore destroyed. Hmmm, that same old story again, eh!? The Manhattan Project cost around thirty billion dollars (almost one third of a trillion dollars in 2021 values) so does it really make any sort of sense that they would spend numerous man-years of time and billions of dollars on constructing a 'successful' nuclear weapon, use it once, and then simply destroy all record of its plans and diagrams? Surely they would have all been preserved for posterity and historical record, if for no other purpose?

Actually, this story is totally reminiscent of the story that NASA told a few years ago, when some retired investigators were searching for the original film footage and telemetry data tapes of the Moon landings. They were told that NASA had lost them – all seven hundred cartons of them plus 13,000 reels of telemetry data! As it turns out, NASA had actually 'erased and re-used the tapes in the 1980s,' apparently because of a 'major recording tape shortage' at the time. Oh, that's OK then – a major recording tape shortage eh? What an utterly bizarre phenomenon! NASA spent only around $100 billion in all, on the Apollo project, so why would anyone need a permanent record of it? And as for keeping a permanent memento of supposedly the greatest human achievement ever, both in technological and symbolic terms... well, that is just tough. Seriously though, how could such a thing be allowed to happen? Surely these were priceless historical records of the utmost importance? But apparently not to NASA, which of course also happens to be very convenient for them.

But back to the nuclear weapons scenario. Concerning the failure to test the first nuclear device allegedly used in war, it was stated that there were several reasons for not testing a *'Little Boy'* type of device. Primarily, there was so little uranium-235 as compared with the relatively large amount of plutonium which, it was expected, could be produced by the Hanford site reactors. Additionally, they told us that the weapon design was so simple that it was only deemed necessary to undergo laboratory tests with the gun-type assembly. Whereas unlike the implosion design, which required sophisticated coordination of shaped explosive charges, the gun-type design was considered almost certain to work.

I would ask you to please read the last two sentences again closely, since the second contradicts the first. In the first sentence, the weapon design is simple, and the only thing that needs to be tested is the 'gun-type assembly.' In the second sentence though, notice that this switches around. The gun-type assembly is certain to work, so by implication it needs no testing at all. But yet the implosion design is now sophisticated. Talk about frying the brain with techno-speak! In reality, neither statement is to the point. This super-weapon was allegedly the first of its kind, as we know and was further alleged to be the first nuclear device to be used. Its construction costs were huge, soaring into the tens (if not hundreds) of billions, in 1940s dollars – and that is a huge amount of money – yet allegedly it required the coordination of all of the best physicists in the West, including several 'imported' from Germany via Operation Paperclip. So how could the weapon's design have been 'simple'?

In fact, it did not simply require 'coordination of explosive charges,' it required the first ever chain-reaction fission explosion, which up until that point in time was merely just theoretical. The notion that this would not be tested in the field is ludicrous in the extreme. Plus, if these weapons required no testing before their first use 'in anger,' why did all the later examples need to be tested by the hundreds, in various parts of the world, both over and underground – to great fanfare – as well as creating extreme inconvenience for those many, poor unfortunates that were forcibly ejected from their islands and homelands to make way for these 'tests?' As is often the case with the fairy stories we are told by the Elite, there is nothing coherent in their statements whatsoever, when examined thoroughly.

Next, consider the difference between the *'Gadget'* and *'Fat Man.'* The *'Gadget'* was the nickname for the test device exploded at Trinity and yet *'Fat Man'* allegedly exploded at Nagasaki twenty-four days later. Is it not incredible that the technology was deemed perfect first time around or alternatively had been refined in a matter of a mere three weeks?

It is obvious from contemporary photographs, that the *'Gadget'* was much bigger than the *'Fat Man'* bomb, so in the intervening three weeks, the *'Gadget'* would have to have been made much smaller and simpler to fit within the shell of *'Fat Man,'* whilst providing the same result. Truly a miracle of modern science, would you not agree?

And this also raises the question that if *'Fat Man'* was already built and was smaller and simpler, then why was the *'Gadget'* so large, cumbersome, and complex? They were both built at the same time – they had to have been, and although they announced that they had been working on this project for many years, in truth they had been working on it for only one year.

The first enriched uranium did not arrive at Los Alamos until the June of 1944 and it is unfeasible to believe that they began constructing this weapon minus uranium. Plus, this totally contradicts their statements about the building of *'Jumbo.' Jumbo'* was the container built just in case the Trinity *'Gadget'* failed to detonate correctly, and which would then recover the plutonium as they apparently did not have enough for a second test – although it was not even used at Trinity. So how did they have enough for the *'Fat Man'*? From whence did that extra quantity suddenly appear?

"By the time it (Jumbo) arrived, the reactors at Hanford produced plutonium in quantity, and Oppenheimer was confident that there would be enough for a second test." Wikipedia™

This is the official explanation for why *'Jumbo'* was not used at Trinity. But this indicates that at the time of the Trinity test, they still were unsure whether the new plutonium would be used in a second test, or in a bomb destined for Japan. They were also not sure that the amount of plutonium was sufficient, and Oppenheimer actually had to state publicly that he was 'confident that there would be enough for a second test,' which indicates that there was a huge question around this issue at the time. Regardless, it certainly indicates that *'Fat Man'* had not been filled at that time. Should we assume that it had been already been built and left empty, just in case the Trinity test was successful, and Truman subsequently ordered its immediate delivery to Japan?

Even if they had enough plutonium from Hanford to fill *'Fat Man,'* they would have had to ship-in the plutonium, fill *'Fat Man,'* calibrate it, load it, etc., all in less than three weeks. In fact, they announced that they did it in about nine days, since *'Fat Man'* left Kirtland on 26th July. It beggars belief that this would be undertaken in such indecent haste since it makes absolutely no sense whatsoever to 'rush' work on such an allegedly unstable device – it would be very poor practice and more pertinently, extremely dangerous. In addition, there was no necessity at all for haste as Japan had already been defeated, was desperately trying to surrender and there was no reason to bomb them further, much less to bomb them with the first ever nuclear weapons. Incidentally, in addition to all the foregoing, before the Trinity test was undertaken, there had been a 'rehearsal' around two months earlier. In this rehearsal, 108 tonnes of high explosive were detonated in exactly the same location and suspiciously, this 'conventional' explosion was spiked with radioactive isotopes and gamma ray producers, an action which was even fully admitted by the directors of the project.

Exactly how could a conventional explosion be a 'rehearsal' for a nuclear explosion? What specifically would be the point of that? Nuclear explosions are allegedly a completely different 'animal' to conventional explosions, so the latter cannot possibly be a rehearsal for the former. This is akin to training for a one mile run by swimming one mile. It is not possible to learn anything about a nuclear explosion by making a conventional explosion, the only way to rehearse a large nuclear explosion is by running a similar, smaller nuclear explosion. The spiking with radioactive isotopes is another quite obvious clue that they were attempting to make a conventional explosion look like a nuclear one. So, who are they attempting to fool? It could only possibly be we, the masses, the ones they believe to be idiots. Unfortunately, regarding this belief, they are probably 95%+ correct.

Of course, this is comparable to the way that they often perform 'drills' around contemporary false flag events. Remember that there were all sorts of drills occurring around the events of 9/11, simultaneous with the actual event? And exactly the same with the 7/7 event in London, where drills identical to the actual event were happening on the same day, as well as other similar incidents such as Sandy Hook and the Boston marathon bombing and many other wholly manufactured catastrophes of recent years. Even the Trinity test, had its own strange rehearsal two months earlier, with the conventional blast being made to resemble a nuclear blast. Why would they do this? Well, when we see pictures of the Trinity test, it is of course impossible to know whether they are from the test event in May or the 'real' event in July as neither were time-stamped. Be honest, could you tell the difference between a 'nuclear' explosion and a large conventional explosion, on sight? No, in any case it is not actually possible because there is no such thing as a nuclear explosion. There are only large conventional blasts and faked nuclear blasts, photo-shopped or faked in large film studios such as Lookout Mountain, California.

They admit that the plume (mushroom cloud) from the rehearsal blast was visible more than sixty miles away, and Major Shields said it looked 'beautiful,' so why are there no photos of it? Why can we not compare the two? Well, one reason is that they undertook this rehearsal at night, in the wee small hours, at 4.37am. Why would they do that? Why would they want to make things more difficult than they need be by messing around with this 'incredibly dangerous' stuff in the pitch darkness, out in the middle of a desert?

The above picture is allegedly from the Eniwetok blast, during Operation Sandstone, but it looks suspiciously fake to me. What has caused the bright light apparently emanating from the actual blast to only show in a diagonal line down towards the bottom left of the picture? Surely that has to be incorrect? What is the light source for that line of brightness? If it is the blast itself (which it must be), then surely the whole picture should have been bright in the same way and not just some random portion of it? Another problem with this photograph is that they have totally forgotten to fake a surge in the sea. There should be a large circular tidal wave moving outwards from the explosion, but there is not. There is a large spout of water moving upwards, but no water moving outwards. The semicircular whitish line below and around the explosion, may appear to be the ocean surge but that is just the atoll itself.

Another problem is indicated by the little clouds simply remaining in situ and not responding to the blast at all. Could this be because the blast at this point is limited solely to the mushroom cloud? Well, the answer is 'no,' because the explosion would travel extremely fast through the air, much faster than a waterspout could ever form and by the time the waterspout had formed to that extent, the shockwave in the air should have reached those nearest clouds and completely obliterated or dispersed them or at least stretched them, so that they pointed at the event.

We are also told this by the mainstream account...

"Observers watching from ships in the lagoon saw a brilliant flash and felt the radiant heat."

If they felt the radiant heat, then they also were treated to a large dose of radiation, since the two waves would travel together. We are not shown what these observers were wearing, but in the Bikini Atoll publicity photos, one of the sailors was shirtless. Also, to which lagoon are they referring? The one in the same atoll? That surely cannot be. But it was reported that the sound took forty-five seconds to reach them, so they were about 15 km (9 miles) away and as the atoll is about 25 km (15 miles) in diameter, they were allegedly

The Falsification of Science

in a lagoon in the same atoll! So, referring back to the above explosion photograph, the observers were inside that circle were they!? At the Trinity test explosion, the soldiers were even closer, watching from only six miles away and we know that those soldiers were unprotected, as this is apparent from the photos. Had either of these blasts been real, that would have been fatal for most of the observers – according to nuclear weapons proponents' own facts and figures and pronouncements about the dangers of radioactivity – but this was plainly not the case as no mass deaths were reported. And they should also have learned their lesson from Trinity in 1945 and have known not to repeat it at Eniwetok three years later in 1948.

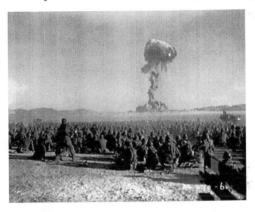

Here is a Trinity photo and obviously the explosion is pretty feeble compared to the way the later ones appear. Trinity was about 20 kilotons, while Able and Baker in the Bikini Atoll were only slightly bigger at 23 kilotons and yet Baker allegedly looked like this...

What a difference three whole kilotons makes, eh? I believe that both pictures are very poor forgeries, each obviously created without regard to the other and totally uncoordinated.

The picture above is also a possible fake, too.

Not only are these photographers possibly even closer to the blast, but they also have no protective clothing. Had these events had been real, they would have suffered a huge dose of gamma rays, which are devastating to the eyes and distance makes less of a difference with gamma rays than with other particles since gamma rays are photons. They travel at the speed of light, which means that they would travel the entire distance to those photographers in about 0.00001 of a second and in that miniscule time period, they would lose no energy, so standing 5, or even 20 miles away does not help at all in any way and Gamma rays have such a high energy that most goggles would not be effective anyway, even if they were wearing them, which is impossible to tell for sure from the picture. Closing the eyes would not help either, since the rays would instantly penetrate the eyelids as though they did not exist. The photographers should be looking through double and triple shielded binoculars, but instead, we are

The Falsification of Science

meant to believe that they are just standing right there with their eyes wide open, happily 'snapping' away.

In fact, we were given similar kinds of misdirection with the later Moon landings and the Kennedy assassination. In the example of the Moon landings, we had to watch grainy, ghostly secondhand images, filmed from flickering tiny, black and white monitors. And in the case of the Kennedy assassination, we had only 32nd degree Freemason and CIA asset, Abraham Zapruder's poor quality film, shot with a Bell and Howell Zoomatic, although it is now known that professional cameramen were standing at the kerbside filming the motorcade as it passed and presumably capturing all the drama. I would venture to ask where did all that footage go? Undoubtedly destroyed in order to promote Zapruder's brilliantly but nevertheless 100%, faked 'home movie,' as 'proof' of their view of events. For far more extensive details about the faking of the Zapruder film and the background to the Kennedy assassination, see my book, *JFK – A Very British Coup.*

The above photo is taken from the 'Lookout Mountain' webpage.

So, it transpires that the military did have some pretty high-tech camera technology after all and therefore had no need to rely upon amateur cameramen positioned around fifteen miles away. Of course, deliberately using grainy low-tech pictures protects their sordid little secrets as much as possible as very few people bother to check out all the true facts – your humble author and others excluded of course! The implausible low-resolution photographs are far more difficult to analyse and critique, and that of course is their purpose.

There are more problems with the alleged time of the Trinity test. Apparently it was originally scheduled for 4.00 am local time and as sunrise in central New Mexico on the 16th July would have been at about 6.00 am, why would they schedule an important test in the pitch darkness? And additionally, if the test was indeed scheduled for 4.00 am, why were all the cameramen there? They could not possibly have obtained any decent images of the event, beyond the initial flash. But in actuality, the detonation actually allegedly occurred at 5.29 am, still a good thirty minutes before local sunrise time. So how did they manage to obtain all the daylight images? It is all elaborate, but not that clever, fakery and so easy to pick huge, gaping holes in the nonsense spouted by the mainstream. Of course, a whole tissue of lies is so difficult to maintain and remain foolproof and this is the distinct advantage we researchers have when investigating these issues.

So, the above picture was taken a half-hour before sunrise, was it? The Moon must have been a good deal brighter in those far-off days, in that case. This should not even be twilight and although human eyes do not respond too well to twilight, cameras tend to appreciate it even less. Also, these cameras would have been saturated with radiation – of which there would have been plenty – and that destroys film,

The Falsification of Science

of course. Therefore, nothing about the story we have been told makes any sense whatsoever.

And please do not imagine for a moment that the above scene is being lit by the explosion, as that is not the case. If it were indeed lit by the explosion, the shadows would be directly behind them, but we can see that the shadows are to the left of the soldiers in the picture. Plus, according to contemporary reports, the detonation only *"lit the mountains for one or two seconds."* This picture depicts a scene several minutes beyond that point. The length of the shadows actually proves this is not at sunrise, and much less *before* sunrise. The seated soldier in the middle foreground of the photo above is about ½ inch tall on my computer screen and his shadow is about 1½ inches long, an increase of 300%, which means that the sun is about 18° above the horizon. That would indicate a time of around 8.00am. William Laurence, the New York Times reporter at Trinity, wrote this about the event...

"A loud cry filled the air. The little groups that hitherto had stood rooted to the earth like desert plants broke into dance, the rhythm of primitive man dancing at one of his fire festivals at the coming of Spring."

Well, just maybe, the 'loud cry' he heard was the communal cry of pain as hundreds of soldiers and other observers were having their eyes permanently blinded by gamma rays and having their fertility permanently destroyed, too. Conveniently and fortunately for them though, that did not happen. What is also very strange is that in the movies of the period preceding the actual blast, there were many signs and notices ordering the soldiers to keep quiet about what they were about to witness. What would the purpose of this be? The test was not a secret, how could it have been? It could be seen for hundreds of miles, over most of New Mexico and it was in many of the newspapers in the West later the same day, and of course they were publicly boasting about it within a matter of weeks.

Hiroshima was bombed only three weeks later, and the full nature of the test was made public at that time. So why all the warning signs to the soldiers on the ground that were apparently scattered liberally around the observation points? I would suggest that they were being ordered to keep quiet about the ongoing fakery that was taking place. Many probably knew or sensed that the explosion was no different to a conventional explosion, and that fact was what needed to be kept secret. This also neatly explains why all the soldiers had no problem being so close to the event without protective clothing of any kind. Protective clothing is obviously unnecessary even close to a conventional blast. Armed with this information, the films from the event begin to make sense. As 'proof' of the Trinity story, the authorities spread the downright lie that some of the radioactive fallout reached places as far away as Indiana, in the process allegedly erasing Kodak film there in at least one instance. But it is interesting to note that although film was destroyed in Indiana some weeks later, yet film in cameras only a few miles away was unaffected. In other words, fallout ruined film thousands of miles away, but gamma rays spared all the 'local' cameras. Quite the miracle really.

"I was staring straight ahead with my open left eye covered by a welder's glass and my right eye remaining open and uncovered. Suddenly, my right eye was blinded by a light which appeared instantaneously all about without any build-up of intensity. My left eye could see the ball of fire start up like a tremendous bubble or nob-like mushroom. I dropped the glass from my left eye almost immediately and watched the light climb upward. The light intensity fell rapidly hence did not blind my left eye, but it was still amazingly bright." This is allegedly an eyewitness account taken directly from Wikipedia™

The 'light intensity falling rapidly' would be totally irrelevant. Your eyes would either be destroyed in the first two seconds or they would not. But nevertheless, the story makes no sense in any other way, either. This person used a welder's glass to cover only one eye? How stupid do they think we are? He was a lawyer, not a welder, so if he had taken the trouble to purchase and bring the welder's glass with him, why would he use it on only one eye? If we are going to believe anything whatsoever about this story, then we must assume that he was advised by scientists to buy the glass and use it. So, he either would do that, or he would not. It is impossible to conceive of a situation in which he would he use it on one eye only, since that would

The Falsification of Science

just be inviting a bout of extreme pain and permanent blindness. The only way to interpret this fake story is to assume that he wished to provide us with accounts of both the 'intense light' and the 'bubble.' He therefore invented a truly asinine story that allowed him to do both, seeing one event with one eye and one with the other.

There are so many other issues that I could take-up with regard to the fakery involved in nuclear test photographs. Indeed, there are so many probable fakes. Having made the point now however, it is far more productive to move on to other issues surrounding all the fakery of the atomic tests.

But what about radiation fallout, you may ask? As part of the synthetically manufactured atomic hysteria in 1947, twenty-six young men who worked closely with these alleged 'weapons of mass destruction' were 'critically contaminated' by plutonium. However, in 1980, a medical survey was conducted of the twenty-four who were still alive and who had lived with plutonium inside them for thirty-three years. This survey concluded that they had all lived normal lives and only two of them had died in the intervening period – one was killed in a traffic accident and the other due to a similarly irrelevant reason. This information was taken from an article published in the UK's *Financial Times* in 1980.

Let us next examine the above photograph. This is taken directly from the *Lookout Mountain* website and is depicted in relation to *'Operation Greenhouse'* which was a nuclear test series from 1951 at

Eniwetok Atoll which is comprised of several small islands in the remote South Pacific region. It is extremely interesting that *Operation Greenhouse* was actually scripted. I could understand the event being recorded maybe, but scripted? Why would a nuclear test need to be scripted? A fictional event would probably require a script but not a real one. Real events do not need scripts. This prominently placed photograph on a governmental website also proves the maxim that the Elite always follow and that is 'hiding evidence in plain sight,' also known as 'revelation of the method.' Whether you believe this or not, it is common practice. The Elite, governed only of course by their slavish devotion to the ancient mystery school teachings, firmly believe that by subtly showing us what they are doing, that this exonerates them of their actions, no matter how utterly insidious or psychopathic these may be.

In August 1945, the world's media published information, provided to them by the United States military, that two medium-sized ports of no strategic value in western and southern Japan had been completely destroyed by 'atomic' bombs killing hundreds of thousands of Japanese men, women, and children civilians. The end result was that Japan surrendered one week later and WWII ended without any loss of face. The conditions for accepting the surrender stipulated by the Americans, were that Japan should keep quiet about the 'atomic' bomb being a hoax and that the US should be allowed to place military bases in Japan indefinitely. In return the US agreed that the Japanese ruling elite, the Emperor, and his family could remain in situ.

But, because no atomic bombs have ever exploded – anywhere, it was all one huge hoax. Proof? There are no public records of anyone in Japan ever having been killed by *nuclear* weapons and *radiation*. Of course, murdering unarmed men, women and children is pure terrorism and is totally unconnected with real, military wars between the armies of mutually antagonistic nations. And the fact is that no one ever protested about these 'terror weapons' being used at the time. It was only much later when 'nuclear' weapons allegedly began to proliferate in the 1950s and 1960s, that huge worldwide protests began.

The Falsification of Science

In truth, the two Japanese cities were simply destroyed by being razed by carpet, napalm firebombing and 'only' a comparatively few Japanese died. Military propaganda and censorship created the impression that Hiroshima and Nagasaki had been 'atom' bombed and the beneficiaries of this were primarily the New World Order and its proponents in the Western world. They wanted an excuse to hold the world to ransom with the threat of a weapon so terribly destructive that it was capable of destroying the entire world, many times over, and secondly the complicit Japanese elite themselves also benefitted.

The Japanese blamed the war on a few of their military leaders who were subsequently executed and passed a law that made war 'illegal' for Japan in order that precious resources, instead of being pumped indiscriminately into the military, could be used for peaceful purposes, and generally improving the living standards of their people. And very quickly, Japan became one of the richest countries in the world. By the 1970s it had become a dominant world force in electronics and car manufacturing, to name but two markets only.

The more we read and hear about nuclear weapons, it seems to me that the USA, the UK, Israel, France and the rest of the 'nuclear nations' are simply 'testing' people in order to see just how much sheer nonsense they can get away with. There are major anomalies and lots of 'red flags' everywhere. Nuclear weapons are a 'military secret,' which is rather convenient for them, and they just drip-feed us small amounts of information from time to time to keep our fears of Armageddon nicely 'topped-up.' All nuclear tests, also conveniently, take place at often secret, remote locations away from the prying eyes of the public and the media, and quite bizarrely, in my opinion, usually in total darkness. In fact, the media just publishes only what they are specifically instructed to report anyway. There is never any actual *evidence* presented, of anything significant whatsoever.

Everyone associated with the nuclear weapons hoax since the 1940s, which began with the Manhattan Project, are in my opinion, criminals guilty of complicity in major fraud. Fraud is deliberate deception undertaken in order to secure unfair or unlawful gain and complicity is the participation in a completed criminal act of an accomplice, or a partner in the crime who aids or encourages other perpetrators of that crime, and who shared with them, an intent to act to complete the crime. Wilfully scaring billions of people with fake nuclear weapons is also a serious crime and the criminals, those still alive who continue to perpetrate this insidious hoax at any rate, should be brought to justice.

According to Wikipedia™ there are at least three types of nuclear fission, that is, splitting the nuclei of atoms to produce radiation or release energy...

As previously stated, the first, peaceful, non-explosive type of fission was discovered by the German physicist Otto Hahn in 1938, in a laboratory and it produces pure energy in the form of heat at nuclear power plants of various types, where neutrons slowly split atoms of 80-100 tonnes or more of uranium oxide under controlled conditions. The neutrons are either slowed down by water under high pressure or drawn up by special control rods, so no exponential reactions can develop inside the reactor. After a couple of years or so, the nuclear fuel is used up (ie. it has transformed into other types of atoms) and refuelling is then necessary. Exactly how the nucleus of the atom is split and becomes two other types of atoms is not fully understood, but one thing is clear – no solid mass 'm' is transformed into pure energy 'e' as per Einstein's formula $e=mc^2$, in which 'c' equals the speed of light. The radiation phenomenon had been discovered forty years earlier in radium that decays (fissions) producing ionising radiation – gamma rays. Very small amounts of radioactive radiation is produced at nuclear power plants. The radioactive material is only contained in the ash of the used fuel.

A second type of fission is one that occurred in nuclear power plants such as Fukushima or Chernobyl, during a 'meltdown.' The fuel then melts and sinks to the bottom of the combustion chamber and after a while the fission ceases by itself.

Nuclear research is carefully controlled by specific, vested interests, such as the US Departments of War and Energy that promote the

The Falsification of Science

third, allegedly deadly, secret, fake, explosive type of fission as used in 'nuclear weapons.' This was invented by Robert Oppenheimer under total military secrecy between 1942-5, and the results of his research were quite simple to fake using mass- propaganda techniques, such as media hype and the falsification of photographs and moving films etc.

We are told that until activation is desired, the military fission is kept 'sub-critical,' that is, in a state of neutrality. In the case of a uranium bomb, this is achieved by keeping 61 kilos of pure uranium, solid metal fuel in a number of separate pieces, each below the critical size, in which state, no destructive fission can occur at all. In order to produce the explosion, the pieces of uranium are compressed together rapidly when neutrons come together at almost the speed of light. This process lasts for nanoseconds only and instantly vapourises any innocent people that just happen to be in the way, at the time.

This is abject nonsense and in reality, cannot possibly happen. They tell us that simply by suddenly compressing two pieces of solid metal together, that they will explode. Metal can explode in this way can it? Imagine that! I think not, somehow. Sad to say, people will actually believe anything they are told by scientists, assuming it is said with authority and published by a 'credible' source.

"Scientists have discovered that...people will believe absolutely anything as long as it is preceded by the words 'Scientists have discovered that...'"

This third, destructive type of fission was and is pure pseudoscience and fantasy. There is no scientific evidence that explosive fission exists, except within highly suspect, secret military documents.

To repeat for the avoidance of any doubt at all...two small metal pieces cannot become one by compressing them together, it is a scientific and physical impossibility and another example of the indiscriminate breaking of the laws of physics...and even if they could, they would not detonate destructively. Solid metal cannot simply explode, under any circumstances any more than 'nothing' can – in the case of the alleged 'Big Bang.' Nuclear fission is only possible under moderated and controlled conditions by using neutrons to produce energy in the form of electricity and heat.

Not even the International Atomic Energy Agency (IAEA) can explain how fission, explosive or not, works, how to instigate an atomic explosion or how to prevent overheating of a nuclear power plant when cooling fails. The IAEA considers that regulating atomic energy safety is a national responsibility, even though it is itself an international agency and its ideas regarding the safety of nuclear power plants and their commissioning and operation therefore reads like satire.

The 1940s/50s fake atomic bombs were all US and USSR joint propaganda. The deception was of course, fully approved by the US presidents Roosevelt and Truman and their close 'friend,' the mass murdering Soviet dictator Stalin, with provisions to keep it secret forever by making it illegal and punishable by death to be a whistle-blower. There is little doubt that the leaders of both countries, the USA and USSR, agreed from the beginning in 1945, to falsify all information about atomic bombs to keep the masses in a constant state of fear of total annihilation. During the cold war that followed, from 1945 to 1990, or even subsequently, no one was ever killed by nuclear weapons. Of course, as well as faking the original weapons, subsequent 'improvements' were also faked, so ongoing close yet secret cooperation continued to be necessary. This cooperation was later extended to include the fake space race that began in the 1950s (as per the previous chapter on NASA and Outer Space).

We have been threatened for over 75 years with annihilation by nuclear weapons and yet at the same time have also been told that nuclear power is essential for our energy needs.

And over and over again it is asserted that plutonium is the most dangerous element in the world and simply just being in contact with it will kill you. Plutonium is extracted from uranium which is mined in the earth and by adding neutrons into the nucleus of uranium atoms, plutonium is thereby formed.

Since it decays by emitting alpha particles from its nucleus, plutonium is classified as an unstable isotope and the energy emitted from it is called nuclear radiation. Is this radiation harmful though? When atomic particles leave the nucleus, the atom becomes ionised and these charged particles make up an electric current only if they are moving in a magnetic field.

Nuclear radioactivity does not constitute an electric current and the radiation is not harmful as are the moving ionised charged particles in microwave and x-ray emissions. The energy escaping from the nuclear radiation is actually minute and yet we are told that uranium and plutonium are so powerful that their half-life decay rate will last hundreds of millions or even billions of years.

We are also told that by harnessing this radiated energy we can derive power from it. The first nuclear power plants were constructed to develop bombs using plutonium. The idea was that energy created by the release of atomic particles would form a chain reaction and cause a massively potent explosion creating bombs more powerful than anyone could imagine. But this never has happened.

The same idea was extended to the sun. It is taught that the sun is powered by the magic nuclear transformation or decay of hydrogen into helium deep inside its interior. This convective process has never been shown to occur. In fact as seen in sunspots the Sun is cooler on the inside, while out in the corona it is much hotter, meaning the Sun is powered externally rather than internally. The Sun is an orb of plasma-storing charged particles and releases them in an electrical drift current. It is not a 'nuclear bomb' as is alleged by science. There is no such entity.

Nikola Tesla proved that electric power is available everywhere in unlimited quantities from the natural electromagnetic field of the Earth and that we have no need of fuel or anything else for energy. But free power is an anathema to the Control Grid. Power companies use the Tesla Coil of alternating currents for power at little cost to them but charge consumers an exorbitant rate.

The first nuclear power plant was built in 1951 near Arco, Idaho as an experiment and in 1957 the first operating plant was built in Shippingport, Pennsylvania. There are now 64 so called nuclear power plants in the USA alone and a total of 104 reactors. Most were built in the late 1970s and early 1980s and were said to last around 40 years. Why forty years? Indeed forty years has now passed? These facilities were built alongside rivers, lakes, and oceans because water is needed for conversion into steam. The steam it is said, is needed to drive turbines and to cool the reactor cores.

Very few uranium enhancement facilities are still around. There is only 0.7% uranium in uranium-235 and when it is enhanced to 4% it is known as 'enriched uranium.' The uranium used for its atoms is known as 'depleted uranium' and is only 0.3% uranium-235. Since we now have laser technology and DEWs (direct energy weapons) making bombs is obsolete, especially ones that never worked in the first place. But there is a lot of depleted uranium just sitting around, seemingly 'wasted,' so the military came up with the bright idea of dropping it on Afghanistan and Iraq. Then someone thought of selling it as a weed killer and it became Monsanto's *Roundup*™, which is devastatingly poisonous to all lifeforms but certainly not radioactive – and which brain dead people now spray all over their gardens and farmers all over their fields and by default thereof, all of their crops, which we consumers subsequently eat.

Back in the 1940s it was said that Russia had developed a nuclear bomb and so therefore it became necessary to build bigger and better nuclear bombs than those they possessed, which then of course resulted in an ongoing escalation as each side engaged with the other in truly scary brinkmanship.

Bikini Atoll is one of the Marshall Islands in the East Indies in the South Pacific. Twenty-three offshore nuclear tests are said to have taken place there, but firstly of course the natives had to be evicted. These blasts are supposed to create not only tremendous winds but also 100,000° Fahrenheit heat and since the explosion was under the water it would also create a tsunami. But upon examining 'before' and 'after' pictures of Bikini Atoll, the grass-roofed huts, palm trees and docks have remained miraculously untouched.

The Falsification of Science

As a matter of fact, this inconceivable amount of heat would have, at the very least, turned the whole island into magma. Temperatures of 'merely' 1,300° Fahrenheit will easily produce lava.

We were also deceptively told from the 1940s onwards, that particles in a nuclear bomb travel at a rate of greater than one mile per second and that anyone in the vicinity will become blind, have their reproductive organs damaged beyond repair and yet in all probability would die instantly.

In reality, everything at the extreme temperatures claimed would be vapourised! How ludicrous it all is? The islanders eventually returned but were evacuated again because it was said that strontium-90 (a radioactive isotope) was found in crabs. But if so, why does the US military still maintain a presence there and why are crops still grown there if it is all so contaminated by nuclear fallout? Why are diving tours still being conducted there and why in all the alleged photographs of these tests, do the boats out on the water near the blast site remain unscathed? Could it perchance be because the photos were all poorly constructed fakes? Why is it that the island is not only thriving, but also that the coral reef there is such a diverse and well-populated underwater habitat?

A Few More Pertinent Questions

The Trinity test in New Mexico was alleged to be even larger than the Bikini tests. So why did this blast not generate a crater? Why did Robert Oppenheimer and his Generals openly pose for a picture a mere few weeks afterwards at ground zero? That area should have been highly radioactive according to nuclear 'science.' The onlookers who were allowed to watch said they felt the heat from the blast, but this also means that they would have been severely affected by the radiation too. Why were they totally unconcerned by this and why did none of them suffer any ill effects, whatsoever?

The Bravo bomb test in Nevada was said to be the biggest of all. So why were so many soldiers allowed to watch? Why did they not have protective clothing or goggles or any form of bodily defences? Their eyes should have been fried and they themselves rendered sterile! They should also have ALL contracted a cancer of some type. The blast cloud is supposed to have reached 60 miles into the air and travelled dozens of miles on the land surface and yet some soldiers are even seen to be without shirts – not that a shirt would have provided any kind of protection whatsoever. But of course this was merely a conventional detonation of dynamite and not a so-called 'atomic bomb' and therefore the soldiers knew that and were not concerned. It was all a publicity stunt!

So, if nuclear bombs **are** real, why is all this obvious, transparent fakery, lies and obfuscation necessary? Why have the cities of Nagasaki and Hiroshima experienced no measurable increase in cancer rates since the 1940s and why have the entire areas not been abandoned as heavily nuclear-contaminated sites? Both are now thriving tourist towns with zero excess background radiation at all. We submit, that this is because the bombs deployed there and also in Tokyo and dozens of other Japanese cities in 1945, were simply incendiary firebombs. These were insidiously dropped on innocent Japanese civilians for a further nine months after the Japanese emperor had first attempted to sue for peace – and on many subsequent occasions in that time period. Instead of Japan's hoped-for peace and acceptance of its surrender offer, all it received was nine more months of lethal firebombing, costing several million totally innocent lives, sacrificed as they were on the altar of political expediency.

Nuclear bombs allegedly incinerate everything within a radius of tens of miles so why were all the non-wooden buildings in Hiroshima and Nagasaki still standing with little structural damage afterwards?

Why has Chernobyl in Russia remained a haven for wildlife if radiation is so deadly and why has Three Mile Island in Pennsylvania remained open if it was supposedly leaking copious amounts of radioactive elements for a long period and causing cancer en-masse in the adjacent populations?

We were assured that the radiation fallout from the Fukushima disaster in Japan would contaminate the entire Pacific Ocean and

The Falsification of Science

even contaminate the west coast of the US, causing widespread cancer among the residents of coastal cities, plus many other radiation-associated health problems. This never happened. In fact your author spent two weeks in Los Angeles in early 2014, three years after the Fukushima disaster, which would have allowed plenty of time for the Fukushima radiation to drift across the Pacific and using a reliable, portable Geiger counter, proved that all readings fell within the normal 'safe' range, despite extensive testing in various different locations.

And another question springs to mind. Why were there no electricity blackouts in Japan after Fukushima exploded, when only one power plant remained open subsequently? Nuclear power is supposed to account for one third of their energy, but this had no discernible, detrimental effect on the Japanese power supply.

We are also told that nuclear power plants, such as the one at San Onofre in Southern California leaks, but beachgoers have been using the beaches right beside it for many years and no one is known to have suffered any ill effects.

Indian Point, in upstate New York, is alleged to be leaking radioactive material into the drinking water of New York City. Why has no one reportedly died from this? Both Diablo Canyon and Oyster Creek are among many other nuclear plants which are said to continually leak. Maybe they are all just steam plants after all! Yucca Mountain in Nevada is supposedly a huge nuclear waste storage facility, but in reality, is it simply just a huge, deep underground military base (DUMB)?

Maybe because it is all sheer fakery and deception? Yes, that would indeed explain all the above anomalies.

In fact the possible nuclear scam is second only to the fake NASA moon landing scam. Both have received upwards of $100 billion and both have openly claimed to have lost and/or erased records of significant technological achievements. Not at all suspicious and just a coincidence then, do you not agree?

Galen Winsor was a hands-on nuclear expert in the fullest sense of the phrase. Before irrational radiation protection rules were imposed, he and his colleagues directly handled used fuel. Since they needed to touch radioactive materials to accomplish their mission, they could not maximise distance or use shielding. So instead, they limited their exposure time and depended on just one out of three of the triple protection means learned by all radiation professionals, 'time, distance and shielding.'

Winsor and his colleagues knew enough about the material that they were handling to prevent most skin burns, but they had a job to do and did not allow a desire to lower doses below the level of immediate risk to impede their successful accomplishment.

During his more than 30 years of professional involvement in handling nuclear materials, Winsor stubbornly refused to change his habits. He considered the used fuel pool at the Morris, Illinois recycling plant to be his personal 'warm swimming hole,' and he gave talks during which he licked uranium dioxide off of the palm of his hand and he once filled a two litre bottle from a used fuel pool and kept that water on his office desk from which he took a daily drink.

Winsor was no fool. He was well aware of the real behaviour of the materials that he measured and in his opinion, imposition of unreasonably tight rules associated with radiation protection has been simply a cost-increasing strategy. He passed away a few years ago, in his 80s, and his death was apparently from natural causes.

Winsor frequently pointed-out the monetary value of the irradiated material that some people insist on calling high level waste. He asked the vitally important questions, *"Who owns the plutonium?"* and *"How much is it worth?"* He also recognised that using it beneficially threatens a number of powerful interests.

Unfortunately, Winsor's message did not receive widespread attention in the 1980s when he gave his many lectures on the subject. He did not live in the Internet era and did not have access to tools such as blogs and YouTube™ and thus has his legacy been successfully suppressed.

Workers at Los Alamos regularly have plutonium in their bodies. Some have had it for over 50 years, so why have they not died in extreme agony?

However, to conclude this chapter, here is proof of a persistent hoax deliberately perpetrated by the British government in the 1950s onwards. Although this does not provide proof of an absolute **total** hoax re nuclear weaponry, it does at least offer up proof that governments in general are not beyond any hoax they deem necessary and it is therefore no great leap of faith to assume that the whole nuclear weapons situation is nothing more than elaborate 'smoke and mirrors.'

This article (transcribed below in its entirety) appeared in the British mainstream newspaper, *The Independent* on Thursday 24th March 1994 under the headline below. 'Patriotic scientists'?! Amazing isn't it how the mainstream apologists can turn any negative into a positive with a well-chosen phrase or two!? . . .

News > World

Britain's H-bomb triumph a hoax: Patriotic scientists created an elaborate and highly secret bluff to disguise dud weapons, Peter Pringle reports from New York

"Last week the United States announced it was extending its moratorium on nuclear testing – including the testing of British nuclear weapons in the Nevada desert – for another year, until September 1995. That is after the non-proliferation treaty is due for renewal, in May 1995, and could mean the end of British nuclear tests for good.

If so, it will be the final chapter in a story that goes back to an elaborate bluff in the late 1950s over the nation's early H-bombs. Declassified documents from the Public Records Office reveal scientists claimed they had successfully tested three bombs when, in fact, two were duds. The third made a big bang but it was not an H-bomb.

John Hamer

For almost four decades, 'Short Granite,' 'Purple Granite' and 'Orange Herald' have been the official codenames of British H-bombs tested in the Pacific in 1957. Reported by newspapers of the day as evidence of Britain's triumphant entry into the Elite club of H-bomb nations, which then only included the United States and the Soviet Union, the two 'Granite' tests were of H-bomb design, but they fizzled, and 'Orange Herald' was a massive A-bomb.

The bluff was so successful that even defence chiefs not directly involved were kept in the dark, and the then prime minister, Harold Macmillan, was also misled.

The two 'Granite' bombs used hydrogen isotopes instead of the uranium and plutonium fuel used in the older A-bombs, but the devices were duds, according to an obscure report by the National Radiation Protection Board based on the newly declassified figures. The Aldermaston bluff is also confirmed by the authors of a new US book on British nuclear weapons history, published this week in Washington.

A handful of academics in Britain and the US, including Professor Norman Dombey of the University of Sussex and his co-author, Eric Grove, formerly of the Royal Naval College, have suspected the bluff for some time. They thought all three bombs might have been H-bomb attempts using a formula adopted earlier by the Russians – who also failed to produce a big bang first time around.

Now the record confirms their suspicions but shows the formula for the 'Granite' bombs was the same as that used by the Americans, the so-called 'two-stage H-bomb' invented and successfully tested by Edward Teller and Adam Ulam. The explosive yield expected from the Teller-Ulam design was at least in the megaton range – the equivalent of 1 million tons of TNT, more than 70 times bigger than the bomb that destroyed Hiroshima. The two 'Granite' bombs produced less than one third of a megaton.

One of several remarkable aspects of the hoax is that it was carried out as an act of supreme patriotism by probably no more than a dozen scientists led by Sir William Penney, the director of the nuclear weapons factory at Aldermaston. In 1954, Churchill had ordered him to make a bomb

The Falsification of Science

in the megaton range for use in the Blue Steel and Blue Streak missiles, which were later cancelled. As work on the bombs proceeded, public protest over fallout from the US bomb tests was growing and the scientists at Aldermaston knew their time for aerial tests was short.

In the 1957 tests, which took place on the Malden Islands in May, Sir William Penney was concerned that the two-stage H-bomb design in the 'Granite' bombs had been so hurriedly put together that they would not work, so he inserted 'Orange Herald,' the older, proven uranium bomb as a 'fall-back,' and that is the one British journalists watched being detonated and wrote up as though it was an H-bomb.

The new documents show that Britain's first real H-bomb, code-named Grapple X, was not detonated until November the same year on Christmas Island. The yield of Grapple X was 1.8 megatons, far bigger than even the Aldermaston scientists had predicted.

The new US book is volume five in a series called the Nuclear Weapons Databook produced by researchers at the Natural Resources Defence Council, a group of highly respected environmental researchers in Washington. One of the authors, Stan Norris, says the British bluff did not altogether surprise him. 'It was a period when nuclear weapons symbolised global power; nations sacrificed a lot to become a member of the club,' he says.

What has surprised Dombey and Grove is that the scientists were able to extend their hoax into the uppermost reaches of the British defence establishment.

And what has surprised all the researchers is the extent of the help given to Aldermaston by the Americans. The extraordinary story puts more dents in the already battered idea of Britain's nuclear deterrent being 'independent.'

In 1958, after Britain with its relatively meagre resources at Aldermaston had proved it could build an H-bomb on its own, President Eisenhower successfully urged the US Congress to amend the Atomic Energy Act so that the free flow of nuclear weapons information that existed during *the war but which was halted immediately the war was over, could be resumed.*

From previously classified documents of the US Atomic Energy Commission, Norris and his co-authors discovered that the US simply handed over blueprints of nuclear warheads and these then went into production at Aldermaston, under US supervision. Within three years of the 1958 amendment to the US Atomic Energy Act, American scientists had handed to Britain 16 blueprints for nuclear weapons designs. The number of blueprints handed over since then is not known, but the flow continued into the Reagan administration."

In addition to the above information, there is so much censorship on the internet nowadays regarding the nuclear hoax that the only conclusion to draw is that the powers that be have much to hide on the topic. How do I know this for a fact? So many of the links to evidence that were saved during the early part of my research for this chapter have now mysteriously become 'broken.' Of 24 saved links, 22 of them were found to be non-existent anymore and the other two were very innocuous pieces indeed and entirely unusable as credible evidence! That cannot all be pure coincidence, surely?

And so, in conclusion, I now rest my case on this particular matter! I firmly believe that given all the copious amounts of evidence presented above it is perfectly safe to state that nuclear weapons do not exist!

Chapter 8

Health, Food and Medicine

Health

Germ Theory – A Totally False Paradigm

"Doctors are men who prescribe medicines of which they know little, to cure diseases of which they know less, in human beings of whom they know nothing." Voltaire

"We need to re-direct our perspectives of microbes and see them in a new light. In terms of bacteria, for example, we need to appreciate them as bodily inhabitants who assist us in such ways as protecting us from other organisms (e.g. fungi), assisting in digestion and metabolism of food, synthesising vitamins, and helping to eliminate waste materials." Dr Paul Goldberg

"Hygienists object to the germ theory of disease because germs do not cause disease. They may be present in disease processes, and they may complicate a disease with their waste products which can be very toxic at times, but the germ or virus alone is never the sole cause of disease." Dr Virginia Vetrano

My thanks once again go to Shannon Rowan for her hugely significant contributions to this chapter.

In the medical schools of all 'Western' countries today, doctors are taught an utterly monstrous lie. A lie which is actually the basis of all modern medicine and which thus renders its entire premise totally false. This lie is commonly known as 'germ theory' and the scientist credited with postulating it was Louis Pasteur, also credited with finding a cure for rabies although it probably does not exist as such and is simply another made-up disease (more of this later).

Once again, it is important to point out that the clue to the deception here is in the description of this particular 'sleight of hand.' So-called 'germ theory' is exactly that, yet another insubstantial theory that has somehow mutated into a kind of concrete fact, accepted without question by the whole of the medical and pharmaceutical industries and its many detractors are ruthlessly vilified and ridiculed by the protective medical 'mafia.'

Germ Theory does not account for or explain many diseases, such as cancer, heart disease, diabetes, and several degenerative and chronic conditions. Therefore, there is clearly a gap in the conventional wisdom of how our bodies work and why we become ill.

Incidentally, is it not also significant that those who oppose the germ theory are now branded as 'germ theory deniers' by the medical establishment and the wholly corrupted mainstream media? This is standard practice in line with those who argue against the validity of the 'holocaust™' and vaccines, to name but two examples. The wholesale usage of these 'labels' is intended to emphasise the validity of the conventional views whilst subtly denigrating those who seek only to propound the truth, often against insurmountable odds.

Pasteur has been fêted by the medical profession for making some of the most important discoveries of all time, yet when the historical evidence is examined, it leads inescapably to the conclusion that Pasteur was merely an incompetent fraud. Not only did he NOT understand the processes with which he experimented and wrote about, but most of what he is credited with discovering was plagiarised from several other scientists as detailed in the 1940s book *'Pasteur, Plagiarist, Imposter'* by R.B. Pearson.

Both Pasteur and a contemporary of his, Antoine Beauchamp, were experimenting with the process of fermentation. The prevailing theory was that fermentation was a simple chemical reaction, but

The Falsification of Science

Beauchamp's experiments clearly demonstrated that fermentation was a process brought about by microorganisms in the air. However, Pasteur continued to insist for some time after Beauchamp's discovery that fermentation was a process that did not require oxygen because it was a lifeless chemical reaction (known as spontaneous generation). It actually took Pasteur many years to finally grasp the concept that the fermentation of sugars is caused by a yeast fungus, a living organism. And when he finally did comprehend and write about these concepts, he presented them as his own discoveries, giving no credit at all to Beauchamp. So at the very least, he was a definitely a plagiarist.

Throughout their careers, Pasteur and Beauchamp individually continued to experiment with microorganisms. Pasteur continued to adhere rigidly to the principle of 'monomorphism,' the belief that all microbes and bacteria have only one form. Whereas Beauchamp was able to prove, however, the existence of 'pleomorphism,' the crucial fact that microbes can alter their form to appear as different entities. This discovery was eventually confirmed by many scientists that succeeded Beauchamp, including Gunther Enderlein. In his experiments, Enderlein discovered that every living cell contains two distinct kinds of microorganisms known as endobionts (which means 'inside life'). These microorganisms exist within the cell, cannot be removed from it, and play an important role in cellular health. The state of a person's health is determined by the stage of development of these organisms and Enderlein further discovered that all microbes that live permanently in our bodies go through three stages...

The Primitive Stage (microbe), the Middle Stage (bacteria), and the End Stage (fungus).

Subsequently, other scientists were later able to confirm that there was actually a fourth stage which occurs only after extreme toxicity and in which the fungus goes through a transformation, mutating into what we know as a 'virus.'

So, most of the diseases in modern society today are not caused by the alleged 'pathogenic bacteria' that enter our bodies from outside us, as was emphatically suggested by Pasteur, and in fact disease occurs as these endobionts are transformed from the microbe stage to more virulent forms of life and the state of development of these organisms depends upon the state of the medium in which the organism lives. In other words, the microbes which live in our cells and assist the cells in maintaining a healthy state will mutate into bacteria, fungus, and viruses when the tissues of our bodies in which they live, change to provide a medium for their growth. They gradually become better able to survive harsh conditions, but it is these conditions which actually cause the problem.

The entire process may be summarised thus...

Primitive phases live in a strong alkaline pH environment

Bacterial phases live in mild alkaline pH environment

Fungal forms live in a medium acid pH environment

Viral forms live in a strong acid pH environment

These primitive organisms can live in our bodies in the microbe stage indefinitely, and do not cause disease, but rather perform a restorative function. Bacteria and other germs consume dead tissue. That is their function. Experiments have demonstrated that were we to place a fresh, raw steak that still has active live enzymes in it, and a cooked steak outside in the open air, it is the cooked steak only that will become infested with maggots.

Microorganisms cannot survive in living tissue. It is only when the tissue becomes dead that they move in to do their job. This is exactly what happens in a garden compost pile, the waste food and vegetable matter along with bacteria, is placed there and the bacteria decompose the food remnants into soil. Everything that exists on this Earth eventually biodegrades back to its original, basic chemical composition. (Incidentally, this process also takes place on a much wider, non-biological scale and is what is known as 'entropy,' the phenomenon by which all matter eventually breaks down over time and returns to its original chemical constituents. e.g. The process of iron turning firstly to rust and then iron oxide powder and the general 'wear and tear' and breakdown, experienced by all things.)

The Falsification of Science

Most of the germs which enter our bodies from the outside, we are told, are quickly dealt with by the 'immune system,' but in reality sickness is nothing to do with microorganisms invading our bodies. The microbes already exist within us.

"The medical myth of the immune system came about in the early 1970s to explain why some people are susceptible to disease/cancer, whereas others are not. The immune system has been presented to us as a type of army in our body whose job is to eventually destroy the malignant cancer cells and malignant germs. There is no such thing as a malignant cancer cell. They exist when one is suffering from cancer: they do not cause cancer. Cancer is caused by a biological shock... There is also no such thing as malignant germs; they are active in the reparation phase of all disease... The medical myth of an immune system is another dogma invented by Modern Medicine to explain why the Germ Theory of Disease dogma does not work." James McCumiskey, *'The Ultimate Conspiracy: The Biomedical Paradigm'*

So in reality, whether or not we get sick and die has very little to do with so-called 'catching diseases' but has everything to do with whether or not we keep our bodies free of the dead matter upon which these microorganisms feed.

"There are many microorganisms that must be permanently present in order for us to live. For example, E-coli is actually NEEDED in our guts, but an overabundance of it can be created by any form of toxicity and bacteria are produced to deal with the problem. It is the overabundance of toxins which actually cause the problem. This then is the real cause of illness. Simply put, if we were to completely sterilise our bodies, we would die. It is the same too, with the environment in which we live. It is all a question of balance, so the 'more the merrier' when it comes to microorganisms in terms of variety. A variety will keep things nicely 'in check' so any particular one does not overly flourish and create a problem. Take this garden analogy, for example... if there is a sufficient variety of insect and plant life, then one insect cannot takeover and destroy everything. Some plants may well be eaten but not the majority, so life will continue. This is how our bodies and our guts should be viewed, as an ecosystem that thrives better on diversity. But it is also true that the

stronger and healthier plants will more easily survive bug infestations and what makes them stronger and healthier is all dependent on nutrients in the soil, and enough, but not too much, water and sunlight. And toxic exposures in whatever form will also sicken, weaken, and destroy them readily." Shannon Rowan, 'alternative' healthcare expert, geopolitical researcher, and author

As with that other 'great deceiver' of the 19th century, Charles Darwin, at the end of his life, Pasteur ultimately admitted that his theory was a total fraud. He confessed that it was not the germs that were the issue, but the medium in which they lived. And yet, his false, so-called 'theory' is now the basis of the whole medical model of disease and healing.

Pasteur finally realised that germ theory was flawed, and so he abruptly reversed his position, acknowledging that germs were not the primary cause of disease. But it was Koch's Postulates that proved to be the ultimate proof of this premise, providing a clear basis for the scientific testing of the theory. The German scientist, Robert Koch, (1843-1910) a bacteriologist, physiologist and one of Pasteur's contemporaries, specified the correct testing procedure, which to this day is still considered a valid endorsement of the true scientific method.

Koch proposed four postulates to prove whether or not a bacterium is pathogenic and thereby the cause of a specific disease…

1. The pathogenic microbe can be observed in the body fluids of a host suffering from the disease. This pathogenic organism is not present in a healthy host.

2. The pathogenic microbe can be isolated from the diseased host and cultured in the laboratory.

3. The cultured pathogenic microbe causes the same disease when introduced into another host.

4. The microbe can be re-isolated from that experimentally infected host.

Viruses on the other hand, do not actually exist as it is not possible to 'isolate' them. And furthermore, according to current scientific standards one must, in order to prove existence of a virus, isolate it, and photograph and biochemically characterise it. Bacteria do exist and can live outside the body whereas no virus has ever been proven to exist, let alone 'live' outside the body. What we refer to as a 'virus' is simply an RNA sequence observed under electron microscopes and even then it is not readily identifiable as is often seen in various cases of people supposedly 'carrying' the same virus in their cells. This is instead a by-product of the cells themselves and it is actually cell debris which cannot be broken down.

Highly significantly, a Freedom of Information (FOI) request to the British government in November 2020, enquiring as to whether the Covid-19 'virus' had ever been isolated, was responded to with an emphatic 'no,' thus proving beyond a shadow of doubt that the Covid-19 virus does not in fact even exist.

"In relation to bacteria you can isolate bacteria and culture them, the first two of Koch's postulates. However you cannot inject say tubercular bacteria or cholera bacilli into somebody and cause tuberculosis or cholera. The Germ Theory of Disease does not work for viruses because they do not exist; it also does not work for bacteria because you cannot cause the same disease by injecting the bacteria that allegedly case that disease into another person." James McCumiskey

Because not all four of Koch's postulates are met when describing disease pathology (namely numbers 3 and 4 as neither has ever been demonstrated for any so-called 'infectious' disease to date) bacteria and viruses can be said with absolutely certainty to not be the direct cause of disease. There are so many fraudulent photographs, allegedly of viruses, in existence but there are two ways to detect the fraud. By reading the caption for the photo and by examining the photo. The caption usually states that the photos show the cell which supposedly contain the virus.

"The captions do not even claim to have isolated the virus, yet they are being fraudulently and criminally misrepresented to the general public

as being direct pictures of viruses.... These photographs show cells and contained in the cells are typical cellular substances of all types. These structures are well known and serve functions such as transport inside and outside the cells. These are fraudulently claimed to be viruses by scientists...All viruses of a particular type always have the same size and shape. If the pictures show a particular type of virus but with the virus particles having different shapes, you then immediately know that they are fraudulent." 'The Ultimate Conspiracy: The Biomedical Paradigm' by James McCumiskey

The scientific method of confirming that a specific bacterium is the cause of a disease means satisfying all four postulates, not just one, two or three.

Scientists know that specific bacteria are not found in every case of a specific disease. The eminent Canadian physician, Sir William Osler (1849-1919) found that the diphtheria bacillus is absent in 28 to 40% of cases of diphtheria. Green's Medical Diagnosis says that tubercle bacilli may be present early, more often late, or in rare instances be absent throughout the disease condition and indeed, the first postulate is not fulfilled in tuberculosis, diphtheria, typhoid fever, pneumonia, or any other disease, in fact.

And nor is the first postulate fulfilled, because it is a medically-accepted fact that so-called 'pathogenic' bacteria are found in the bodies of humans and animals which exhibit no symptoms of any disease whatsoever. Also, specific bacteria are repeatedly found when the disease they are alleged to cause, is totally absent.

The third postulate is not fulfilled either. Neither Pasteur nor any of his contemporaries or successors have ever induced disease by the ingestion of airborne bacteria. Contagion has never been proved by airborne means and the poisons or proteins may only cause sickness if injected and even then, not in every instance.

Koch's third postulate originally specified that introducing germ cultures in a healthy body or organism will always produce signs and symptoms of the disease. This again is not so and thus in order to protect germ theory it was revised to... *"Introducing germ cultures in a*

The Falsification of Science

susceptible body or organism produces signs and symptoms of the disease." But again, although germs are present in illness, that is no evidence that they have *caused* the disease. The condition of the host is of primary importance in the production of illness.

"The 74-year old German hygienist Max von Pettenkofer wanted to prove his theories on the true causes of cholera. He got a cholera culture from the Koch Institute without revealing his true intention. On the 7th October 1892 he drank a millilitre of these cholera bacilli. He didn't get sick nor did he get cholera. His assistant repeated the same experiment on the 17th October 1892. He got slightly sick but recovered rapidly. He did not get cholera. The Bio-Chemical Society of Toronto, Canada carried out similar experiments in which pure cultures of typhoid, diphtheria, pneumonia, tuberculosis, and meningitis were consumed in large quantities by a group of volunteers. No ill-effects were observed." 'The Ultimate Conspiracy: The Biomedical Paradigm' by James McCumiskey

Disease is a process of physiological and biochemical changes within the body, producing certain signs and symptoms which we label incorrectly as specific diseases. Diseases are categorised as communicable or infectious, but it is not true that any disease, per se, is transmitted from one person to another.

Disease has many different causes; most frequently it is due to the inadvertent introduction of toxins into the body by various methods.

Eating GMO chemical-laden, highly processed 'foods,' drinking aspartame-poisoned or heavily sugared 'soft' drinks, overindulgence in alcohol (becoming 'inTOXICated'), taking harmful drugs (especially chemical pharmaceutical 'medicines'), being injected with heavy metal-containing 'vaccines,' breathing carcinogenic polluted air, and being exposed to excessive amounts of electromagnetic radiation, as well as not getting adequate sleep, rest or exercise all contribute to toxic overload in our bodies.

The fundamental causes of disease can be also described as the result of 'enervation.' Enervation is body depletion and / or exhaustion due to the diminution of nervous energy when the body has expended more than it is capable of regenerating in a timely fashion. The general energy level diminishes, extreme tiredness and functional efficiency then ensues, and we develop a state known as 'toxicosis' as the body becomes saturated with toxic matter. Toxicosis often manifests as a disturbance of the blood and tissue fluids, and the accumulation of toxic by-products of the metabolic processes. In recent years, studies of biochemical pathology have revealed this disturbance within the homeostatic mechanism of the body, caused by the accumulation of toxic substances.

"The Bio-Chemical Society of Toronto conducted a number of very interesting experiments in which pure cultures of typhoid, diphtheria, pneumonia, tuberculosis, and meningitis germs were consumed by the millions in food and drink by a group of volunteers. The results proved that no ill effects whatsoever, were experienced by a single one of the volunteers." 'The Germ Theory Re-examined' by Bob Zuraw and Bob Lewanski, in *'Vegetarian World,'* Volume 3, Number 11, September-November 1977

The diagnoses and treatments (in the form of Big Pharma drugs) provided by today's doctors inevitably make the patient sicker by simply poisoning our systems.

All the preceding information here is based upon sound scientific studies, and documents, yet anyone who chooses to deny the effectiveness of, or refuse allopathic medical treatments is regarded as being somehow mentally unstable or in dire need of psychiatric help. And should parents refuse to allow their child to be given these modern medical so-called 'cures,' they are looked upon as guilty of gross negligence at best and 'child abuse' at worst, and risk having their children removed from them—simply for not agreeing with an incorrect interpretation of scientific data.

But perhaps the most stunning and revelatory aspect of the entire 'germ theory' paradigm is that without it, an entire, corrupt, fraudulent industry would almost cease to exist. The modern (Rockefeller) allopathic, petroleum-based drug industry is based almost entirely upon the 'curing' of diseases allegedly caused by bacteria and viruses. Take away the invalid theory of their cause and what is left? Ailments that can be treated extremely simply, and without great expense to the patient, but of course that in no way benefits the giant, mega-rich drug cartels or indeed the current medical paradigm.

To me, this speaks volumes as to the reason why germ theory is so popular amongst conventional medical practitioners and especially of course, the drug industry. Their entire livelihoods and a huge monolithic industry are almost entirely dependent on it.

"Actually bacteria are our symbiotic partners in both health and disease. They serve a useful role. As scavengers they make harmless or remove undesirable substances within our bodies. They also elaborate certain of our body needs. That is, they help build complex organic compounds from simple ingredients. A notable example of this is the production of vitamin B-12 in our intestines." Dr T.C. Fry

Fry also wrote... *"'Infection' is no war in which the body is fighting invaders. The bacteria that come to these sites are symbiotic and help the body in elaborating dead cells and tissues for expulsion. They are partners in the clean-up process. When this has been accumulated the bacteria disappear and the wound heals. Infection... is a body-cleaning process for a body burdened with toxic materials."*

So please, for the sake of your health and that of your family, choose to reject the current, false medical paradigm and in doing so, you will avoid ALL of the so-called 'plagues' of the 20th and 21st centuries, including but not exclusively, cancer, AIDS, SARS, Ebola, Swine Flu, Bird Flu, Covid-19 and indeed any other future engineered 'pandemics.'

Bearing in mind the above facts, it also immediately renders the entire concept of vaccines, as a preventative for disease, as being totally irrelevant, superfluous – and most importantly, fraudulent. And given that simple fact, how can vaccines possibly be claimed to be, as they most certainly are, the saviours of mankind due to the eradication of so many diseases? Indeed the whole premise is shown up to be exactly what it is... utter fraud and deception designed to extract trillions of dollars per annum from governments and the world population in general.

Vaccines

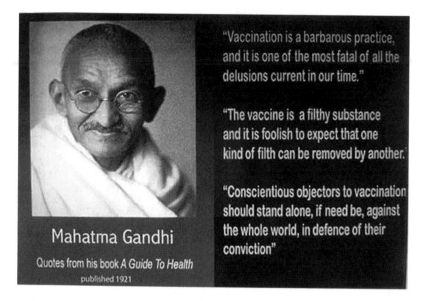

We are so fortunate to live in times where most endemic diseases have been eradicated and we are free to live out our long lives relatively untroubled by the thoughts of succumbing to such horrors as smallpox, poliomyelitis, tuberculosis, typhoid, anthrax, whooping-cough (pertussis) or measles.

Of course the main contributory factor to achieving this state of affairs has been the advent of safe and effective vaccines which stimulate our natural defence mechanisms into providing us with a resolute barrier against the ravages of these former killer diseases.

…Or so says the medical establishment, but if only this statement were true.

How can these 'proven' FACTS actually be disputed you may ask? Sadly, we have to remind you that the power of persistent, overwhelming propaganda and lies never ceases to amaze, does it not? The fact is that not only have vaccines *never* been the great saviour they have been proclaimed to be to our grandparents, parents, ourselves and now our children and grandchildren, but in actuality, the exact opposite is true. Vaccines have been proven over and over again not only to be utterly useless in disease prevention, even assuming germ theory to be valid, but they are known to be killers and purveyors of disease, misery, death, and destruction on a large scale.

But how can this be? We have been told all our lives by people who profess to 'know' about these things that we simply *must* have vaccinations if we want to stay healthy. It surely is simply just impossible that something so basic, so familiar to us, so known as a fact to be true, could be a huge deception and lie, is it not?

Is it really so impossible to believe though?

"If I had a child now, the last thing I would allow is his/her vaccination." Retired vaccine researcher

"Almost everyone who promotes vaccination is paid to do so. The supporters of vaccination have a personal interest in promoting vaccination. On the other hand, just about everyone who questions vaccination does so at great personal cost. Vaccination is big business and many of those who promote it, and make money out of it, do everything they can to protect an intellectually vulnerable but enormously profitable exercise." Dr Vernon Coleman MB ChB DSc FRSA

The British Medical Association is in effect a 'wholly-owned subsidiary' of Big Pharma, despite its facade of independence and recently, its former chairman, Sir Sandy Macara (significantly but unsurprisingly, recently 'outed' by new General Medical Council disclosure rules as a Freemason) called for the highly controversial MMR (measles, mumps, and rubella) triple vaccine to be made compulsory.

Big Pharma has been increasing its influence and control over medicine for decade after decade and by funding, cajoling, incentivising, and bribing corrupt politicians and senior medical professionals, it has in effect made its own laws and in doing so has become 'above the law.' Now it constantly lobbies for mandatory vaccinations so that every child, in fact *everyone*, must be injected with its venomous poisons. This is merely the latest stage in Big Pharma's war on the human body, wholly designed to cause still more disease, death, and indeed drug sales, by devastating the body's natural healing mechanisms.

These horrendous cocktails of destruction ALL contain toxic heavy metals and chemicals such as thimerosal (mercury), aluminium, aspartame, food dyes and preservatives, plus DNA from animal tissue, aborted human foetuses, and foreskins from mutilated (aka 'circumcised') infants, and foreign proteins in the form of live or dead bacteria and casein (milk protein). Small babies and toddlers with their immune defences still developing, are now given in some cases twenty-five vaccines, including single-dose multiple vaccine combinations, before the age of two. One simply cannot imagine the havoc it is wreaking on their tiny, immature bodily defences. No wonder that once virtually unheard of problems such as Cot Death (Sudden Infant Death Syndrome or SIDS), SBS (Shaken Baby Syndrome), and Autism are absolutely rife and increasing exponentially year on year in the Western world.

"I believe that the whole vaccination story is one of the great modern scandals of our time. The entire medical profession has been bribed and most doctors know very little about vaccination but simply follow the establishment line, never question what they are told by the drug industry, dismiss all critics of vaccination as dangerous lunatics and get very rich by promoting mass vaccination programmes which have never been proven to be safe or effective." Dr Vernon Coleman MB ChB DSc FRSA

Dr Andrew Wakefield, a British vaccine expert was indeed removed from the medical register for publishing absolute proof and refusing to retract his view that vaccinations were a huge contributory cause in Cot Death and Autism cases. This is what happens when the 'little man' stands up to these bullies. He is ruthlessly destroyed.

*"The world today has 6.8 billion people. That's heading up to about nine billion. Now if we do a **really great job on new vaccines**, health care, reproductive health services, we could lower that by perhaps 10 or 15 percent. But there we see an increase of about 1.3 billion."* Bill Gates, founder of Microsoft, prominent eugenicist, and Elite 'gofer,' speaking in 2010

The Falsification of Science

What more proof do we need that vaccines and 'healthcare' in general are being actively used for depopulation and in line with the insidious tenets of Agenda 21 / 2030?

As staggering as it may seem, the Big Pharma octopus has within the grip of its many tentacles, the entire mainstream medical profession and industry. It controls what is taught in the medical schools, the drugs that doctors prescribe and for what reason, and it also dictates governmental health policy through its lobbying and a range of techniques that include massive political contributions and bribes to doctors such as all-expenses-paid trips to 'conferences' in exotic places that are nothing more than simply a 'free vacation.'

Mainstream Big Pharma-controlled medicine is a huge morass of corruption and self-interest made even worse by the extraordinary levels of ignorance, incompetence and sadly, corruption among those who are paid large sums to be our so-called medical professionals. Take the example of Dr Paul Offit at the Children's Hospital of Philadelphia, who was paid at least $29 million from his share of royalties for Merck's *Rotateq* vaccine, which provably causes severe diarrhoea – often leading to death, in infants. Despite this outrageous conflict of interest, he used his position with the US *Centers for Disease Control and Prevention* (CDC) to ensure that childhood vaccination with **his** vaccine became compulsory. The authors of an article exposing Dr Offit said:

"Clearly, based on the distribution of income rights outlined, Paul Offit had a greater personal interest in Rotateq's commercial success than any other single individual in the world and more than any other individual in the world, he found himself in a position to directly influence that success. Unlike most other patented products, the market for mandated childhood vaccines is created not by consumer demand, but by the recommendation of an appointed body called the ACIP. In a single vote, ACIP can create a commercial market for a new vaccine that is worth hundreds of millions of dollars in a matter of months."

But surely Dr Offit was removed from the medical register for corruption and had his medical licence revoked? Unfortunately not.

Nothing whatsoever happened to him and this is just one small example and far from an isolated instance. Even had he been stripped of his medical practitioner's licence, would he actually care, with $29m sitting in his bank account and no doubt he would have been simply handed a senior position with Merck, regardless, due to the 'revolving door' policy that operates between leading medical professionals and the drug companies?

However, should a registered medical practitioner even dare to attempt to treat people with proven natural remedies, they would firstly receive a severe warning and censure, and then should they persist in this heinous practice, they will very quickly be hauled before the medical boards (GMC in the UK and AMA in the USA) and have their name forcibly removed from the medical practitioner's registers. This is not mere supposition; this has happened in literally thousands of well-documented cases down the years and decades. In fact my namesake, but no relation, Dr Ryke Geerd Hamer, now sadly deceased, was actually imprisoned for two years? His 'crime' was recorded as 'Agitation against Orthodox Medicine' and all because his alternative approach to curing cancers in so-called 'terminally ill' cancer patients actually worked! Of course, no one is allowed to practice anything which detracts from drug company or the medical profession's profits.

As another example amongst many of this trend, Dr Jennifer Daniels author of '*The Lethal Dose*' had to flee the country and is now living as an exile in Panama as she feared imprisonment and loss of her liberty in the US, subsequent to losing her licence through helping patients heal naturally. Her biggest 'crime' was advising patients in her practice to quit pharmaceutical drug regimens after they had experienced adverse reactions.

The level of ignorance and deceit amongst medical professionals is truly astounding. Doctors in general have no real idea what happens 'behind the scenes' in the hierarchy of medicine and so they absolutely believe their own propagandised education which leads to the killing and maiming of innocents on a monumental scale, worldwide. How frightening and disgusting it is to report that the medical

The Falsification of Science

profession itself the largest cause of death in America with around 800,000 on average dying in US hospitals every year from unnecessary surgery, medication and other errors, the effects of the drugs and vaccinations given to 'help' them, and infections picked up in hospital, all whilst following basic 'standard of care' practice.

However, at the risk of being totally unfair to American healthcare professionals, this also holds true of Western 'civilisation' as a whole, including the UK. Indeed in the 1980s, there was a strike by doctors in Sweden which lasted around a month and during that period *the death rate actually dropped dramatically*. This is absolutely, provably true. And in other examples of the medical profession 'striking' ...

"Whenever doctors strike, death rates go down. In 1976 in Bogota, Columbia, doctors went on strike for 52 days, with only emergency care available. The death rate dropped by 35%. In 1976 in Los Angeles County, a similar strike by doctors resulted in an 18% drop in the death rate. In another strike by doctors in Israel, a 50% decrease in mortality occurred in 1973 during a one month stoppage." 'The Ultimate Conspiracy: The Biomedical Paradigm' by James McCumiskey

"Prescription drugs kill some 200,000 Americans every year. Will that number go up, now that most clinical trials are conducted overseas – on sick Russians, homeless Poles, and slum-dwelling Chinese – in places where regulation is virtually non-existent, the FDA doesn't reach, and 'mistakes' can end up in pauper's graves?" 'Deadly Medicine,' Vanity Fair magazine, 2011

"In an airline industry, the evidence from scheduled airlines is that the risk of death is one in 10 million. If you go into a hospital in the First World, the risk of death from medical error is one in 300." James McCumiskey

As people's scepticism about the efficacy of vaccination programmes is at last beginning to gain momentum, the medical mainstream is fighting back aggressively with such blatantly totalitarian tactics as ...

"Local councils could boost immunisation rates among children by making it part of the preparation for going to school. Parents could be

asked if their child has been immunised. If they haven't, this could be done by the school nurse as a matter of routine when the child visited their new school before enrolment." Charles Waddicor, writing in the Local Government Chronicle in the UK

And this extremist nonsense, which could have been lifted verbatim from the pages of Huxley's *'Brave New World'*…

"There also seems to be no way to avoid the conclusion that we need to fight fire with fire—fear with fear. We have to make parents more scared that their children will contract serious preventable infectious diseases than they are about the false fears surrounding vaccines. And I need to emphasise—parents should be more scared of this." Steven Novella, writing in *'Science-based Medicine'*

I am not entirely sure what is 'science-based' about that, but nevertheless that is where the quote originated.

"The bloodline families (Elite) and the higher levels of the Big Pharma pyramid couldn't give a damn about protecting the health of the population. Like almost everything else in this crazy world we need to reverse their statements to see their true motivation.

They want to cull the global population and reduce human numbers dramatically and there can be few more effective ways to access the body than compulsory vaccination. Once these laws are passed anything goes with regard to vaccine content because you will see the demands increase for another kind of immunity—immunity from prosecution for drug companies who kill and brain-damage people with their witches brew.

Those they don't kill directly they plan to do so indirectly by devastating the immune system and opening people to death by other means that the human defences would otherwise eliminate. This is why the number of vaccines and combinations go on being increased for children under two while the immune system, and indeed the brain, is still developing and getting up to speed. They want their immunity to disease and their potential for clear thought dumbed-down and ideally eliminated." David Icke, geopolitical researcher, 2009

The Falsification of Science

This actually came to pass in early 2011 and the legislation is now in place protecting these murderers from liability.

Strong stuff, but nevertheless absolutely true. Vaccinations are nothing less than part of an elaborate Elite plan to reduce the population of the world to their own stated 'desirable' levels whilst having the happy (for them) side effect of their victims paying them utterly obscene amounts of money to facilitate their own deaths. Their other purpose of course being to make people sick and even more reliant on prescription drugs.

The 'safety' of vaccines is actually a complete myth. Under the 1986 *National Childhood Vaccine Injury Act*, the Vaccine Adverse Reporting System was established. Annually, it reports about 11,000 serious vaccine reactions, including up to 200 deaths and many more permanent disabilities. However, worryingly, the FDA estimates that **only 1% of serious adverse reactions are actually reported**. And worse still, several medical school students testified before the US Congress that they are instructed *not to report* these incidents.

According to the US National Vaccine Information Centre, only one in forty New York doctors reported adverse vaccine reactions or deaths and yet international studies state that there are up to *10,000* US SIDS (Sudden Infant Death Syndrome or 'cot death') deaths annually and at *least* half of them, but likely much more than that, are from vaccines.

Another American study revealed that 3,000 children in the US die annually from vaccines and that the measured incidence of non-reporting of problems in America suggests that annual adverse vaccine reactions, in fact number from 100,000 to one million!

Since 1988, the US government's *National Vaccine Injury Compensation Program* has paid families of affected children $1.2 billion in damages. It should also be noted that in settling vaccine damage suits, drug companies impose 'gagging' orders as part of the deal, to keep vital information from the public and additionally, insurers refuse to cover adverse vaccine reactions because of the high potential liability they would face. This of course distorts the true figures in a distinctly downward direction.

Medical literature documents significant numbers of vaccine failures for measles, mumps, smallpox, pertussis, polio, bacterial meningitis, and pneumonia. In 1989, the Middle Eastern country of Oman experienced a widespread polio outbreak six months after completing a population-wide immunisation program. In Kansas in 1986, 90% of 1,300 reported pertussis cases were 'adequately vaccinated,' and 72% of Chicago pertussis incidents in 1993 were also similarly vaccinated.

Between the years 1850 to 1940, well before widespread vaccination programmes, the British Association for the Advancement of Science reported a 96% decrease in childhood diseases solely due to improved nutrition, sanitation, and hygiene practices. By 1945, US medical authorities noted a 95% drop in deaths from the leading childhood infectious diseases (diphtheria, pertussis, scarlet fever, and measles) again well before mass-immunisation programmes began. This then, is the *real* reason for the decline in mortality, disingenuously attributed wholly to vaccines.

In the late nineteenth century, the Smallpox vaccine was not only made mandatory in the UK, but this was rigorously enforced by the police too. Anyone who refused vaccines, either for themselves or for their children were ruthlessly hauled before the courts and often sentenced to prison terms and even 'hard labour.' This was even true of parents whose refusals to vaccinate were based upon the fact that their older children had already suffered damage (and frequently death) from the ongoing inoculation programme.

And then, something quite remarkable happened. The English midlands city of Leicester unilaterally decided to halt the compulsory mass vaccination programmes, the results of which were totally unexpected and stunning (for most). It quickly became apparent that far from this event causing even further widespread misery and death as had been widely predicted, the smallpox cases in Leicester actually began to drop dramatically, until within a few short years, smallpox had virtually disappeared whilst it continued to rage unabated in the rest of the country. Of course this period in time also corresponded to the period whereby improved sanitation and especially diet, were becoming the norm, thereby strongly suggesting that it was in fact the vaccine that was causing the disease that it was actually alleged to prevent.

Eventually of course, the compulsory vaccine programme was ended, grudgingly, yet fortunately for many thousands who would no doubt have succumbed had it continued unabated. But of course it made no difference to the overall situation or the general public's wholly engineered perception that vaccination is beneficial to health.

A recent World Health Organisation (WHO) report found that Third World disease and mortality rates had no direct correlation with immunisation programmes but were closely related to hygiene and dietary standards. There is no evidence whatsoever that links vaccines with the decline of infectious diseases, apart of course from disingenuous statistics produced by the drug companies themselves and their often highly paid apologists. Indeed proper hygiene, clean water and most importantly of all, a healthy, nourishing diet are proven to be far more effective in this regard.

Although some vaccines may stimulate antibody production, there is categorically no evidence whatsoever to suggest that this assures immunity. In 1950 the British Medical Council published a study that found no correlation of any kind between antibody count and disease incidence. Natural disease protection involves many bodily organs and systems and artificially producing antibodies cannot achieve that.

Research also shows how squalene adjuvants, which are rife in vaccine serums, actually causes harm to the human body, effectively poisoning and weakening it, contrarily to the stated purpose, making it susceptible to numerous illnesses and diseases ranging from the merely irritating, to actually life threatening – and of course, frequently death itself. In addition, the 'herd immunity' notion of mass immunisation's effectiveness is now totally discredited, not least by the rebuttal of 'germ theory.' In fact the exact opposite is closer to the truth and evidence shows that fully vaccinated populations have experienced numerous epidemics in the past. But as an aside, is it not significant that labelling us all as being part of a 'herd' speaks volumes as to how our 'rulers' view us!?

Furthermore, vaccine effectiveness remains scientifically unproven because no double-blind studies are ever conducted to test the theory. Significantly, recent disease outbreaks have affected more vaccinated children than unvaccinated ones and the common practice of 'one size fits all' is worthless, if not downright dangerous, as it allows new-born babies to be given the same dosage as a twenty-year-old or a one hundred year-old and suffers additionally from dubious quality-control practices.

Shockingly (but maybe not surprisingly knowing what we now know about its modus operandi) the FDA absolutely refuses to act preventatively against vaccines. In fact, individual vaccine batches have almost never been recalled even when proven to be associated with severe adverse reactions or deaths. Instead, they are administered under the assumption that all recipients respond the same way, regardless of age, size, gender, ethnicity, genetics, or any other characteristics one can name. There are even reports that suggest that 'bad' vaccine batches that are reluctantly withdrawn are then surreptitiously sold-on or generously 'donated' to Third World countries.

A recent study reported in the *New England Journal of Medicine* found that a significant number of Romanian children receiving polio vaccine actually contracted the disease. One dose alone raised the polio risk eightfold.

"Unbeknownst to most doctors, the polio-vaccine history involves a massive public health service makeover during an era when a live, deadly strain of poliovirus infected the Salk polio vaccines, and paralysed hundreds of children and their contacts. These were the vaccines that were supposedly responsible for the decline in polio from 1955 to 1961! But there is a more sinister reason for the 'decline' in polio during those years; in 1955, a very creative re-definition of poliovirus infections was invented, to 'cover' the fact that many cases of 'polio' paralysis had no

poliovirus in their systems at all. While this protected the reputation of the Salk vaccine, it muddied the waters of history in a big way." Dr Suzanne Humphries, *'Smoke, Mirrors and the 'Disappearance' of Polio,'* 17th November 2011

New research may reveal other unknown hazards, but public safety will not be addressed adequately until government health officials act ethically, report accurately and adequately protect their populations from vaccines they never should allow in the first instance. How much chance is there of that coming to pass under the current medical regime? The answer to that question is 'absolutely none at all.'

Medical institutions insist that childhood diseases are extremely dangerous – or so they would have us all believe. This is absolutely false. *Centers for Disease Control* (CDC) data shows a 100% pertussis recovery rate during the period 1992-94. One Cincinnati Children's Hospital infectious diseases expert said at the time... *"The disease was very mild, no one died, and no one went to the intensive care unit."*

He also said that nearly always, childhood infectious diseases *"...are benign and self-limiting. They usually impart lifelong immunity, whereas vaccine-induced immunisation (even when achieved) is only temporary."* In fact, vaccines can increase vulnerability later in life by postponing illnesses better-tolerated in childhood, until adulthood when death rates (though still low, relatively speaking) are far higher than would normally be the case. Most importantly, nearly all, common diseases are rarely dangerous and in fact, strongly contribute to the development of robust, healthy adult, natural defence systems when they are most needed i.e. in adulthood. Additionally, it is commonly known amongst more 'honest' medical practitioners that children who did not contract measles in childhood have a higher incidence of skin diseases, degenerative bone and cartilage issues and tumours, whilst ovarian cancer is higher among childhood mumps-free adult women. The human 'immune system' allegedly benefits greatly from common childhood infectious diseases, so why do we even have, let alone tolerate dangerous and unnecessary vaccine programmes effectively hampering our bodies' natural development process? Freedom from the normal childhood ailments may well be harmful to us later in life.

So, in conclusion, it is my firm belief that the dangers of childhood disease are greatly exaggerated by Big Pharma and consequently the medical profession, to scare parents into having their children vaccinated with dangerous compounds.

What about polio though? Surely that is a classic example of a vaccination programme eradicating a disease? Was it not completely conquered or made almost non-existent by the mass-immunisation programmes in the US and Europe in the late 1950s? Unfortunately, the real facts say not, despite the mass hysteria and propaganda generated at the time, to that effect. In 1955, when Jonas Salk's polio vaccine was introduced, polio was considered the most serious post-war public health problem. One year after the vaccine was widely deployed, six New England states in the US reported sharp rises ranging from more than double in Vermont to a 642% increase in Massachusetts. Other states also were so badly affected that Idaho and Utah to name but two, halted immunisations due to a greatly increased incidence of adverse reactions and soaring death rates.

In his 1962 congressional testimony, Dr Bernard Greenberg, head of Biostatistics at the University of North Carolina, reported sharp polio *increases* from 1957 to 1959 but this was completely covered-up and suppressed at the time by a Public Health Service whitewash in order to protect the reputation of vaccinations. In 1985, the CDC actually reported that 87% of US polio cases between 1973 and 1983 were caused by the vaccine!

But perhaps of more significance than the actual vaccine, the decline in polio was due mainly to the reduction in the widespread use of the ubiquitous (at the time) insecticide and deadly poison, DDT and this is the reason that India and other countries where DDT is still manufactured and used by its citizens, still has high incidences of 'polio' which really just describes 'paralysis' and indeed this is exactly how insects are killed by the chemical. It simply paralyses them. And the huge irony is that at the time, DDT was sold as a way to 'combat polio' since it was assumed that houseflies carried the disease and since DDT killed the flies they reasoned that it would prevent infection through fly population control. But in reality this actually exposed the children (and adults) to the primary cause of polio. Parents were even encouraged to lace their children's sandwiches and clothing with the deadly stuff, through a mass advertising campaign, as incredible as that may seem to us now!

Furthermore, misdiagnoses, deceptive reporting and cover-ups suggest that the actual number of vaccine-associated paralytic poliomyelitis cases "...*may be 10 to 100 times higher than that cited by the CDC.*" In 1977, even Jonas Salk himself, admitted that mass inoculations had caused most polio cases since 1961. In fact the Salk vaccine proved highly dangerous and truthful information about it was hugely suppressed. The truth is that incidences of the disease were already dramatically declining when mass-immunisations were commenced. In Europe, the declines even occurred in countries that rejected the vaccine, proving that it was never needed in the first place. To repeat, by far the largest contributory factors to the decline of polio (as with most others) from the turn of the twentieth century to the 1950s were major improvements in hygiene, nutrition, and general living standards, plus of course the progressive elimination of DDT from the environment.

But maybe, just maybe...

"Polio has not been eradicated at all. And will not be eradicated. Polio has been renamed...Why have Acute Flaccid Paralysis and Polio been put together? Because Acute Flaccid Paralysis is a catch-all name for what looks like polio, and what you call polio when the crucial polio viral tests don't show polio VIRUSES at all...While the cases of 'polio' have gone down, cases of polio which didn't return a positive virus test, and which are now called 'Acute Flaccid Paralysis' have skyrocketed. Nice little bit of magicians' sleight of hand...A parent who saw that WHO website page, wouldn't know that Acute Flaccid Paralysis was simply polio of old, which covered the same syndromes and symptoms, caused by a large variety of viruses, as well as various toxins. A parent looking at the WHO website, would think to themselves, Acute Flaccid Paralysis must be some other 'valid' disease in its own right...All the kids who used to be on clumsy iron lungs, are now on high tech iron lungs and renamed under the autoimmune moniker called 'Transverse Myelitis' and no doubt other creative

titles to spread the decoys around." Hilary Butler, *'Polio and Lemmings,'* May 2011

"Doctors around the world are being faced with children catching the diseases they have been vaccinated against. Rather than diagnosing these children correctly, professionals have discovered that the doctors are giving the diseases new names. This suggests a cover up is going on and the vaccinations we are all being told are safe and effective are in fact completely useless. Vaccinations are now being given to children to keep them safe from every disease known to man. There appears to be a vaccination for everything from polio to a broken fingernail. However, many professionals now believe that the vaccinations are actually causing the diseases they are supposed to prevent." Vactruth.com 25th April 2012

The same is true for other diseases, including the fake, massively-hyped non-epidemic of so-called 'Swine Flu' of early 2009 and the pathetic, repeated, now probably long-forgotten by the time you read this, attempts of the medical establishment in league with the mass media to resurrect the scare yet again in the winter of 2010/11. The World Health Organisation (WHO) and CDC admitted that most cases were mild, unthreatening, and generally passed without any form of treatment, let alone worth risking highly dangerous and unnecessary vaccines. In other words, it was just another strain or variant of regular flu. Does this ring any bells perchance regarding the year 2020?

But does the lack of an initial adverse reaction prove that vaccines are quite safe, really? Far from it, unfortunately. Documented long-term health problems arising from the ingestion or injection of vaccines include arthritis, chronic headaches, skin-rashes indicative of disease, non-healing skin lesions, seizures, autism, anaemia, multiple sclerosis, dementia, and cancer. And ingredients common to all vaccines are also a serious issue. For example, squalene adjuvants are a biological time bomb that can harm or even destroy human health, long-term.

Other ingredients are known toxins and carcinogens, including thimerosal (a mercury derivative), aluminium phosphate, formaldehyde, phenoxyethanol and numerous gastrointestinal toxins like liver toxicants, cardiovascular and blood toxicants, and reproductive toxicants. Chemical ranking systems rate many vaccine ingredients among the most hazardous substances known to man, even in microscopic doses. And if a vaccine containing vial breaks in a medical setting, the mandated clean-up protocol follows the same as any other HAZMAT (hazardous material) removal procedure including evacuating the area and clean-up crew wearing HAZMAT suits. Truly staggering and criminal, I'm sure you agree!

"Millions of children (and adults) are partaking in an enormous crude experiment and no sincere, organised effort is being made to track the negative side effects or to determine the long-term consequences." Dr Bart Classen, founder, and CEO of *Classen Immunotherapies*

Dr Classen's research named vaccines as the cause of 79% of type-1 diabetes cases in children under ten years old. The sharp rise in numerous other diseases may also be linked with mass-immunisation programmes. California's autism rate has soared by 1500% in the last 25 years alone in line with the increase in vaccinations, whilst In the 1990s in Britain, MMR vaccine usage (for measles, mumps, and rubella) increased and at the same time autism rose sharply in an entirely correlated fashion. The January 2000 edition of the *Journal of Adverse Drug Reactions* reported that no adequate testing was done, so in a truly caring, ethical society, the vaccine would never have been licensed for use.

Meanwhile, the Big Pharma-compliant and controlled *Autism Society* blithely stated that... *"Autism is a complex developmental disability that typically appears during the first three years of life and is the result of a neurological disorder that affects the normal functioning of the brain..."*

That may well be so, but what causes the 'neurological disorder' in the first place and why is it increasing exponentially in direct correlation with mandatory vaccine usage? And in a similar vein, could the exponential increase in dementias of various kinds in the elderly and now even the middle-aged, be related to vaccine-induced brain damage. Again, the incidence of dementia seems to have increased in line with the huge increases in vaccinations of various kinds. This is at very least, a possibility worth considering, but of course one which never will be publicly investigated, for obvious reasons.

According to the CDC and National Vaccine Information Centre, one in every 54 US children now develops autism and in fact some US sources now quote a figure as high as one in 10, in some deprived urban areas. Tens of millions are affected worldwide, making it more common in children than paediatric cancer, incurable type-1 diabetes and AIDS combined. In the early 1940s, prior to mass immunisations, autism was so rare that few doctors ever encountered it. Today it is truly a 'genuine' global pandemic, unlike several others that could be mentioned!

Long-term vaccination reactions have been suppressed and ignored in spite of the alarming correlation between their use and the rise of autoimmune and other diseases. Vaccines are not administered for our protection; they are administered 'for profit' and other nefarious purposes such as depopulation and reducing our ability to fight disease. Avoiding them ALL is absolutely essential to protecting and maintaining human health and wellbeing.

Had I known forty years ago what I now know today, my children would never have been subjected to vaccinations. I have met many people with similar views to myself who told me that their children were never vaccinated at all and they, almost without exception, had never had a single day's illness in their lives.

Natural solutions and healthy diets have proven many times more effective than allopathic (pharmaceutical) medicines in the treatment and prevention of disease. During the 1840s US cholera outbreak, homeopathic hospitals recorded a 3% death rate compared to 50-60% in conventional ones. It is just as true today if not even more so and recent epidemiological studies show that natural remedies are far superior to vaccines in preventing diseases. They are safe, effective and toxin and side effect-free, yet most health insurers will not even cover them. Something is very, very wrong with this picture.

Vaccination history shows documented instances of deceit, portraying vaccines as 'mighty disease conquerors,' when in fact vaccines have had little or no discernible impact on disease prevention or have even delayed or reversed pre-existing disease declines. Conflicts of interest are the norm in the vaccine industry and government agencies such as the FDA and CDC are grossly overloaded with corporate officials who often return to highly-paid positions in commerce and industry provided of course that they place the profit considerations of those corporations over public health and safety issues.

In November 2000, concern over this and adverse reactions prompted the *American Association of Physicians and Surgeons* to pass a unanimous resolution at its 57th meeting calling for a moratorium on mandatory childhood vaccinations and for doctors to insist on "*...truly informed consent for their use.*"

"*It is clear...that the government's immunisation policies are driven by politics and not by science. I can give numerous examples where employees of the US Public Health Service...appear to be furthering their careers by acting as propaganda officers to support political agendas. In one case...employees of a foreign government, who were funded and working closely with the US Public Health Service, submitted false data to a major medical journal. The true data indicated the vaccine was dangerous; however, the false data indicated no risk.*" Dr Bart Classen, founder and CEO of *Classen Immunotherapies*, told US Congress in 1999

In addition, Dr Classen stated that... "*Four letters from the FDA/ Public Health Service...clearly revealed that the anthrax vaccine...*" intended for US military personnel was approved "*...without the manufacturer performing a single controlled clinical trial.*" Honest and ethical trials are essential to determine safety and effectiveness and failure to conduct them proves devastating to the health and well-being of recipients. In any case, all vaccines are unsafe, and some are even extremely dangerous. In fact, multiple vaccinations for all US military personnel practically ensure damage to their systems and severe health problems later in life. Amongst US veterans there is a massive percentage who are damaged both physically and psychologically by the military's insistence upon a truly mind-blowing array of mandatory vaccines. And these vaccines, by the way, are also regarded by some as largely culpable for the incredibly high suicide rate amongst US ex-service personnel of both sexes, as these people are callously abandoned by their government after sacrificing both their physical and psychological welfare, purportedly in the 'service' of their country.

"One compelling little 'factoid' about the military vaccination program is the large number of military personnel who developed 'Gulf War Syndrome' after having received the anthrax vaccine but had never stepped foot outside of the US and had not in fact been to the Gulf or seen combat yet were given (by military medical officials) the diagnosis of 'Gulf War Syndrome.' Many of these victims were so significantly injured that they suffered partial paralysis and were wheelchair bound for life as a result. (the paralysis part is interesting, since that was initially, I believe, the PRIMARY diagnostic requirement for 'polio' back in the day, but one of the reasons that polio incidence decreased when it did, was that the paralysis aspect was no longer necessary to the diagnostic criteria.) So it's really all about labelling when it comes to disease stats, and of course skirting the vaccine injury issue and laying blame elsewhere." Shannon Rowan, 'alternative' healthcare expert, geopolitical researcher, and author

It is also unfortunately an undisputable fact that many senior public health officials approve dangerous vaccines on unsuspecting recipients and profit handsomely for their efforts by way of incentives, bonuses, lucrative job offers with the corporations and 'brown envelope' payments.

All vaccines are akin to biological weapons that weaken or destroy the human body's natural defences. They usually fail to protect against the very diseases they are designed to prevent and indeed are often proven to directly cause them.

"The medical authorities keep lying. Vaccination has been a disaster on the immune system and it actually causes a lot of illnesses. We are actually changing our genetic code through vaccination...a hundred years from now we will know that the biggest crime against humanity was vaccines."
Dr Guylaine Lanctot, *Medical Post*, December 1994

The Falsification of Science

"There is no evidence whatsoever of the ability of vaccines to prevent any diseases. To the contrary, there is a great wealth of evidence that they cause serious side effects." Dr Viera Scheibner, internationally known leading expert on adverse vaccine reactions.

Her analysis concluded that *"Nonetheless, immunisation programs proliferate because the profit potential is enormous despite growing numbers of reputable scientific figures citing concerns."*

Currently, over two hundred new vaccines are being developed for everything from birth control to curing cocaine addiction. Many of them are in clinical trials using human 'guinea pigs,' thus knowingly and deliberately, seriously jeopardising their health and safety. Many participants in vaccine trials have been and continue to be killed and seriously injured.

New delivery systems are also being developed that include nasal sprays and genetically engineered fruits containing vaccination doses. With every person and country in the world a potential customer of these poisons, health and safety considerations are being suppressed and / or ignored for the sake of Big Pharma's profits. Unless somehow this utter madness is ended, the harm to our descendants, society, and the human race in general will ultimately be catastrophic.

And the new gene-editing or 'recombinant gene therapy' vaccines being created for Covid-19, that actually reprogramme our DNA to 'behave differently' as a way to 'combat the virus.' This will literally turn us into genetically modified humans and likely in untold (or undisclosed) or even unforeseen ways. Yet the overall rhetoric emanating from world leaders is the 'necessity' for the majority of the world's population to be injected with this new type of experimental vaccine. And already we see the trend being set (December 2020) with some airlines claiming they will not allow unvaccinated passengers once (any) Covid-19 vaccines are available.

"ALL vaccines are causing immediate and delayed, acute and chronic, waxing and waning, impairments to blood flow, throughout the brain and body. This IS causing us all to become chronically ill, sick, and causing brain damages along a continuum ranging from clinically silent

to death. This is causing ischemic 'strokes.' In some respects, this is also 'ageing.' Since the damages are microscopic, we cannot see them as they occur. However, we can now see the neurological aftermath of these damages—within hours and days of vaccination—all vaccinations.

If you place your hand on a hot stove element, you will be burned. If you do not experience pain and you cannot see the burn, then you will not learn that touching hot stove elements is harmful.

All vaccines have been causing 'burns' to body and brain. The brain has no pain receptors. You will not feel the pain. You can, however, see the footprints of these 'burns' immediate and delayed, from each vaccination. The evidence was before our eyes all along. We simply did not appreciate what these 'burns' meant, let alone that they were emerging after each vaccination. The 'burns' are largely to internal organ systems. We can ALL now see the damaging effects of these 'burns' with our own eyes.

As a physician, it is my sworn duty to cause no harm. As a human being, it is my duty to watch over my fellow beings. As an educator, it is my responsibility to teach awareness and understanding. As a scientist, it is my duty to separate cause from coincidence. As a Christian, it is my value to do unto others, as I would have others do unto myself. As a man, it is my responsibility to stand up to power, with truth and understanding, when those that wield power are in error.

My statements are not the words of a zealot. These are the words of integrity, couched with understanding that has the potential to reside in every one of you. Seek, and you shall find. Knock, and the door shall be opened unto you. I have sought. I have knocked. The door has been opened. I have found the truths I was seeking. The answers have not come from my own understandings. The answers are simply self-evident—'res ipsa loquitur'—the thing speaks for itself.

All vaccinations cause brain damage, disease, chronic illness, aging, and death—'res veritas loquitur'—the truth speaks for itself. If you do not seek, if you do not knock, if you do not look, if you trust your own understandings, then 'caveat emptor'—buyer beware.

The Falsification of Science

We have answers. We have solutions but we cannot provide solutions or treatments to a medical system and model that denies it is sick.

My training has taught me how to translate anecdotal clinical observations into empirically sound clinical measurements in medicine, neurodevelopment, and human physiology. This is especially so in dealing with functions of the human brain and behaviour.

…The truth about all vaccinations causing harm is now available for your understanding. The truth is frightening, disheartening, alarming, and now self-evident.

We have made a global medical mistake based on a lack of knowledge and understanding. We have translated forest fires for 1% of the population into chronic brush fires for the entire population. The brush fires are chronic illness and disease, not least of which are the neuro-developmental disorders." Dr Andrew Moulden, MD. PhD

And just to cap it all, in early 2011, the US Supreme Court ruled that vaccine producers i.e. Big Pharma, **can no longer be held to account for ANY damage to anyone's health, whether proven or not to be caused by their hideous, deadly cocktails.**

"The Supreme Court showed the world today that there is nothing supreme or noble about it and that it is as corrupt and cruel as most other governmental institutions. In a 6-3 vote, the high court ruled for Wyeth, saying they could not be sued for vaccine damages. Wyeth is now owned by Pfizer Inc. The US Supreme Court ruled that federal law shields vaccine makers from product-liability lawsuits in state courts seeking damages for a child's injuries or death from a vaccine's side effects. The trial case was a lawsuit by the parents of Hannah Bruesewitz, who suffered seizures as an infant after her third dose of a diphtheria-tetanus-pertussis (DTP) vaccine in 1992. The US Supreme Court ruled on Tuesday February 22nd, 2011, sustaining the federal law that shielded vaccine manufacturers from desperate parents who seek damages for serious health problems suffered by their children. Today's children are hit with more shots in a day than most of us were hit with in our entire childhood. No doubt certain pharmaceutical madmen fantasize having a permanent tubular hook-up with every child receiving constant (24/7) chemical injection and the

Supreme Court would obviously go along with that. Pharmaceutical terrorism and medical madness is alive and well in this world of ours and is part of the backbone of our modern civilization and the legal system has totally bought into it even though they understand nothing about medicine and the consequence of supporting the madness of pharmaceutical companies." Mark A. Sircus, director of the *International Medical Veritas Association*, 22nd February 2011

Natural health solutions have been proven time and time again over the years to be far more beneficial to the human body. Why is it then that many governments in conjunction with health authorities are trying their best to make vaccinations mandatory?

The answer is simply that they wish to maximise their profits from us, maim us or keep us in a permanent state of unhealthy limbo, which quite frankly, disempowers us both physically and mentally as well as making us even more dependent on their often lethal drugs, too. If we can understand that governments and the profit-driven, unethical, genocidal, Elite owned and run corporations that control them and pull their strings are only interested in two things, then the picture becomes much clearer. The only two things in which the entire corporatocracy are interested in are…absolute maximisation of profits to the detriment of ALL else and also the complete and absolute control of the great majority of mankind by whatever means they deem necessary.

It is vitally important that everyone should understand that nothing, but nothing else matters at all. Not you, your health or wellbeing, your education, your career, your life, your future, your wider family, your friends, or children, absolutely nothing, nil, zero, zilch, nada, zip, nought, nothing at all.

Once we are able to comprehend that one simple fact, everything else falls into place. It is a little similar to completing the outside, straight edges of a jigsaw puzzle in that one then has a 'framework' to operate within and reference points from which one can proceed to gain an even greater understanding of how the world really works by slowly and methodically inserting all the pieces into their correct places

The Falsification of Science

and revealing the absolute truth of how the medical paradigm really functions.

Spanish Flu 1918/9

The so-called 'Spanish Flu' pandemic, responsible for the deaths of around 50 million people worldwide, 1n 1918/19, was not 'flu' at all, it was a simple, easily treatable chest 'infection.' The fledgling 'wonder-drug' aspirin played a significant role in all those eminently preventable deaths, but the real question is this; was the pandemic a case of pharmaceutical genocide perpetrated by the Elite to further their stated population reduction agenda or was it simply a case of misdiagnosis and/or prescription error, compounded by a huge corporation's desperation to put profits before people? We will let the reader decide for themselves. Here are the facts...

In 1899 aspirin was first produced and was patented by the German pharmaceutical company Farbenfabriken Bayer in 1900. Indeed, Bayer is still one of the 'Big Pharma' companies (as Bayer AG) today. In the first decade of the 20th century its strenuous worldwide marketing efforts had left few places in the civilised world lacking aspirin. In the United States, Bayer's giant factory produced aspirin under American management and after Bayer executives were charged with violating the 'Trading with the Enemies' Act in August 1918, copious numbers of advertisements were produced that re-encouraged the lost confidence in aspirin.

The world has believed for over a century now, that a new and virulent 'flu virus' appeared from nowhere and killed millions worldwide in 1918. However, two reports, one published in 2008 and the second in 2009, have now laid that particular myth to rest for good. But of course this information has been largely unreported, indeed suppressed. It just would not 'do' for the general population to be made of such information—especially facts which cause serious doubt in the efficacy of anything as lucrative as the market for a 'household name' drug.

The first report came as a press release on 19th August 2008, from the American National Institute of Allergy and Infectious Diseases (NIAID)...

"Bacteriologic and histopathologic results from published autopsy series clearly and consistently implicated secondary bacterial pneumonia caused by common upper respiratory tract bacteria in most influenza fatalities. People were killed by common bacteria found in the upper respiratory tract. The 20 to 40 million deaths worldwide from the great 1918 Influenza Pandemic were NOT due to 'flu' at all or even a virus, but to pneumonia caused by massive bacterial infection."

The NIAID press release did not however, address the actual cause of the bacterial infections, but further, follow-up research by Dr Karen Starko certainly did. This research is quite categorical in its implication of aspirin as the real culprit, dovetailing with the NIAID research on pneumonia from massive bacterial infection and goes much further toward also explaining the extremely rapid deaths in young people.

Mortality was caused in this instance by two overlapping syndromes; an early, severe acute respiratory distress condition, which was estimated to have caused 10%-15% of deaths and a subsequent, aggressive bacterial pneumonia 'super-infection', which was present in the majority of deaths.

In examining reports of those who died, two distinct groups were discovered to be apparent, based on a very distinctive timeframe from health to death...

1. People who died of pneumonia from an alleged bacterial infection became sick and deteriorated at varying rates from there to death, and...

2. People who died so astoundingly fast that those deaths became a classic part of the frightening legend of the 1918 'flu' – people perfectly well in the morning and dead before the afternoon was out.

It has subsequently been discovered that In both groups, ***aspirin was the likely cause.***

In the case of the first group, pneumonia, it was the use of aspirin that allowed the bacterial infections to take hold. However this is another stance supporting Germ Theory, so if one is questioning that, this relationship can also be explained simply by observing that the toxic effects of the drug are enough to cause severe damage to the lungs, which is still admitted, even by the medical establishment today. There is even a specific disease condition known as 'AERD' or 'aspirin exacerbated respiratory disease' which manifests in asthma, nasal polyps, and other sinus issues. Incredibly (or not, knowing the absurdity of modern medical dogma), the 'cure' recommended for lung and sinus symptoms arising from aspirin use is 'desensitisation to aspirin' by *"increasing doses of aspirin until their body no longer reacts to the drug... This treatment includes a lifetime continuing dose of aspirin."* Excerpt from the article, '7 Warning Signs Aspirin is Destroying your Lungs and Sinuses,' The Hearty Soul, April 2016

So in the case of AERD, higher doses of aspirin are recommend for treating a disease caused by taking aspirin? Talk about fighting fire with fire! And this is recommended even though it has been admitted that high doses of aspirin can also cause 'pneumonitis' described by the Mayo Clinic as manifesting in shortness of breath, cough, and fatigue and is caused by an irritation to the lungs, which it states can occur from a myriad of factors, including chemotherapy drugs, radiation therapy and **aspirin overdose.** Untreated it can result in pulmonary fibrosis which can result in heart failure and death.

Aspirin use in children and teens, especially (and perhaps ironically) if used to treat flu-symptoms (or chickenpox) can result in a rare but horrific condition called 'Reye's disease' which is characterised by swelling in the liver and brain. Signs of the disease may include confusion, seizures, and loss of consciousness and if left untreated can prove fatal. The Mayo Clinic instead recommends parents give their children acetaminophen (i.e. Tylenol) for fever. Even though acetaminophen can cause fever (of course!) and pulmonary eosinophilia and is now linked to a likely cause of autism! Other known side effects (or simply 'effects') of using this drug include bloody stool and urine, bruising, jaundice, diarrhoea, vomiting, stomach cramps... and the list goes on.

Aspirin has a similar but even longer list of known possible effects of taking the drug (at normal recommended doses), which is nearly 60 symptoms long! Some of which include abdominal cramping, chest pain, convulsions, bloody urine, fever, fainting, seizures, hyperventilation, irregular heartbeat, sweating, drowsiness, haemorrhage, and loss of consciousness.

Some conventional medical sources still admit that high doses of aspirin can also cause pneumonia, but this particular effect is no longer as prominently listed or searchable as it once was (another item which seems to have gotten lost down the Internet 'memory hole' in recent years) and which is not included on current warning labels.

But many doctors at the time of the 'Spanish Flu' were most definitely relating pneumonias to the use of aspirin.

The salicylates, including aspirin and quinine, were almost the sole treatments given by most conventional doctors and it was quite common for them to lose 60% of their pneumonia patients.

Aspirin directly or indirectly was the cause of the loss of more lives than was the alleged infection itself. The majority of 'infection' or 'flu symptoms,' it should be noted, were manifest in respiratory complaints (and in some cases, fever) but not necessarily presented as pneumonia, which mostly seemed to arise after conventional treatment for the more mild symptoms arose. Aspirin caused harm in two ways. Firstly, through its indirect action derived from the fact that aspirin was taken until prostration resulted and the patient developed pneumonia. And for the second group which died so precipitously, their symptoms were consistent with aspirin overdose and with extraordinarily rapid deaths from it.

Another report noted that...

"The disease was a veritable plague. The extraordinary toxicity, the marked prostration, the extreme cyanosis, and the rapidity of development

stamp this disease as a distinct clinical entity heretofore not fully described. Salicylate toxicity is often overlooked because another condition is present, the dose is thought to be trivial and the symptoms (hyperventilation, vomiting, sweating, headache, drowsiness, confusion, dyspnoea, excitement, epistaxis, vertigo, pulmonary oedema, and haemorrhage) are nonspecific. In 1918, differentiating progressive salicylate intoxication from infection pathologically or clinically, the dyspnoea lasts from a few hours to a day followed by respiratory failure, circulatory collapse, convulsions, and death."

To summarise, just before the 1918 death spike, the widespread use of aspirin was still in its infancy and was unfortunately being recommended in doses now known to be potentially toxic and to cause pulmonary oedema and may therefore have contributed to the overall mortality. Young adult mortality may be explained by their willingness to use the new, recommended therapy and especially so the presence of youth in regimented circumstances, such as the military. The apparent lower mortality of children may well have been as a result of lower aspirin dosages . The most influential source of paediatric medicine in 1918 recommended hydrotherapy for fever, not salicylate; however, its 1920 edition condemned the practice of giving "*coal tar products*" ie. Rockefeller medicine's pharmaceuticals, in full doses for the reduction of fever. Varying aspirin recommended dosages may also have contributed to the marked differences in mortality between cities and even between military bases.

But another compelling reason not to be overlooked for young military men succumbing to the 'Spanish Flu' in higher numbers than the population at large, was that they were also the guinea pigs for the Rockefeller Institute's newly manufactured vaccines produced between 21st January and 4th June 1918. These vaccines were subsequently forcibly administered to millions of soldiers in the battlefields of Europe, but also soldiers at home in the United States such as those at Fort Riley, Kansas, among others.

According to author and researcher Eleanora McBean, in 'The Swine Flu Expose,' American soldiers during WWI commonly proclaimed, *"more soldiers were killed by vaccine shots than by shots from enemy guns."*

McBean also quoted a report from US Secretary of War, Henry L Stimson who *"confirmed that seven soldiers had dropped dead in an army camp after being vaccinated and that there had been 63 deaths and 28,585 cases of hepatitis as a direct result of the yellow fever vaccine during six months of WWI. This yellow fever vaccine was one of 14 to 25 shots given to US soldiers."*

Another researcher into the Spanish Flu/vaccine link, Dr Frederick Gates detailed an experiment where soldiers were injected with three doses of bacterial meningitis vaccine – a serum derived from horses.

"Reactions… Several cases of looseness of the bowels or transient diarrhea were noted. This symptom had not been encountered before. Careful inquiry in individual cases often elicited the information that men who complained of the effects of vaccination were suffering from mild coryza, bronchitis, etc., at the time of injection.

Sometimes the reaction was initiated by a chill or chilly sensation, and a number of men complained of fever or feverish sensations during the following night.

Next in frequency came nausea (occasionally vomiting), dizziness, and general "aches and pains" in the joints and muscles, which in a few instances were especially localized in the neck or lumbar region, causing stiff neck or stiff back. A few injections were followed by diarrhea.

The reactions, therefore, occasionally simulated the onset of epidemic meningitis and several vaccinated men were sent as suspects to the Base Hospital for diagnosis." Dr Gates, from the report, 'Antimeningitis Vaccination and Observation on Aggultinins in the Blood of Chronic Meningococcus Carriers'

The so-called 'Spanish Flu' was only later labelled as such after Spain (a country, significantly, that was not involved in WWI and thus had no need to spread propaganda) heavily reported on a 'terrible flu' afflicting Europe during the war. The Spanish press itself had actually called it 'French flu.' And it should be noted that that the time, 'influenza' or 'flu' was a catchall label used to describe a disease of unknown origin. It was only much after the fact that the medical establishment made attempts to discover the microbe responsible for the 'outbreak,' even in recent years blaming it on the H5N1 ('Bird Flu') virus, in an attempt to spread fear about this particular virus during the last so-called 'pandemic' in 2009.

After the alleged amazing 'success' of vaccines on WWI soldiers (many of whom were already susceptible to disease from living in extremely unsanitary 'disease conditions' in the trenches and other crowded areas without access to clean food and water), these same vaccines were pushed on the public at large after fear propaganda was widely spread regarding infectious disease-ridden soldiers coming home from the war and bringing all manner of potential pathogenic 'germs' with them.

These poor 'shell-shocked' victims were not simply made insane by their experiences in the theatre of war, but suffered post-vaccination encephalitis, which it turns out was likely the real cause of their 'shell-shock.'

As with the current modern-day 'plague' of Covid-19, holding us all captive at the hands of the medical mafia, it is probable that there were multiple causes for the Spanish Flu symptoms and especially flu related deaths, such as, vaccine damage, disease conditions (malnutrition and poor sanitation), possibly new electromagnetic pollution exposures (electrification of towns and increasing radio transmissions), and of course use of new experimental pharmaceuticals. Which brings us back to the role of aspirin in the high mortality rates now wrongly attributed to a virus.

In February 1917, Bayer lost its American patent on aspirin, thereby opening up a lucrative drug market to many manufacturers. Bayer fought back with a sustained advertising campaign, emphasising its own version of the brand's 'purity,' just as the epidemic was reaching its peak. The New York Times reported that...

"Aspirin packages were produced containing no warnings about toxicity and few instructions about use. In the fall of 1918, facing a widespread

deadly disease with no known cure, the surgeon general and the United States Navy recommended aspirin as a symptomatic treatment and the military bought large quantities of the drug. The Journal of the American Medical Association suggested a dose of 1,000 milligrams every three hours, the equivalent of almost 25 standard 325-milligram aspirin tablets in 24 hours. This is about twice the daily dosage generally considered safe today."

Dr Karen Starko's research clearly demonstrated that... *"Aspirin advertisements in August 1918 and a series of official recommendations for aspirin in September and early October immediately preceded the death spike of October 1918. The number of deaths in the USA increased steeply, peaking first in the Navy in late September, then in the Army in early October and finally in the general population in late October."*

One single sentence in her work, stands out as being extremely significant in my view... *"Homeopaths, who thought aspirin was a poison, claimed few deaths."*

That sentence alone speaks volumes about the millions of deaths caused by Bayer and the pharmaceutical industry of the day, as indeed it does today. Homeopathy threatened pharmaceutical industry profits (as it also does to this day – hence the massive propaganda campaign against natural remedies) and worse still, the homeopathic doctors criticised coal-tar/petroleum based synthetic drugs in general, the very basis of the pharmaceutical industry.

Aspirin and the other coal-tar products are condemned as causing great numbers of unnecessary deaths and the omnipresent aspirin is the most pernicious drug of all. Its deceptive malignancy is partly concealed by its fast pain-relieving quality. In several instances, aspirin weakened the heart, depressed the vital forces, increased the mortality in mild cases and made convalescence slower. In all cases it masked the symptoms and rendered immeasurably more difficult the selection of the correct curative remedy. Apparently aspirin bears no curative relation to any disease and strictly speaking it ought to be removed from sale as unsafe. The alleged blood thinning capability of aspirin, meaning that it is in widespread use for heart-attack patients, is also grossly overstated. Whilst aspirin does not actually thin the blood, it does in fact inhibit the clotting process, an effect, which can be extremely dangerous and often fatal due to uncontrolled haemorrhaging in some users. However this truth is completely misrepresented by all producers of aspirin in order to maintain the huge annual profits engendered by this silent killer. It is widely thought to be directly responsible for at least 6000 deaths each year in the USA alone, not all through misuse or from incorrect dosages.

A disturbing side effect of aspirin is also that it causes the lining of the stomach and intestinal walls to breakdown after prolonged use, leading to irreparable tissue damage and the severe degradation and weakening of the digestive tract. Doctors today still advise angina and heart attack sufferers to take an aspirin every day in complete and utter disregard of this fact. My own mother who passed-away in 2011, had suffered with angina for twenty years and completely unknown to the rest of the family until shortly before her sudden death, significantly from a ruptured colon, had been taking two aspirins a day for that entire time period—all under her 'ignorant' doctor's directives. Incidentally, this is also the same doctor that prescribed an ongoing regimen of twenty-three different drugs for my father—until his death in 2016. Both my mother AND father, I believe, were sadly two of the hundreds of thousands of victims of doctor 'error,' each year in the Western world. But when one considers how dangerous most 'legal' drugs are, then these figures are actually rather unsurprising.

But back to the main thrust of the story. Perhaps the most shocking aspect of it however, is that using only natural (therefore un-patentable and unprofitable) substances, homeopaths saved the lives of almost everyone who turned to them rather than general practitioners, during the 1918 'massacre of innocents,' meaning that in effect, millions died for no reason at all—if corporate greed and criminal recklessness is excluded, of course. This also threatened to expose the fact that the new coal-tar based synthetic drugs (derived of course from Big Oil) the basis for huge investment, were in fact medically disastrous—and that had to be prevented at all costs.

John Hamer

In 1918, The Rockefeller Foundation (shortly afterwards inextricably connected to Bayer) used the Spanish flu epidemic and the media, that it already controlled by this time, to commence a 'witch-hunt' and propaganda campaign against all forms of natural medicine that were not covered by its patents, the full force of which still continues and is being felt, to this day.

"The Rockefeller Foundation was the front organisation for a new global business venture.... This new venture was called the pharmaceutical investment business. Donations from the Rockefeller Foundation went only to medical schools and hospitals, which had become missionaries of patented pharmaceutical drugs, developed by a new breed of companies that manufactured patented, synthetic drugs." From *'Rockefeller Medicine Men'* by Richard Brown.

Also, from the same source... *"These newly discovered natural molecules had only one disadvantage; they were non-patentable. Thus, already in its first decades of existence, the pharmaceutical investment business faced a mortal threat. Vitamins and other micronutrients promoted as public health programmes would have prohibited the development of any sizable investment business based on patented synthetic drugs. The elimination of this unwanted competition from micronutrients and other natural therapies became a question of survival for the young pharmaceutical investment business.*

To promote public acceptance of his 'new medicine' as the philanthropic umbrella of the newly created pharmaceutical investment industry with patented drugs, the Rockefeller-controlled media used the Spanish flu epidemic of 1918, to start a campaign against all forms of non-patented medicine and discredit them as 'unscientific.' Within the next 15 years, essentially all medical schools in the US, all influential hospitals and, most significantly, the 'American Medical Association' became part of this strategy to align the entire health care sector under the control of the pharmaceutical investment business."

This witch-hunt, with Bayer in a leading role, has a long and violent history and continues today whereby Big Pharma is now resorting to draconian measures, with the banning of all herbal remedies across the EU, which commenced in April 2011, with the attempt to criminalise homeopaths and all natural practitioners including midwives, with the banning of IV vitamin C because it is a powerfully effective yet gentle treatment for cancer, in December 2010.

Given Big Pharma's epic yet highly lucrative battle against nature, suppressing the truth of what happened during 1918 and the pharmaceutical industry deliberately killing millions in the name of obscene profits, natural healing becomes even more important.

In 1999, the *Centers for Disease Control* claimed that it had reconstructed the 'virulent' 1918 'flu' virus. *"CDC researchers and their colleagues have successfully reconstructed the influenza virus that caused the 1918-19 flu pandemic, which killed as many as 50 million people worldwide."*

Whatever the CDC 'reconstructed,' it was certainly not whatever killed millions of people in 1918-19. The CDC appears unaware or more likely is totally ignoring the fact that the NIAID has proven beyond reasonable doubt that it was not a flu or even a virus, (even ignoring the fact that germ theory is fallacious) but common upper respiratory bacteria and massive infection combined with an insidious and incorrectly used drug that killed millions. Bayer was also possibly unsurprisingly, responsible through its vaccine division, Baxter, for sending out to 18 countries, a seasonal flu vaccine in 2009 that allegedly contained 'live' avian virus, (or so we are told, but whatever the ingredient was, it proved lethal), in effect whether accidentally or by design, a bioweapon which could have killed millions itself. And, had it not been for a single vigilant laboratory technician discovering that the vaccines were lethal through injecting ferrets all of which died, Bayer's vaccines could have potentially initiated a 'pandemic' such as nature has never before seen, which had it happened would no doubt have been ascribed to natural causes of some kind. In 1918, the convergence of a toxic drug, massive corporate advertising and government, military, and medical pressure to use the drug aspirin and the massive, forced inoculations, led to millions of deaths.

Despite Bayer's 2009 seasonal vaccines containing a bioweapon and there being no reasonable explanation, Bayer was never charged, not even with negligence, but instead was immediately selected by the WHO to produce the H1N1 vaccines. Those vaccines were promoted strongly by the government and the CDC, even as the CDC itself acknowledged that the H1N1 vaccines were predicted (surreptitiously) to maim or kill approximately 30,000 people in the USA alone.

So, why does the CDC persist in this virulent 'flu epidemic' myth, despite abundant evidence to the contrary? Disregarding motives, one might only suggest some effects of projecting the terrifying idea that millions of people died from the unproven-to-exist virus alone...

1. Distraction from the hard reality that natural treatments were the only effective ones during the 1918 health crisis.

2. The hiding of Bayer, aspirin, and the industry's role in the deaths.

3. The ongoing sales of billions of dollars in anti-viral drugs and vaccine development.

4. Increased financial power to the pharmaceutical industry to control media and influence governments.

5. Increased illnesses and deaths from chemical, pharmaceutical agents in line with the depopulation agenda.

6. Fearful dependence on 'expert medical authority' and complex, expensive solutions to save people.

7. Surrender of unlimited authority to government regulatory agencies to 'protect' the public from natural products.

8. Use of the spectre of millions of deaths as the justification for the removal of human rights in order to 'protect' the public. Exactly as we are seeing during the 2020 'Covid-19 crisis.'

The Falsification of Science

9. Enhancement of the pharmaceutical industry's move toward an uncontested global monopoly over health

10. The industrialisation, commercialisation, and militarisation of disease.

11. An open door to the use of 'pandemic emergency' to justify martial law and the extreme restriction of freedoms.

The truth, that medical authorities using Bayer aspirin (and generic aspirin) killed so many millions of people, that those deaths became one of the most terrifying events in human history, fundamentally threatens a global multi-trillion dollar industry built around unassailable 'medical expertise' and their use of synthetic drugs. But the fuller truth contains something even more dangerous to the pharmaceutical industry than simply the millions of deaths they directly or indirectly caused, because something else of great significance occurred at that time too. Non-industrial medicine ie. those using natural, organic substances, actually *saved* further millions of people from death. The events of 1918 unintentionally instigated a worldwide trial, facilitating a comparison of the millions of people who were treated by pharmaceutical medicine with those treated by natural medicines. The first group died in unimaginable numbers whilst of the second group, virtually all of them survived. More than anything else, this is the most important lesson to be learned from this dark period in medical history.

However, what is blatantly obvious is the spectacular failure and toxicity of one of the industry's oldest and purportedly most trusted 'standards,' aspirin. We can also add to that, Bayer's extensive global advertising of its toxic product, government and media influence and the fact that it increased deaths to a level never before seen. Bayer's product, political and financial power, and media influence, combined to produce so many millions of deaths that it actually merits comparison with the 'Black Death' of the mid-fourteenth century which decimated the world's population. Forty to fifty million deaths during 1918 and 1919 versus twenty-five to seventy-five million during the course of many years of the bubonic plague.

But that single, synthetic pharmaceutical product has by now far outstripped the number of deaths of the most infamous disease in history since the deaths have not yet ceased and are continuing to be of epidemic proportions. The numbers of those who have died from aspirin poisoning alone over the hundred years since this huge tragedy, are absolutely incalculable, including as previously related, that of my own mother.

As previously stated, it was natural health practitioners who saved lives during 1918, whilst aspirin killed in indiscriminately and in vast numbers But today Big Pharma in league with complicit, 'bought and paid for' governments is actively and vigorously suppressing access to natural health products which are perfectly safe and saved lives during 1918, in favour of predictably lethal vaccines and pharmaceutical products, the need for which is based on an immensely misleading, yet ubiquitous myth. What does this tell us about the insidious methods of the Big Pharma companies, who operate with impunity, hand in glove with corrupt politicians of all affiliations, whilst carrying out their genocidal yet highly profitable policies?

In light of the 1918 deaths from Bayer's pushing of aspirin onto an unsuspecting population and Bayer and other manufacturers' vaccines (synthetic, even genetically engineered) an involuntary reassessment of natural versus synthetic treatments has been instigated.

"A confluence of events created a 'perfect storm' for widespread salicylate toxicity. The loss of Bayer's patent on aspirin in February 1917 allowed many manufacturers into the lucrative aspirin market. Official recommendations for aspirin therapy at toxic doses were preceded by ignorance of the unusual nonlinear kinetics of salicylate (unknown until the 1960s), which predispose to accumulation and toxicity; tins and bottles that contained no warnings and few instructions; and fear of 'Spanish influenza,' an illness that had been spreading like wildfire." Dr Karen Starko

Given the role that it played in the millions of 1918 deaths, a further, more in-depth assessment of aspirin is probably expedient. Bayer aspirin was one of the earliest of drugs from the pharmaceutical industry, dependent on the oil industry and has become the most commonly used staple, trusted 'cures.' As such, it represents a good example of the displacement of natural treatments by synthetic drugs. But *is* it actually safe?

During 1918 repeatedly, first-hand medical accounts point to aspirin as the source of pneumonias. *"I had a package handed to me containing 1,000 aspirin tablets, which was 994 too many. I think I gave about a half dozen. I could find no place for it. My remedies were few. I almost invariably gave Gelsemium and Bryonia. I hardly ever lost a case if I got there first, unless the patient had been sent to a drug store and bought aspirin, in which event I was likely to have a case of pneumonia on my hands."* Dr J.P. Huff, Olive Branch, Kentucky

Aspirin was the first of the non-steroidal anti-inflammatories, others not becoming available until 1955 when Tylenol™ was first marketed. They are the most commonly used drugs on the market, sold both with and without prescriptions. For Bayer and the entire pharmaceutical industry, they are the absolute foundation stones of their profitability *and yet they all routinely kill.*

"Over 100,000 people are hospitalized for internal bleeding and of those, 16,500 die every year. And these values are considered 'conservative.' Also the figures only include prescription NSAIDs used to treat only arthritis and only in the United States. If prescription and over the counter NSAID-related hospitalisations and death rates were counted for not only arthritis, but for all conditions and throughout the world, the figures would no doubt be enormous. Taking those figures and applying them over the many years that this class of drug that has been available since the early 1970s and the numbers would be horrific. And yet, no study to date has attempted to quantify these figures." 'Toxic and Deadly NSAIDs,' an investigative report by Roman Bystrianyk

Another important observation is that most people receive no warning signs that these drugs are causing them internal damage before they end up in hospital with a serious medical condition and approximately 10% of these hospitalisations end in death. Considering that aspirin is still being highly recommended by the vast majority of medical practitioners to reduce the incidence of heart disease we must also consider the catastrophic levels of gastrointestinal damage being caused.

It has been discovered that no particular dose of aspirin between 75 mg and 300 mg daily currently used in the treatment of heart ailments, is completely free of risk of causing intestinal bleeding. Even very low doses of aspirin reportedly caused gastric bleeding in volunteers. Some 10,000 episodes of bleeding occur in people aged 60 and over each year in England and Wales alone and it is estimated that around 90% of those 10,000 episodes could be associated with and directly ascribed to aspirin usage.

Unfortunately, the risk of hospitalisation and death is not the only problem caused by taking these types of drugs. Other studies have also indicated that the risk of congestive heart failure (CHF) while using NSAIDs is also quite substantial. One author suggested that the number of deaths could be similar to those that are evident with gastrointestinal bleeding. If so, the numbers of deaths attributed to NSAIDs would increase dramatically from the already large figure of 16,500.

It was also discovered that recent use of NSAIDs by elderly patients doubles the odds of being admitted to hospital with an episode of CHF. The estimated relative risk for first admission with heart failure and the risk of this outcome was increased substantially by NSAID use in those with a history of heart disease.

NSAIDs, particularly generic aspirin are truly a silent epidemic that has caused a tremendous amount of pain and unnecessary, premature deaths. Public knowledge of this tragedy is virtually non-existent, with an enormous amount of critical information primarily existing within the sanctuary of medical libraries and thus being unavailable to the public in general. Big Pharma still markets and promotes worldwide sales of these toxic substances and governmental agencies have done nothing at all to alert the public or even medical practitioners, many of whom remain totally unaware of these facts.

The conservative estimate of 20 to 50 million deaths during 1918 has long been attributed to a virulent new virus but the NIAID has now

The Falsification of Science

clearly stated that common upper respiratory bacteria was responsible, not a new virus. There was no new deadly virus but there was *something* new in 1918 and that was toxic aspirin, being used in totally inappropriate, dangerous dosages. And of course the heavy use of new experimental vaccines, forced on soldiers and also pushed on a fearful unsuspecting public via coercion and in some cases mandates.

Based on the primary role that aspirin and vaccines played in the millions of 1918 deaths versus the survival of those who avoided it in favour of natural treatments, it would suggest that 1918 was not a plague caused by a virus, but directly by the pharmaceutical industry and is another perfect example of the iatrogenic effect, or 'death by medicine.' And given the scale of the deaths, it was without question, the greatest medical catastrophe in human history, exceeding even the 'Great Plagues' of the Dark and Middle Ages.

Whilst the events of 1918 expose the extreme toxicity of aspirin, it also reveals something even more profound; the continuing abject failure of the oil-based synthetic, pharmaceutical drug industry to treat disease and the persistent Elite propaganda directed at we, the masses, in order to cover up and obfuscate this fact.

The overall conclusion must therefore be that 'Spanish influenza' was not the cause of the 1918/19 deaths. Whilst focusing on medical evidence indicating that aspirin overdose is the most reasonable explanation for the terrifying rapid deaths and aspirin use appears responsible for the conditions that led to lethal pneumonias, there is an instigating factor prior to the use of aspirin itself and that is Bayer's desire for profits to the detriment of all else. As millions of people died in this reckless quest, how much money did Bayer make because of the 1918 'flu?'

This question is extremely pertinent and should be answered. However, we would not advise 'holding our collective breath' on any answers being forthcoming any time very soon. As always, the Elite spider web of deceit and chicanery will prevent anyone from investigating these issues too closely and anyone who does get too close...well, we will leave that to the reader's own imagination.

HIV/AIDS

Almost forty years ago, at the beginning of the 1980s, a small group of homosexual men in San Francisco began dying of a strange, never before seen disease.

However, it turns out that there was nothing too mysterious about these deaths, either then or now. The combination of three cultural revolutions, drug, sexual and gay, took a heavy toll on these, its most ardent practitioners. They took part in rampant, anonymous, soulless gay sexual intercourse on an almost industrial scale, copiously administered pre-emptive antibiotics, thinking that this would keep them healthy (useless against viruses, which do not exist anyway, but deadly effective against gut-friendly bacteria which are vital to overall health) and imbibed recreational drugs with no consideration for their long term effects.

The main drug used by gay revolutionaries beginning in the 1970s nightclub scene is commonly known by its slang term 'poppers' which describes a chemical psychoactive class of drugs called alkyl nitrites. The nitrite is taken by inhalation after 'popping' glass ampoules wrapped in fabric between the fingers to release the chemical. The drug's effects are a relaxing of smooth muscles including the throat and anus, the latter, which facilitates anal sex by relaxing the sphincter. This is in part why it became so popular in the gay party scene. The other reason is that the rush or 'high' created also enhanced the experience of recreational sex. This same drug is known for its abuse on airlines amongst the '5-mile-high club' because of the access flight attendants had to this drug, once included in airline first-aid kits until the abuse led to health problems in airline staff and was subsequently withdrawn.

The inhalation of this chemical substance has a profound detrimental effect on the lungs and can cause a myriad of respiratory ailments such as those common in AIDS sufferers. Additionally it can cause pronounced lesions around the nose, mouth, lips, and face; the type of lesions considered hallmark symptoms of AIDS patients.

The Falsification of Science

Headaches and allergic reactions are other known effects of using poppers.

In fact some of the first doctors, when faced with pressure to diagnosis this 'new disease' as first GRID (Gay Related Immune Deficiency) and later AIDS, recognised that the symptoms they were seeing in these groups of gay men did not in fact describe a new disease at all but lined up perfectly with the physical effects resultant of excess indulgence in the nitrite drugs used recreationally by these patients.

Other health issues related to the gay revolution lifestyle derived from lack of sleep (due to excess partying), malnutrition (due to lack of regular meals, sometimes caused by deliberate starvation related to eating disorders associated with body image issues like anorexia or bulimia, or as a result of the extreme partying lifestyle). The use of other recreational drugs in conjunction with nitrites and 'uppers' such as amphetamine drugs to stay awake for jobs the day after partying, also all contributed to disease manifestation (arising from disease conditions, including poor sanitation in some cases arising from drug abuse and orgies) that came to be known by the catchall label of AIDS.

Also noteworthy is the profound effect that extreme promiscuity had on the psyche of participants in the 'gay revolution' of the times. It was then considered a rite of passage for gay men to have as many sexual partners as possible in order to in effect prove their 'allegiance' to the cause and to display their 'pride' in being homosexual. But so much disconnected or 'soulless' intimacy can be very emotionally damaging, which can also lead to physical illness from a holistic perspective. So it is not homosexuality or gay sex per se, which was to blame for the serious illnesses originating in this particular subculture during the late 1970s and early 1980s, but the related physically and emotionally taxing lifestyle. Since the Germ Theory of disease is not provable, thus STD (sexually transmitted disease) is not the true cause of any manifestation of illness and not then the origin of the AIDS 'epidemic.'

Just how a handful of 'AIDS' cases specific mainly to the San Francisco male homosexual demographic reached global 'pandemic' levels next affecting Sub-Saharan Africans, then IV drug users and blood donors, and finally copulating heterosexuals and most tragically newborn babies is, as usual, mired in political and industry corruption.

When Ronald Reagan took Presidential office in 1981, he had a mandate to downsize government and the Centers for Disease Control and Prevention (CDC) was an obvious target. The *'War on Cancer'* declared by Nixon in 1971, had little to show for all the money spent. The CDC had been acutely embarrassed in 1976 when it attempted to make believe that five soldiers with a mild dose of flu were a potential national 'swine flu' epidemic. A subplot of this charade was their attempt to seize on the completely coincidental outbreak of pneumonia among some visitors to an American Legion convention in Philadelphia. The CDC used this as an excuse to rush out a vaccine that killed hundreds of people, which is hundreds more than did the flu itself. So-called 'Legionnaire's Disease' as it transpired, turned out to be attributed to a known microorganism, and not in fact related to the swine flu, but the fact remains we do not actually know if this supposed 'disease' is unique or if there are other causes to the supposed 'outbreak' such as aspirin use.

So it was very fortunate for them that in 1981, as potential budget cuts were mooted and imminent, that the CDC received a report about five young homosexuals dying of what was being referred to as immune deficiency disorders. If a 'new' deadly disease could be discovered and promulgated, it could give the CDC a whole new lease of life and raison d'être. The more dangerous and terrifying the better and ideally something a little more deadly than simply 'the flu' this time, would fit the bill nicely.

'Gay Related Immune Deficiency' (GRID) as an initial name for the new disease, was soon discarded. Besides being too 'politically incorrect,' it did not sound too threatening to the general (straight) population and thus would not threaten a large enough portion of the population to be profitable. So it was eventually replaced with

The Falsification of Science

'Acquired Immune Deficiency Syndrome' (AIDS). The French scientist Luc Montagnier later 'discovered' (but significantly never isolated or proved its existence by accepted scientific criteria) the Human Immunodeficiency Virus (HIV) and failed American cancer researcher Robert Gallo co-discovered it (if you can call finding it a year later 'co-discovering it'). Gallo was later investigated for misconduct but was eventually cleared).

So now today, we have AIDS as the disease and HIV as the 'undisputable' cause of it. A press conference by the CDC subsequently launched the idea to the general public, and thus the under-threat CDC was saved (hoorah!) and a new multi-billion dollar profit centre for the medical cartel was created.

In the years since the CDC announcement, every scientist or doctor who has tried to question the official story, rather than being applauded for practicing good investigative and stringent science, has instead been attacked and dismissed. That same old MO again – it never fails! But now with forty plus years of history behind it, it is becoming more and more obvious that certain facts do not stand up to scrutiny with the official story.

The filmmaker Brent Leung made a stunning documentary revealing the full story behind the Aids/HIV scam, called *'The House of Numbers.'* This film was absolutely savaged by the controlled mainstream media film critics (why would it not be?) with one review in particular, likening Leung to someone 'who would question gravity.' Oh, the irony! In my view and for what it is worth, had the media been complimentary and demanding answers from the medical establishment, then I for one, would have seriously questioned whether or not the film was anti-AIDS or subtle disinformation. If there is one thing that one can virtually guarantee, that is that when the mainstream media sharks attack (or indeed defend) anything, there *must* be an ulterior motive for it and you can usually bet your house that the exact opposite of their pronouncements will prove to be truth, upon further, in-depth research.

In the film, Leung totally debunks HIV testing. But it transpires that HIV testing is almost irrelevant anyway as the WHO has provided a definition for AIDS which lists simple symptoms to use for diagnosing AIDS **without testing**...

"The WHO AIDS surveillance case definition was developed in October 1985 at a conference of public health officials including representatives of the CDC and WHO in Bangui, Central African Republic... For this reason, it became to be known as the 'Bangui definition for AIDS.' It was developed to provide a definition of AIDS for use in countries where testing for HIV antibodies was not available." Wikipedia™

It stated the following:

Exclusion criteria

Pronounced malnutrition

Cancer

Immunosuppressive treatment

Inclusion criteria with the corresponding score	**Score**
Important signs	
Weight loss exceeding 10% of body weight	4
Protracted asthenia	4
Very frequent signs	
Continuous or repeated attacks of fever for more than a month	3
Diarrhoea lasting for more than a month	3
Other signs	
Cough	2
Pneumopathy	2
Oropharyngeal candidiasis	4

Chronic or relapsing cutaneous herpes	4
Generalised pruritic dermatosis	4
Herpes Zoster (relapsing)	4
Generalized adenopathy	2
Neurological signs	2
Generalised Kaposi's sarcoma	12

The diagnosis of AIDS is thereby established when the score is 12 or more.

Although it was moderated nine years later with the instruction that testing should really be undertaken, it was instrumental in kick-starting the supposed AIDS epidemic in Africa. How convenient it is though, to create a disease from nothing in this fashion by artificially constructing a score-based system from arbitrary, unconnected symptoms.

But even with testing, it is quite easy to say that there is more HIV in one place than another, as the tests are interpreted differently in different countries. At one point in the film Leung straddles the US/Canadian border and comments, *"No other disease behaves differently when you cross a border."*

He also visits South Africa to examine the 'epidemic' for himself. It is difficult to say what is more shocking about Leung's visit to a poor indigenous village, the ignorance, and superstitions that people have regarding AIDS. It is also highly significant that these villages tend to have no access to clean water, necessitating their inhabitants to drink sewage-contaminated water and the horrendous malnourishment and starvation rampant in those parts is also a significant contributory factor to causing disease

Leung interviewed several scientists and doctors in the film and they essentially fell into two distinct groups. Those sceptical about the AIDS story included, among others, Kary Mullis, incidentally the inventor of the PCR test, which has now recently come to media prominence, who shared a 1993 Nobel Prize in chemistry, Joseph Sonnabend, a physician who was involved with AIDS research and treatment since the very beginning and James Chin, an epidemiologist at the WHO for five years, whose characterisation of that agency's statistics on the AIDS epidemic in Africa gave the movie its name. And then there was also Peter Duesberg, who was a cancer researcher until he was recently rebuked and subsequently ostracised for questioning the official line on HIV.

On the other side of the debate were, among others, Robert Gallo, Luc Montagnier and Dr Anthony Fauci (whose name will appear prominently in the final chapter of this book!). This group, speaking in defence of the AIDS syndrome comprised a curious mixture of the detached and mildly irritated as they related their weak arguments asserting that the virus exists, that HIV causes AIDS, everyone is at risk and that anyone simply not agreeing with their viewpoint is an idiot.

However, when questioned more closely and intensively, they unanimously conceded that there are gaps in the knowledge of how HIV works, but nevertheless contradicted each other and in one instance, the subject actually contradicted *himself*. None of them were able to define AIDS in a simple and consistent phrase, were able to explain satisfactorily or convincingly the mechanism by which HIV works, directly address the issue of the HIV virus never having been isolated (and thus proven to exist) or were willing to address the problem of so many deaths attributed to AIDS that in fact were caused by the incredibly toxic drugs administered to cure it.

Leung himself avoided taking sides, rather playing the annoying devil's advocate, persistently asking probing questions. As the film progressed however, it became obvious that the answers given were totally inadequate and would not satisfy anyone with anything even resembling a semi-functioning brain.

The film also related the story of a girl by the name of Lindsey Nagel. She was originally a Romanian orphan adopted by Steve and Cheryl Nagel, a couple from Minnesota. Having been tested for HIV in Romania which proved negative, she was then tested again upon arriving in America and this time the test was positive. This is hardly

The Falsification of Science

surprising once it is appreciated that the tests can and do vary so much from country to country (and from test manufacturer to test manufacturer).

Of course, in the beginning, even though their daughter showed no signs of illness, the Nagels followed their paediatrician's recommendations to treat her with antiretroviral drugs, which at the time meant high dosage AZT (the most common anti-AIDS drug). For months, the Nagels watched on helplessly as their initially healthy daughter deteriorated, becoming sicker and sicker. Among other things, her growth became stunted. Of course all symptoms were ascribed to her supposed HIV infection and not to the drug used to 'treat' it.

Then, after nearly two years of this utter horror show, the Nagels became aware of Peter Duesberg's dissenting view through a relative who read an article about his work. They became intrigued and wrote to Duesberg, who replied immediately, telling them to take Lindsey off the antiretroviral drugs or they would kill her. They did so and for that reason alone, Lindsey is still with us today – and completely healthy with no symptoms of 'AIDS.'

The paediatrician responsible for prescribing the poison that almost killed poor little Lindsey actually received an award in 2005 (what a surprise!) for her 'leadership in treating HIV patients' and in a subsequent interview about the award, had this to say...

"We started on AZT (Retrovir) for a child who was adopted, and the parents said it was a poison and they called Peter Duesberg, the man who wrote a book claiming that AIDS isn't caused by HIV and they pulled the child from my care."

She obviously though conveniently omitted from her statement the fact that Lindsey was on the cusp of death until Duesberg's timely intervention! However, others were not so fortunate...

"There was nothing you could do years ago. Most children back then did not live past seven to twelve years old. And it was hard; these were children that you got attached to. It was really hard. All we could do was

provide some supportive care and treat their opportunistic infections. We had many deaths, ten to twelve in 1994."

The doctor quoted above also went on to state that children stand a better chance of survival now, implying that the treatment has improved, but does not mention the fact that this is only because the dosage of retroviral drugs has been **greatly lowered**. However, these drugs are still nonspecific, toxic, and eventually still do kill most of those who take them.

Of course, Lindsey Nagel was not the only one who benefitted from ending her intake of deadly AZT, this was also happening in Africa too.

"Recently CNN dispatched a reporter to the West African country of Gambia to do a story on Gambian President Yahya Jammeh, who announced in January that he has discovered a cure for AIDS. President Jammeh has come up with a recipe consisting of seven herbs and spices that is administered to an HIV-positive patient once a day…

…Mr. Sow, according to his own testimony, has been HIV-positive since 1996 and had been taking antiretrovirals for the past four years until he volunteered to try President Yahya Jammeh's new treatment. After only four weeks, he gained 30 pounds and felt like a new person. He feels cured and has no more 'HIV symptoms.'…Mr. Koinange is sceptical, though, not so much about the witness's honesty (in fact, he interviewed a lot of patients who made similar statements) but of the scientific basis of the treatment, as the government refuses to provide any medical records that might back the patients' claims.

But the patient, if one will listen carefully, may have provided a full explanation of the efficacy of the dictator's dream-potion. He has been HIV-positive since 1996. What a strange virus this is, that threatened to ravage North America so many years ago, as once a healthy person caught it (which could happen quite easily through normal sexual contact, we were told), the incurable and unstoppable disease AIDS always set in, inevitably killing the patient. Today in North America, the only people who are killed by AIDS are the same people who have always been killed by it, i.e., severe drug abusers and/or homosexual men engaging in a

The Falsification of Science

certain sexual practice (usually both). So the AIDS 'disease' has moved on to a new market, Africa, where millions of people are supposedly infected with the HIV virus and are going to start dying any day now. Yet, Ousman Sow had the virus from 1996 until 2003, seven years, before he started taking the antiretrovirals to 'save' his life. It was then, I would be willing to bet, that he started really experiencing his 'HIV symptoms.' Within only four weeks of ceasing the antiretrovirals, he regained lost weight and felt well again.

The cure that our witch doctor has inadvertently found may be nothing more than getting 'HIV positive' patients off their antiretrovirals."
James Foye

The treatment for HIV has always been non-specific, DNA destroying drugs. In a supreme irony, the prophecy of a destructive epidemic became on a small scale, self-fulfilling, as tens of thousands died from the very drugs that were supposed to cure them. Of course, they 'officially' died from the disease and not the drugs themselves. All of the defenders of the HIV/AIDS orthodoxy are paid, directly or indirectly by government (i.e., they work for the government, or a university that is subsidised by government or a pharmaceutical company whose AIDS drug business largely depends on people believing what the government says about AIDS and whose drugs are paid for by the government). Dissenters like Peter Duesberg, are shut out and disenfranchised, attacked or both.

"HIV does NOT cause AIDS. HIV does not cause anything. This is a staggering statement given the hype and acceptance by the scientific establishment and through them, the public that the HIV virus is the only cause of AIDS. HIV is a weak virus and does not dismantle the immune system. Nor is AIDS passed on sexually. There are two main types of virus. Using the airplane analogy, you could call one of these virus strains a 'pilot' virus. It can change the nature of a cell and steer it into disease. This usually happens very quickly after the virus takes hold. Then there is the 'passenger' virus which lives off the cell, goes along for the ride, but never affects the cell to the extent that it causes disease. HIV is a passenger virus!

John Hamer

AZT was developed as an anti-cancer drug to be used in chemotherapy, but it was found to be too toxic even for that! AZT's effect in the 'treatment' of cancer was to kill cells—simple as that—not just to kill cancer cells, but to kill cells, cancerous and healthy. The question and this is accepted even by the medical establishment, was: would AZT kill the cancer cells before it had killed so many healthy cells that it killed the body? This is the drug used to 'treat' HIV. What is its effect?

It destroys the immune system, so CAUSING AIDS. People are dying from the treatment, not the HIV. AIDS is simply the breakdown of the immune system, for which there are endless causes, none of them passed on through sex. That's another con which has made a fortune for condom manufacturers and created enormous fear around the expression of our sexuality and the release and expansion of our creative force.

What has happened since the Great AIDS Con is that now anyone who dies from a diminished immune system is said to have died of the all-encompassing term, AIDS. It is even built into the diagnosis. If you are HIV positive and you die of tuberculosis, pneumonia, or 25 other unrelated diseases now connected by the con men to 'AIDS,' you are diagnosed as dying of AIDS. If you are not HIV positive and you die of one of those diseases you are diagnosed as dying of that disease, not AIDS. This manipulates the figures every day to indicate that only HIV positives die of AIDS. This is a lie.

AZT is the killer. There is not a single case of AZT reversing the symptoms of AIDS. How can it? It's causing them, for goodness sake. The AIDS industry is now worth billions of pounds a year and makes an unimaginable fortune for the drug industry controlled by the Rockefellers and the rest of the Global Elite." David Icke, geopolitical researcher, and author. 2007

Interesting is it not, that the AIDS industry also refers to those who try to expose the truth as being 'deniers' which has shades of the other great 'denials,' those of the 'holocaust™ industry'—and also as previously related, vaccines and germ theory? So in effect they are psychologically likening doctors who try to tell the world the truth

The Falsification of Science

about what is really happening, with 'Nazi sympathisers' and 'conspiracy nuts.'

"I reported that in early 1987 I had received a telephone call from a researcher for a TV company who had told me that his company (Thames TV) was planning a documentary about AIDS. 'What do you think about AIDS?' he asked me. I told him that I thought that the threat had been exaggerated by some doctors, a lot of politicians and most journalists. The researcher was silent for a moment or two. I could tell by the silence that he was disappointed. It wasn't quite what he'd hoped to hear.

'We're planning a major documentary,' he said. 'We want to cover all the angles. Haven't you got anything new to say about AIDS?' 'I don't think AIDS is a plague that threatens mankind,' I insisted. I then pointed out that I believed that the evidence about AIDS had been distorted and the facts exaggerated. 'We really wanted you to come on to the programme and talk about some of the problems likely to be caused by the disease,' persisted the researcher. 'I'm happy to come on to the programme and say that I think that the dangers posed by the disease have been exaggerated,' I told the researcher.

The researcher sighed. 'Quite a few doctors have said that to me,' he said sadly. 'But it really isn't the sort of angle we're looking for.' Very gently I put down the telephone. I didn't expect to hear from the researcher again and I didn't. His company produced a networked television programme about AIDS that appeared on our screens a short time after that conversation. And I suspect that most of those who viewed it went to bed believing that AIDS is the greatest threat to mankind since the Black Death. That was by no means an isolated incident. The facts about AIDS were carefully selected to satisfy the public image of the disease—and to satisfy those with vested interests to protect." Dr Vernon Coleman MB ChB DSc FRSA, British former 'TV Doctor,' dropped from the networks because of his noncompliant views

And whilst we still argue and debate the issues, people are literally dying to know the truth.

ADHD and ADD

Many young students, now used to such modern educational devices as the Internet and computers, are classified by the dinosaurs of the education system in conjunction with Big Pharma as having ADHD 'Attention Deficit Hyperactivity Disorder' or ADD 'Attention Deficit Disorder.' These 'abnormal children,' for some strange reason, just cannot seem to handle outdated, centuries-old educational systems since they now have access to much more interesting methods of self-learning. So, what do the authorities do with these 'misfits?' They drug them to dumb them down and make them compliant and unresponsive whilst labelling them as a 'problem child' so that they can handle sitting in an uncomfortable wooden chair and reading uninteresting so-called 'facts' out of dingy old books, all day long, whilst listening to some boring old 'wrinkly' (aka anyone over 30), droning on to them about the latest scientific and historic propaganda.

Dr Leon Eisenberg, the 'father' of ADHD, said just before his death that ADHD *"…is a prime example of a fictitious disease."* He passed away at the age of 87 in 2009 and was a prominent figure in the field of child psychiatry who during the 1950s and 60s conducted medical studies of children with developmental problems, including some of the first rigorous studies of autism and attention deficit disorder. As described by the *'British Medical Journal'* (BMJ) Dr Eisenberg *"…transformed child psychiatry by advocating research into developmental problems."*

Early in his medical career, in the mid-1950s, Leon Eisenberg became fascinated with the childhood mind. Wanting to know more, he broke free from the shackles of the Freudian psychoanalytic dogma that dominated child psychiatry at the time to conduct ground-breaking biologically based research of childhood developmental problems. This research included the first randomised clinical drug trials in child psychiatry.

"I think what Leon brought to the field was a different way of thinking—thinking out of the box," said David DeMaso, chairman of psychiatry at Boston Children's Hospital and professor of psychiatry

and paediatrics at Harvard Medical School. *"He was thinking in terms of biology, of evidence based treatment, way before anybody else. His was a bio-psychosocial model at a time when psychoanalytical thinking was the norm."*

But was he correct in believing that ADHD was and is, 'fictitious'?

Nowadays, if parents physically chastise naughty children, social workers will soon remove them from parental care, and they may even end-up behind bars for their 'troubles.' But please do not worry, the government actually pays parents now to give them the kind of mind-altering drugs that if sold on the streets would constitute a serious criminal offence. The drug *Ritalin*™, the most common form of treatment for ADHD and ADD is almost literally, chemical castration for the brain.

Many years ago, when I was at school, we had our fair share of students who always handed in their homework on time, always got top marks for everything, always did their work neatly, presented it in clean and tidy exercise books, and were primarily, from so-called, 'middleclass' families. They also went on to attend university in an era when only two percent of all students in the UK achieved that distinction and are in all probability now, on the cusp of 70 years of age, drawing their healthy pensions following their retirement from highly lucrative careers. I have no problem with this. No doubt they worked hard to support their families, brought up their children to be equally 'good' citizens, led thoroughly productive lives, and they now deserve to relax a little and enjoy the fruits of a lifetime's hard work—as does everyone of that age, regardless of background or achievements.

They are what I call 'conformers,' and there is nothing wrong with that to be fair. They probably married people who also handed in their homework on time, got top marks, did their work neatly, and also came from similar middleclass backgrounds. And doubtless their children and probably now their grandchildren also handed in *their* homework on time, got top marks and presented their work neatly in clean, tidy exercise books. Barring the unexpected, they now watch these grandchildren develop and achieve good grades in their tests and exams. They no doubt (until recently!) also go on annual cruises, safari holidays and attend golf club and reunion dinners.

I was one of the lucky ones too, despite definitely not emanating from a middleclass background. I went to a 'good,' well-respected school, a Grammar school no less, with mainly excellent teachers and good 'discipline.' We knew that any, even minor transgression would result in a form of corporal punishment being meted-out, not that I am in any way advocating physical punishment as a solution to child behavioural problems, you understand.

It was not so much the fear of the terrible retribution, ever lurking below the surface that persuaded us to behave, but first and foremost the respect we had learned at home, from our parents. They had also worked hard, and my father for example served with the British Army during his 'National Service,' in India in the late 1940s—and both my grandfathers had patriotically volunteered for service in World War I. Upstanding citizens, all!

Secure in our elitist, well-regarded Grammar school, we were also aware that not every student was the same. The children from the 'other' school in our fairly remote rural area were generally not so well-regarded. They smoked at bus stops, habitually stole from the local shop, and adopted what could be described as a more 'liberal' attitude to their compulsory school uniform, and some dispensed with the sheer inconvenience of it altogether, as well as growing their hair long and generally being rebellious in many other aspects of their behaviours. These were the children with whom our 'respectable' parents had warned us about mixing. They had no worries about handing *their* homework in on time because they never bothered to do any. Their idea of fun in their resultant, copious amounts of spare time was hanging out on the street with their mates causing trouble in a variety of inventive ways, and the best they could ever hope for in life was an existence of low paid, unskilled, repetitive work, and eventually producing equally troubled, neglected offspring (often to different mothers—or fathers) and occasional encounters with the police of varying severity.

The Falsification of Science

I am generalising and exaggerating greatly above of course in case anyone takes me to task for snobbery of some kind!

But back to being a little more serious, there has always been social division in the world. The standards and principles of the 'haves' are passed down to their offspring, just as the standards and principles (or lack of them) of the 'have-nots' are passed to *their* offspring. Children are hugely influenced by the behaviour and the attitudes of their parents. It is all part of the rich tapestry of life here on Earth.

ADHD is virtually unknown in the UK Asian communities. A fact which raises an interesting question. Why then is it so common in benefit-dependent white families?

This is the moment where I stick my head firmly above the parapet and probably commit some kind of criminal offence or at very least commit a form of 'political incorrectness.' It seems to me that ADHD/ADD is much less a 'disease' than the lack of parental care and ensuing discipline, that causes it. A good friend of mine, a very experienced and successful psychotherapist said that in his opinion, ADHD/ADD is a 'social/council housing disease.' I believe he is correct. Most children who are diagnosed with ADHD are almost all from families to whom parental responsibility—and the part it plays in successful child rearing—is an alien concept. However, I do concede that it is maybe less a socio-economic issue than it is to do with parents being present with their children. And nowadays, because of technologies both distracting parents and being used as a babysitter for kids, there are a LOT more children with this so-called disorder and this does cover a wide range of economic or 'class' groups.

The 'politically correct brigade' will doubtless all have collective heart failure at the mere mention of this idea, but the plain fact is that the children of reasonably well-to-do and/or responsible parents are almost never on Ritalin™. Instead, their parents tend to talk to them, guide them, spend quality time with them and they monitor their progress at school as growing, developing human beings. But I do concede that this is not *always* the case. They also usually instil in them a sense of respect, for themselves and others, a sense of self-esteem and self-discipline, and above all, the idea that the way to be the best is to do your best. This is what is commonly known as *taking an interest* and *good parenting*. All it requires is a little common sense, time, and dedication to one's own flesh and blood.

So it would appear that it is now the kids from the families whose mothers or fathers are unwilling or unable to engage in good parenting or to provide their children with the correct guidance that begin their formative years on prescription drugs and there is a terrible price to pay for this short-term solution. Children who are now drugged with Ritalin™ (the so called 'chemical cosh') do not learn and are robbed of their ability (and their right) to develop. It does not take a genius to work out that this is storing-up huge problems for the future and a recipe for promulgating the exact same problems down through the generations yet to come.

There are no two ways about it. Prescribing Ritalin™ to children is legalised and government sponsored drug dealing, as well as legalised child abuse. Cannabis and a dozen other illegal drugs have exactly the same effect and yet taking these drugs can land you in court and result in social services removing your children into care. The government and their apologists continually lecture us on the pitfalls of smoking and drinking (and taking *certain* drugs), yet they allow our children to be pumped full of mind-altering drugs of which Ritalin™ is by far the most popular and market leader. If it *were* cannabis or amphetamines they were prescribing, there would be an outcry. Yet Ritalin™ in many respects, is worse than these.

The Government's official health policy on ADHD is tragically flawed. It is a perfect example of what happens when politicians take the line of least resistance in a desperate attempt to ease the financial burden on the NHS and other government services. Our education and healthcare systems are already stretched to breaking point, so it is far cheaper and simpler to dole out the pills rather than spend the time giving the children the more traditional tried and tested guidance. Of course, the government recognises that you cannot legislate against parents who simply cannot be bothered to practice effective childcare, so in the absence of good parenting, the quick and

The Falsification of Science

easy solution is 'pill-popping' for seven or eight year olds – or even younger ones.

Naughty children have been around since the dawn of humanity. The cure has always been proper parenting and firm discipline and in America, this is called 'tough love.' In the modern world, it is often claimed that parents are just too busy, a hypothesis which does have some merit. We work much longer hours for less money (relatively speaking) than we have for many years and I can therefore understand that parents find it increasingly difficult to devote the ideal amount of time required, to their children's mental welfare. And on their downtime parents are sucked into virtual worlds, addicted to their digital gadgets and social media. If they pay any attention to their children at all, it is through the screens of their phones while taking cute pictures to immediately post to Fake-book and see how many 'likes' they can get. These children soon learn that they need to be louder and bolder and 'cuter' in order to get attention from 'mummy' at the playground or elsewhere. Children are competing with digital technology for their parents' attention. Ultimately the addicted, distracted parents (also exhibiting 'ADD' symptoms) pass along the same addiction to their children especially when providing them with tablets and iPhones as substitute babysitters. And in most families these days, both parents working fulltime has been socially engineered to become the 'norm,' primarily because of the extra tax benefits it brings to governments and not as glibly stated by them, 'sex equality.' But also of course, because one income households are no longer adequate financially to provide for the whole family.

The inconvenient truth however, is that the ability to give birth to babies is nothing special, as truly wonderful, and almost miraculous as it may be. Half of the population can do it, and most of them plan their pregnancies in line with their ability to bring up their offspring responsibly, taking into consideration such factors as finances and the time and resources they reasonably expect to have available. These usually tend to be the people who handed their homework in on time and exercised a degree of self-discipline etc. etc. They plan their families sensibly and are more likely to make sure their children do *their* homework and hand it in on time too.

There is certainly more than just a grain of truth in the theory that Big Pharma has identified certain conditions and reinvented them as diseases so they can sell more drugs. (see the section later in this chapter on invented diseases) The number of prescriptions for Ritalin™, given by British doctors alone and who themselves are now often too busy to care, has risen by more than 50% in just five years. From 420,000 to 650,000! This leads to another wholly related problem and that is that children will soon learn that the answer to any problem is to 'pop a pill.' Is that really the message we want to send to our children? Nevertheless, the good news for Big Pharma is that the ADHD drugs market is now worth more than £12bn ($16bn) per year.

Turning children into sedated zombies however is only part of the problem. There are often disturbing side-effects for children, such as depression, which then leads to more drugs to counter these symptoms. Another side-effect is a loss of appetite, almost certainly a result of the child not doing normal childhood things, such as playing outside in the fresh air. Children instead, are now seated in their school classrooms, quietly sedated 'zombies' or glued inextricably to their computer screens playing mindless, repetitive 'games' that require no concentration at all, into the wee, small hours day after day, month after month.

Schools simply cannot cope with the ever-growing numbers of undisciplined children from undisciplined homes, and so legal sedation is the quick and easy solution. The upside of course though, is that when these kids eventually leave school, there is a ready supply of labourers, garbage collectors and cleaners, etc. All very convenient of course for servicing modern cities, where removing the trash and cleanliness is a very important job, but the big drawback is that it also totally stifles creativity and originality in the children it is supposed to be helping. Remember, these kids are not mentally ill, despite the propaganda, they just have not had the same chance in life as some of their peers or may simply be more physically oriented, preferring physical over mental stimuli.

The Falsification of Science

And physically motivated kids are at an increasing disadvantage in today's world where due to liability fears, (so we are told, though it may just be an excuse to further drug and keep children captive, chained to desks and screens); time outdoors engaged in physical activities is being increasingly curtailed at schools the world over. Specifically at some schools, children whilst outside during their meagre, allotted break/recess times are no longer allowed to **run**. Running, apparently, is now 'too dangerous' in this overly risk-averse world and may cause a child to have an accident!

And other children get headaches from exposure to computer screens and have extreme difficulty focusing whilst test taking in this manner. And since test scores now count more than any other grading methodology, these kids are mislabelled as lacking intelligence or having a 'learning disorder.' Some kids, forced into learning in highly-toxic electrically-polluted wi-fi saturated environments, are unable to cope with these EMF exposures but are treated as disciplinary cases instead of with compassion when they complain. Many of these sensitive kids, especially in their teens, end up committing suicide through sheer desperation, when teachers and parents casually dismiss their pain and protestations.

It is also important to understand that any overworked GP or medic will also naturally opt for the line of least resistance. All that is needed to deal with this situation is a couple of lines in the child's online medical record and the filling-out of a prescription. That way at least, the parents of these children will not be bothering him/her again, anytime soon.

The government in the UK now claims that 5% of children suffer from ADHD/ADD. This is something that was virtually unheard of twenty years ago. In the US it is considered to be as high as 10% yet in France, Germany, and Italy, where family ties are stronger and closer than in either the UK or the US, where families go out together, play, dine and entertain together, ADHD is almost unknown. What does that tell us? The real problem is that ADHD is being treated as a disease, which it most certainly is not, but it is a great strategy for selling even more drugs.

In fact ADHD/ADD has now become the Orwellian 'newspeak' for any form of childhood behavioural problem. All human beings, including all children, are different. Some children are naturally talented in certain areas, for example in music and the arts or sciences, whilst some children of all educational persuasions are naturally boisterous and more physically orientated (as previously mentioned). Children finding it hard to concentrate for set periods of time, should have their energies and talents directed elsewhere, somewhere where they will be allowed to realise their hidden potential. It is a fundamental mistake to assume that our schools should be turning out attentive, scholarly 'clones,' or obedient robots. But that is exactly the direction in which our educational systems are headed.

That children need discipline is as true today as it ever was, particularly with the new temptations furnished by the Internet and social media. It is not possible to administer even a gentle slap on the legs anymore, because of all the political correctness that prevails and pervades every aspect of our lives. I am not referring to the brutal beatings of Victorian times of course—all agree that would amount to abuse—but the traditional short, sharp shock is a tried and tested strategy as a gentle but firm reminder to a child of their responsibilities, in some circumstances. It is also reflected throughout the animal kingdom, where adults often give their offspring a gentle cuff to bring them back into line. Undisciplined kids grow up to be undisciplined adults and undisciplined adults cannot form cohesive or productive social groups—and in turn produce equally undisciplined children. And so on and on the destructive cycle continues.

There is now also a Big Pharma instigated new description for temper tantrums, which are henceforth to be known as 'Disruptive Mood Dysregulation Disorder.' And disobedience, which all children experiment with from time to time as they test boundaries as well as their parents' patience, is simply part of life and growing-up. We all did it on occasion, except it is now known as 'Oppositional Defiant Disorder.' The cure for both these truly *horrendous* conditions is also now the prescription drug, Ritalin™, which is to repeat myself, chemical castration for the brain.

The Falsification of Science

Despite all of the above, it is too easy to parrot the line of *'I blame the parents.'* That would be totally unfair and has the wrong emotional connotations, if only because parents who themselves did not have a 'proper' upbringing cannot be expected to cope with the onslaught of their own unruly kids. *They* need help too, but definitely not in the form of drugs.

In the uber-politically correct third decade of the 21st century, parents have become afraid of their own authority because of the ever-present spectre of social services, the new 'SS.' Here again, is the dichotomy... gently slap your naughty child and social workers will take them away from you—you could even go to prison. But it is absolutely fine to give them the kind of drugs that if sold on the streets would constitute a serious criminal offence.

In the UK, ADHD/ADD is now classed as a disability, enabling parents to claim state benefits. Carer's allowances and disabled child tax credits have, in a depressingly large percentage of cases, turned children into cash cows. This is of course a really clever ploy on the part of Big Pharma, in effect convincing the government to pay for these drugs with taxpayer's money. Very astute—and nice work if you can get it. Some parents even receive benefit-funded cars as a result of 'having to cope' with an ADHD-afflicted child. Unfortunately, there are many parents out there who do not give a damn about their children and are only too happy to exploit them, just as there are women who get pregnant simply so that they can upgrade from a social-housing, one bedroomed flat to a social-housing two or three bedroomed house, or who even become pregnant in an attempt to avoid prison sentences. But in a sense who can really blame them? They are only exploiting 'loopholes' in the system, just like their richer counterparts who have large scale tax avoidance and evasion off to a 'T,' or the criminal banksters who routinely 'fleece' we ordinary folk at every given opportunity.

In the UK in 2018 there were 57,000 families receiving extra benefit payments for ADHD children. As a result, our beautiful, precious, vulnerable, innocent children are being turned into licensed junkies. So what, if some of them are badly behaved or over boisterous, surely this is not the solution in any caring society? But the harsh truth is that it is far from a caring society any longer – if indeed it ever has been. What will be next do you imagine? Pre-frontal lobotomies for kids available at *Tesco*™ or *Safeway*™ next to the Botox injections counter or the 'instant chemical castrations' counter? Or two children's lobotomies for the price of one on Wednesdays?

Or just maybe... parents could try raising their voices ever so slightly occasionally (in a constructive way), making steady eye contact with their child and saying firmly, *"turn that off, put it down, and do your homework and if you don't do it, then no more computer, Xbox, or mobile phone – and you are also grounded, for a week!"* But they would have to mean it and stick to it. That is presuming of course, that as parents, they can even be bothered to deal with the hassle that would no doubt in and of itself, cause.

But just when I thought that the debate about ADHD could not get any more heated or contentious, along came a perfectly reasoned and rational explanation for our obsession with ADHD. A recent study suggests that the cause may be even more simple, and the solution may have been staring us in the face all this time!

Could it possibly just be that all those kids with ADHD might just be too young and immature to be in that particular classroom in the first place. The fact is that babies born in August are more likely to be at risk of being diagnosed with the accursed ADHD than those born in September. This makes perfect sense (assuming any school year group runs from 1st September through to 31st August the following year – as is the case in the UK). After all, of the 500,000 kids diagnosed, researchers discovered that it is the youngest kids in any given class that are more at risk of being labelled as suffering from ADHD. What this means, in simple language, is that the younger kids do not have the same emotional maturity as the older kids in any given class. Worse, teachers, parents and psychologists have been making a dreadful mistake – they have been judging the youngest by the higher standards of the oldest and therefore often the higher achievers.

The Falsification of Science

Think about this. In any class of say 4 or 5 year olds (or any age group for that matter) there will be almost one whole year's difference between those who turned say, 6-years old in September and those who will not turn 6 until the following August. As time goes by the relative difference will obviously become smaller as a proportion but it will be particularly marked in the 4-8 age range for example, where one whole year is a huge proportion of a young child's entire lifetime.

This means that some children are being misdiagnosed, even discriminated against, because of their age and the difference in ability to learn which ageing brings naturally. If this is really the case, then the whole ADHD phenomenon is an absolute scandal (even assuming it was not before)!

The number of boys thought to suffer from ADHD who were born in August is 4.5% compared to 2.5% for those born in September. Likewise, 2.9% of girls born in August were diagnosed as positive, compared with only 1.8% born in September. Obviously, other factors have a role to play such as nature and family environment, but the age gap discrepancy is just too great to ignore, and researchers have confirmed that the youngest children in a school year are more likely to be diagnosed with the condition. Thus 'developmental immaturity' may be a significant factor in the diagnosis. However, lest we forget, these children are all receiving medication in the form of a dangerous drug!

When the statistics were broken down by age, researchers found that adolescents (12-16 year olds) born in August were not at an increased likelihood of being diagnosed. These figures are consistent with previous studies in the US and Canada and seem to imply that increasing age and maturity reduces the impact of birth month on ADHD diagnoses. But please remember, it is predominantly the very young who are being prescribed the Ritalin™.

Symptoms of ADHD/ADD include impulsivity, fidgeting, arguing and excessive talking. I believe that this is a normal part of childhood development and the obsession with 'pigeonholing' behaviour and labelling any behaviour that falls outside certain predetermined parameters, as 'unusual,' is the main culprit here. These symptoms often continue into adulthood and are just simply a product of the diversity inherent in human life. But the possibility that some children may be diagnosed simply because they are younger and less mature is frankly, quite scary and smacks of criminal deception and fraud.

These findings have been published in the '*Journal of Paediatrics*' and emphasise the importance of considering the age of a child within a year/grade when diagnosing ADHD and prescribing medication. It has long been accepted that the month in which a child is born has also been found to affect educational achievement, with older pupils generally achieving better grades than their younger peers.

I have always believed that as human beings, we are all often very different and some are just different in a different way, if that makes sense? The large numbers of children who have been diagnosed with ADHD suggest that this difference is more 'normal' and that sufferers should be more accepted. Indeed, ADHD/ADD diagnosis reeks of an exercise in being grossly unfairly judgemental and using false parameters to justify the prescription of totally unnecessary drugs—and that surely has to be totally reprehensible!?

But try convincing the drug companies of that. They are not interested in children's welfare in any way, shape, or form. All they care about is their financial 'bottom-line,' and there is no doubt that the so-called, wholly invented 'diseases' of ADHD and ADD contribute significantly to that.

Cancer

"Everyone should know that the 'war on cancer' is largely a fraud." Linus Pauling, Nobel Laureate

"I keep telling people to stop giving money to 'cancer research' because no one is frigging looking for a cure. We have several and they have been carefully hidden away from public view...this is a multi-billion dollar

The Falsification of Science

per year industry and a 'cure' would put a lot of people out of work." Geraldine Phillips, cancer research worker, 2011

"The chief, if not the sole, cause of the monstrous increase in cancer has been vaccination." Dr Robert Bell, former Vice President, International Society for Cancer Research

Dr James Watson won a Nobel Prize along with Dr Francis Crick for discovering and describing the double helix shape of the DNA molecule at Cambridge University in the early 1950s and during the early 1970s he served two years on the US National Cancer Advisory Board. In 1975, he was asked his thoughts about the American National Cancer Programme. Watson declared, *"It's a bunch of shit."* Blunt and crude though his assessment may be, it also happens to be true.

Cancer, that 'life-threatening disease' and ruthless killer of countless millions of mothers, fathers, sons, and daughters in the last 100 years or more, is relatively easy to cure and even easier to prevent.

I am acutely aware of the emotive subject that cancer has become and do not make this glib-sounding statement lightly, but with due deference to the millions who have lost loved ones and / or suffered terribly and had their own lives cruelly cut short for what amounts to no reason at all, unless of course you consider the vast, unimaginable profits made by the purveyors of this great criminal racket, for criminal racket is exactly what it is.

In 1953, a United States Senate investigation reported in its initial findings that there was the strong suspicion of an ongoing conspiracy to suppress and destroy effective cancer treatments. The Senator in charge of the investigation died suddenly in unexplained circumstances, the usual MO (modus operandi or operating method) in these cases, which was obviously very convenient for those with much to lose from his revelations. As a result of his death, the investigation was subsequently, suddenly disbanded without further ado and was never resumed. Unsurprisingly, the good Senator was neither the first nor the last of literally hundreds if not thousands of strange, unexplained deaths involving people in positions to threaten the interests of those running the Elite controlled cancer programmes and indeed the Elite controlled anything else. Ethical people who attempt to disrupt the flow of profits into the Elite's coffers have to be silenced one way or another, after all.

But this is only the small tip of a very large and extremely dangerous iceberg. In 1964, the FDA spent millions of dollars to suppress and bury an 'alternative' cancer treatment which had cured hundreds, if not thousands of cancer patients according to well-documented sources. It became apparent and was later disclosed that in the subsequent court proceedings, the FDA had falsified the testimony of witnesses, to suit its own ends. The FDA lost the court case because the jury found the defendants innocent and recommended that the substance be evaluated, objectively. In fact it never was evaluated but instead all the evidence was totally suppressed and then conveniently 'lost.'

For many years (and still to this day), the American Medical Association (AMA) and the American Cancer Society (ACS) coordinated their own 'blacklists' of cancer researchers who were regarded as threats to their cancer monopoly and who were to be singled-out for smear campaigns and ostracised by the mainstream. One investigative reporter declared the AMA and ACS to be "... *a network of vigilantes prepared to pounce on anyone who promotes a cancer therapy that runs against their substantial prejudices and profits.*" The ACS, believe it or not actually makes political donations! A 'charitable organisation' that makes political donations? What does this tell us about them and the system within which they operate?

In the late 1950s, it was learned that Dr Henry Welch, head of the FDA's Division of Antibiotics, had secretly received $287,000 (a colossal sum in those days) from the drug companies he was supposed to regulate. In 1975, an independent government evaluation of the FDA still found massive 'conflicts of interest' among the agency's top personnel.

And In 1977, an investigative team from the prominent newspaper '*Newsday*' found serious 'conflicts of interest' at the National Cancer

The Falsification of Science

Institute (NCI) and in 1986, an organised coverup of an effective alternative cancer therapy, orchestrated by NCI officials, was revealed during Congressional hearings. The list goes on and on and once again, I strongly suggest to the reader that they should undergo their own research on this topic and not simply take my word for it. It is extremely simple. Just key into a search engine, the phrase 'alternative cancer treatments,' or 'cancer has been cured,' for example, plus any other similar or relevant phrase. However please be aware that there is now so much Internet censorship, especially through the chief culprit in this regard, Google™ that many useful websites which contain this information are consigned to total anonymity by the simple expedient of being suppressed by search engines. Some do 'slip through the net' though.

The cancer 'industry' now has a more than 70-year history of vast corruption, incompetence and organised terror against its many detractors and a shameful track record of suppression of cancer therapies which are actually beneficial. Millions, if not billions of people have suffered terrible torture and death because those in charge took bribes, had closed minds to the innovative, or simply were afraid to do what was obviously and ethically correct. Instead, corporate, and individual greed and the desire of the few to profit from the many, as always, take precedence.

"The doctor's union (AMA) the cancer bureaucracy (NCI) the public relations fat-cats (ACS) and the cancer cops (FDA) are conspiring to suppress a cure for cancer ... It would be easy for any Congressional committee, major newspaper, television network or national magazine to confirm and extend the evidence presented here in order to initiate radical reform of the critical cancer areas -- the hospitals, the research centres, the government agencies, and especially state and local legislation regarding cancer treatment.

But that will not happen without a struggle. Neither Congress nor the media desire to lift the manhole cover on this sewer of corruption and needless torture. Only organized, determined citizen opposition to the existing cancer treatment system has any hope of bringing about the long-needed changes. I expect the struggle to be a long, difficult one against tough, murderous opposition. The odds against success are heavy. The vested interests are very powerful..." Barry Lynes, *'The Healing of Cancer'*

There is a veritable mountain of overwhelming evidence and examples which support the theory of collusion between activities of Western governments, especially the United States, along with other prominent members of the 'medical Elite' to prevent an effective cancer treatment being promulgated.

Surgery is a massive shock to the system, uses carcinogenic anaesthesia and increases the risk of cancer in the resultant scar tissue. It has value only where the threat to life processes is immediate, as in digestive obstruction etc. The routine removal of every malignant or sometimes even benign lump, surrounded by the body with a defensive shield, can be virtually a death sentence, especially in the elderly.

Chemotherapy involves the use of extremely toxic petrochemical drugs originally derived from the highly toxic chemical weapon nitrogen mustard also known as 'mustard gas' famous for its deadly use in battle during WWI. Oddly enough, in 1942, two Yale doctors (Goodman and Gilman) after researching the poisonous effect of mustard gas on WWI soldiers, decided since the gas seemed to destroy normal white blood cells (and notably lead to cancer) in these exposed victims maybe it could also destroy cancer cells. So they experimented on a patient (simply referred to as J.D.) with advanced lymphoma who had several serious tumour growths and found that nitrogen mustard significantly shrunk the tumours. Never mind J.D. still died 6 months later, the reduction in tumour size was enough to declare the procedure as a 'success' and bring the drug, with minor modifications, to market. Chemotherapy, despite its questionable origins, is still used today in the hope, which is often never realised, of killing the disease before killing the patient. The drugs are designed to kill all fast growing cells, cancerous or not, and to systematically poison all cells caught in the act of division. The effects include hair loss, violent nausea, vomiting, diarrhoea, cramps, impotence, sterility, extreme pain, fatigue, extreme cognitive impairment or 'brain fog' (aka 'chemo-brain'), cancer and death. According to the government's own figures, around 2% of chemotherapy recipients are still alive after 5 years. The term 'alive' is used here in its literal sense, i.e. not yet clinically dead. One of chemotherapy's less well-known side-effects is pneumonia. Many cancer patients die of this after undergoing chemo treatment and their cause of deaths are not recorded as 'cancer.' In this way, death statistics can easily be manipulated to demonstrate that they are 'winning the war' against cancer.

"Toxic chemotherapy is a hoax. The doctors who use it are guilty of premeditated murder. I cannot understand why women take chemotherapy and suffer so terribly for no purpose." Oncologist, Channel 4 TV, 2010

Radiotherapy likewise is equally, if not more deadly. One person who chose to have treatment with the radiation machine turned off altogether was the British Grand National winning jockey Bob Champion. Convinced by the early detectors, in spite of feeling well, that he was "... *likely to die of cancer of the lymph gland,*" he decided that he did not relish the thought of a treatment that "... *could have ruined his lungs,*" let alone the rest of him. He eventually survived the alternative treatment and the 'lymphoma.' His doctor, the 'cancer specialist,' Ann Barrett, declared that "*He is the only patient in my experience who has come through this disease and achieved such a high degree of physical fitness afterwards. His recovery is even more remarkable when you consider that he refused to have the conventional treatment!*" Or not?

The plight of the ever-increasing number of parents of child cancer victims facing 'radiotherapy' was well illustrated in October 1993 "... *after learning of the appalling side-effects of radiotherapy... her anxious mother has opted to take her to America for private treatment... I've been told the radiotherapy will cause brain damage knocking forty points off her I.Q... Her growth would be stunted... she would need hormones to help her growth and sexual development. It is also likely she would be sterile.*'" Further associated 'delights' include bone and nerve damage, leading to amputation of limbs, severe burns and of course, death, at a future time, from cancer and leukaemia due to the highly carcinogenic effects of the huge doses of radiation.

"Chemotherapy and radiotherapy will make the ancient method of drilling holes in a patient's head, to permit the escape of demons; look relatively advanced... the use of cobalt... effectively closes the door on cure."
A cancer researcher who wished to remain anonymous

The 90/95% death rate within a five year period has not stopped the cancer industry from carrying out the same procedures, day in, day out, for decades with the same deadly, inevitable results. Temporarily suppressing, with the scalpel, drug or radiation the symptoms of cancer does nothing for the victim's chances of survival.

Adding gross insult to injury, the treatment involves massive doses of carcinogens and super-poisons. The patient is subject to a regime diametrically opposed to that which is needed for survival. Succumbing to cancer is an acceptable form of suicide for those who have lost the desire to live, this loss being a major factor in the development of the disease in the first place. The great tragedy and scandal is in cases where the victim has a strong determination to live and fight but is then destroyed by the assault from the lethal, useless treatment and not by the cancer itself.

So why are the vast majority of doctors against alternative cancer treatments and why would they actively encourage us to undergo known-to-be-dangerous treatments such as chemotherapy, radiotherapy, and surgery instead of using natural cures?

Unfortunately, doctors are against 'alternative' treatments because from the first day of Elite-controlled medical school, they are brainwashed into believing that disease can only be effectively treated by those methods proscribed by Big Pharma. They most certainly will have been led to believe that there are no cures for cancer, when in reality there are several, none of which will enhance the profits of Big Pharma or sustain the payments on a senior hospital consultant's Aston Martin. Additionally they operate under the severely inhibiting paradigm that food is good enough to keep you alive but not sufficiently good enough to keep you healthy or heal you when you are sick.

The Falsification of Science

Most cancer drugs cost in the region of $40,000 per annum *per patient*. In the US this is payable either by the individual or by their health insurer (assuming they are adequately insured) whereas in the UK this is paid by the NHS (National Health Service). However, whichever way, the fact is that this is the amount paid into the coffers of Big Pharma, per person, per annum and when you consider the number of people worldwide who suffer from and die from cancer each year, I am sure you can do the maths. What incentive is there for any organisation whose first responsibility is always to maintain a profit for its shareholders and owners, to discover a cure? (Turkeys 'voting for Christmas' springs readily to mind.) I submit that there is no reason at all and this is the true cause of the utter failure (despite the eloquent hype) of Big Pharma in their self-styled 'war on cancer.'

We are even deceived by the so-called professionals in such seemingly beneficial activities as 'cancer screening programmes.' For example, mammograms, heavily promoted as being an integral part of the early detection of breast cancer, provably achieve nothing other than to irradiate the breast and in many cases actually *cause* the cancer it is supposed to be detecting.

Most doctors believe not only that what they were taught in medical school *must* be true, but they also believe that what they were *not* taught cannot be important and as a result of this are unable to comprehend anything that falls outside of their area of knowledge. Most doctors are still thinking 'inside the box' when it comes to cancer and doctors who do think for themselves instead of regarding their learning as gospel and treat the actual cause of disease rather than the symptoms are regarded as 'quacks' and are subjected to huge pressure, ridicule and threats to conform. One of the FDA's modus operandi is to raid the offices of alternative thinkers and practitioners, destroying their medical records and often all their equipment and putting them in jail.

Additionally, some doctors (especially in the US) are afraid of expensive, time-consuming lawsuits and their insurer could well refuse to pay out if they use alternative treatments of any kind. Their medical boards may fine them and even revoke their licence to practice or strike them from the medical register, effectively disbarring them from medical practice forever. Peer pressure is a huge issue too. After all, doctors are only human, and their colleagues will not be slow to publicly ridicule them if they use alternative treatments or are seen to be using or endorsing 'non-conventional' medicine.

"Doctors will continue to fail with cancer until they buck the training and accept that a patient is not some collection of malfunctioning cells but a human out of homeostasis. We have cultures alive today who don't get cancer. No stress, no speed cameras, no mobile phones, no Iraq War. Don't get me wrong, I truly believe 21st century civilisation has much to commend It, but there are downsides. We're a toxic society and that includes the medicines. If cancer is striking 1 in 3 of us, that means something is going fundamentally wrong and we're either going to be honest about it or continue canoeing down that long river in Egypt called De-Nial, splurfing down the ratburgers until the meat wagon comes to collect us." Philip Day, health researcher

Cancer Research UK spends £170 million, annually, on 3,000 research scientists whose brief is to *avoid* any research into holistic, naturopathic, nutritional treatments; therapies which provide the ONLY means to successfully treat a cancer victim.

"Using the guise of 'established' medical science, many widely accepted studies are disseminated through medical journals and accepted as the ultimate authority by many. In the case of Professor Sheng Wang of Boston University School of Medicine Cancer Research Center, his cancer research was found to be misconducted, fraudulent and contain altered results. What is unsettling is the fact that his research had been previously accepted and used as a cornerstone from which to base all subsequent cancer research." Andre Evans. Activist Post, 19th October 2011

"The American Cancer Society was founded by the Rockefeller family to act as a propaganda outlet and public relations tool to suck-in money and help promote pharmaceuticals for cancer 'therapy.' Gary Null did a fantastic exposé on who and what the ACS is, in a series of articles about 10 years ago and he often retold his experiences on the radio in coming to realise what a fraudulent outfit the ACS actually is. People are simply

The Falsification of Science

giving aid and comfort to Big Pharma when they support the ACS." Ken Adachi, political researcher, May 2011

However, not all studies are fraudulent, but when the motivation for these doctors and professors is financial, it turns the current medical paradigm into a war zone. As a consumer, it is vitally important that you undertake your own research on the harsh side-effects of traditional cancer treatment methods such as chemotherapy.

There is much evidence that there are in existence literally hundreds of alternative cancer treatments which really do work. Some are even of sufficient potency or are fast acting enough to effectively treat a cancer patient who has been deemed to be 'terminal' by his/her doctor. As untold millions are pumped into the fake cancer industry that thrives on provably fraudulent research, it is important to remember that free, alternative health options do exist. Utilising natural sweeteners, vitamin D therapy and eliminating artificial sweeteners such as aspartame in its many guises, are extremely simple ways to effectively prevent cancer and potentially begin reversing it. Additionally reducing or eliminating exposure to wireless (microwave) radiation and avoidance of chemical containing personal care and cleaning products are very potent cancer prevention exercises, as all of these exposures have been proven carcinogenic. Even as recently as 2018 the US National Toxicology Program, after a $30m FDA funded decade-long study into the correlation between microwave radiation from regular mobile phone use, found *"clear evidence of cancer"* from exposure to this form of radiation. (Notably the FDA has thrown out the results stating that 'animal studies have no bearing on humans,' even though it was the FDA which requested the animal studies in view of the fact that it is considered scientifically unethical to experiment on humans before testing on animals, and that animal testing for safety of products used by humans is the current scientific standard.)

It is not my intention here to relate those cures to the reader as this is outside the remit of this book. It is obviously desirable that everyone become familiar with a few different working methods of *preventing* the disease rather than trying to affect a cure at the eleventh hour, so to speak and these preventative and curative strategies are all available in abundance on the Internet. However, even should the worst happen, and you are unfortunately diagnosed with cancer of some kind then it is still not too late to adopt the 'cure rather than prevention' approach in 90% of cases and this is true even in cases where traditional cures have been attempted and apparently failed.

Autism

Controversy has raged for years over whether mercury received through vaccines is sufficient to cause harm to children. Virtually all studies absolving mercury-containing vaccines of safety deficiencies have been conducted by vaccine insiders with a financial stake in the outcome, rendering them in effect, worthless.

Chronic neurological disorders, especially autism, have increased rapidly during the past four decades in correlation with increases in vaccines and total mercury and aluminium exposure (plus a huge increase in wireless radiation exposure). In July 1999, the CDC, the American Academy of Paediatrics and vaccine companies agreed to remove mercury from all childhood vaccines 'as soon as possible,' but at the time of writing it still remains in literally dozens of licenced vaccines, many of which are *still* given to infants.

The CDC claims that there is 'no convincing evidence' that vaccines cause significant neurological damage or autism and cites a number of studies claiming to exonerate mercury from blame. But of course, they would say that, wouldn't they?

"The studies they reference are all deeply flawed, and as was the case with decades of 'tobacco epidemiology,' were deliberately manufactured to hide the truth," Jim Moody, director of *SafeMinds*

Data from the CDC's Vaccine Safety Datalink first revealed an association between mercury and brain damage, but these findings were later suppressed, and the data were manipulated to exonerate thimerosal. This scientific manipulation was first revealed through documents obtained under the Freedom of Information Act, leading to a best-selling book, '*Evidence of Harm*' and to a published retraction of any 'no cause' interpretation by the study's lead author Thomas Verstraeten (who by then had left the CDC for the vaccine manufacturer, Glaxo Smith Kline.) The most recent study of VSD data by Young et al, made additional findings that vaccine mercury not only caused autism but several other neurodevelopmental disorders. Despite criticism from the Institute of Medicine and Congress, the CDC still refuses to grant access to VSD to private researchers and the Justice Department refuses to permit petitioners in vaccine courts, access to these crucial data.

As part of my research for this particular topic, I conducted an Internet search by use of the phrase, 'autism links to vaccines' and received literally hundreds of links the vast majority of which led to articles or comments about how any links of autism to vaccines were as a result of, 'fraudulent research,' 'now discredited research,' 'false hypotheses,' or 'fake news,' plus a whole wide variety of other disinformation. I found it rather significant that all of these websites had either, government, CDC or other mainstream medical connections. Most 'natural health' websites and blogs tell it how it really is. For example...

"As for the safety of vaccines, I am afraid the media have perpetuated their usual propaganda, which is not based on scientific fact. In truth, there have been NO safety studies published on vaccines. I repeat: NONE. Scientists have tried, sure, but nothing exists currently that proves the safety of any vaccine. The American Association of Paediatrics (AAP) published what was supposed to be a safety study, but they would have been better off not to; it turned out that the 'study' was riddled with fraudulent data and a scandal broke as a result, which is now in full bloom!

Some years ago, the US government was forced to create a Vaccine Court which, to date, has paid out about $2 billion to vaccine-injured kids' families. The causal link between some childhood vaccines and autism is now so well established that the Vaccine Court has had to pay out big awards. This has been discussed in the US Congress and is no longer deniable – scientifically or clinically.

However, it is not necessarily true that autism is always caused by vaccines. Two other causes are well documented in the medical literature. One cause of autism is brought about in a transmission of Lyme bacteria to the foetus in the mother; the other frequent cause of autism comes from mercury toxicity in fish, water sources, or from dental amalgam fillings. The extreme form of that is Minimata Disease, which first occurred in Japan due to mercury-contaminated water. (This is all reported in the mainstream medical literature).

With this in mind, we can clearly see that autism is not an inherited disease as some would have us believe. There may be evidence to the contrary someday, but currently we don't have any." Helke Ferrie, *'vitalitymagazine.com'* 2016

…all of which sounds much more plausible to me than the pathetic excuses and obfuscations, routinely spouted by the mainstream medical mafia and their apologists in order to propagandise we, the masses.

Big Pharma has been under pressure to remove mercury from all its vaccines for more than two decades now, and eventually did so, to a limited degree.

"A great example of this type of deception is the story of mercury (thimerosal) in vaccines. As parents became more hesitant and vocal during the late 1990s, the CDC began a highly publicized campaign announcing that they were 'phasing out' mercury from U.S. childhood vaccines. By 2001, they declared that the manufacturing of thimerosal-containing vaccines had ceased. In reality, some of these vaccines were changed to 'low-mercury.' Because childhood vaccines were deemed "essentially" mercury-free at this point, health officials were quick to assert that autism was not linked to mercury as autism rates continued to increase. What they didn't tell the public was that during the so-called phase-out period, the CDC began recommending mercury-containing influenza vaccines to pregnant women. They also added four doses of the pneumococcal vaccine (PCV), which has a high aluminum content, to the childhood immunization schedule in 2000, and two doses of the aluminum-containing hepatitis-A vaccine in 2005. These increases

led to a 25 percent increased uptake of aluminum for babies by 2005. Additionally, in 2001, the CDC began recommending pregnant women receive the aluminum-containing pertussis vaccine (Tdap), despite the fact that studies show that aluminum crosses the placenta and accumulates in foetal tissue.

Perhaps the most shocking recommendation is the influenza vaccine, as the vast majority of these vaccines contain 25 micrograms (mcg) of thimerosal per dose. Unsuspecting parents who take their children to be vaccinated according to schedule will most likely submit their children to nineteen doses of mercury-containing influenza vaccines between six months and eighteen years of age. In other words, today's children receive nearly as much mercury as prior to 2001 and they receive 25 percent more aluminum. Although the FDA and CDC vehemently deny a correlation, could this be why one in two children in the U.S. is chronically ill, one in six suffers from a neurodevelopmental disorder and one in thirty-six is autistic?"

https://www.westonaprice.org/health-topics/vaccinations/aluminum-in-vaccines-what-everyone-needs-to-know/

What further evidence do we need that there is a conspiracy to maintain the status quo and continue to poison and kill on an industrial scale? The very fact that governments remain complicit and do not intervene also speaks volumes and adds fuel to the fire of conspiracy. It is obviously in everyone's interests (except the general populace of course) that things remain as they are, and that the cull should continue unabated.

Dental Amalgam

The official position of the British Dental Association (BDA) is that *'silver'* fillings are safe and that amalgam (containing mercury) is an appropriate material to shore-up decaying teeth. They also state that there is no proven connection with adverse health conditions. This actually conflicts directly with a statement by the Health and Safety Executive which states... *"Mercury forms a large number of organic and inorganic compounds. Mercury vapour and almost all of these compounds are highly toxic. Less hazardous substitutes should be used whenever possible."*

Over 14,000 scientific papers worldwide have been published which suggest that mercury is a toxic material and should therefore *not* be used in fillings. However, more recently the BDA has released a statement saying that more research needs to be carried out to ensure safety. The UK Department of Health actually advises all dentists not to remove or replace amalgam fillings in pregnant women!

It is almost universally agreed that mercury can produce serious side effects. It has become increasingly apparent that a large minority of dentists are no longer happy placing amalgam fillings in patients' mouths or subjecting nursing staff or themselves to further potential toxicity.

"About 3% of the population are estimated to suffer from 'mercury sensitivity.'" The British Dental Association (ie. the Elite-controlled and run British Dental Association)

That statement would be hilarious if it was not so serious. Who do they really think they are kidding? Everyone is 'sensitive' to mercury in much the same way as we are 'sensitive' to Sarin nerve gas, concentrated hydrochloric acid or a bullet to the head. Mercury, as they well know, is a deadly, cumulative poison causing all manner of unpleasantness to our bodies. Amalgam based dental fillings contain not only mercury but other highly toxic metals which produces vapour as our teeth naturally grind together in the chewing process, and that slowly poisons us.

Good health is impossible to achieve in the face of constant, low-level mercury poisoning, which certainly occurs in people with mercury based amalgam fillings. As an example, thyroid abnormalities cannot be corrected until amalgams are removed. From the time the mercury is placed in your mouth, until the day you die (or you lose the teeth) chewing causes small quantities of mercury vapour to escape. It then enters your bloodstream and is delivered to all parts of your body, including your brain.

The Falsification of Science

In California, dentists are required by law to inform patients of mercury risk. Several European countries have outright bans, and the German government reimburses victims for mercury removal. It is commonplace for dental associations to deny the dangers, due to their well-founded fear of being beset with lawsuits from people they have damaged over many decades. Mercury fillings, unfortunately, are still widely used in the UK and are still vehemently defended by many dentists. I leave you to guess their motives in this gross deception.

"The ADA owes no legal duty of care to protect the public... If you are a dentist still using mercury amalgam, be careful. If you tell your patient that it is harmful, you already know that the ADA will come after you. But... if you don't tell your patient, you might be sued for not providing informed consent." American Dental Association in a letter to dentists. This says it all, quite succinctly.

The California State Board of Dental Examiners recently published a warning that mercury is a known toxin that has been shown to escape into the body. The US Environmental Protection Agency classifies mercury-filling material, once removed from the mouth, as a toxic waste that must then be carefully handled in special containers and buried in toxic waste sites.

"When I trained as a dentist some years ago we had no training in nutrition or in the safety of materials. In fact, we were told once mercury is mixed into the alloy to make amalgam it is perfectly safe.

However, not long after working in general practice I realised that the suction unit has a filtration device that captures the chunks of amalgam drilled out of a tooth and that the dental nurses would empty this into a jar containing x-ray fixative chemicals at the end of the day so that it could not gas-out mercury.

Once the jar was full it would be collected to go away as a hazardous waste material as we are legally not allowed to dispose of it down the sink or in the garbage. That material is identical to what I was drilling and placing in teeth yet once out of the mouth was a toxic waste product. How could that be?" Rachel Hall, former National Health Service dentist in the UK

Mercury toxicity has been linked to the following ailments;

Severe headaches and migraine

Fatigue

Poor Concentration

Irritability

Myalgic Encephalomyelitis (M.E.) or 'Chronic Fatigue Syndrome'

Multiple Sclerosis

Alzheimer's disease

If you have amalgam fillings, my strong advice would be to find a dentist who is capable of and willing to remove them and replace with bio-friendly alternatives. As you have probably gathered by now, our controlled 'healthcare' system does not care about you or your family's health in the slightest. We need to start fighting back against this insidious group as soon as we can, in every way that we can.

Energy-Saving Lightbulbs

These lightbulbs (Compact Fluorescent bulbs or CFLs) are strongly promoted as being a huge contributor to the 'green' economy, even to the extent that it is now no longer possible to buy the standard lightbulb we have been used to all our lives. However, when we examine the reality, it can be seen to be somewhat different to the rosy picture painted of the eco-friendliness of this abomination.

Firstly, eco-friendly or healthy they most certainly are not. CFLs are filled with mercury and furthermore they also emit UV radiation when activated. An average CFL bulb contains 5mg of mercury and considering the fact that ingesting even the tiniest amount of mercury can be very harmful and as the US OSHA (Occupational Safety and Health Administration) points out; the permissible level of mercury vapour is 0.1 milligram per cubic metre. A CFL contains 50 times that amount and even more significantly, the threshold limit value for skin contamination is just 0.025mg per cubic metre, 200 times the amount within one CFL!

Should one of these 'eco-friendly' bulbs break open, anyone within the same room will be exposed to a massive risk of contamination. A study published on the 6th July 2011 showed that once broken, a CFL continuously releases mercury vapour into the air for months and depending on how well ventilated the room is, can well exceed human safety levels.

The effects of exposure to mercury vapour can be damage to central and peripheral nervous systems, lungs, kidneys, skin and eyes in humans and damage to immune function and perhaps most alarmingly, the release of methylmercury, a chemical compound formed in the environment from released mercury that can cause brain damage. This is not exaggeration at all, but simply stated fact.

CFLs operate in the 24-100 kHz frequency range, classified as Intermediate Frequency 5 (IF5) by the World Health Organisation. This in itself gives rise to fears of biological damage as it has been demonstrated that disturbances in this range cause sufficient interference to raise severe health concerns. In California, a recent study of cancer clusters linked the increased risk of cancer with teachers who taught in classrooms where the GS (Graham Stetzer, a means of measuring dirty electricity) was above 200 GS units. Ironically, a lightbulb which is designed to be eco-friendly and less hazardous is classified as emitting 'dirty electricity,' yet the traditional incandescent lightbulb does not.

Also despite the wild claims of CFL's promoters, questions also arise over its life expectancy. There have been claims suggesting that these bulbs can last up to 10,000 hours whereas 5,000–8,000 hours would probably be closer to the truth. However, it is impossible to accurately determine a specific lifespan due to the large variety of different conditions existing in each individual home. However, what is known for certain is that by switching the CFL on and off continually, it unsurprisingly, dramatically reduces its lifespan. Even that Elite propaganda dispenser *Wikipedia* states, '*In the case of a 5-minute on/off cycle the lifespan of a CFL can be reduced to close to that of incandescent lightbulbs'.*

And disposal of such toxin-containing products is not regulated in most countries, and even where this is the case, this regulation is not monitored. It is common for any type of lightbulb consumer to dispose of spent bulbs in the bin and not treat them (as CFLs should be treated) as hazardous waste. Public refuse bins often fill with several bulbs, many of which are broken, putting the waste collector at highest risk for excess mercury exposures.

CFLs also take around two to three minutes to fully 'warm up' and therefore should light be required purely for a short period, for example for quickly entering a room to retrieve an item, far more energy will be used than with conventional bulbs. The US Energy Star programme suggests that *fluorescent lamps should be left on when leaving a room for less than 15 minutes to mitigate this problem.'* Is life not already complicated enough without all this to remember?!

A CFL 26-watt bulb is the equivalent of a 100-watt incandescent bulb and costs around £2.50 ($3.30) with a life span of an average of 7,000 hours without factoring in the five-minute on/off cycle. Comparatively speaking, an incandescent bulb costs just around 60p (80 cents) with an average life-span of 800 hours, so according to these figures we would spend around £4.00 ($5.30) more per lifecycle using incandescent eco-friendly, clean bulbs. In effect this then means that we would save a whole £4.00 ($5.30) over a period of one year.

But what cannot be excluded from any considerations of the 'value' of this abomination are the severe, potential health risks and the enormous power surges required by CFLs when switching them and also the cost of the quick on/off cycle. These figures are typically always absent from any study on costs and efficiency, creating as usual, a totally false impression for widespread consumption by the masses.

In addition, the current price of CFLs reflects their almost exclusive manufacture in China, where labour costs are dramatically less and that will surely change, as the Chinese economy grows and achieves world domination as is widely expected by most commentators.

The Falsification of Science

Governments cannot be trusted to control the cost either, as they certainly wish to maximise the import duty on any item. The more expensive the bulb, the more duty they collect is simple mathematics. A good example of this tactic is how we were strongly encouraged to buy diesel cars, as the fuel was much cheaper, but now that sufficient numbers of motorists have been convinced by the 'benefits' of diesel, we find that the tax on diesel has been miraculously increased and thus we now pay considerably more for diesel than petrol (gasoline)! The game is rigged, and the table is most definitely 'tilted.'

Sunscreen

Here is a humorous aside which I wrote for and published on my website falsificationofhistory.co.uk a few years ago . . .

How to Get Away With Mass Murder

Killing stupid people is lots of fun, if you are a sick psychopath that is. But unfortunately, there is such a thing as the justice system, and it tends to go after you if it notices that you are killing people. But do not worry, there is a solution to this, and it is very simple . . . outsmart the justice system. I will show you how this can be done, and how you can achieve a high body count in the dystopian nightmare that is our modern Western civilisation.

Step 1

Find something that is dangerous but is not yet illegal. *"How do I find something that is dangerous but not yet illegal,"* you may wonder? Luckily for you, this is very simple. 'Independent' scientists are constantly looking for things that are killing people, because they are born with a conscience and like to save lives. Look at their discoveries and put your money into whatever it is that they are worried about. For example, let us say that they discovered that microwave radiation is dangerous. Invest in microwave technology, and fund studies that pretend that everything is safe. This allows you to buy time before your technology is banned due to public outrage.

Step 2

Step 2 is for the real geniuses amongst you. It is one thing to just kill people but it's another to get paid for this task by the people you are killing. However, if you are really clever, you can make your weapon mandatory for the public, whilst getting paid to kill people. Vaccines are one example, and another is fluorescent lightbulbs. If you have the idea to ban traditional lightbulbs, this will force people to buy fluorescent lightbulbs. But how would you get traditional lightbulbs banned? That's easy, simply look for something that the masses all have an irrational fear of. A good example would be terrorism. Tell them that traditional lightbulbs can be used to build homemade bombs and if no-one believes that tell them instead that you can help to prevent 'climate change' by banning traditional lightbulbs.

Step 3

So how do you kill people with fluorescent lightbulbs? Encourage them to snort up the mercury? That would be one way, but my suggestion to you is to be patient. A study from 1982 found that exposure to fluorescent lighting at work was associated with a 2.1 relative risk of malignant melanoma, aka deadly skin cancer. In those exposed for more than 10 years, the risk becomes massive. Note, this is 1982, when fluorescent lighting was still relatively rare. How many people have died from skin cancer since then and whose deaths could have been prevented? Countless numbers without a doubt, so what we do now is sell our lightbulbs and wait to get rich.

Step 4

Now, skin cancer deaths are increasing, people are beginning to get worried, and are looking for a scapegoat. Our plan is succeeding. You may be a little disappointed. Perhaps you say *"I've been selling lightbulbs for ten years, spent millions to get the traditional lightbulbs banned, and still my body count has not reached 6 digits yet!"* Well, do not worry just yet, because now your body count will definitely increase. The reason for this is that we are going to blame something healthy for the deaths that we caused. Imagine if asbestos companies, when it was discovered that asbestos causes mesothelioma had said, *"mesothelioma*

The Falsification of Science

is caused by eating fruit!" They would have managed to kill far more people by just telling them a lie than by selling them something deadly. Well, this is what we are going to do too! Your body makes a hormone in response to exposure to sunlight, and the hormone is called vitamin D. It has many beneficial effects, but one of the main effects is that it prevents cancer. You can also get vitamin D in pill form. In studies where women were given vitamin D, they had a 77% reduced risk of getting cancer.

In other words, if you can make people deficient in vitamin D, you can make them get cancer. And how do we make them deficient? Easy, we have to keep them out of the sun! Remember that skin cancer epidemic we created? That is going to come in very handy for the next step in our plan! What we are going to do now is tell people that the reason that they are now all getting skin cancer is not from our fluorescent lightbulbs at all, but because they sit around in the sun all day!

Let's look at the facts here. Each year more than 2,500 people die from skin cancer in the UK. Sounds like a lot? Well, not when we consider the total amount of cancer deaths. In 2016, there were more than 166,000 cancer deaths in the UK. But actually most of the skin cancer deaths, almost 2,000 each year, are from malignant melanoma.

These skin cancer deaths are provably not caused by the sun. How do I know this? Swedish scientists did a study comparing people who work in an office; work indoors but not in an office; and people who work outdoors. This is what they found... There were 1,364 cases of malignant melanoma in office workers, compared to an expected 1,043. For indoor, non-office workers, they found 2,426 cases versus an expected number of 2,583. But here is the best part... In outdoor workers they found only 916 cases versus an expected number of 1,065! The people exposed to sunlight the most had a lower melanoma risk than the people not exposed to sunlight. Another study found that melanoma is more common on covered parts of the body than on those exposed to the sun.

So let's ignore the melanoma deaths for a moment. That leaves us with 500 skin cancer deaths per year in the UK, compared to a total number of cancer deaths of 166,000. That is about one sixth of the number of people that die in the UK every year from road accidents. But it gets even better. About 193,000 people die every year in the UK from cardiovascular disease and vitamin D deficiency also raises the risk of cardiovascular disease. It also reduces the risk of Diabetes. I think you get the point... If we can convince people to stay out of the sun, we can cause slaughter on an unimaginable scale!

So here is what we will do... We will start a campaign to 'prevent skin cancer,' by telling people to stay out of the sun or if they have to be in the sun, to cover their exposed skin with 'factor 50' sunblock, which will in effect, do the same job. It is very easy. We just have to set it in motion by raising the alarm about an increase in skin cancer. Then society will do much of our work for us. They will rally behind our goal, because they are scared. Doctors will tell their patients with a tan to stop sitting in the sun every day. Mothers will keep their children indoors, to watch TV and eat fast food instead of going outside where the sun will 'kill them.' People are so stupid; I promise you they will swallow our story! Before you know it, you will have campaigns like these everywhere!

But of course not everyone will be convinced to stay out of the sun, there are always a few rebellious ones. There is a solution for that, however. We will tell them that if they must go outside, they should apply a strong sunblock. The type of radiation that creates vitamin D is within the UV-B wavelength. However, people will still want to get their tans, thus, we cannot block all radiation, but now we have scammed them so much, let's scam them a little more! Let's sell them sunblock that blocks UV-B but does not block UV-A! We will tell them that UV-B causes skin cancer, when in fact it is mostly UV-A that is responsible!

Step 5

By now there are not many mass murderers left who can compete with our body count. Idi Amin, Pol Pot, Stalin, Mao-Tse Tung—all

The Falsification of Science

mere amateurs! But we have not even started yet. People now avoid the sun, and thus stay inside more, where they are exposed to the fluorescent lighting we convinced them to install in their homes. Also many workplaces leave fluorescent lighting on despite the fact that it is the middle of the day and thus, skin cancer death figures keep increasing! Because people are now seriously vitamin D deficient thanks to us, heart disease and diabetes and cancer in general are going up too! This is where most of our score is now coming from. But people are catching on now. They tell us, *"I have avoided the sun for the past twenty years but now I've got melanoma, and my stupid sister who goes to the beach every day is fine, what's going on here?"* Now we tell them our next scam and that is we are really sorry, but it is now already too late. They had sunburn as kids, and now they are destined to die from skin cancer. They will simply have to accept their fate.

We will tell them to please, think of their children. Do not let them play outside, they might start to enjoy their existence on this planet, and we cannot risk that. Keep your children inside at all costs and let them play FIFA 2020 instead of real football. Let them watch the Discovery and History channels so that they will realise that this is for their own good and keep them restricted. If they get depressed, give them Prozac. If they have too much energy, it is not because they are not free to roam anymore, it is because they are sick, and their sickness is called ADHD. You can cure it by growing your son a pair of breasts through the use of Ritalin™ and other such 'wonder-drugs.' We will tell them that peer-reviewed, double-blind super-scientific studies done by people with IQs a thousand times higher than theirs have shown that growing your son breasts cures his ADHD. If they disagree, we will call them 'far-right populist, scientifically illiterate demagogic conspiracy theorists' and while they are busy trying to figure out what the hell that means, we will be off to live in luxury in some distant 'banana republic' that does not have an extradition treaty.

... Seriously though, sunscreen, sunblock—whatever you choose to call it, is very far from the benevolent preventative measure it is purported to be by those who deceive us for financial gain at every opportunity. Since people first began wearing sunscreen lotion routinely, in the last 30-40 years, in order to protect the skin against the sun's allegedly harmful rays, the incidence of malignant melanoma has soared exponentially. Co-incidence? Once again I will let you, dear reader, decide for yourself.

Invented Diseases

How is it possible to 'invent' a disease – or a syndrome to be more accurate? The answer is quite simple. When you have the financial muscle and the powerful propaganda machine that only extreme wealth can buy, it is fairly straightforward to convince millions that several hitherto unrelated symptoms, when grouped together, comprise a new illness, disease, or syndrome. Indeed this is exactly the case with AIDS/HIV but as that issue has been covered in detail in an earlier section we will concentrate on other examples.

Psychiatry itself admits that, "*...it has not proven the 'disease theory' or the cause or source of a single mental illness it has classified and the theory that a chemical imbalance in the brain causes mental illness has been thoroughly discredited by the psychiatric industry itself,*" mainly because as critics pointed out that as there is no test for a chemical imbalance, it is complete nonsense. They say that psychiatry is the 'original pseudoscience, medical fraud and completely made up.' I believe they are correct.

When one delves beneath the surface in the specialty of psychiatry, what is uncovered is so ludicrous it is difficult to believe that it is really true. Prominent psychiatrists from all over the world gather annually for a meeting at which new diseases are invented. There are no objective findings that establish the diagnosis of these diseases. These new diseases are included in the Diagnostic and Statistical Manual of Mental Diseases. Potential new diseases are discussed at these meetings and new diseases are voted in or out by a show of hands, believe it or not. Truly laughable were it not so serious.

Among recent new 'diseases' voted in by psychiatry, are 'social anxiety disorder' (everyone who ever feels uncomfortable in a social setting has this 'disease') and 'mathematics disease' (anyone who has ever struggled with a maths problem has this disease). 'Gender identity disorder,' 'passive-aggressive disorder,' 'disorder of written expression' and 'sexual disorder' are all other examples of invented diseases that will follow the individuals tagged with these ridiculous diagnoses for the remainder of their now-blighted lives. Naturally, all these completely fake diseases have a psychoactive drug, which supposedly ameliorates the 'condition,' complete with a usually impressive price tag of course.

This may indeed be laughable but unfortunately it has serious consequences. When a child is diagnosed with depression, as they often are these days, the child is often placed on a potent antidepressant drug. The manufacturer of one of these leading drugs knew for many years that the drug caused loss of the ability to control violent behaviour, thus increasing violence towards self (suicidal tendencies) and others (mass murders). This information is of course, completely covered up because it negatively impacts upon sales of the drug. Every teenager involved in the many recent school-shooting rampages (those that were not 'false flag' attacks at least) was taking an antidepressant prescription drug of some kind. There is of course a major, fundamental flaw in a society more concerned about sales of drugs than the welfare of its children, but this we know already.

Another subtle ploy is to artificially set a limit on the incidence of certain parameters within the human body and then declare that anything above (or below) that limit is a cause for concern and/or a danger to our wellbeing. By that method it then becomes a simple task to produce a drug that 'cures' that imbalance and make even more millions for Big Pharma.

High cholesterol is just such a similar scam.

Cholesterol

"Somewhere along the way however, cholesterol became a household word—something that you must keep as low as possible or suffer the consequences. You are probably aware that there are many myths that portray fat and cholesterol as one of the worst foods you can consume. Please

understand that these myths are actually harming your health. Not only is high cholesterol most likely not going to destroy your health (as you have been led to believe) but it is also not the cause of heart disease." Dr Joseph Mercola, 10th August 2010

One certainly needs cholesterol. It is present not only in the bloodstream, but also in every cell in the body, where it helps to produce cell membranes, hormones, vitamin D and bile acids that assist in the digestion of fat. Cholesterol also helps in the formation of memory and is vital for neurological function.

The liver makes about 75 percent of a body's cholesterol and according to conventional medicine, there are two types:

High-density lipoprotein or HDL: This is the 'good' cholesterol that helps to keep cholesterol away from the arteries and remove any excess from arterial plaque, which may help to prevent heart disease.

Low-density lipoprotein or LDL: This 'bad' cholesterol circulates in your blood and accumulates in arteries, forming a plaque that narrows arteries and renders them less flexible (a condition called arteriosclerosis). If a clot forms in one of these narrowed arteries leading to your heart or brain, a heart attack or stroke may result.

Please understand that the total cholesterol level is not an indicator of one's heart disease or stroke risk. Health officials in the United States urge everyone over the age of 20 to have their cholesterol tested once every five years. Part of this test is your total cholesterol or the sum of your blood's cholesterol content, including HDLs, LDLs, and VLDLs.

The American Heart Association recommends that total cholesterol should be less than 200 mg/dL, but what they do not tell you is that 'total cholesterol level' is just about worthless in determining your risk for heart disease, unless it is above 330. In addition, the AHA updated their guidelines in 2004, lowering the recommended level of LDL cholesterol from 130 to LDL to less than 100, or even less than 70 for patients at 'very high risk.'

The Falsification of Science

In order to achieve these outrageous and dangerously low targets, you typically need to take multiple cholesterol lowering drugs. So the new guidelines instantly increased the available market for these dangerous drugs. And now, with the advent of the testing of children's cholesterol levels, they are increasing their potential market even more. So, another zero on Big Pharma's bottom line profits there then.

Actually, you may be surprised to hear that all cholesterol is actually neither 'good' nor 'bad.' Now that good and bad cholesterol has been defined, it has to be said that there is actually only one type of cholesterol!

"Notice please that LDL and HDL are lipoproteins—fats combined with proteins. There is only one cholesterol. There is no such thing as 'good' or 'bad' cholesterol. Cholesterol is just cholesterol. It combines with other fats and proteins to be carried through the bloodstream, since fat and our watery blood do not mix very well. Fatty substances therefore must be shuttled to and from our tissues and cells using proteins. LDL and HDL are forms of proteins and are far from being just cholesterol. In fact we now know there are many types of these fat and protein particles.

LDL particles come in many sizes and large LDL particles are not a problem. Only the so-called small dense LDL particles can potentially be a problem, because they can squeeze through the lining of the arteries and if they oxidise, otherwise known as turning rancid, they can cause damage and inflammation. Thus, you might say that there is 'good LDL' and 'bad LDL.' Also, some HDL particles are better than others. Knowing just your total cholesterol tells you very little. Even knowing your LDL and HDL levels will not tell you very much." Ron Rosedale MD, a leading anti-aging doctor in the USA

The idea that cholesterol is 'bad' has been very much ingrained in most people's minds due to the propaganda expounded by the pharmaceutical cartel, but this is a very harmful myth that needs to be eliminated as soon as possible.

Dr Rosedale further points out, *"First and foremost, cholesterol is a vital component of every cell membrane on Earth. In other words, there is no life on Earth that can live without cholesterol. That will automatically tell you that, in and of itself, it cannot be evil. In fact, it is one of our best friends. We would not be here without it. No wonder lowering cholesterol too much increases one's risk of dying."*

Vitamin D is a much-neglected source of wellness and general health and what most people do not realise is that the best way to obtain vitamin D is from exposure to sun on one's skin. The UVB rays in sunlight interact with the cholesterol on the skin and convert it to vitamin D.

Therefore, if cholesterol levels are too low it will be impossible to use the sun to generate sufficient levels of health-giving vitamin D. As vitamin D is a major influence in the prevention of many diseases, particularly cancer, this obviously contributes to higher incidences of these diseases, which I am sure is no coincidence.

Essentially, HDL takes cholesterol from the body's tissues and arteries and brings it back to the liver, where most cholesterol is produced. If the purpose of this was to eliminate cholesterol from the body, it would make sense that the cholesterol would be dispatched back to the kidneys or intestines so it could be removed. But instead it goes back to the liver. *Why should this be?* The reason is because the liver can reuse it.

"It is taking it back to your liver so that your liver can recycle it; put it back into other particles to be taken to tissues and cells that need it," Dr Rosedale explains. *"Your body is trying to make and conserve the cholesterol for the precise reason that it is so important, indeed vital, for health."*

If cholesterol levels drop too low this can have a devastating effect on the body in so many ways. Remember, every single cell needs cholesterol to thrive including those in the brain. Perhaps this is why low cholesterol wreaks havoc on the psyche. One large study conducted by Dutch researchers found that men with chronically low cholesterol levels showed a consistently higher risk of having depressive symptoms. (Anyone feel a long, intensive course of suicide-enhancing antidepressants coming on?) Seriously though, this could well be

The Falsification of Science

because cholesterol affects the metabolism of serotonin, a substance involved in the regulation of moods.

On a similar note, some Canadian researchers found that those in the lowest quarter of total cholesterol concentration had more than six times the risk of committing suicide as did those in the highest quarter. Dozens of studies also support a connection between low or lowered cholesterol levels and violent behaviour. Lowered cholesterol levels may lead to lowered brain serotonin activity, which may, in turn, lead to increased violence and aggression. And one analysis of over 41,000 patient records found that people who take statin drugs to lower their cholesterol as much as possible may have a higher risk of cancer, while other studies have linked low cholesterol to Parkinson's disease.

Probably any cholesterol level reading under 150 is too low—an optimum would be around 200. How strange then that doctors tell us that cholesterol needs to be *under* 200 to be healthy—or maybe not?

In 2004, the US government's National Cholesterol Education Program panel advised those at risk of heart disease to attempt to reduce their LDL cholesterol to specific, very low, levels. Before 2004, a 130 milligram LDL cholesterol level was considered healthy but the updated guidelines however, recommended levels of less than 100, or even less than 70 for patients at very high risk. It is worth noting that these extremely low targets often require multiple cholesterol-lowering drugs to achieve.

Fortunately, in 2006 a review in the '*Annals of Internal Medicine*' found that there is insufficient evidence to support the target numbers outlined by the panel. The authors of the review were unable to find research providing evidence that achieving a specific LDL target level was important in itself and found that studies attempting to do so suffered from major flaws. Several of the scientists who helped develop the guidelines even admitted that the scientific evidence supporting the 'less than 70' recommendation was not very strong. So how did these excessively low cholesterol guidelines come about? As if we cannot guess!

Eight of the nine doctors on the panel that developed the new cholesterol guidelines had been taking money from the drug companies that manufacture statin-based cholesterol lowering drugs—the same drugs incidentally for which the new guidelines suddenly created a huge new market in the United States. Coincidence? Probably not, but I will allow the reader to decide.

Now, despite the finding that there is absolutely no evidence to show that lowering one's LDL cholesterol to 100 or below is good for health, what do you think the American Heart Association still recommends to this day? Nothing less than keeping your LDL cholesterol levels to less than 100. And even better for Big Pharma, the standard recommendation to get to that level always includes one or more cholesterol-lowering drug.

If you are personally concerned about your cholesterol levels, taking a drug should be your absolute last resort. The odds are very high indeed that you do not need useless, expensive drugs to lower your cholesterol levels.

According to recent data from *Medco Health Solutions Inc.*, more than 50% of health-insured Americans are using drugs for chronic health conditions and cholesterol lowering medications are the second most common variety among this group, with almost 15% of chronic medication users taking them (high blood pressure medications—another vastly over-prescribed category, were first in the list). This is true as you would imagine for the rest of the Western world, its healthcare systems managed and controlled as it is by Big Pharma's aggressive profit targets.

"Some researchers have even suggested that the [cholesterol lowering] medications should be put in the water supply." Business Week magazine, 2008

Indeed, cholesterol-lowering drugs are some of the most insidious on the market and please believe me when I say that once again, that is despite the fact they are up against some pretty stiff competition for that particular 'honour'!

The Falsification of Science

Statin drugs take effect by inhibiting an enzyme in the liver that is needed to manufacture cholesterol. What is so worrying about this is that when we mess around with the extremely delicate workings of the human body, the major risk is putting the body's natural cycles out of balance, causing a chain reaction of knock-on effects. Big Pharma's answer to that scenario of course is to keep prescribing more and more drugs to counter the side-effects of the previous one. Can you even imagine the internal disruption caused by this state of affairs with some patients taking in excess of twenty different medications, each to combat the side-effects of the others? My own father's situation springs readily to mind. Additionally, do you think that ALL the different combinations of drugs it is possible to have prescribed by your local 'death-dealer' have been tested, even cursorily? The answer has to be 'no' as the permutations would run into many millions if not billions.

"Statin drugs inhibit not just the production of cholesterol, but a whole family of intermediary substances, many if not all of which have important biochemical functions in their own right," Enig and Fallon

Statin drugs deplete the body of Coenzyme Q10 (CoQ10), which is beneficial to heart health and muscle function. Because doctors do not as routine, inform people of this risk and advise them to take a CoQ10 supplement, this depletion leads to fatigue, muscle weakness, soreness and eventually heart failure.

Muscle pain and weakness is actually the most common side-effect of statin drugs, which is thought to occur because statins activate the gene which plays a key role in muscle atrophy.

They have also been linked to:

An increased risk of nerve damage

Dizziness

Cognitive impairment, including memory loss

A potential increased risk of cancer

Depression

Liver problems

Motor Neurone (Lou Gehrig's) disease

With all of these risks to consider, one would hope that statin drugs were effective. Well, unfortunately this is highly improbable.

Most cholesterol lowering drugs can effectively lower cholesterol numbers, but this does in no way make for a healthier individual and there is certainly no available evidence that says that they may help prevent heart disease.

Food and Nutrition

"Most of what sits in our stores is not really food as we have known it. It is a stew of sorts; chemicals, additives, flavourings, colourings, enhancers, preservatives, aspartame, neotame, and stuff we can't even identify along with residual hormones, vaccines, antibiotics, herbicides, and pesticides. It has been irradiated, sprayed with viruses, and now covered in ammonia. Reading any label for content makes one think you would be just as well off if you drank floor cleaner and it most likely might be a lot tastier although just as empty of nutrition." Marti Oakley, July 2011

"If people let the government decide what foods they eat and what medicines they take, their bodies will soon be in as sorry a state as are the souls of those who live under tyranny." Thomas Jefferson, 1778

"There is no right to consume or feed children any particular food. There is no generalised right to bodily and physical health. There is no fundamental right to freedom of contract." US Department of Health and Human Services and FDA, 2010

And there in a neat little nutshell, is everything we need to know about how much these organisations actually care about the welfare of we, the 'common herd.'

It is an inescapable fact that our food (especially fruit and vegetables, but meat products too) now contains around 50-60% less nutrients and vitamins, by quantity, than was the case as recently as 40 or 50 years ago. What this means in effect is that we have to consume 50-60% more in quantity of the same foodstuffs to achieve the same nutritional benefits as in the past. The reason for this is that modern farming and cultivation methods and food production processes have become such that the only way to compete and survive in the cut-throat world of commercial food, is to foster and encourage faster growth of both animate and inanimate food and this has a knock-on, detrimental effect on its nutritional value as well as other downsides brought about by the widespread use of artificial growth hormones, nonorganic fertilisers and pesticides.

The quantities we need to consume have therefore increased hugely and this is partly responsible for the obesity epidemic we see today. I say 'partly,' because this does not quite tell the whole story. In addition to consuming ever greater quantities of foodstuffs, it is also the case that much of what we now eat can be classified as 'processed food,' that is we no longer tend to buy the basic raw, unadulterated products in order to make meals as our forebears as recently as two generations ago, would have routinely done. Instead, we rely on the vast supermarket chains kindly provided for our convenience by 'Big Food,' where most of us now tend to do our weekly or monthly 'buy-in.' And by far the majority of food products in supermarkets fall under the heading of 'processed food.'

Processed food, whilst undeniably 'convenient' in that it is easily and quickly prepared and served, in these days of the instant fix, unfortunately contains many nasty additives, preservatives and colourings which are known in some cases, to have a severe detrimental effect on human health and contribute hugely to the obesity epidemic. The quantity of processed food sold and served these days thus contributes to the overall malaise and generally poor health and obesity prevalent today in the Western world.

Aspartame

One of the worst culprits of all food additives is the substance, Aspartame. This is present in much of our processed foodstuff today including *most* carbonated drinks (fizzy pops and sodas) and especially in those labelled as 'diet,' 'lite,' 'zero' and 'low-sugar' or 'sugar-free' products, as a sweetener.

The story of how this abomination, Aspartame came to be present in our food and drink is a real eye-opener and I am sure you will agree, nothing less than criminal activity.

In the late 1970s the head of GD Searle, one of the pharmaceutical giants, was a certain Donald Rumsfeld, he of 9/11 infamy amongst many other dubious US government-sponsored atrocities. It is well documented that he wanted to get Aspartame, discovered by accident in a laboratory in the 1960s, on to the world markets as a potentially huge money-spinner. When I say huge I mean, bigger than you or I could ever imagine. The Aspartame market is worth billions and billions per annum if not trillions, so please do not think I am exaggerating too much.

The substance went through the normal channels of short-term testing by the FDA to determine its safety and efficacy as a sweetener before being accepted and was thus rejected due to the health dangers it was found to exhibit. All this took place towards the end of Jimmy Carter's reign as president of the USA, so Rumsfeld as head of a large corporation, duly poured millions of company dollars into Ronald Reagan's presidential campaign coffers which no doubt 'helped' in seeing him elected as the next President. To cut a long story short, as soon as Reagan was installed as President, Aspartame was miraculously approved by the FDA without further tests and rushed on to the market as a sweetener in 1981, within days of Reagan taking office. Just a coincidence though, I'm sure.

"If Donald Rumsfeld had never been born think of how many millions of people the world over would not suffer headaches and dizziness. Thousands blind from the free methyl alcohol in Aspartame would have sight, and there would be much fewer cases of optic neuritis and macular

The Falsification of Science

degeneration. Millions suffering seizures would live normal lives and wouldn't be taking anti-seizure medication that won't work because Aspartame interacts with drugs and vaccines. Think of the runner, Flo Jo, who drank Diet Coke and died of a grand mal seizure. She, no doubt, would still be alive. Brain fog and memory loss, skyrocketing symptoms of Aspartame disease, would not be epidemic.

Millions suffer insomnia because of the depletion of serotonin. Think of Heath Ledger. He took that horrible drug, Ambian CR for sleep, which makes the optic nerve and face swell and gives you terrible headaches. Plus, he drank Diet Coke (which destroys serotonin) *and took other drugs and died of polypharmacy.*

Since Aspartame has been proven to be a multi-potential carcinogen, would Farrah Fawcett still be alive?" Hesh Goldstein, 'The Rumsfeld Plague' 2009

Aspartame triggers an irregular heart rhythm (arrhythmia) and interacts with all cardiac medication. It damages the cardiac conduction system and causes sudden death. Thousands of athletes have fallen victim to it. Simply drink a can or bottle of diet drink, perform some vigorous exercise and you too could suddenly die, just as thousands of others provably have since 1981.

As the phenylalanine in Aspartame deletes serotonin, it triggers all kinds of psychiatric and behavioural problems. The mental hospitals are full of patients who are nothing but Aspartame victims. If the FDA had acted ethically, the revoked petition for the approval of Aspartame would have been signed by FDA commissioner Jere Goyan and the mental hospitals would house probably 50% less victims. Jere Goyan would never have been fired at 3.00am by the Reagan transition team to overrule the Board of Inquiry. Instead, FDA commissioner Goyan would have signed the revoked petition into law and the FDA today would still be Big Pharma's adversary instead of being their whore.

For over a quarter of a century there has been mass poisoning of the public in over one hundred countries of the world by Aspartame because Donald Rumsfeld, as he put it, *"called in his markers."* The

Aspartame industry has several bought-and-paid-for front groups and professional organisations to defend them and 'push' it to the very people upon whom it can inflict most harm. How surprising.

If Aspartame had not been approved, Motor Neurone (Lou Gehrig's) Disease, Parkinson's and other neuro-degenerative diseases would not be killing-off people in record numbers. Michael J. Fox, a *Diet Pepsi*™ spokesman for several years, would never have contracted Parkinson's Disease at the ridiculously early age of 30. He would probably still be making films and enjoy robust health. Aspartame interacts with 'L-dopa' a Parkinson drug and another, 'Parcopa' actually contains Aspartame but the pharmaceutical company refuses to remove it despite strong, proven evidence linking it to Parkinson's disease. It would be hilarious if not so serious.

Is it not criminal that there is not even a warning for pregnant women? Aspartame triggers every kind of birth defect from autism and Tourette's Syndrome to cleft palate and is an abortifacient (a drug that induces foetal abortion).

It is of course perfectly normal for young girls to look forward to marriage and children, yet many drink diet drinks or consume Aspartame products not realising that Aspartame is an endocrine disrupting agent, stimulating prolactin, which is a pituitary hormone that stimulates milk production at childbirth, changes and inhibits the menstrual cycle and causes infertility. Many women go through life never knowing why they cannot have children and Aspartame could even be accused of destroying marriages because it causes male sexual dysfunction and prevents the female arousal response.

Clear evidence demonstrates that Aspartame can cause every type of blood disorder from a low blood platelet count to leukaemia. And because it can also precipitate diabetes type II, this disease has recently become epidemic. To make matters worse, it can stimulate and aggravate diabetic retinopathy and neuropathy, destroy the optic nerve, cause diabetics to go into convulsions and interact with insulin. Diabetics lose limbs from the free methyl alcohol; professional organisations such as the American Diabetes Association push and defend this poison because they take large sums of money from the manufacturers. How many millions would not now have diabetes if Rumsfeld's actions had been prevented?

Aspartame (NutraSweet/Equal/Spoonful/E951/Canderel/Benevia, etc.) along with High Fructose Corn Syrup (HFCS) and Monosodium Glutamate (MSG) are responsible for the epidemic of obesity the entire world over. Why? This is because they instigate a craving for carbohydrates and cause great toxicity in the liver.

An FDA report lists 92 symptoms of Aspartame poisoning, from unconsciousness and coma to shortness of breath and shock. Medical texts list even more: *"Aspartame Disease: An Ignored Epidemic,"* H.J. Roberts, MD and *"Excitotoxins: The Taste That Kills,"* by neurosurgeon Russell Blaylock, M.D. There is simply no end to the horrors triggered by this literally addictive, excito-neurotoxic, and genetically engineered, carcinogenic drug. This chemical poison is so deadly that Dr Bill Deagle, a respected physician once said it was worse than depleted uranium because it is found everywhere in food. He is absolutely correct.

The formaldehyde converted from the free methyl alcohol embalms living tissue and damages DNA according to the 'Trocho' Study done in Barcelona in 1998. Even though this devastating study clearly demonstrates how serious a chemical poison Aspartame is, the FDA has turned a blind eye to it and it continues to flourish everywhere. When the Monsanto attorney Michael Taylor was appointed as Deputy Commissioner to the FDA by President Barack Obama, it simply became nothing more than Monsanto's Washington Branch Office. Even before the 'Ramazzini' studies showing Aspartame to be a multi-potential carcinogen were published, the FDA knew that this was the case. Their own toxicologist, Dr Adrian Gross, even admitted that it violated the Delaney Amendment because of the brain tumours and brain cancer it provably causes. Therefore, no allowable daily intake ever should have been established. Aspartame causes all types of tumours from mammary, uterine, ovarian, pancreatic, and thyroid to testicular and pituitary. Dr Maria Alemany, who undertook the Trocho Study, commented that Aspartame could possibly kill *"200 million people,"* and damaging DNA could even destroy humanity itself.

Dr James Bowen told the FDA over 30 years ago that Aspartame is mass poisoning the USA and likewise more than 100 countries of the world. No wonder it is known colloquially as 'Rumsfeld's Plague.'

Big Pharma indeed knows only too well all about Aspartame and its effects and yet they still allow it to be added to drugs, including the ones used to treat the problems caused by Aspartame in the first place. People are so sick from Aspartame and yet this truly criminal organisation continues to sell these dangerous pharmaceuticals at outrageously inflated prices.

Dr H.J. Roberts said in one of his books that we should *"charge Aspartame with killing children."* We are talking about a drug that changes brain chemistry. Today's children are indeed 'medicated' rather than 'educated.'

Donald Rumsfeld has actually foisted death and disability onto consumers simply to make money. Charles Fleming used to drink about around ten 'diet' drinks per day and then used Creatine (Creatine is found in vertebrates where it facilitates recycling of adenosine triphosphate (ATP), the energy currency of the cell, primarily in muscle and brain tissue. Wikipedia™) on top of this, which interacts with Aspartame and is thus considered the actual cause of death. Yet his widow, Diane Fleming, remains in prison in Virginia convicted of his death, despite being the very one who tried to get her husband to stop using these dangerous products containing Aspartame in the first place.

The list goes on and on. At least six *American Airlines*' pilots, who were heavy users of Aspartame have died, one of them actually in flight, drinking a Diet Coke™. So pilots too are sick and dying from Aspartame and when we fly our life is literally in their hands.

In the Persian Gulf, diet drinks sit on pallets daily in temperatures of over 100°F for as long as eight weeks at a time before the Allied forces drink them. Aspartame converts to formaldehyde at 86°F and it interacts with vaccines and damages the mitochondria or heart of the human cell, and the whole molecule breaks down to become a brain tumour causing agent.

Doctors at mortuaries carrying out autopsies know immediately when Aspartame has been the cause of death, because when the skull is opened, they can instantly smell the methanol in the brain tissue and cavity. However it is more than their career is worth to overtly name Aspartame as a cause of death. The hands of physicians are well and truly tied. Most are clueless that a patient is using Aspartame and even that Aspartame is deadly and the drugs they prescribe to treat the Aspartame problem will probably interact negatively and may even contain Aspartame themselves. Those that are aware of the situation also know that there is nothing they can do in the face of strong financial and political pressure.

This is the world for which Donald Rumsfeld is personally responsible.

Of course we will never be told this by our compliant media, who are in any case owned by the same small group of Elite people at the tops of the pyramids. The game is rigged folks. It's total war and we are the enemy under siege.

Genetically Modified Organisms (GMOs)

The concerted drive to inflict genetically modified food upon us by food corporations such as Monsanto (now absorbed by the gigantic

German conglomerate Bayer) with the compliance of the highly corrupt FDA, in the name of profit, is nothing but a disgrace and a sham. Listening to their spokespersons quoting rigged and heavily massaged trial results and lying about the safety of these substances, is no different to listening to the worst excesses of Soviet propaganda during the 'Cold War.'

Make no mistake about it, GMOs are nothing short of deadly, to humans and animals alike and the worst news of all is that they have already infiltrated much of the food chain. According to these Big Food, Elite-controlled corporations and their media poodles, GM foods are the answer to the world food crisis and a cheap, efficient way of feeding the starving millions worldwide. Of course they really do care about the starving millions — we believe them. Of course we do.

The truth is that the reason that genetically modified foods are being produced in the first place is so that the corporations who produce them can patent the genetics and thus control the market for them. Can you imagine owning the sole rights to the genome for a carrot or a tomato or even a strain of cattle? This is a very dangerous situation and very much 'the thin end of the wedge' in the sense that allowing a corporation to own a genome means that they can in effect corner a market and basically charge whatever they then wish for that product. That is really what GM is all about and why it is so insidious. It also gives them the power to make demands for payment for seeds containing *their* genetic blueprint, as opposed to the natural genome, found on other private farms or lands.

There have already been court cases in the US and worldwide where Monsanto have illegally sent their 'agents' into private farmland to gather samples of seeds / plants bearing their genome blueprint, carried in by birds or the wind from neighbouring farms. In all cases the judges found in Monsanto's favour, declaring that the seeds were 'Monsanto's property' and were being grown illegally. And of course the very fact that those Monsanto agents were trespassing on private property in order to make their discoveries, is entirely of no

The Falsification of Science

consequence to the highly corrupt legal system. This is not satire by the way, this is the absolute truth.

In addition Bayer (and other corporations') GM seeds contain what has become known as the 'terminator gene.' This means that unlike in bygone days when farmers / growers could save the seeds from one year's crop in order to propagate the following years crop, they are now forced to buy from the Corporations again and again, year after year as the ability of the plants to naturally produce seeds has been 'deleted' from the genome. In my humble view, although this is perfectly legal, it is absolutely, totally immoral, and callous profiteering on a grand scale.

Here is a case against Monsanto that made the headlines...

"For 40 years Percy Schmeiser grew oilseed rape on his farm in the Canadian province of Saskatchewan. Each year, he would sow each year's crop with seeds saved from the previous harvest. In 1998 Monsanto took Schmeiser to court. Investigators employed by the company had found samples of its GM oilseed rape among Schmeiser's stock. Monsanto's lawsuit alleged that the farmer had infringed on the firm's patent. It even stated that Schmeiser had obtained Monsanto seeds illegally; going so far as to suggest that he might have stolen them from a seed house.

The corporation later admitted that Schmeiser had not obtained the seeds illegally but said that wasn't important. What did matter, Monsanto argued, was that it had found some of its canola plants in the ditch along Schmeiser's field (note that the plants were not found in Schmeiser's fields); that meant that the farmer had violated the firm's patent.

The judge agreed with Monsanto, ruling that 'the source of (GM) oilseed rape...is not really significant for the issue of infringement.' In other words, it was irrelevant how the patented canola plants got on Schmeiser's land. It could have happened as a result of cross-pollination or by seed movement caused by wind. (The latter is the biggest cause of contamination involving GM crops, and the farm next to Schmeiser's did grow Monsanto's crop.) The judge told Schmeiser that all his seeds, developed over almost half a century, were now the property of Monsanto. The Ecologist, May 2004

This is so typical of the immoral behaviour of the Elite corporations and their paid-for dupes in the legal profession. Profit is everything and ordinary, hard-working, decent people and their families count for nothing other than the fact that they are considered fair game for extortionate practices or at best regarded as 'profit centres' to be exploited to the maximum.

Worse still, it is a well-documented fact that thousands of farmers in India are committing suicide due to Bayer's hideously unethical and immoral business tactics. In early 2011 it was estimated that at least **300,000** Indian farmers had committed suicide by 2014 and during the previous 15 years, due to Monsanto's business practices causing their livelihoods to virtually evaporate. Powerful GM lobbyists however are stating that GM crops have transformed India's agriculture, providing greater yields than ever before.

"So who is telling the truth? To find out, I travelled to the 'suicide belt' in Maharashtra state. What I found was deeply disturbing—and has profound implications for countries, including Britain, debating whether to allow the planting of seeds manipulated by scientists to circumvent the laws of nature.

For official figures from the Indian Ministry of Agriculture do indeed confirm that in a huge humanitarian crisis, more than 1,000 farmers kill themselves here each month. Simple, rural people, they are dying slow, agonising deaths. Most swallow insecticide—a pricey substance they were promised they would not need when they were coerced into growing expensive GM crops.

It seems that many are massively in debt to local money-lenders, having over-borrowed to purchase GM seed. Pro-GM experts claim that it is rural poverty, alcoholism, drought and 'agrarian distress' that is the real reason for the horrific toll. But, as I discovered during a four-day journey through the epicentre of the disaster, that is not the full story.

In one small village I visited, 18 farmers had committed suicide after being sucked into GM debts. In some cases, women have taken over farms from their dead husbands—only to kill themselves as well. Latta Ramesh, 38, drank insecticide after her crops failed—two years after her

The Falsification of Science

husband disappeared when the GM debts became too much. She left her ten-year-old son, Rashan, in the care of relatives. 'He cries when he thinks of his mother,' said the dead woman's aunt, sitting listlessly in shade near the fields.

In village after village, families told how they had fallen into debt after being persuaded to buy GM seeds instead of traditional cotton seeds. The price difference is staggering: £10 for 100 grams of GM seed, compared with less than £10 for 1,000 times more traditional seeds. But GM salesmen and government officials had promised farmers that these were 'magic seeds'—with better crops that would be free from parasites and insects.

Indeed, in a bid to promote the uptake of GM seeds, traditional varieties were banned from many government seed banks. The authorities had a vested interest in promoting this new biotechnology. Desperate to escape the grinding poverty of the post-independence years, the Indian government had agreed to allow new bio-tech giants, such as the U.S. market-leader Monsanto, to sell their new seed creations.

In return for allowing western companies access to the second most populated country in the world, with more than one billion people, India was granted International Monetary Fund loans in the Eighties and Nineties, helping to launch an economic revolution. But while cities such as Mumbai and Delhi have boomed, the farmers' lives have slid back into the dark ages.

Though areas of India planted with GM seeds have doubled in two years—up to 17 million acres—many farmers have found there is a terrible price to be paid. Far from being 'magic seeds,' GM pest-proof 'breeds' of cotton have been devastated by bollworms, a voracious parasite. Nor were the farmers told that these seeds require double the amount of water. This has proved a matter of life and death.

With rains failing for the past two years, many GM crops have simply withered and died, leaving the farmers with crippling debts and no means of paying them off. Having taken loans from traditional money lenders at extortionate rates, hundreds of thousands of small farmers have faced losing their land as the expensive seeds fail, while those who could struggle on faced a fresh crisis.

When crops failed in the past, farmers could still save seeds and replant them the following year. But with GM seeds they cannot do this. That's because GM seeds contain so-called 'terminator technology,' meaning that they have been genetically modified so that the resulting crops do not produce viable seeds of their own. As a result, farmers have to buy new seeds each year at the same punitive prices. For some, that means the difference between life and death.

Take the case of Suresh Bhalasa, another farmer who was cremated this week, leaving a wife and two children. As night fell after the ceremony, and neighbours squatted outside while sacred cows were brought in from the fields, his family had no doubt that their troubles stemmed from the moment they were encouraged to buy BT Cotton, a genetically modified plant created by Monsanto.

'We are ruined now,' said the dead man's 38-year-old wife. 'We bought 100 grams of BT Cotton. Our crop failed twice. My husband had become depressed. He went out to his field, lay down in the cotton and swallowed insecticide.' Villagers bundled him into a rickshaw and headed to hospital along rutted farm roads. 'He cried out that he had taken the insecticide and he was sorry,' she said, as her family and neighbours crowded into her home to pay their respects. 'He was dead by the time they got to hospital.'

Asked if the dead man was a 'drunkard' or suffered from other 'social problems,' as alleged by pro-GM officials, the quiet, dignified gathering erupted in anger. 'No! No!' one of the dead man's brothers exclaimed. 'Suresh was a good man. He sent his children to school and paid his taxes. He was strangled by these magic seeds. They sell us the seeds, saying they will not need expensive pesticides, but they do. We have to buy the same seeds from the same company every year. It is killing us. Please tell the world what is happening here.'

Monsanto has admitted that soaring debt was a 'factor in this tragedy.' But pointing out that cotton production had doubled in the past seven years, a spokesman added that there are other reasons for the recent crisis, such as 'untimely rain' or drought, and pointed out that suicides have

The Falsification of Science

always been part of rural Indian life. Officials also point to surveys saying the majority of Indian farmers want GM seeds – no doubt encouraged to do so by aggressive marketing tactics.

During the course of my inquiries in Maharastra, I encountered three 'independent' surveyors scouring villages for information about suicides. They insisted that GM seeds were only 50 per cent more expensive – and then later admitted the difference was 1,000 per cent. (A Monsanto spokesman later insisted their seed is 'only double' the price of 'official' non-GM seed – but admitted that the difference can be vast if cheaper traditional seeds are sold by 'unscrupulous' merchants, who often also sell 'fake' GM seeds which are prone to disease.)

With rumours of imminent government compensation to stem the wave of deaths, many farmers said they were desperate for any form of assistance. 'We just want to escape from our problems,' one said. 'We just want help to stop any more of us dying.' Cruelly, it's the young who are suffering most from the 'GM Genocide' – the very generation supposed to be lifted out of a life of hardship and misery by these 'magic seeds.' Here in the suicide belt of India, the cost of the genetically modified future is murderously high."
Andrew Malone, UK Daily Mail, 3rd November 2008

The above tragic report calls to mind the famous fairy tale 'Jack and the Beanstalk,' in which peasant Jack, foolishly sells all his family's wealth (in the form of a cow) to a traveling con artist who convinces him to buy 'magic beans' promising him riches beyond his imagining. Of course the beans instead bring near ruin to his family and a life threatening conflict with the giant (not unlike the GM food giant, Monsanto) who lives atop the monstrous beanstalk sprouted from the beans. Fortunately in the most popularized version of the story Jack defeats the giant, stealing his riches and toppling him to the ground after cutting down the beanstalk, and lives with his mother, in opulence, happily ever after. But what ending will the people in India and around the world ultimately have to their own version of this story? One can only hope that Jack will reign triumphant and the GM giant will eventually fall.

Dr Arpad Pusztai a Hungarian-born protein scientist spent most of his working life at the Rowett Research Institute in Aberdeen, Scotland – a total of 36 years. He was considered to be the world's foremost expert on plant lectins and has authored 270 scientific papers and 3 books on the subject.

However, in 1998 Pusztai publicly announced that the results of his extensive, meticulous research showed conclusively that feeding genetically modified potatoes to rats substantially harmed them, leading to his summary dismissal from the institute. The resulting controversy over his dismissal and the attempts to invalidate the conclusions of his research became known as the 'Pusztai Affair.'

The rats fed on the genetically modified potatoes showed significant intestinal damage and harm to their general health. These effects were not observed in control rats fed on unmodified potatoes, or unmodified potatoes mixed with snowdrop lectin. The team concluded that the effects observed were a result of the genetic modification, not the snowdrop lectin.

Dr Pusztai commented, *"We had two kinds of potatoes – one GM and the other non-GM. I had expected that the GM potato, with 20 micrograms of a component against the several grams of other components, should not cause any problems. But we found problems. Our studies clearly show that the effects were not due to that little gene expression, but it depended on the way the gene had been inserted into the potato genome and what it did to the potato genome."*

In early 2009, I was fortunate enough to briefly meet Arpad's wife, also a scientist, Dr Susan Bardocz at a conference in London. She told me that what *actually* happened, although it was obviously distorted in the media, was this…

As soon as the report on the trials was published, at the headquarters of a large corporation primarily concerned with the production and distribution of GM foods, alarm bells began to ring and so the head of the corporation involved (no prizes for guessing its name but it begins with 'M' and ends with 'onsanto') contacted the then US President Bill Clinton to demand some sort of government intervention in order to limit the damage done to GM foods by this worrying turn of events. According to Susan Bardocz who was told this by the director of the institute, Clinton immediately contacted the British PM, Tony Blair and Blair immediately put pressure on the director of the Rowett Institute to summarily dismiss Dr Pusztai from his position.

In addition to this, the Rowett Institute's director Philip James, who had initially supported Pusztai, suspended him and used misconduct procedures to seize the relevant data. His annual contract was not renewed and Pusztai and his wife were banned from speaking publicly. Phone calls to his office were diverted and his research team was disbanded. Initially the Rowett Institute claimed that they were not performing any research on GM crops but later the Institute claimed that Pusztai had voluntarily retired and apologised for his 'mistake.' According to another version of the story, the experiments had never been performed in the first place and then, yet another version emerged whereby a student had accidentally confused control data with experimental data. This is typical of the modus operandi. Confuse the issue as much as possible so that the whole story becomes a total mess, and no one really knows the actual truth anymore.

Dr Bardocz also said that for the next ten years, Arpad's life became unbearable. His work, not just the recent work he had done with GM food, but all of his past research going back 40 years was totally discredited. He was as a result, unable to obtain any sort of position anywhere in the world and all the stress he suffered culminated in him having a stroke in late 2008, severely affecting his health and his life thereafter.

Scientists who tell the truth revealed by their genuine research projects are really not at all welcome in the world of our Elite masters. The truth about most issues, not just health is far too dangerous to their position of power to be allowed out in the open and if it should accidentally leak out then a huge, no expense spared propaganda machine immediately cuts in to discredit the source and cover-up or eliminate the process by which it was leaked. This can take the form of anything from a simple lie at one extreme, to mass murder at the other. Either way, it is never a problem.

These silent killers (GMOs) are deadly, and do not discriminate. They target babies, the elderly, teenagers, young adults, middle-aged housewives, and businessmen alike. They poison livestock, pets and wildlife and the people behind them deny complicity in the carnage. These killers are the seemingly beneficial, killing-fields of genetically modified (GMO) crops and the people behind them are the US and other Western governments, the Rockefellers, Monsanto, Dow, DuPont, Syngenta and Bayer Crop Science.

Eugenics is a dirty word, yet particularly applicable to America's GMO killing fields and their inception. In 1974 Henry Kissinger drafted the controversial NSSM-200, called *"the foundational document on population control issued by the United States government."*

"According to NSSM-200, elements of the implementation of population control programs could include the legalization of abortion; financial incentives for countries to increase their abortion, sterilization, and contraception-use rates; indoctrination of children; mandatory population control, and coercion of other forms, such as withholding disaster and food aid unless an LDC implements population control programs. NSSM-200 also specifically declared that the United States was to cover up its population control activities and avoid charges of imperialism by inducing the United Nations and various non-governmental organizations to do its dirty work." Human Life International, 2008

In 1970, Henry Kissinger also said, *"Control oil and you control nations; control food and you control the people."*

How do you control food? By consolidating agricultural interests into what was to be termed agribusiness, creating genetically modified organisms out of heritage seeds with funding from the Rockefeller Foundation, patenting the new seeds and making sure that these new seeds are force-fed to US farmers as well as the rest of the world. By holding the patents on these seeds and requiring farmers to purchase new seeds every year, the control is complete. Also, by controlling how these GMO seeds are created, other more sinister uses come to

The Falsification of Science

mind. But firstly it is necessary to convince the world of your good intentions. This is accomplished through constant, persistent lies, deception, and a modicum of controlled-media manipulation. By promising farmers that this technology was safe and would result in increased yields at less cost, they were more than happy to comply. The fact that in most cases this claim was false had yet to be proven by the innocent farmers that believed the lie and by the time independent studies started revealing that GMO is harmful, it was too late, and agri-business was on its way to fulfilling its purpose. That is to make as much money as possible by spreading GMO seeds as far as possible and thus gaining control of the population via food.

"In what should amount to a wildly imaginative narrative created by an overzealous science fiction aficionado, the following agencies, their connections, and past actions are real, nonetheless.

Imagine, if you will, a world in which health sciences, disease control, cancer research, bioweapons research, vaccine development, biotechnology, food and agriculture, national defence, and chemical companies all work together under the military. Then imagine if you will that a certain chemical company under the guise of a life sciences operation, produces an herbicide/defoliant for military use so destructive and highly toxic that contact with it causes cancer, diabetes and birth defects. And then that same chemical/life sciences company partners with a funding corporation whose team members include the ex-partner of the inventor of the world's first completely synthetic organism, which was recently unleashed in the ocean and has since turned its ever-hungry sights on human flesh. Then imagine that same company with a monopoly on our food supply...

Sound like an episode out of the Twilight Zone or Dr Who? Well, it's not. It's history, and it's documented." Barbara H Peterson, Farm Wars, June 2011

And if that all is not frightening enough, consider the implications of the current push in 2021 to inject the majority of the world's population with a vaccine, which uses gene-editing technology not unlike that which is used to create GM seeds. A vaccine, which will in effect genetically modify humans by reprogramming human DNA (to combat the 'virus,' so we are told). Which begs the question, if we become GM humans, who then owns us? Could we not, after being altered thus, become property of the vaccine manufacturers or whoever owns the patent for these new genes? This is not at all out of the realm of possibilities in our increasingly dystopian world.

High Fructose Corn Syrup (HFCS)

Another product found in much of our processed foods is HFCS which has been added since the early 1970s. In effect a corn-based sweetener, this product is largely responsible for the high incidence of Diabetes type 2 prevalent in much of Western society today. Today, food companies use HFCS, a mixture of fructose and glucose, because it is inexpensive, easy to transport and keeps foods moist. And because HFCS is so sweet, it is also cost effective for companies to use small quantities of HFCS in place of other more expensive sweeteners or flavourings.

For the above reasons, HFCS is not going away any time soon. If HFCS is one of the first ingredients listed on a food label, my advice would be to not eat it.

HFCS is a highly refined, artificial product. It is created through an intricate process that transforms corn starch into a thick, clear liquid. White sugar and HFCS are not the same, but corn-grower industry advocates say that they are, and yet nutritional science studies say there is a huge difference between the two. They say that HFCS is definitely *'worse than sugar,'* for us.

This manufactured fructose is sweeter than sugar in an unhealthy way and is digested differently, in a bad way. Research has shown that HFCS goes directly to the liver, releasing enzymes that instruct the body to then store fat. This may elevate triglyceride (fat in blood) levels and elevate cholesterol levels. This fake fructose slows fat burning and causes weight gain. Other research indicates that it does not stimulate insulin production, which usually creates a sense of being full, therefore, people are artificially induced to eat much more than they should. Also, indications are that the important chromium

The Falsification of Science

levels are lowered by this sweetener which contributes to diabetes type 2.

So we have another product that would seem to be harmful in many ways that is allowed indiscriminately into our foodstuffs. Why are we not warned? Why are we not informed about it so that we can make healthy lifestyle choices for ourselves and our families? Once again it would appear to be that profit comes before health. It also strongly indicates to me that the people in control of our food and thus our health are deliberately deceiving us into disease and obesity for the sake of their own bank balances.

Monosodium Glutamate (MSG)

MSG is an amino acid that adversely affects every organ in the body and is a highly addictive substance. There is much evidence that as with HFCS, MSG is also similarly responsible in large part for the incredible epidemic of obesity and diabetes type 2 from which the Western world is currently suffering.

MSG appears in a vast range of processed, tinned, and packaged foods from supermarkets and can often be added separately to food in restaurants, cafeterias, and institutes such as hospitals and care homes as a 'flavour enhancer.' So in effect we are not safe from it, anywhere.

Staggeringly, it is not compulsory for manufacturers of food to specify the exact quantities of MSG that are contained in their products despite the fact that a single 12gm (¾oz) dose of MSG has been found to be lethal to a 2kg (4½lb) rat, whilst much smaller doses have been shown to cause massive brain damage. Additionally MSG is now allowed to be sprayed onto crops and it may also become airborne as a result of this activity.

Study after scientific study has shown MSG to be capable of damaging a multitude of organs and soft tissues and also artificially enhance appetites in a variety of consumers, leading in turn as a by-product of this, to obesity. In other experimental studies using rodents, the food additive has been successfully used to generate diabetes on a large scale.

Perhaps more worryingly, MSG has been proved to pass through the placenta of rodents into the unborn foetus, the implications of this being quite serious for humans. Human foetal development has also been shown to be jeopardised by high quantities of MSG in other studies.

"Children undergoing perinatal brain injury often suffer from the dramatic consequences of this misfortune for the rest of their lives. Despite the severe clinical and socio-economic significance, no effective clinical strategies have yet been developed to counteract this condition. This review describes the pathophysiological mechanisms that are implicated in perinatal brain injury. These include the acute breakdown of neuronal membrane potential followed by the release of excitatory amino acids such as glutamate and aspartate. Glutamate binds to post-synaptically located glutamate receptors that regulate calcium channels. The resulting calcium influx activates proteases, lipases, and endonucleases which in turn destroy the cellular skeleton. Clinical studies have shown that intrauterine infection increases the risk of periventricular white matter damage especially in the immature foetus. This damage may be mediated by cardiovascular effects of endotoxins." From the scientific paper, 'Perinatal brain damage from pathophysiology to prevention'. Jensen A, Garnier Y, Middelanis J, Berger R. Eur J. 22nd September 2003

There are few chemicals that we as a people are exposed to that have as many far-reaching physiological effects on living beings, as Monosodium Glutamate. MSG directly causes obesity, food addiction, diabetes, triggers epilepsy, destroys eye tissues and is toxic to many organs. Considering that MSG's only reported role in food is that of 'flavour enhancer' is that use worth the risk of the myriad of physical ailments associated with it? Does the public really want to be tricked into eating more food and faster by a food additive?

MSG is entering our bodies in record amounts with absolutely no limits. The studies outlined in this report often use a smaller proportional dosage than the average child may ingest daily.

The Falsification of Science

Consider the children of the world who routinely consume MSG in great quantities in their school cafeterias, hospitals, restaurants and homes. They deserve foods free of added MSG, a substance so toxic that research scientists use it purposely to trigger diabetes, obesity and epileptic convulsions in animals. Perhaps we will see a reduction in obesity, diabetes and other diseases once the excess MSG threat to our health has been removed. We *can* stop the slow, deliberate poisoning of mankind for profits if we all act together.

If you undertake an Internet search for MSG, be prepared to be bombarded with lies and propaganda. There are literally thousands of websites out there trying to fool you into believing that it really is harmless.

Fluoride

Perhaps the most deadly of all the poisons we are deliberately and deceptively being forced to ingest, fluoride also comes complete with the biggest hype of these toxins by far.

There is actually no such chemical name as 'fluoride.' The name stems from the gas 'fluorine' and from the use of this gas in various industries, such as aluminium smelting and the nuclear industry, where toxic by-products containing fluorine molecules are created. One example of this is sodium fluoride which is the most common and is a hazardous waste by-product of aluminium production. Used for a long time as rat and cockroach poison, it is also an ingredient of anaesthetics and psychiatric drugs as well as the deadly Sarin nerve gas.

As we all know, fluoride is used in 99.9% of proprietary toothpastes and is present in the drinking water (tap and bottled) of much of the First world. Advertised and promoted as protecting teeth, it is unfortunately much more sinister than that. Historically, this substance was quite expensive for the worlds' premier chemical companies to dispose of, but in the 1950s and 60s, *Alcoa* and the entire aluminium industry, with a vast abundance of the toxic waste at hand, somehow sold the FDA and the US government on the insane (but highly profitable) idea of buying this poison at a 20,000% mark-up and then injecting it into our water supply as well as into the world's toothpastes and dental products.

Consider also that when sodium fluoride is injected into our drinking water, its level is approximately one part-per-million but since we only drink ½ of one percent of the total water supply, the hazardous chemical and the chemical industry not only has a free hazardous waste disposal system, but we have also paid them for the privilege!

Independent scientific evidence over the past seventy plus years has shown that sodium fluoride shortens our life span, promotes various cancers and mental disturbances and most importantly, reduces intelligence and makes humans docile, unquestioning and subservient. There is also increasing evidence that aluminium in the brain is a causative factor in Alzheimer's Disease, and evidence points towards the fact that it has the ability to 'trick' the blood-brain barrier by imitating the hydrogen ion, and thereby allowing this deadly chemical, access to brain tissue.

Honest scientists with no ulterior motive who have attempted to 'whistle-blow' on sodium fluoride's mega-bucks propaganda campaign have consistently been given a large dose of professional 'blacklisting' in return and thus their valid points disputing the current

The Falsification of Science

vested interests never have received the recognition they deserve in the national press. 'Just follow the money' as the saying goes, to find the source and you will find prominent, Elite families to be the major influence in this absolutely scandalous situation.

In 1952 a slick PR campaign forced the concept of 'fluoridation' through public health departments and various dental organisations worldwide. This slick campaign was far from the objective, scientific, stringently researched programme that it should have been. But as with all these situations, where money is concerned – and billions upon billions of dollars of it in this case – there were no barriers to be seen. It has continued in this same vein even until the present day and sodium fluoride use has now become the ubiquitous, de facto standard.

"There is a tremendous amount of emotional, highly unscientific 'know-it-all' emotions attached to the topic of sodium fluoride usage – but I personally have yet to find even ONE objective, double-blind study that even remotely links sodium fluoride to healthy teeth at ANY AGE. Instead, I hear and read such blather as '9 out of 10 dentists recommend fluoride toothpaste' etc. etc. etc. Let me reiterate; truly independent (unattached to moneyed vested interest groups) scientists who've spent a large portion of their lives studying and working with this subject have been hit with a surprising amount of unfair character assassinations from strong vested-interest groups who reap grand profits from the public's ignorance as well as from their illnesses.

Do you have diabetes and/or kidney disease? There are reportedly more than 11 million Americans with diabetes. If it is true that diabetics drink more liquids than other people, then according to the Physician's Desk Reference these 11 million people are at much higher risk drinking fluoridated water because they will receive a much deadlier dose because of their need for higher than normal water consumption. Kidney disease, by definition, lowers the efficiency of the kidneys, which of course is the primary means in which fluoride (or any other toxic chemical) is eliminated from the body. Does it not make sense that these people shouldn't drink fluoridated water at all? Cases are on record (Annapolis, Maryland, 1979) where ill kidney patients on dialysis machines died because they ingested relatively small amounts of sodium fluoride from unwittingly drinking the 'fluoridated' city water supply? Will adequate warnings be given to people with weak kidneys, or will the real cause of such deaths be covered up in the name of 'domestic tranquillity?'" A. True Ott, August 2000

It is also worth contemplating the fact that in the USA, all toothpaste tubes / packs come with a warning as standard. *'Warning – harmful or fatal if ingested. If swallowed, please consult a doctor immediately.'* A small clue as to its capabilities there, I would have thought. How often have you accidentally or absent-mindedly swallowed a toothbrush-full of toothpaste? More importantly, your children and grandchildren love the taste of their special toothpastes with added artificial flavour calculated to make them want to brush their teeth more often. How many times might they have swallowed a toothbrush-full over the years and what pernicious harm must it be doing to their immature little bodies?

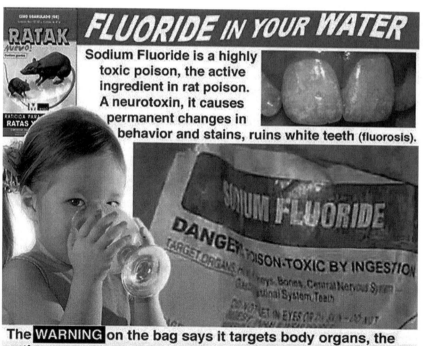

The Falsification of Science

So once again we have a situation where vested interests spend millions, rake in favours, or provide financial incentives to get their way and in doing so guarantee themselves billionaire status many times over.

The following letter is also worthy of note...

"It appears that the citizens of Massachusetts are among the 'next' on the agenda of the water poisoners. There is a sinister network of subversive agents, Godless intellectual parasites, working in our country today whose ramifications grow more extensive, more successful and more alarming each new year and whose true objective is to demoralise, paralyse and destroy our great Republic—from within if they can, according to their plan for their own possession. The tragic success they have already attained in their long siege to destroy the moral fibre of American life is now one of their most potent footholds towards their own ultimate victory over us.

Fluoridation of our community water systems can well become their most subtle weapon for our sure physical and mental deterioration. As a research chemist of established standing, I built within the past 22 years 3 American chemical plants and licensed 6 of my 53 patents. Based on my years of practical experience in the health food and chemical field, let me warn; fluoridation of drinking water is criminal insanity, sure national suicide. DON'T DO IT!!

Even in very small quantities, sodium fluoride is a deadly poison to which no effective antidote has been found. Every exterminator knows that it is the most effective rat-killer. Sodium fluoride is entirely different from organic calcium-fluoro-phosphate needed by our bodies and provided by nature, in God's great providence and love, to build and strengthen our bones and our teeth. This organic calcium-fluoro-phosphate, derived from proper foods, is an edible organic salt, insoluble in water and assimilable by the human body; whereas the non-organic sodium fluoride used in fluoridating water is instant poison to the body and fully water soluble. The body refuses to assimilate it.

Careful, bona fide laboratory experimentation by conscientious, patriotic research chemists and actual medical experience, have both revealed that instead of preserving or promoting 'dental health,' fluoridated drinking water destroys teeth before adulthood and after, by the destructive mottling and other pathological conditions it actually causes in them and also creates many other very grave pathological conditions in the internal organisms of bodies consuming it. How then can it be called a 'health plan?' What's behind it?

That any so-called 'doctors' would persuade a civilised nation to add voluntarily a deadly poison to its drinking water systems is unbelievable. It is the height of criminal insanity! No wonder Stalin fully believed and agreed from 1939 to 1941 that, quoting from Lenin's 'Last Will'... 'America we shall demoralize, divide, and destroy from within.'

Are our Civil Defense organisations and agencies awake to the perils of water poisoning by fluoridation? Its use has been recorded in other countries. Sodium fluoride water solutions are the cheapest and most effective rat killers known to chemists; colourless, odourless, tasteless; no antidote, no remedy, no hope; instant and complete extermination of rats. Fluoridation of water systems can be slow national suicide, or quick national liquidation. It is criminal insanity – treason!!" Dr E.H. Bronner *(a nephew of Albert Einstein)* Research Chemist, Los Angeles, January 1952

The public outcry at the time (now sadly forgotten) was such that the addition of fluoride into public water supplies was abandoned for a year. However, the populace have a very short memory and the topic was resurrected again shortly thereafter, this time with little to no resistance whatsoever.

Fluoride, sad to say, has exactly the opposite effect on teeth to that which is promoted to us; strong healthy teeth. There is now, especially in America, an absolute epidemic of dental fluorosis with up to 80% of children in some cities being affected. The first visible sign of excessive fluoride exposure according to the US National Research Council are brownish flecks or spots, particularly on the front teeth, or dark spots or stripes in more severe cases.

The Falsification of Science

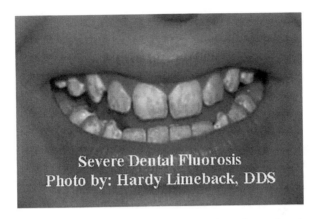

Severe Dental Fluorosis
Photo by: Hardy Limeback, DDS

What is much less known to the public is that fluoride also accumulates in bones.

"The teeth are windows to what's happening in the bones." Paul Connett, Professor of Chemistry at St. Lawrence University, New York.

In recent years, paediatric bone specialists have expressed alarm about an increase in stress fractures among young people in the US. Connett and other scientists are concerned that fluoride, linked to bone damage in studies since the 1930s, may be a contributing factor.

In 1944 a severe pollution incident occurred downwind of the E.I. DuPont de Nemours Company chemical factory in Deepwater, New Jersey. The factory was then producing millions of pounds of fluoride for the Manhattan Project whose scientists were racing to (allegedly) produce the world's first atomic bomb.

The farms downwind in Gloucester and Salem counties were famous for their high-quality produce. Their peaches went directly to the Waldorf Astoria Hotel in New York City; their tomatoes were mainly bought by Campbell's for soup. But in the summer of 1944 the farmers began reporting that their crops were blighted. They said that poultry died after an all-night thunderstorm and that farm workers who ate produce they had picked would sometimes vomit all night and into the next day. The horses looked sick and were too stiff to work, and some cows were so crippled that they could not stand up; they could only graze by crawling on their bellies.

The account was confirmed in taped interviews with Philip Sadtler (shortly before he died) of Sadtler Laboratories of Philadelphia, one of the USA's oldest chemical consulting firms. Sadtler had personally conducted the initial investigation of the damage.

Kidney disease is another hallmark of fluoride poisoning. Multiple animal studies have found that fluoride levels as low as 1 part per million (ppm) which is the amount added to most fluoridated water systems, cause kidney damage. And a Chinese study found that children exposed to slightly higher fluoride levels had biological markers in their blood indicative of kidney damage.

It has also been found that fluoride impairs proper thyroid function and debilitates the endocrine system. Up until the 1970s, fluoride was used in Europe as a thyroid suppressing medication because it lowers thyroid function. Many experts believe that widespread hypothyroidism today is a result of overexposure to fluoride.

Since fluoride is present in most municipal water supplies in North America and in much of Western Europe, it is absurd to even suggest that parents avoid giving it to their young children. How are parents supposed to avoid it unless they install an expensive 'whole house' reverse-osmosis water filtration system? And even if families install such a system, fluoride is still found in all sorts of food and beverages, not to mention that it is absorbed through the skin every time people wash their hands with or take a bath or shower in fluoridated water.

There simply is no legitimate reason to fluoridate water. Doing so forcibly medicates an entire population with a carcinogenic, chemical drug. There really is no effective way to avoid it entirely and nobody really knows how much is ingested or absorbed on a daily basis because exposure is too widespread to calculate. But political pressure and bad science have continued to justify water fluoridation in most major cities, despite growing mountains of evidence showing its dangers.

Is it surprising that people now, after years of ingesting this toxin, have little to no interest in the world around them and the fate awaiting them and their children and grandchildren? After all, what's more important than who wins the game this weekend or the latest adventures of our favourite soap characters? All as planned of course.

A study pre-published in December 2010 in the journal '*Environmental Health Perspectives*' confirmed that fluoridated water definitely causes brain damage in children. The most recent among 23 others pertaining to fluoride and lowered IQ levels, the study so strongly proves that fluoride is a dangerous, brain-destroying toxin that experts say it could be the one that finally ends water fluoridation. I have severe doubts about that. The 'gravy train' must go on unhindered – at all costs.

I would strongly urge any dentists or doctors reading this to undertake their own research into the efficacy and effects of fluoride and not just parrot the propaganda relayed to them by their own Elite-controlled professional organisations.

However, it is also up to us all to take a stand and say 'no' to mass water fluoridation. If someone wishes to voluntarily imbibe fluoride, then let him/her do so in the genuine belief it is beneficial, I have no problem with that, but to forcibly administer this poison to an unsuspecting and unwilling populace without prior knowledge or consent is a criminal activity. We are not prisoners... not just yet anyway, but we are creeping ever closer to that undesirable state, that is for sure.

Contaminated Drinking Water

What is actually in our water other than fluoride? Is it pure or does it contain impurities and toxins? If we take heed of the water companies, they will tell us that water is absolutely pure due to being filtered at source. Wrong! This is totally untrue.

Have you ever come across a bottle of prescription tablets or a medicine bottle that you no longer needed or perhaps had passed its expiration date? You probably disposed of the substance by flushing it down the toilet, down the sink waste disposal unit or by throwing it in the waste bin?

It is extremely concerning that environmental contaminants originating from industrial, agricultural, medical and common household substances, ie., pharmaceutical waste, cosmetics, detergents and toiletries are being disposed of into the water systems of the world. A variety of pharmaceuticals including painkillers, tranquilisers, antidepressants, antibiotics, birth control pills, oestrogen replacement therapies, chemotherapy agents, anti-seizure medications etc., are finding their way into the water table via human and animal excreta from mass disposal into the sewage system.

Flushing unused medications down the toilet and pharmaceutical residue from landfills has a tremendously detrimental impact on groundwater supplies and thus drinking water. Agricultural practices are a major source of this contamination and 40% of antibiotics manufactured are fed to livestock as growth enhancers. Manure, containing traces of pharmaceuticals is often spread on land as fertiliser from which it can leach into local streams and rivers and thence into the water table. Conventional wastewater treatments, filtration and recycling as commonly used by our water companies does not even come close to being effective in eliminating the majority of pharmaceutical residues.

The prevalence of pharmaceuticals in water is nothing new. In fact, it is reasonable to assume that as long as pharmaceuticals have been in use, they and their metabolites have contributed to overall environmental contamination. What is new is our ability to detect trace amounts of these contaminants in water; hence, we are finding pharmaceuticals in water because we are finally able to detect them.

According to an article published in the December 2002 issue of *Environmental Health Perspectives*, the number of pharmaceuticals and personal care products (PCPs) released into the environment each year is roughly equivalent to the amount of pesticides used each year.

During 1999-2000, the US Geological Survey conducted the first nationwide investigation of the occurrence of pharmaceuticals, hormones and other organic contaminants in 139 streams from 30 states. A total of 95 contaminants were targeted including antibiotics, prescription and non-prescription drugs, steroids and hormones, 82 of which were found in at least one sample. In addition 80% of streams sampled were positive for one or more contaminant. Furthermore, 75% of the streams contained two or more contaminants, 54% had greater than five, while 34% had more than ten and 13% tested positive for more than twenty targeted contaminants. There is no valid reason to believe that this is not the case elsewhere in the world. And do you think that the situation has improved or grown worse in the last twenty plus years, since then?

Pharmaceuticals have since been found in treated sewage effluents, surface waters, soil and tap water. Antibiotics and oestrogens are only two of many pharmaceuticals suspected of persisting in the environment either due to their inability to naturally biodegrade or continued prevalence as a result of continuous release.

Recent monitoring studies fail to address one question; are the levels of pharmaceuticals in the environment significant? At first glance, one would say 'no' since levels found in the environment are six to seven orders of magnitude lower than therapeutic doses in spite of the fact up to 90 percent of an oral drug can be excreted in human waste, both solid and liquid. However, low and consistent exposures would not likely produce immediate acute effects but rather subtle impacts such as behavioural or reproductive effects that could very well go unnoticed.

In addition, concern remains over the increasing practice of artificial recharge of groundwater with sewage effluent where pharmaceuticals have been found to percolate into the groundwater. Some common pharmaceutical contaminants are known to persist for more than six years in the subsurface or groundwater.

The bad news is that conventional water and wastewater treatment methods allow many classes of pharmaceuticals to pass through into our drinking water supplies unchanged and untreated.

So, what is the true risk assessment of pharmaceuticals and other similar contaminants in water? Do they present a health threat to any humans, animals (or even plants) being exposed to them? Many scientists are concerned about long-term, chronic and combined exposures to agents designed to cause a physiological effect in humans and believe we should be very concerned about aquatic ecosystems where sperm levels and spawning patterns in aquatic organisms have been clearly altered in environments heavily polluted with a class of hormone-altering pharmaceuticals known as endocrine disrupters.

With a growing and ageing population as well as increased reliance on drug treatments and the development of new drugs, the problem with pharmaceutical contamination promises only to increase.

Since we (almost) all drink tap and bottled water routinely, every single day in one form or another would it therefore not be unreasonable to conclude that we are all being systematically 'poisoned' on an ongoing basis? Worse still, could any of these substances be cumulative in nature, making the impact on our health and that of our children even more significant? As long as the water corporations continue in denial over this issue, I would suggest that the health of every one of us is at serious long term risk, all through that supposedly most benign of all substances, water.

Junk Food

Many years ago there was a famous story about a man in the USA who accidentally left a McDonald's™ hamburger in a jacket pocket for a year and when he finally got around to wearing the jacket again, to his surprise, the burger looked and smelled exactly as it did when first bought, twelve months earlier. This took place in 1990. Since then he has bought a McDonald's™ hamburger every year and saved the results. He now has a veritable 'museum' of McDonald's™ burgers, complete with bread bun, French fries and associated relishes going back thirty years and has labelled and dated them. Interestingly, none of them have decomposed – at all. From the one bought in 1989 to the one bought in 2020, they still look virtually identical.

How can this be? Why do fast-food burgers and fries not decompose like 'regular' food? The obvious answer is partly that they are infused

The Falsification of Science

with so many chemicals that even mould and bacteria will not eat them. This is partially true but not the entire explanation.

Many processed foods do not decompose and are not consumed by moulds, insects or even rodents. For example, try leaving an open tub of margarine outside in your garden and see if anything bothers to eat it. It will not!

Potato crisps (chips) can last for decades. Frozen pizzas are remarkably resistant to decomposition and even some highly processed sausages and meats can be kept for years and they will never decompose.

With meats, the primary reason why they do not decompose is their high sodium (salt) content. Salt is a great preservative, as our ancestors have known for thousands of years. McDonald's™ meat is absolutely loaded with sodium – so much so that it qualifies as 'preserved' meat, without even considering the chemicals you might find in it. However, the real question should be 'why do the bread buns not turn mouldy?' That is the real issue, since regular, nutritious bread begins to grow mould within days. What could possibly be in McDonald's™ hamburger buns that would ward off microscopic life for almost three decades?

Unless you have qualifications in chemistry, you probably cannot even read the ingredients list for these burgers. McDonalds'™ website states that these frankly, scary ingredients are contained in their 'bread' buns...

"Enriched flour (bleached wheat flour, malted barley flour, niacin, reduced iron, thiamin mononitrate, riboflavin, folic acid, enzymes), water, high fructose corn syrup, sugar, yeast, soybean oil and/or partially hydrogenated soybean oil, contains 2% or less of the following: salt, calcium sulphate, calcium carbonate, wheat gluten, ammonium sulphate, ammonium chloride, dough conditioners (sodium stearoyl lactylate, datem, ascorbic acid, azodicarbonamide, mono- and diglycerides, ethoxylated monoglycerides, monocalcium phosphate, enzymes, guar gum, calcium peroxide, soy flour), calcium propionate and sodium propionate (preservatives), soy lecithin."

How delicious. Our old friend HFCS makes an appearance in there as does sugar, partially-hydrogenated soybean oil (real heart disease inducing fodder) and the long list of chemicals such as ammonium sulphate and sodium propionate.

However, the truly shocking part about all this is...

"In my estimation, the reason nothing will eat a McDonalds™ hamburger bun (except a human) is because it's not food! No normal animal will perceive a McDonalds™ hamburger bun as food and as it turns out, neither will bacteria nor fungi. To their senses, it's just not edible stuff. That's why these bionic burger buns just won't decompose. There is only one species on planet Earth that's stupid enough to think a McDonalds™ hamburger is food. This species is suffering from skyrocketing rates of diabetes, cancer, heart disease, dementia and obesity. This species claims to be the most intelligent species on the planet and yet it behaves in such a moronic way that it feeds its own children poisonous chemicals and such atrocious non-foods that even fungi won't eat it (and fungi will eat cow manure, just FYI)." Mike Adams, Natural News

Do you cook your chicken with *dimethylpolysiloxane*, an antifoaming agent made of silicone? How about *tertiary butylhydroquinone*, a chemical preservative so deadly that just five grams can kill you?

These are just two of the ingredients in a McDonalds™ '*Chicken McNugget™.*' Only 50% of a McNugget™ is actually chicken. The other 50% includes corn derivatives, sugars, leavening agents and completely synthetic ingredients.

The Organic Authority helpfully transcribed the full ingredients list provided by McDonalds...

"White boneless chicken, water, food starch-modified, salt, seasoning (autolyzed yeast extract, salt, wheat starch, natural flavouring (botanical source), safflower oil, dextrose, citric acid, and rosemary), sodium phosphates, seasoning (canola oil, mono- and diglycerides, extractives of rosemary).

Battered and breaded with...water, enriched flour (bleached wheat flour, niacin, reduced iron, thiamine mononitrate, riboflavin, folic acid,) yellow corn flour, food starch-modified, salt, leavening (baking soda, sodium acid pyrophosphate, sodium aluminium phosphate, monocalcium phosphate, calcium lactate), spices, wheat starch, whey, corn starch. Prepared in vegetable oil (Canola oil, corn oil, soybean oil, hydrogenated soybean oil with TBHQ and citric acid added to preserve freshness). Dimethylpolysiloxane added as an antifoaming agent."

Just in case you may think I am singling out McDonalds™ for some unfair treatment here, let me just say that I have no particular 'axe to grind' on this issue. I would include burgers and chicken burgers and any other 'delicious snacks,' not just from McDonalds™, but from also, take a bow...Burger King™, Wendy's™, Wimpy™, Kentucky Fried Chicken™ et al. (With sincere apologies to any other manufacturer of this vile filth masquerading as nutrition that I have omitted to mention – you know who you are!)

Until we, as a species, stop buying this dreadful garbage, they will continue to poison us and infect us with disease, while we pay them for the privilege. Let us all let them know that we are not prepared to suffer this any longer by boycotting all such establishments.

However, perhaps the biggest danger we face is not from the Elite corporations who seek to abuse our bodies and health in the name of profits, but from an organisation, truly Orwellian in scope which actively seeks to prevent us from gaining access to healthy eating alternatives. Let me introduce you to...

Codex Alimentarius

Codex Alimentarius is the United Nations/European Union plan to destroy organic farming and to eradicate the complementary and natural healthcare industry which is obviously a threat to the health, welfare and profits of Big Food and Big Pharma.

There would appear to be an almost total lack of awareness (or even interest) with regard to the implications of this pernicious global Commission, particularly amongst those most affected by the excesses of this restrictive legislation. In the words of the National Health Federation, the aims and objectives of Codex Alimentarius are to...

- Allow only low potency supplements that will do nothing positive for one's health.

- Allow all or most foods to be genetically modified.

- Make beneficial supplements unavailable or sold by prescription only.

For many people, this agenda is so outrageous, they cannot believe such goals are achievable; yet this may well very soon be reality, if the Codex Alimentarius Commission continues to disregard input from those who offer a counter perspective to the combined forces of Big Food and Big Pharma.

For the past five years the European challenge to Codex has been led by Dr Robert Vererk, Executive Director of the Alliance for Natural Health and Scientific Advisor to the National Health Federation. Yet despite the efforts of the ANH, the NHF and the Dr Rath Foundation the Codex agenda lumbers ever closer to the EU statute books.

In April of 2005, the ANH mounted a legal challenge to the Codex Commission; Justice Leendert A. Geelhoed, the European Union Advocate General, referred to the arbitrary powers of the Codex-supporting EU legislation as being *"about as transparent as a black box."* The subsequent 12th July 2005 ruling of the International Court of Justice in Luxembourg followed the 4th July Rome meeting of Codex when the 85 countries present ratified the restrictive guidelines for dietary supplements. In 2005, six days after London was awarded the 2012 Olympics and just five days after the London 7/7 bombings, there was little mention of the ICJ ruling in the British media. Another so-called 'good day to bury bad news' it would seem.

Amongst the most disturbing component contained within Article 6 of the EU Directive, is that it strictly prohibits information about

The Falsification of Science

diseases being treatable by nutrients and calls for future supplement dosage restrictions. Such is the power and influence of Big Pharma.

Today the EU pays an annual fine of $150 million to maintain its ban on US hormone-fed beef. Clear evidence that WTO rules put *free-trade* interests of agribusiness above national health concerns. Meanwhile, a flood of new GMO products are surreptitiously being introduced into EU agriculture. Monsanto (now Bayer), Syngenta and other GMO multinationals have already taken advantage of lax national rules in new EU member countries such as Poland to get the proverbial GMO 'foot-in-the door.' Pro-GMO governments, such as that of Angela Merkel in Germany, abdicate any responsibility by claiming they 'are only following WTO orders,' which is exactly the line taken by the FSA and strangely reminiscent of another organisation in history coincidentally from the same part of the world, whose followers used exactly the same defence.

Powerful agribusiness multinationals such as Bayer, Dow Chemicals and DuPont are working through the WTO-backed Codex Alimentarius Commission in their determination to overrun national or regional efforts to halt the march of GMO.

NHF & ANH resistance to Codex will potentially be little more than an inconvenience to the prime movers behind this pernicious global agenda... unless and until such time as there is a wider realisation that the organic farming and natural health industries may soon be little more than a distant memory.

Meanwhile, Big Pharma and Big Food have a very well prepared strategy to ensure the success of the Codex agenda. But our challenge is to demonstrate that no corporate strategy can be effective against the universal desire to retain the basic human right to food and health freedom.

Contrary to popular belief, Codex Alimentarius is neither a law nor a policy. It is in fact a functioning body, a Commission, created by the Food and Agricultural Organization and the World Health Organisation under the direction of the United Nations. The confusion in this regard is largely due to statements made by many critics referring to the 'implementation' of Codex Alimentarius as if it were legislation waiting to come into effect. A more accurate phrase would be the 'implementation of Codex Alimentarius guidelines,' as it would more adequately describe the situation.

Codex is merely another tool in the chest of the Elite whose goal is to create a one-world government (the oft promoted NWO) in which they will wield complete control over humanity. Power over the food supply is essential in order to achieve this. As will be discussed later, Codex Alimentarius will be 'implemented' whenever guidelines are established, and national governments begin to arrange their domestic laws in accordance with the standards set by the organisation.

The existence of Codex Alimentarius as a policy-making body has roots going back over a hundred years. The name itself, Codex Alimentarius, is simply Latin for 'food code' and directly descended from the 'Codex Alimentarius Austriacus,' a set of standards and descriptions of a variety of foods in the Austria-Hungarian Empire between 1897 and 1911. This set of standards was the brainchild of both the food industry and academia and was used by the courts in order to determine food identity in a legal fashion.

Even as far back as 1897, nations were being pushed toward harmonisation of national laws into an international set of standards that would reduce the barriers to trade created by differences in national laws. As the Codex Alimentarius Austriacus gained traction in its localised area, the idea of having a single set of standards for all of Europe began to engender support too. From 1954-1958, Austria successfully pursued the creation of the Codex Alimentarius Europaeus (the European Codex Alimentarius) and almost immediately the UN directed FAO (Food and Agricultural Organisation) sprang into action when the FAO Regional Conference for Europe expressed the desire for a global international set of standards for food. The FAO Regional Conference then sent a proposal up the chain of command to the FAO itself with the suggestion to create a joint FAO/WHO programme dealing with food standards.

The Falsification of Science

The very next year, the Codex Alimentarius Europaeus adopted a resolution that its work on food standards be taken over by the FAO. Then in 1961, it was decided by the WHO, Codex Alimentarius Europaeus, Organisation for Economic Cooperation and Development (OECD), and the FAO Conference to create an international food standards programme known as the Codex Alimentarius. In 1963, as a result of the resolutions passed by these organisations two years earlier, Codex Alimentarius was officially created.

Although created under the auspices of the FAO and the WHO, there is some controversy regarding individuals who may or may not have participated in the establishment of Codex. Some anti-Codex organisations have asserted that Nazi war criminals, Fritz ter Meer and Hermann Schmitz in particular, were the principal architects of the organisation. Because many of these claims are made with only indirect evidence or no evidence at all, one might be tempted to disregard them at first glance. However, as the allegations gain more and more adherents, Codex has attempted to refute them. In the 'Frequently Asked Questions' (FAQ) section of its website, the question, 'Is it true that Codex was created by a former war criminal to control the world food supply?' Is answered by stating;

"No. It is a false claim. You just need to type the words 'Codex Alimentarius' in any search engine and you will find lots of these rumours about Codex. Usually the people spreading them will give no proof but will ask you to send donations or to sign petitions against Codex.

Truthful information about Codex http://www.codexalimentarius.net is found on the Internet —there is nothing to hide from our side—we are a public institution working in public for the public—we are happy if people want to know more about our work and ask questions.

There is an official Codex Contact Point http://www.codexalimentarius.net/web/members.jsp in each member country, who will be pleased to answer your questions on Codex."

But, as one can see from the statement above, Codex's response does very little to answer this question beyond simply disagreeing with it.

While it is true that many individuals who make this claim provide little evidence for it, the presentation of the information does not necessarily negate its truthfulness. In fact, Codex offers its own website as a source for accurate information about the organisation;' yet, beyond the FAQ section, there is nothing to be found that is relevant to the 'war criminal' allegations; furthermore the Codexalimentarius.net website is virtually indecipherable, almost to the point of being completely useless.

In the end, this response raises more questions than it answers. This is because Codex, if it wished to do so, could put these rumours to rest by simply posting a list of the individuals and organisations that funded or played an integral role in its creation. However, it does nothing of the sort. Beyond mentioning the FAO and the WHO, we are completely unaware of who or how many other individuals and organisations participated in the creation of Codex Alimentarius.

However to repeat, Codex does nothing to dispel the allegations besides simply disagreeing with them and the connections are not at all implausible. Codex is very secretive about its beginnings, as evidenced on its website where it only states that it was created at the behest of the FAO and the WHO. It is highly unlikely that such an organisation would be created without the assistance, input, and even funding of privately owned international corporations. Thanks to both the anti-Codex community and Codex Alimentarius itself, there is no evidence that documents which individuals or corporations were involved in its establishment. Nevertheless the fact remains that Codex is an extremely sinister organisation, with an utterly pernicious agenda.

Medicine

The Rockefeller Medical Paradigm

"Our current system of drugs-and-surgery conventional medicine will bankrupt any state or nation foolish enough to depend on it. No nation

The Falsification of Science

that bets its future on pharmaceuticals and chemotherapy is going to win that bet. They will all collapse in the end because you can't create a healthy nation by drugging your population into a state of health. As long as Big Pharma dominates health care and it currently runs the medical journals, medical schools, hospitals, and even the FDA so you will never have a health care system that has any interest whatsoever in teaching people how to be healthy. When profits come from sickness, the corporations always find new ways to keep people sick." Mike Adams

Medical totalitarianism is undoubtedly with us and has permeated our entire mainstream health regime.

Probably one of the most callous and insidious ways by which we are being covertly attacked as a species is not only via the food we eat, but also through our totally corrupt healthcare system which has been infiltrated and is largely controlled by 'Big Pharma' the large pharmaceutical cartels and to a lesser but nevertheless significant extent, 'Big Food' the giant food conglomerates.

To say that our healthcare system has failed us in the past, continues to fail us in the present and is constantly being manipulated to fail us in the future, is a contender for understatement of the millennium. Make no mistake about it; general human health is not improving one iota, despite the so-called 'medical advances' of the last century or more. It is an undisputable fact that the incidence of all the newer, major destructive diseases is increasing exponentially as time goes by, despite the near-eradication of certain formerly 'killer' diseases.

Fifty years ago, one person in fifty would have been expected to contract cancer during their lifetimes and now that situation has deteriorated to the point where we are lucky if it is as low as 1 in 3 and decreasing yearly. This is also true of many diseases such as Diabetes types 1 and 2, Alzheimer's, Multiple Sclerosis, and the many invented 'mental' syndromes that have only been prevalent for a decade or so, ADHD, Bipolar Disorder and the like.

Why should this be? Does anyone ever really try to answer that question? I believe that this question is sidestepped and avoided at all costs by the Elite-run medical establishment because they know the answer to the question already, but would very much rather that you did not, if at all possible.

For example, consider this for a moment; a typical medical doctorate takes about 4 to 6 years of study to achieve, depending on the exact course undertaken, at any Western medical education establishment or University. One would naturally expect that any education or teaching programme about the workings of the human body should include a rather finely detailed study of human nutrition, its effects on all the bodily organs and functions and how to help *prevent* disease and other ailments by way of a balanced diet and vitamin consumption. Is it not more than a little strange then that our standard medical education *does not include one single lecture in six whole years of study, on the subject of nutrition or disease prevention?*

Western medicine however, is at its foundation, wholly a Rockefeller creation. The Rockefellers, of course, are one of the most rich and powerful families of the Elite. Behind their spurious facade of philanthropy, they are power-hungry tyrants intent on owning the entire world and depopulating it through eugenics-based programmes such as forced sterilisation, water fluoridation, abortions and vaccinations, to name just a few of them. They created and still control the United Nations, the World Health Organisation, the Council on Foreign Relations, the Trilateral Commission, Planned Parenthood and many, many other organisations that totally influence and dominate our current culture to a large extent.

Despite the world dominance of allopathic medicine today, even just one hundred and thirty years ago the situation was very different, so let us cast our minds back to how we arrived at this dire situation. How did Western medicine and the giant conglomerate of multinational pharmaceutical corporations, aka 'Big Pharma,' become the mainstream medical system in the US, Europe and all other 'First world' nations?

In the late 19th century, John D. Rockefeller, a man quoted as saying *"competition is a sin,"* was the head of the Rockefeller family and had become extremely rich through the extraction of oil from the ground.

The Falsification of Science

Subsequently, after searching for ways to capitalise even further upon his surplus oil, he hit upon the idea of using coal tar – a petroleum derivative – to create substances that affect the human mind, body, and nervous system. These eventually became the drugs with which we are now so familiar and used ubiquitously by the pharmaceutical cartels.

So, Rockefeller used some of his vast oil profits to buyout part of the massive German chemical organisation, I.G. Farben and then embarked on a truly evil plan – evil from the point of view of a free and healthy humanity that is, but nevertheless brilliant from a purely business perspective.

Rockefeller realised that there were many different types of doctors and healing modalities in existence at that time, including chiropractic, naturopathy, homeopathy, holistic medicine, herbal medicine and more, but he desperately needed to eliminate these competitors to his medical paradigm. He therefore hired a man called Abraham Flexner to submit a report to Congress in 1910 which concluded that there were far too many doctors and medical schools in America, and that all the natural healing modalities which had existed for hundreds and even thousands of years were nothing more than 'unscientific quackery.' It therefore called for the standardisation of medical education, whereby only the newly formed, Rockefeller allopathic-based AMA be allowed to grant medical school licences in the US.

The rest as they say, is 'history' and sadly, the US Congress almost immediately acted upon these deceptive conclusions and made them law. And so, incredibly, allopathy became the standard mainstream modality, even though its three chief methods of treatment in the 1800s had been bloodletting, surgery, and the injection of toxic heavy metals such as lead and mercury to supposedly cure disease. It should be noted that hemp was also demonised and criminalised not long after this, not because there is anything inherently dangerous about it, quite the contrary, but simply because its efficacy was viewed as a huge threat (as both medicine and fuel) to the Rockefeller drug and oil industries, respectively.

But this sad story does not end there. Rockefeller and another member of the Elite, Andrew Carnegie used their tax-exempt Foundations, from 1913 onwards, to offer huge grants to the 'best' medical schools all over America – but on the strict proviso that only an allopathic based curriculum be taught, and that their own hand-picked agents be allowed to sit on the Boards of Directors of those institutions. They then proceeded to systematically dismantle the curricula of these schools by removing any reference or mention of the natural healing power of herbs and plants, or of the importance of diet to health. The result of which is a system which to this day creates doctors who are, totally ignorant of nutrition and utterly disregard the idea that what we eat can actually heal or conversely, harm us.

Two decades subsequent to these events, another law was passed that further entrenched Rockefeller medicine in the US. The Hill-Burton Act of 1946 gave hospitals grants for construction and modernisation, on the strict condition that they provide free healthcare to anyone in need, without discrimination of any kind. Although on the surface this would appear to be a positive, the very real downside was that once people had become dependent on this system for their healthcare needs – especially those permanently on the pharmaceutical industry's pills and medicines – the system very quickly switched from being free into one that had to be paid for, and therefore the Rockefellers found themselves with millions of new, lifelong customers.

The bitter truth is that, in general now, whenever we visit our doctor we are regarded as a potential market for the pharmaceutical industry's products and of course, for Big Pharma, there is no financial incentive to heal you, because put simply and bluntly, a patient cured is a customer – and therefore revenue – lost. However, Big Pharma continues to target us, even if not currently one of its billions of customers by using insidious propaganda in an attempt to convince us that we are ill (for example with psychiatry's ridiculous list of fictitious diseases), and as previously related in this chapter. Also pregnant women (for example) are a particularly lucrative target. They are sold on the ideas of intravenous fluid bags, foetal monitors, ultrasound

The Falsification of Science

(dangerous radiation for a vulnerable baby), a whole smorgasbord of various drugs, the totally unnecessary episiotomy, induction, and as a final, parting shot (literally), a Caesarean delivery, whether or not it is medically necessary.

And of course, many of their drugs are derived from plant compounds, but because nature itself cannot be patented and sold, Big Pharma has no interest in natural cures. So, what they do instead is engage in biopiracy – research natural compounds, copy them (or modify them slightly) in a lab, then in effect steal and attempt to patent them. And if they are successful in receiving a patent, they then market their pill as a 'wonder drug' whilst simultaneously (through fake scientific research and statistics) suppress and demonise the original plant as being worthless. Ironically, John D. Rockefeller used, and the British Royal Family still uses, homeopathy as their own preferred method of medical treatment.

Dr Barbara Starfield published a study in the year 2000 that found there were 225,000 iatrogenic (allopathic doctor caused) deaths in the US every year. However, this only included direct deaths, but when all indirect deaths are factored in too, as per Dr Gary Null's 2011 report 'Death by Medicine,' the actual figure is close to 800,000 per year. In other words, over 2,000 per day on average and a truly staggering 8 million people over the course of a decade.

Null's report concluded that... *"It is evident that the American medical system is the leading cause of death and injury in the United States..."*

The famous, ancient Greek physician Hippocrates wrote... *"Nature heals. The doctor's task consists of strengthening the natural healing powers, to direct them, and especially not to interfere with them."*

How ironic it is then that the oath that all doctors are required to take prior to commencing their careers and being licenced as a practicing physician is known as the 'Hippocratic Oath' and named for the above physician.

Here it is, in all its glory...

"I swear by Apollo Healer, by Asclepius, by Hygieia, by Panacea, and by all the gods and goddesses, making them my witnesses, that I will carry out, according to my ability and judgment, this oath, and this indenture.

To hold my teacher in this art equal to my own parents; to make him partner in my livelihood; when he is in need of money to share mine with him; to consider his family as my own brothers, and to teach them this art, if they want to learn it, without fee or indenture; to impart precept, oral instruction, and all other instruction to my own sons, the sons of my teacher, and to indentured pupils who have taken the Healer's oath, but to nobody else.

I will use those dietary regimens which will benefit my patients according to my greatest ability and judgment, and I will do no harm or injustice to them. Neither will I administer a poison to anybody when asked to do so, nor will I suggest such a course. Similarly I will not give to a woman a pessary to cause abortion, but I will keep pure and holy both my life and my art. I will not use the knife, not even, verily, on sufferers from stone, but I will give place to such as craftsmen therein.

Into whatsoever houses I enter, I will enter to help the sick, and I will abstain from all intentional wrong-doing and harm, especially from abusing the bodies of man or woman, bond or free. And whatsoever I shall see or hear in the course of my profession, as well as outside my profession in my intercourse with men, if it be what should not be published abroad, I will never divulge, holding such things to be holy secrets.

Now if I carry out this oath, and break it not, may I gain for ever reputation among all men for my life and for my art; but if I break it and forswear myself, may the opposite befall me."

You would, dear reader, be forgiven for thinking that this must be some kind of sick joke. I refer specifically to the fact that every allopathic doctor on the planet, despite undertaking this oath probably breaks it several times every day. Truly Orwellian stuff, I am sure you agree.

Living a healthy natural lifestyle in a supportive healthy environment is our best defence against any disease and most natural medicine is designed to treat the body holistically, taking into account multiple factors involved in the disease process and not simply treating us part by part, as if a machine, and as is the case in a mechanical, allopathic approach. Likewise, traditional Chinese medicine defines all disease as stagnation and treats sickness as an imbalance to be rectified, unlike allopathic medicine which 'cures' one ailment, often by directly causing another. What is the point of transferring an imbalance in one area into an imbalance in another area? None, unless of course it is done in order to continuing profiting from sickness, which of course is Big Pharma's entire raison d'étre. This is no real healing paradigm.

Please do not misunderstand me however. The vast majority of doctors are extremely well-trained, professional, thoroughly competent, and knowledgeable in the area of physical, bodily issues such as fractures, sprains, wounds, and muscular problems, in other words mending physical 'broken' bodies. I would not hesitate to seek their assistance should I find myself in need of such a remedy but the track record of first world doctors in treating or preventing general bodily ailments and diseases is nothing short of scandalous and a disgrace to any so-called civilised society.

We have become so conditioned and propagandised to regard pharmaceutical solutions to health issues as being the 'norm' or the 'only way,' that we now regard the treatment or prevention of disease through nutrition and balanced vitamin rich diets as a form of 'quackery' or witchcraft. Indeed this is how Big Pharma attempts to portray natural health solutions, putting them on a par psychologically with the mediaeval practices of using leeches and the 'bleeding' of patients.

"...doctors and health 'experts' have astonishing gaps in knowledge that should be considered basic health information in any first world nation. Parents, too, lack any real literacy in nutrition and health. That's largely because medical journals, health authorities and the mass media actively misinform them about health and nutrition issues, hoping to prevent people from learning how to take care of their own health using simple, natural remedies and cures." Natural News, 2009

Pharmaceutical companies are certainly not in business to maintain human health at its optimum levels. They are there to make money for their Elite owners and shareholders and maintain health levels at an appallingly low level in order to further their drug sales agendas. They certainly excel at both of these activities. Big Pharma companies are among the richest if not actually *the* richest organisations in the world. In recent research conducted to determine the top 100 wealthiest organisations worldwide, 51 were corporations and *only 49 were countries*, which in itself is truly staggering information. Included in the top 10 were the 'big 4' pharmaceutical companies.

So, here we have a situation where companies that have more disposable income than most countries and are concerned with profits to the detriment of all else, are dictating to the medical profession how to conduct a successful health regime and in the process are making colossal profits based on that philosophy. What is wrong with this picture?

"The medical practice of today is anything but a healing modality. It is geared toward maximum profit generated by those for whom disease is a growth industry. They poison the environment and encourage you to eat bad food, simply because they are invested in your becoming sick." Les Visible, musician and researcher.

Your body never becomes ill because it lacks artificial, pharmaceutical, allopathic drugs. It becomes ill through lack of proper, correct nutrition or because it has become 'poisoned' or infected with an outside agent of some kind, often exacerbated by incorrect nutrition. So why do we rush to dose ourselves with, more often than not, harmful chemicals at the first sign of any problems or anything untoward with our bodies? I suggest that it is because both we and our doctors alike are conditioned from birth by 'the system' engendered by Big Pharma, to do so.

Sadly, I have to remind you if you did not already know this, that the giant, multi-national pharmaceutical companies have completely infiltrated all of the important healthcare organisations from cancer research charities such as the hugely corrupt American

The Falsification of Science

Cancer Society (ACS) to the equally corrupt American Food and Drug Administration (FDA), to the British and American Medical Associations (BMA and AMA) and even medical education establishments across the Western world with the primary intention of deceiving the world about healthcare. Indeed the FDA is provably culpable for allowing highly toxic yet highly profitable, substances into the food chain in what can only be described as at best, irresponsible and at worst, criminal activities.

Our healthcare system does not exist to make you well. It exists to profit from your illnesses and keeps you just well enough to stay alive so it can continue to push drugs on you. The last thing the Elite want is a healthy, well nourished, physically, and mentally strong population able to think clearly and look after themselves and their families. What they actually *do* want is a sickly, malnourished society, totally dependent on them and their poisons and unable to act and think freely for itself, so that they can exploit our helpless situations to the maximum in the cause of reaping their huge annual profits. Unfortunately, over time, this is exactly what they have managed to achieve. As a main thrust of this policy, they also do not want you to understand nutritive disease prevention as there is little or no profit to be had from prevention, but plenty from 'cures.'

When I say 'cures,' please be advised that I use the term very loosely. In actual fact most cures to be derived from pharmaceuticals, only treat the symptoms and not the underlying cause of the problem. For example if you have a severe headache, then Big Pharma says you should take a strong painkiller or one of their migraine formulas. This may well remove the pain but ignores the *extremely important fact* that pain exists for a reason and that is to let us know there is something amiss somewhere. Simply eliminating the pain does not make the reason for the pain go away and could actually be dangerous as it removes the 'alarm call' that pain is designed to be. To use an analogy, imagine your car engine develops a fault and a red warning light appears on the dashboard. Would you remove or cover-up the small red light to cure the fault and declare it fixed when the light can no longer be seen, or would you use that warning signal to actually check under the bonnet (hood) for the real problem?

Almost 100% of pharmaceutical drugs work in this fashion and I include their so-called cancer drugs. All that radiotherapy and chemotherapy does is remove the tumour (sometimes) and kill ALL cells (healthy and unhealthy) whilst destroying one's natural resistance leaving one open to succumb once more to cancer in other areas of the body and many other deadly ailments. Sadly, this is what is happening all the time. Many people are delighted to receive the news that their cancer is in remission shortly after a course of radiation. However, what we are rarely told is that the cancer more often than not, returns with a vengeance a few months or sometimes years later, to take advantage of a thoroughly ravaged and weakened body. Of course even if by luck the cancer does not return, the body's natural defences are often devastated, leaving the victim open to all manner of further diseases.

"As clear-thinking people, natural health consumers sometimes look at the actions of the Food and Drug Administration (FDA) and wonder what planet its decision makers seem to be from. It's like the FDA is living in a completely different world than the rest of us—a world where nutrients are dangerous, but synthetic chemicals are perfectly safe for human consumption. In fact, the idea that FDA bureaucrats and modern medicine promoters are living in a different reality is not far from the truth. In my view, FDA decision makers have no connection with reality. They're simply operating on a system of false beliefs and circular reasoning that justifies their efforts to protect Big Pharma profits by exploiting, misleading and directly harming the public." Natural News

So, Big Pharma, in conjunction with Big Food, the FDA and their corporate media whores (all owned by the same bloodline families when you follow the pyramids to the very top) conspire together to wreak havoc on human health so that they can make billions if not trillions out of our misery, whilst pretending that they are spending *our* millions in the form of charitable donations, to search for cures to diseases that can, in most instances, be treated simply, inexpensively and easily through proper and adequate nutrition—or at worst, natural remedies.

The Falsification of Science

If this chapter alone does not prove the adage that science is being systematically falsified for the personal gain of so many entities, then I am not sure what does!

However, we will now leave this particular topic, with the following quote...

"The nature of the medical establishment today is unsettling, to say the least. Doctors of all kinds have been trained to prescribe double-edged medical 'solutions' to their patients, draining the finances of patients through side-effect ridden pharmaceuticals and invasive surgeries. Mainstream medical science is increasingly being found to be fraudulent, but many still see doctors and medical officials as 'experts' that can do no wrong." Andre Evans. Activist Post 19th October 2011

Chapter 9

The Environment

The Great 'Climate Change' Hoax

The debate about climate change is finished. Over. Climate change has now been categorically proven NOT to exist, as confirmed by one of the world's leading climate authorities and despite what establishment and media talking heads and institutions such as the Royal Society, Wikipedia™, the National Geographic, the Meteorological Office and similarly countless other so-called environmental action groups as well as the controlled mainstream media et al, have to say on the topic. These organisations and others of a similar ilk are little more than mouthpieces for the Elite's scientific mafia and the 'powers that be.'

John Coleman, who was co-founder of the *Weather Channel*, shocked establishment so-called 'academics' by insisting that the theory of manmade climate change was no longer scientifically credible. Instead, what little evidence there is for rising global temperatures points to a 'natural phenomenon' within a developing ecosystem. In an open letter attacking the Intergovernmental Panel on Climate Change (IPCC) he wrote...

"The ocean is not rising significantly. The polar ice is increasing, not melting away. Polar Bears are increasing in number. Heat waves have actually diminished, not increased. There is not an uptick in the number or strength of storms (in fact storms are diminishing). I have studied this

The Falsification of Science

topic seriously for years. It has become a political and environment agenda item, but the science is not valid."

Coleman said that he based many of his views on the findings of the NIPCC, a non-governmental international body of scientists aimed at offering an 'independent second opinion of the evidence reviewed by the IPCC.' The IPCC incidentally is 100% a political organisation and not a scientific one, although it does include many scientists – which one can only assume are fully 'paid-up' members – in more than one sense of the phrase.

He added for good measure... *"There is no significant manmade global warming at this time, there has been none in the past and there is no reason to fear any in the future. Efforts to prove the theory that carbon dioxide is a significant greenhouse gas and pollutant causing significant warming or weather effects have failed. There has been no warming over 18 years."*

The IPCC argues strongly that their research shows it is certain that manmade global warming will lead to extreme weather events becoming more frequent and unpredictable, but perhaps more predictably, big business is *"now fully engaged and actively responding to climate science and data."*

The US, along with the UK and other developed countries, is expected to pledge further actions on climate change very soon, all of which contributes to the ongoing lie.

Climate expert William Happer, from Princeton University, fully supported Coleman's claims. He said, *"No chemical compound in the atmosphere has a worse reputation than CO_2, thanks to the single-minded demonization of this natural and essential atmospheric gas by advocates of government control and energy production. The incredible list of supposed horrors that increasing carbon dioxide will bring the world is pure 'belief' disguised as science."*

Is it not the case that an abundance of carbon dioxide in the atmosphere is beneficial in at least one respect? Which gaseous compound is it again that feeds and nourishes all plant life here on Earth, in much the same way as oxygen sustains all animal life? Please remind me. Oh yes, I remember now...it is called carbon dioxide (CO2), of course. Surely then, is not a plentiful amount of CO2 in the atmosphere in fact a positive thing, leading to a good healthy environment for the growth of plants and trees – you know, those large sprawling, mainly green things that in turn excrete the very oxygen that we need to sustain us and the entirety of the rest of the animal kingdom? Well one would assume so at any rate, but not so according to the lying scaremongers of the completely politically biased and controlled IPCC.

And of course demonising CO2 also does the same to humanity by instilling guilt in we CO2 exhaling 'polluters.' In order to actually achieve the absurd 'zero CO2' emissions 'footprint' goals we would all have to stop breathing (and walking), which is, no doubt what our masters would prefer that we do! The current asphyxiating mask mandates brings us one step closer to that goal also.

The 2010 Inter Academy Council review was launched after the IPCC's hugely embarrassing 2007 benchmark climate change report, which contained alarmist, exaggerated and even wholly false claims that Himalayan glaciers 'could melt by 2035.' However, climate change proponents remain undeterred in their mission, ignoring numerous recent scientific findings indicating that there has been no warming trend at all for nearly two decades.

The 'godfather' of the global warming scam, Al Gore, whom many believe to have been 'given' the role of worldwide promotion of the entire scam in return for being cheated out of the presidency by Bush 'the lesser' due to voting irregularities, predicted among other dire occurrences, the melting of polar ice on a massive scale. These predictions have without exception proved utterly false. In fact, in 2014 – a year that was proclaimed as being 'the hottest ever' in the Earth's history – there were record amounts of ice reported in Antarctica, an increase in Arctic ice, and record snowfalls across the world!

On top of these 'inconvenient truths,' the White House's assertion that 97% of scientists agree that global warming is real has been completely debunked. Several independently researched examinations of the literature used to support the '97%' claim found that the conclusions were cherry-picked and misleading.

More objective surveys have revealed more realistically that there is a far greater diversity of opinion among scientists than the global warming crowd would have us believe.

"A 2008 survey by two German scientists, Dennis Bray and Hans von Storch, found that a significant number of scientists were sceptical of the ability of existing global climate models to accurately predict global temperatures, precipitation, sea-level changes, or extreme weather events even over a decade; they were far more sceptical as the time horizon increased."
The 'National Review'

Other mainstream news sources besides the National Review have also been courageous enough to speak out against the global warming propaganda – even the Wall Street Journal published an op-ed piece in 2015 challenging the Anthropogenic Global Warming (AGW) pseudoscience being promulgated by global warming proponents. And, of course, there are the more than 33,000 American scientists to date, who have signed a petition challenging the climate change narrative and 9,029 of them hold PhDs in their respective fields. But of course Al Gore and his lackeys have also ignored that particular inconvenient truth, too.

Many of the scientists who signed that petition were likely encouraged to speak out in favour of the truth after retired senior NASA atmospheric scientist John L. Casey revealed that solar cycles, known as Grand Solar Maxima and Minima, are largely responsible for warming and cooling periods on Earth – not human activity. But the global warming proponents continue to push their agenda on the gullible public whilst lining their cavernous pockets in the process. If you are still inclined to believe the gospel according to St. Al Gore, on global warming, please consider the fact that since he embarked on his 'crusade,' his wealth has grown from a reported $2 million in 2001 to well over $150 million in 2020 – largely due to investments in fake 'green tech' companies and the effective embezzlement of numerous grants and loans in support of his utter lies and complete distortions of the real facts.

As stated above, 2014 was supposed to have been the hottest year ever recorded. If this were actually true, all the terrible catastrophes predicted would be happening right now, but in fact the exact opposite is happening. The Earth is actually beginning to cool dramatically as we enter the latest Grand Solar Minimum.

And of course the utterly deceptive 'climate change' agenda blends suspiciously neatly into the tenets of Agenda 21 and Agenda 2030, too. Hmmm, strange that.

In that 'hottest year ever' of 2014 there was record sea ice in Antarctica. In fact, a 'global warming' sponsored expedition managed to become stranded in it. Oh, the irony. Arctic sea ice also made a huge comeback in 2014 and the Great Lakes in North America experienced record levels of ice too. Lake Superior only had three completely ice-free months in 2014. You would think that in the 'hottest year ever' that ice would be melting, just as Al Gore said in his ridiculous, highly risible 'hit-piece,' the documentary, *An Inconvenient Truth*' which for the sake of painstaking accuracy really should have been titled, *'An Extremely Convenient Lie.'*

In 2014 we experienced all kinds of cold records. Surely, in reality there should have been all kinds of heat records broken in that good old 'hottest year ever'? 2014 also saw record snowfall in many areas. It was not too long ago that the deceitful, lying 'climate change' crowd said that global warming would cause snow to disappear and that children 'will not know what snow is!' Laughable stuff but consider how many people actually believe this and similar lies and metaphorically bow before the altar of the church of St. Al and his faithful flock of adherents. And these faithful apologists (the believing masses and the 'scientists' promoting the global warming/climate change myth), when confronted with evidence of record cold winters, actually claim that 'global warming' is also causing cooling via its ability to somehow create 'extreme weather' events of all kinds, conveniently blaming any type of weather event on global warming. How logical is this?

The Falsification of Science

You may think that Polar Bears would really be in trouble in 2014, *"the hottest year ever"* but they are actually thriving. And remember the emotive photographs of polar bears apparently 'stranded' on remote ice floes and allegedly in danger of drowning, all designed to gain your emotional support for the fake agenda?

A small point maybe, but a valid illustration of how we are all being conned by these people and their extravagant lies and distortions, is the myth of imminent polar bear extinction. According to Dr Mitchell Taylor who has been involved for decades in the research on polar bears in northern Canada, this is absolutely untrue. They are in fact experiencing their highest numbers for almost sixty years. The famous photograph of the two polar bears standing on an ice floe, implied to be in danger of drowning and used by Gore to promote his infamous *An Inconvenient Truth* was in fact taken by Amanda Byrd who publicly stated, to no avail of course, that those particular bears were in no danger whatsoever. In fact polar bears, far from being in danger of drowning in these situations, are known to be capable of swimming for up to 250 miles at a stretch when seeking-out new food sources.

Al Gore actually predicted that the world's oceans would rise 20 feet by 2100, but the latest data demonstrates that we are currently on track for around two inches! Really Al, really? Indeed most tide gauge meters show no rise in sea level whatsoever, and almost none show any acceleration over the past 20 years.

A few years ago the moose population in Minnesota dropped rapidly and of course the 'Gore worshippers' immediately blamed global warming. That is until a study proved that it was actually wolves that were killing the moose. Wolves have already been removed from the endangered species list and are now themselves endangering other species, so the state of Minnesota instituted a wolf-hunting season, and the moose numbers are now beginning to return to normal once again. Indeed, the years in which the moose populations actually decreased were some of the colder ones.

You may have heard it repeated over and over again that 99% of scientists 'believe' in global warming. Well actually, the exact opposite is true. That figure is derived from a study in which only 75 scientists said that they 'believe' in global warming, whereas more than 33,000 scientists from all disciplines have signed a (largely ignored by the mainstream) petition, stating that they do not believe in catastrophic 'manmade global warming.'

In 2014 NASA launched a satellite that measured CO_2 levels around the entire Earth. They had assumed (hoped?) before the results were analysed, that most of the CO_2 would be emanating from the highly industrialised areas of the Northern Hemisphere but much to their surprise, and no doubt disappointment, but quite naturally, the results proved unequivocally that the vast bulk of it was actually being produced by the rainforests in South America, Africa, and China.

In fact, if the satellite data is examined impartially, 2014 was *not* the warmest year ever. There has provably been no global warming due to any cause, for more than 25 years. The reason the IPCC is able to state that it *was* the warmest year is because they were using the ground weather station data, which is heavily influenced by the 'urban heat island' effect, affecting many of the stations, which are on or close to paved areas, which naturally reflect and magnify heat rays of the sun. Even then, they still had to cherry-pick that data to produce the deceptive figures and it was still only the warmest by only 2/100ths of one degree within a dataset that has a variability of 50% of one degree. The fact that they had to ignore accurate data and fudge sketchy data to justify their figures and therefore their agenda, alone proves, in my considered opinion, that climate change is a hoax.

Another factor that clearly points to global warming being simply an elaborate hoax is that the main purveyors of 'climate change' falsehoods, have lifestyles representing the total opposite of what they preach. They all own multiple large homes and yachts and they fly around the world in private jets pushing their propaganda. Not to mention the fact that many of them actually personally profit from the many carbon taxes and other green energy laws. If they actually

The Falsification of Science

believed what they preached they would surely be leading quite different lives.

The global climate change industry is worth an annual $1.5 trillion, according to *'Climate Change Business Journal.'* That is the equivalent of $4 billion a day spent on really 'vital' stuff such as carbon trading, biofuels, and wind turbines. Or in other words, it is almost precisely the same amount that the *entire* world spends *every year* on online shopping.

It is all pure propaganda unfortunately.

But there is a (not so) subtle difference between these two 'industries' of global warming and online shopping. When we go online shopping, no one is holding a gun to our head. We do it through choice – or even necessity, but we are buying those things entirely of our own volition – either for ourselves or for someone we love. We have paid for them, with our own money, because we have made the personal decision that they will make our lives a little better in some way.

On the other hand, when we buy from the climate change industry, we have no choice in the matter whatsoever. It is already built into the tax element of purchases, utility bills, the cost of petrol, the cost of airfare, and indeed the cost of every product we buy and every service we use. This expenditure is utterly inescapable, yet unlike online shopping we get precisely zero in return. But of course, who *does* benefit from this scam? I do not believe that I need to point it out, yet again.

But in reality it is actually much worse than that. We receive *less* than nothing. We have things forced upon us that we really do not want…

- Those hideous monstrosities known as 'wind farms' (see the following section on renewable energy for a detailed examination). These 'bird-slaughtering' eco-monstrosities blight the landscape, slashing property values and contributing almost zero to the sum total of energy generation whilst making a huge killing for certain vested interests.

- Huge swathes of solar panels where wheat and other crops used to be grown or where we used to walk.

- Letters from our local councils expecting us to be grateful for the fact that we now have to separate our garbage into seven different recycling bins rather than the previous 'one,' whilst notifying us that they are only going to collect our regular rubbish once every two weeks now instead of weekly.

- Cultural Marxism-influenced teachers who fill our children's heads with junk science 'eco-friendly' propaganda such as happened at my youngest child's school when his class were shown the disgraceful propaganda-filled hit piece 'An Inconvenient Truth,' thereby defying the UK High Court direction which had recently decreed that it was NOT to be shown in schools due to its mass of scientific inaccuracies and lies.

- Free parking spaces for electric cars which we do not own, but which we subsidise for richer people who do.

- 'Feel bad' nature documentaries about how it is all the fault of we humans that certain 'things,' including countless numbers of animal species, 'may' soon disappear forever.

- So-called 'energy saving' lightbulbs that take our nocturnal homes back to the kind of sepulchral gloom that Western civilisation thought it had bidden farewell to in the 1890s.

- Yawning gaps where forests used to grow but which have now been cut down and chipped to create biomass for burning in power stations, which used to run more cheaply and efficiently on coal.

...and so on and on, ad nauseum.

And as regards the chief proponents of the scam, here are some of the people who benefit substantially from this $1.5 trillion climate change industry...

The Falsification of Science

- The carbon traders

- The bought and paid for academics

- The 'vulture capitalist' classes feeding greedily on the bloated carcass of renewable energy

- The environmental NGOs

- The environmental consultancies who specialise in providing 'expert' testimony at planning appeals, arguing on the most spurious grounds (aka lying) that the bats and birds and other wildlife in the area are not going to be affected by the wonderful new wind turbine 'farm' they are proposing to build

- The 'sustainability officers' at every level of local government

- The 'green' advisers attached to every business who advise them how to reduce their CO_2 count

- The PR companies that specialise in green awareness

...et al. These people do not deserve a single penny of our money, but they nevertheless, receive plenty of it.

I do not begrudge anyone the right to earn a living, just so long as they are providing someone, somewhere with something they actually need – i.e. a valued product or service. Not a single person working in the climate change industry fulfils this criterion. Not one. If all the above 'jobs' were scrapped tomorrow, the world would not suffer in the slightest and science itself would be all the better for it.

You may argue that there is some kind of 'trickledown' effect as the money we are forced to pay these utter charlatans and oxygen thieves, via various taxes and tariffs, feeds back into the economy. But you could make the same argument were these people paid the same amount of money by the government to perpetually dig holes in the ground and then proceed to simply fill them in again, all of which would be a vastly preferable use of taxpayer money because then these utterly useless parasites would be reminded every day how pointless the 'work' they perform, actually is. Whereas as things are, many, if not all of them suffer under the ridiculous delusion that their 'green' non-jobs are somehow really virtuous, important, and saving the planet.

The section heading refers to the climate change industry as being a hoax. That is simply because, on any objective level, it certainly is. This does not necessarily mean that the scientists and businesses and politicians promoting it are *all* abject liars – just *most* of them, even if it means that in order to keep earning their livings, they have to be dishonest with themselves about something they know in their hearts to be untrue.

Alex Epstein, author of the '*Moral Case For Fossil Fuels*,' clearly states the fundamental problem with the climate change industry...

"Increasing the amount of CO_2 in the atmosphere from 0.03 per cent to 0.04 per cent has not caused and is not causing catastrophic runaway global warming. Dishonest references to '97% of scientists,' equate a mild warming influence, which most scientists agree with and more importantly can demonstrate, with a catastrophic warming influence – which most don't agree with and none can demonstrate."

This succinctly sums-up the situation. If you accept the validity of that statement – and quite frankly, how could you not, it is unimpeachably accurate and verifiable – then it follows that the $1.5 trillion per annum (and growing) global warming industry represents the most grotesque misuse of manpower and scarce resources in the history of the world, simply to benefit a small few powerful, greedy, vested interests.

"...And so you've got the green movement creating stories that instil fear in the public. You've got the media echo chamber – fake news – repeating it over and over and over again to everybody that they're killing their children, and then you've got the green politicians who are buying scientists with government money to produce fear for them in the form of scientific-looking materials, and then you've got the green businesses, the rent seekers and the crony capitalists who are taking advantage of massive

The Falsification of Science

subsidies, huge tax write-offs, and government mandates requiring their technologies to make a fortune on this, and then of course you've got the scientists who are willingly compliant, they're basically hooked on government grants.

When they talk about the 99% consensus [among scientists] on climate change, that's a completely ridiculous and false number, but most of the scientists — put it in quotes, 'scientists' — who are pushing this catastrophic theory are getting paid by public money. They are not being paid by General Electric or Dupont or 3M to do this research, where private companies expect to get something useful from their research that might produce a better product and make them a profit in the end because people want it — build a better mousetrap type of idea — but most of what these so-called scientists are doing is simply producing more fear so that politicians can use it to control people's minds and get their votes because some of the people are convinced, 'Oh, this politician can save my kid from certain doom.' It's taking over science with superstition and a kind of toxic combination of religion and political ideology. There is no truth to this. It is a complete hoax and scam." Patrick Moore, founder of Greenpeace

Have you noticed that the original term of 'global warming' has now disingenuously and surreptitiously been replaced by 'climate change' by most commentators and reporters? This is presumably because it has now been demonstrated beyond all doubt that the Earth is currently entering a period of cooling, aka a 'Grand Solar Minimum,' which is expected to continue for the next several decades. Indeed, this is all part of a natural cycle of warming and cooling that has been ongoing for millions of years.

The utterly ridiculous and easily disproved fantasy of manmade global warming by CO_2 emissions is nothing but a gigantic hoax, a confidence trick designed to extract even more money from we the people, in the form of taxation, to further line the pockets of the already super-wealthy.

There are also other powerful factors at work here. 'Climate change' as we must now refer to it in our increasingly politically correct world, is also being used to justify the ongoing centralisation of power, deindustrialisation, the passing of global laws and an expansion of the surveillance of the world population, all as part of the inexorable march towards the much-vaunted 'New World Order.' This is also another major example of the Hegelian dialectic, 'problem, reaction, solution' (or 'thesis, antithesis, synthesis) technique. Artificially create a problem, wait for a demand for action and then provide the solution you wanted to see implemented in the first place. By that subtle method is the agenda advanced step by step by step.

So, enter Mr A. Gore, fresh from his (planned) defeat at the hands of Bush 'the lesser' in the 2000 presidential election campaign. I can just imagine the scenario right now...

Rothschild: Sorry Al, but Georgie-boy is our choice this time around, but don't worry, we have something even more important lined-up for you. We want you to front this massive con...err, campaign that we have thought of to make us all even more billions. Basically we are going to employ a whole bunch of scientists, corporations, and politicians to falsely promote the fact that the Earth is suffering from some kind of runaway warming and that the only solution that will save mankind will be draconian measures that will ahem, stop it all. Of course this will cost every single person on Earth an absolute 'arm and a leg,' but never mind eh, just think of the money!? What do you say Al? Will you do it Al? You know it makes sense, Al.

Gore: Yes of course, Mr Rothschild, sir. Of course I will, sir – only too delighted Mr Rothschild. When can I start?

So, Gore embarked upon his quest to convince the entire world that the Earth is warming dangerously as a result of excessive CO_2 emissions, all of course funded by the Rothschild millions. One of his first actions was to produce the aforementioned horrendously scientifically inaccurate, replete with lies film, *An Inconvenient Truth,* which became the third most successful documentary in history and unsurprisingly won the 'Oscar' for best documentary. Gore's book of the same name also reached the top of the best-sellers list in America as the public fell for his scam 'hook, line and sinker' and furthermore, as

The Falsification of Science

if this were all not enough, and to add insult to injury, Gore was subsequently awarded the Nobel Peace Prize for this great work of fiction.

Please bear in mind that this is the same Al Gore that Phillip Eugene de Rothschild, referred to thus...

"Bill Clinton has 'full-blown' multiple personality disorder and is an active sorcerer in the satanic mystery religions. This is also true of Al Gore, as well; I have known Clinton and Gore from our childhoods as active and effective Satanists."

Incidentally (as also alluded to above), in the early 200os, the *Inconvenient Truth* 'documentary' was distributed by the government to be shown to teenagers in all schools in Britain *despite* a High Court judge ruling that it was too full of unscientific inaccuracies and gross distortions of the truth to be shown to highly impressionable young people. But the law matters not a jot to the perpetrators of this hoax, of course. Laws are only there to be obeyed when it is expedient for their agenda to do so; otherwise they may be ignored at will. And perhaps even more tellingly, the psychological damage has now been done and the purpose of the film has been served. No matter how much proof of its distortions and inaccuracies may now be forthcoming, the falsehoods it portrays are now firmly established in millions if not billions of human psyches all around the world. This then, of course, renders the refuting of all its false information, much more difficult.

Of course any conscientious person realises that protection of the environment is vital to long-term human and indeed all life's very survival on the Earth, but that is not the real issue here, despite the fact that it is disingenuously portrayed as being so. Climate change 'theory' is based on the complete myth or even downright lie, that global temperatures are inexorably increasing solely due to the levels of CO_2 emitted by human activities, whilst the real truth is that from the 1970s to the early 2000s, the Earth was actually undergoing a short period of naturally induced warming as part of a normal cycle of sunspot activity aka a 'Grand Solar Maximum.' This trend is now rapidly being reversed as we are now entering a sunspot-induced period of cooling, but of course this irritating fact does not figure too prominently in the 'warmists' agenda and they therefore deny, but mainly ignore *this* particular inconvenient (for them) truth.

All their 'warming' predictions are based on totally unscientific premises and selective data derived mainly from the old computing principle of GIGO, 'garbage-in, garbage-out' which roughly translated means that if the initial data input is incorrect then the resultant computer predictions based on that data will also be incorrect.

It is falsely claimed by these grand scale 'con artists' of zero morals, that it is the excessive amounts of the so-called 'greenhouse gas,' CO_2, emanating from vehicle exhausts and industrial processes, especially those of the Third world that is the root of the problem. According to their shills, 'useful idiots' and downright unscrupulous professional liars, this excessive CO_2 is collecting in the atmosphere and preventing the natural escape of heat into space and thus causing a build-up of warm air all around us, that is the issue.

Of course we are now being extolled to become 'carbon neutral' and partake in the 'carbon offset' scam just as many companies that wish to cause pollution are now able to 'buy' carbon credits from those who do not exceed their stated carbon emission limits. What a farce it all really is. So essentially what is being said then is that it is OK to cause pollution as long as you pay for it. And where does this money go, you may well ask? Well, luckily for us all, Gore has set up a company to facilitate it all- *Generation Investment Management*, based in London.

So in effect, Gore and his cronies can buy their carbon offsets from themselves. Better yet, Gore can buy them with the substantial payments he receives from his long-term relationship with Occidental Petroleum. It is so easy to be carbon-neutral. All you have to do is own a gazillion stocks in 'big oil,' start an eco-stockbroking company to make eco-friendly investments, use a small portion of your oil company share's profits to buy some tax-deductible carbon offsets from your own investment firm and you too can save the planet whilst making a fortune and leaving a carbon footprint roughly the size of Godzilla's as at the beginning of the movie when the main characters are all standing in the big toe imprint wondering what this strange depression is in the landscape!

So please, please never, ever submit to the blatant confidence trick now being perpetrated by some airline companies that so thoughtfully and generously invite you to buy your carbon credits when paying for a ticket. This is all nothing but a highly lucrative fraud being perpetrated against us all through the emotional blackmail of 'saving the planet.' At the time of writing, these airline carbon credits are voluntary, but in line with the 'totalitarian tiptoe' it is probably simply the prelude to more and more taxation, even more Orwellian controls and perhaps most importantly to prevent the developing nations (Third world) from gaining a share of the profit 'gravy train' currently being enjoyed almost exclusively by the First world and the likes of Gore and his Elitist cronies.

"In searching for a new enemy to unite us, we came up with the idea that pollution, threat of global warming, water shortages, famine and the like would fit the bill. All of these are caused by human intervention. The real enemy then is humanity itself." Aurelio Pecci, *The Club of Rome*, a part of the Round Table network, 1991

In 2006 an offshoot of the United Nations, *The International Panel on Climate Change* (IPCC) began to inform the world that global warming is the direct result of human-caused carbon emissions. It has since become the de facto truth (along with many other lies exposed in this book) that carbon emissions are warming the planet at an unsustainable rate despite the now proven fact that the Earth is now entering into an intensive period of 'cooling.' However, please note that the IPCC is not a scientific body at all, despite many of its apologists' claims; it is a purely political institution with its own agenda and replete with all the bias inherent in all such organisations.

"The IPCC is not a scientific institution; it's a political body, a sort of non-government organisation of green flavour. It's neither a forum of neutral scientists nor a balanced group of scientists. These people are

politicised scientists who arrive there with a one-sided opinion and a one-sided assignment." Vaclav Klaus, President of the Czech Republic

The IPCC's claim that it represents 2000+ of the world's foremost scientists is a downright unadulterated lie. Its report purportedly had the backing of all these scientists, most of whom are blatantly and deceptively, named within its covers. However, since the publication of this huge fairy-tale masquerading as truth, at least 60% of these named scientists have formally protested at the inclusion of their names within its pages, all of which were used without their express permission and indeed many have demanded that their names be specifically excluded from it – all of which has fallen, of course, upon 'deaf ears' and as usual remained unreported by the compliant media.

As one example (among many) Professor Paul Reiter, one of the world's foremost authorities on tropical diseases, whose name by the way, was one of those included without his express permission, gave an example of the blatant untruths within the report when he said that it states within its pages that tropical diseases such as malaria were far more likely with the advent of global warming, to spread to formerly 'colder' parts of the world than before. This he pointed out is in fact, utter nonsense. Mosquitos actually thrive better in colder temperatures and are abundant within the Arctic Circle in such places as Siberia for example, where there are 13 million cases of malaria reported each year, on average. In fact, he said, he was horrified by the entire report as it was, he said… *"…so much misinformation…virtually without mention of scientific literature by specialists in those fields."*

The IPCC report is simply put, a document replete with lies and distortions compiled to support a predetermined false agenda and designed to fool the public into believing that which is not true. In this at least, it has been a raging success.

"The IPCC like any other UN body is political. The final conclusions are politically driven." Professor Phillip Scott, the Department of Geography, University of London

We are constantly being fed the line that the Earth is now 'warmer than at any time since records began.' Sounds impressively foreboding, does it not? However, upon hearing statements such as these it is worth remembering that records only began in fact as recently as 1914, just over one hundred years ago, at the time of writing. This is yet another example of the depths of deceit and outright propaganda, to which the warmists do not hesitate to stoop in their constant battering of our senses to drive home their messages of abject doom and gloom. As stated previously, the Earth is subject to constant and ongoing fluctuations in the warming and cooling cycle that has been in evidence for hundreds of millennia and a mere one hundred year period taken in isolation, forms no statistical basis whatsoever upon which to base correct, unbiased scientific data.

In fact, it is quite simple to demonstrate using such methods as tree rings and ice cores that temperatures in what scientists refer to as the 'mediaeval warm period,' were considerably higher on average than those of today. (Personally, I blame those highly dangerous, methane exuding ox-drawn carts that proliferated in those times.) And there is also much socio-historical evidence to suggest that life in this period was far more 'comfortable' than today in many respects, despite there being an approximately two-degree higher average temperature than at present and absolutely no evidence of polar icecap melt or of coastlines being inundated by rising seas worldwide. Indeed, were the temperatures to rise by as much as two or even five degrees in the next few years, then far from it being cataclysmic, most 'real' scientists believe that it would be highly beneficial to life on Earth, especially the human variety.

"What has been forgotten in all the discussions about global warming is a proper sense of history. We have this view today that warming is going to have apocalyptic outcomes. In fact, during the 'mediaeval warm period,' the world was even warmer than today, and history shows that it was a wonderful period of plenty for everyone. When the temperatures began to drop, harvests failed, and England's wine industry died. It makes one wonder why there is such a fear of warmth." Professor Phillip Scott, department of Geography, University of London

"Warming fears are the worst scientific scandal in scientific history... When people come to know what the truth is, they will feel

deceived by science and scientists." Dr Kiminori Itoh, environmental physical chemist

Indeed so!

Then in late 2009, came another extremely important nail in the 'warmists' coffin with the exposure of the 'Climategate emails.' Upon reading the 1,000+ emails and 72 leaked documents it becomes immediately apparent why the East Anglia Climate Research Unit, in the UK, would have wished to keep them under lock and key and away from the prying eyes of the badly misinformed public. Here are just a few snippets below as examples of how the 'warmists' have faked and manipulated data to further their own ends as have been revealed in these highly incriminating documents...

"I've just completed Mike's Nature trick of adding in the real temps to each series for the last 20 years (i.e. from 1981 onwards) and from 1961 for Keith's to hide the decline."

"The fact is that we can't account for the lack of warming at the moment and it is a travesty that we can't. The CERES data published in the August BAMS 09 supplement on 2008, shows there should be even more warming: but the data are surely wrong. Our observing system is inadequate."

"Can you delete any emails you may have had with Keith re AR4? Keith will do likewise. He's not in at the moment – minor family crisis. Can you also email Gene and get him to do the same? I don't have his new email address. We will be getting Caspar to do likewise."

"... Phil and I have recently submitted a paper using about a dozen NH records that fit this category, and many of which are available nearly 2K back – I think that trying to adopt a timeframe of 2K, rather than the usual 1K, addresses a good earlier point that Peck made w/ regard to the memo, that it would be nice to try to 'contain' the putative 'MWP,' even if we don't yet have a hemispheric mean reconstruction available that far back...."

"This was the danger of always criticising the sceptics for not publishing in the 'peer-reviewed literature'. Obviously, they found a solution

The Falsification of Science

to that—take over a journal! So, what do we do about this? I think we have to stop considering 'Climate Research' as a legitimate peer-reviewed journal. Perhaps we should encourage our colleagues in the climate research community to no longer submit to, or cite papers in, this journal. We would also need to consider what we tell or request of our more reasonable colleagues who currently sit on the editorial board... What do others think?"

"I will be emailing the journal to tell them I'm having nothing more to do with it until they rid themselves of this troublesome editor. It results from this journal having a number of editors. The responsible one for this is a well-known skeptic in NZ. He has let a few papers through by Michaels and Gray in the past. I've had words with Hans von Storch about this but got nowhere. Another thing to discuss in Nice!"

Although replete with in-jokes and in-house references, the above few examples provide us with an overall flavour of exactly what was going on there. And despite the reluctance of the mainstream media to cover it in full, with a few noteworthy exceptions, this no doubt demonstrates pretty conclusively the institutional deception taking place in order to perpetuate the hoax. How much more evidence do we need in order for us to realise that the whole concept of global warming by human-caused CO_2 emissions is a monumental fraud, carefully designed and perpetrated in order to further enrich multi-millionaires and billionaires?

A week after James Delingpole, in his *Daily Telegraph* blog, coined the term 'Climategate,' to describe the scandal revealed by the leaked emails from the University of East Anglia's Climatic Research Unit, Google was showing that the word appeared across the Internet more than nine million times. But in all those 'acres' of electronic coverage, one hugely relevant point about these thousands of documents has largely been missed. The reason why most political commentators have expressed their total shock and dismay at the bigger picture revealed by the documents, is that their authors are not just any old bunch of academics, indeed their importance cannot be overestimated. What we have here is the small group of scientists who have for years been more influential in driving the worldwide alarm over global warming than any others, not least through the role they play at the heart of the UN's Intergovernmental Panel on Climate Change (IPCC).

Professor Philip Jones, the *Climate Research Unit*'s director, is in control of the two key sets of data used by the IPCC to draw up its reports. Through its link to the Hadley Centre, part of the UK Meteorological Office, which selects most of the IPCC's key scientific contributors, his global temperature record is the most important of the four sets of temperature data on which the IPCC and governments rely – not least for their predictions that the world will warm to catastrophic levels unless trillions of dollars are spent (by us all) to avert it.

Dr Jones is also a critical part of the closely knit group of American and British scientists responsible for promoting that picture of world temperatures conveyed by Michael Mann's so-called 'hockey stick' graph which several years ago turned climate history on its head by showing that, after 1,000 years of decline, global temperatures have recently shot up to their highest level in recorded history.

Given star billing by the IPCC, not least for the way it appeared to eliminate the long-accepted Mediaeval Warm Period when temperatures were higher than they are today, the graph became the central icon of the entire manmade global warming movement.

Since 2003 however, when the statistical methods used to create the 'hockey stick' were first exposed as fundamentally flawed, by an expert Canadian statistician Steve McIntyre, an increasingly heated battle has been raging between Mann's supporters, calling themselves 'the Hockey Team,' and McIntyre and his own allies, as they have ever more devastatingly called into question the entire statistical basis on which the IPCC and CRU construct their flimsy, misleading case.

The senders and recipients of the leaked CRU emails constitute a cast list of the IPCC's scientific elite, including not just the 'Hockey Team,' such as Dr Mann himself, Dr Jones and his CRU colleague Keith Briffa, but Ben Santer, responsible for a highly controversial rewriting of key passages in the IPCC's 1995 report, Kevin Trenberth,

The Falsification of Science

who similarly controversially pushed the IPCC into scaremongering over hurricane activity and Gavin Schmidt, right-hand man to Al Gore's ally Dr James Hansen, whose own GISS record of surface temperature data is second in importance only to that of the CRU itself.

There are three threads in particular in the leaked documents, which have sent a shockwave through informed observers across the world. Perhaps the most obvious, as lucidly put together by Willis Eschenbach (see McIntyre's blog *Climate Audit* and Anthony Watt's blog *Watts Up With That*) is the highly disturbing series of emails which show how Dr Jones and his colleagues have for years been discussing the devious tactics whereby they could avoid releasing their data to outsiders under freedom of information laws. They have indeed presented every possible excuse for concealing the background data on which their findings and temperature records were based.

This in itself has become a major scandal, and not least Dr Jones' refusal to release the basic data from which the CRU derives its hugely influential temperature record, which culminated last summer in his startling claim that much of the data from all over the world had simply become 'lost.' Most incriminating of all are the emails in which scientists are advised to delete large chunks of data, which, when this is done after receipt of a freedom of information (FOI) request, is actually a criminal offence.

But the question, which inevitably arises from this systematic refusal to release their data is – what is it that all these 'scientists' appear to be so anxious to hide? The second and most shocking revelation of the leaked documents is how they show the scientists trying to manipulate data through their tortuous computer programmes, always to point in only the one desired direction – to lower past temperatures and to 'adjust' recent temperatures upwards, in order to convey the impression of an accelerated warming. This rears its head so often (not least in the documents relating to computer data in the 'Harry Read Me' file) that it becomes the most disturbing single element of the entire story. The 'Harry Read Me' file is an infamous example of one of these so-called climate 'scientists' attempts to fudge the figures to 'prove' the existence of manmade global warming through the blatant manipulation of data. This is precisely what McIntyre caught Dr Hansen doing with his GISS temperature record last year (after which Hansen was forced to revise his record) and two further shocking examples have now come to light from Australia and New Zealand.

In each of these countries it has been possible for local scientists to compare the official temperature record with the original data on which it was supposedly based. In each case it is clear that the same 'trick' has been used – to turn an essentially flat temperature chart into a graph, which shows temperatures steadily rising. And in each case this manipulation was carried out under the influence of the CRU.

What is tragically evident from the 'Harry Read Me' file is the picture it provides of the CRU scientists hopelessly 'at odds' with the complex computer programmes they had devised to contort their data in their desired direction, and continually expressing their own frustration at how difficult it was to obtain the results they desperately needed in order to 'prove' their own false data as entirely accurate.

The third shocking revelation of these documents is the ruthless way in which these academics have been determined to silence any expert questioning of their findings – not just by refusing to disclose their basic data but also by discrediting and freezing-out any scientific journal which dares to publish their critics' work. It appears that they are prepared to stop at nothing to stifle scientific debate and indeed totally falsify science in this way, not least by ensuring that no dissenting research should ever be made public.

Back in 2006, when the eminent US statistician Professor Edward Wegman produced an expert report for the US Congress vindicating Steve McIntyre's demolition of the 'hockey stick' chart, he excoriated the way in which this same tightly knit group of academics seemed only too keen to collaborate with each other and to 'peer review' each other's papers in order to dominate the findings of those IPCC reports on which much of the future of the US and world economy may hang. In light of the latest revelations, it now seems even more evident that these men have been failing to uphold those principles, which lie at the heart of genuine scientific enquiry and debate. Already one respected US climate scientist, Dr Eduardo Zorita, has called for Drs Mann and Jones to be barred from any further participation in the IPCC. Even previous supporters, horrified at discovering their betrayal by the alleged experts they revered and cited for so long, called for Jones to step down as head of the CRU.

The former UK Chancellor of the Exchequer, Nigel Lawson, upon launching his new 'think tank,' the *Global Warming Policy Foundation*, rightly called for a proper independent inquiry into the maze of skulduggery revealed by the CRU leaks. But the inquiry actually mooted to be chaired by Lord Rees, President of the Royal Society, also himself a shameless propagandist for the 'warmist' cause, is very far from being what Lawson had in mind. Our hopelessly compromised scientific establishment cannot be allowed to get away with a whitewash of what has become one of the greatest scientific scandals of our age.

And so to conclude, yes, unfortunately it is all a huge hoax, corporate junk science at its finest and a sordid game to convince us that we are in danger of imminent disaster from the slightest warming of the planet and the only way to prevent this will be to give even more money to an already wealthy Elite group who care nothing whatsoever for you and your families, despite maintaining a thoroughly deceptive pretence to the contrary.

HAARP – The High Frequency Active Auroral Research Program

John Hamer

In a nutshell, here is how HAARP works...

Firstly, the facility's transmitters send radio waves upwards into the ionosphere, between 100 and 350km in altitude. Then the resulting heating effect creates irregularities in the electron density there, which in turn allows communications signals, to be relayed off the ionosphere and the amount of electrical charge found in the atmosphere (e.g. in clouds) to be manipulated.

Out of interest, here is what Wikipedia™ says...

"The High Frequency Active Auroral Research Program (HAARP) was is an ionospheric research program jointly funded by the

U.S. Air Force

U.S. Navy

University of Alaska

The Defence Advanced Research Projects Agency (DARPA) and designed and built by BAE Advanced Technologies (BAEAT.) Its purpose was is to analyse the ionosphere and investigate the potential for developing ionospheric enhancement technology for radio communications and surveillance. The HAARP program operated a major sub-arctic facility, named the HAARP Research Station, on an Air Force-owned site near Gakona, Alaska.

The most prominent instrument at the HAARP Station is the Ionospheric Research Instrument (IRI), a high-power radio frequency transmitter facility operating in the high frequency (HF) band. The IRI is used to temporarily excite a limited area of the ionosphere.

Other instruments, such as a VHF and a UHF radar, a fluxgate magnetometer, a digisonde (an ionospheric sounding device,) and an induction magnetometer, were used to study the physical processes that occur in the excited region.

Work on the HAARP Station began in 1993. The current working IRI was completed in 2007, and its prime contractor was BAE Systems Advanced Technologies. As of 2008, HAARP had incurred around $250 *million in tax-funded construction and operating costs. It was reported to be temporarily shut down in May 2013, awaiting a change of contractors. In May 2014, it was announced that the HAARP program would be permanently shut down later in the year. Ownership of the facility and its equipment was transferred to the University of Alaska Fairbanks in mid-August 2015."*

Note that the Wikipedia™ article says 'was' instead of 'is,' and also states that the project has been *"permanently shut down."*

Now, does anyone really believe that the USAF, US Navy, DARPA, et al would spend $250m plus millions more of 'black project' money on a project they would simply write off? Or indeed that massive remote facilities that bombard the atmosphere with microwaves are really only doing so for communication purposes?! After all, do we not have radio frequency towers and the Internet for that purpose?

No, the real reason for these huge complexes that can affect electron density in the ionosphere (and thus affect the charge found in clouds) is for weather manipulation.

There is also a similar facility at the Pierre Auger Observatory in Argentina. According to Wikipedia™ this observatory is, *"an international cosmic ray observatory in Argentina designed to detect ultra-high-energy cosmic rays: sub-atomic particles traveling nearly at the speed of light and each with energies beyond 1018 eV. In Earth's atmosphere such particles interact with air nuclei and produce various other particles. These effect particles (called an 'air shower') can be detected and measured."*

However it has lately become clear that this facility is doing far more than just *"detecting and measuring."*

It is now also worth recalling what Nikola Tesla said about creating earthquakes. He revealed that an earthquake, which drew police and ambulances to the region of his laboratory was the result of a little machine he was experimenting with at the time which *"you could put in your overcoat pocket."* He said, *"I was experimenting with vibrations. I had one of my machines going and I wanted to see if I could get it in tune with the vibration of the building. I put it up notch after notch. There was a peculiar cracking sound. The police and ambulances arrived. I told my assistants to say nothing. We told the police it must have been an earthquake. That's all they ever knew about it."*

Some shrewd reporter asked Dr Tesla at this point what he would need to destroy the Empire State Building and the doctor replied: *"Vibration will do anything. It would only be necessary to step up the vibrations of the machine to fit the natural vibration of the building and the building would come crashing down. That's why soldiers break step crossing a bridge."*

In other interviews, Tesla claimed that the device, properly modified, could be also used to map underground deposits of oil. A vibration sent through the earth returns an 'echo signature,' using the same principle as sonar. This idea was actually adapted for use by the petroleum industry and is used today in a modified form with devices used to locate objects at archaeological digs.

As you may probably imagine, if buildings can be made to resonate with the right frequency and pressure – and with a pocket-sized device, then so too can the ground itself at the right frequency and with a much larger device.

With that in mind, it is somewhat obvious that the technology exists, to allow manmade earthquakes, as well as targeted weather manipulation to induce local or regional droughts, famines, hurricanes, etc.

HAARP is possibly the most dangerous and sinister weapon known to man as well as being probably the least well known to the general public.

It consists of a huge installation of antennae, in Alaska USA and is the largest ionospheric heater in the world (see the picture above) and Is capable of heating a 1,000 square kilometre area of the ionosphere to over 50,000 degrees. It is also a phased array; meaning that it is steerable, despite widespread claims to the contrary by its apologists, and its waves can be directed to a selected target area. By transmitting radio frequency energy up into the skies above us and focusing, it

The Falsification of Science

causes the ionosphere to heat considerably. This heating literally lifts the ionosphere within a 30-mile diameter area thereby changing localised pressure systems or perhaps the route of jet streams. Moving a jet stream is in itself a phenomenal event. The problem being that it is not possible to model the system accurately. Long term consequences of atmospheric heating are unknown. Changing weather in one place can have a devastating downstream effect and HAARP has already been accused of modifying weather for geopolitical as well as other possibly even more sinister functions.

Manipulating the weather and the environment for the purposes of US government sponsored terrorism is definitely not within the discussion remit of the mainstream news. But it would appear that that is exactly what is happening.

HAARP is part of the weapons arsenal of the New World Order under the Strategic Defence Initiative (SDI). From military command points in the US, entire national economies could potentially be destabilised through climatic manipulations. More importantly, the latter can be implemented without the knowledge of the enemy, at minimal cost and without engaging military personnel and equipment as in a conventional war. The use of HAARP, if it were to be applied, and as it most probably has, could have potentially devastating impacts on the Earth's climate.

HAARP is based in Alaska, where not only are we witnessing an increased incidence of earthquakes, but the prolonged eruption of volcanoes. But there are several other locations throughout the world where other, smaller HAARP arrays are located, including Puerto Rico and the significance of this will become apparent later in this section. In 1958, then chief White House adviser on weather modification, Captain Howard T. Orville said that the US Defence Department was looking for ways to *"manipulate the charges of the earth and sky and so affect the weather by using an electronic beam to ionize or de- ionize the atmosphere over a specific area."* Recently, an ice bridge in the Antarctic collapsed and the Wilkins Ice Shelf could be on the brink of breaking away. Of course this is all conveniently blamed on global warming, but is HAARP actually being used to melt ice in an attempt to 'prove' global warming? And is HAARP merely mapping weather patterns, agricultural seasons, and crop cycles, or actually influencing or even creating them?

Responding to US economic, strategic, and geopolitical interests, it could easily be used and indeed has been used to selectively modify climate in different parts of the world, resulting in the destabilisation of agricultural and ecological systems. It is also worth noting that the US Department of Defense has allocated substantial resources to the development of intelligence and monitoring systems on weather changes. NASA and the US Department of Defense's National Imagery and Mapping Agency (NIMA) are working on 'imagery for studies of flooding, erosion, landslide hazards, earthquakes, ecological zones, weather forecasts, and climate change' with data relayed from satellites.

On 26th December 2004, the Indian Ocean earthquake that generated the subsequent tsunami that killed over one quarter of a million people was unleashed on an unsuspecting world. The magnitude of this earthquake was 9.3 on the Richter scale, making it one of the most deadly and destructive in all known history. Wikipedia™ states that the earthquake itself lasted almost ten minutes when most major earthquakes last no more than a few seconds. It caused the entire Earth to vibrate several centimetres and it also triggered earthquakes elsewhere, even as far away as Alaska. The section in Wikipedia™, named 'Tectonic Plates' provides more shocking pieces of information.

"Seismographic and acoustic data indicate that the first phase involved the formation of a rupture about ... 250 miles ... long and ... 60 miles wide, located ... 19 miles ... beneath the seabed – the longest known rupture ever known to have been caused by an earthquake.

As well as the sideways movement between the plates, the seabed is estimated to have risen by several metres, displacing an estimated ... 7 cubic miles ... of water and triggering devastating tsunami waves. The waves did not originate from a point source, as mistakenly depicted in some illustrations of their spread, but radiated outwards along the entire ...

750 miles...length of the rupture. This greatly increased the geographical area over which the waves were observed, reaching as far as Mexico, Chile, and the Arctic. The raising of the seabed significantly reduced the capacity of the Indian Ocean, producing a permanent rise in the global sea level by an estimated 0.1 mm."

In the section entitled 'Power of the Earthquake,' there are even more disturbing statistics.

"The total energy released by the earthquake in the Indian Ocean...is equivalent to 100 gigatons of TNT, or about as much energy as is used in the United States in 6 months. It is estimated to have resulted in an oscillation of the Earth's surface of about 20–30 cm (8 to 12 in), equivalent to the effect of the tidal forces caused by the Sun and Moon. The shock waves of the earthquake were felt across the planet; as far away as Oklahoma, vertical movements of 3 mm (0.12 in) were recorded. The entire Earth's surface is estimated to have moved vertically by up to 1 cm. ... It also caused the Earth to minutely 'wobble' on its axis by up to 2.5 cm (1 in)... or perhaps by up to 5 or 6 cm (2.0 to 2.4 in)."

So, here is an event that expends so much energy it literally shortened the day, caused the Earth to wobble and its surface to raise significantly.

Why would the Elite-controlled, US military-industrial complex want to turn the Indian Ocean into a huge disaster area? It seems inconceivable and pointlessly destructive – at first glance. However, consider this; the Aceh area of Indonesia, the geographical location most badly hit by the earthquake/tsunami is known to be extremely rich in untapped oil resources. Once this fact is understood it all becomes much clearer. At this time, probably significantly, Aceh was in the grip of a devastating civil war, largely unreported in the Western, controlled media and seriously hampering the extraction and distribution of the oil reserves.

It would seem that on the morning of 26th December 2004, despite subsequent denials, the US authorities had foreknowledge of the earthquake (via the extensive early warning system in place) and the probable course and effects of the resulting tsunami. As was the case on 9/11, the automatic warning system stood down, to allow a terrorist act to occur using the HAARP technology, to justify upcoming militarisation of the area with US troops in an area rich with oil. Whether HAARP induced or not, the critical issue would be the stand-down of the automatic earthquake warning system among its subscribing member nations. This is at best criminal negligence and at worst a deliberate act of genocide for monetary and geopolitical gain.

The provable fact that an advance warning was given to Australia and the US military base in Indonesia only, demonstrates conclusively that there was criminal intent involved. The Third world areas affected were left to suffer horrendously whilst oil rich Aceh was simultaneously invaded by over 2,000 heavily armed US marines and two aircraft carriers equipped with dozens of 'state of the art' Cobra attack helicopters. Strange also that the US military base in Indonesia also knew in advance about the planned attack/tsunami. The critical point of foreknowledge, whilst not comprising conclusive proof of the use of HAARP technology, at least makes this possibility realistic. We know that the technology has the capability to create the Asian tsunami, so it is a fairly simple step of logic to deduce that the end goal, i.e. the US invasion of Aceh province would have justified the action in the twisted minds of those self-styled masters of the Universe. Why would it be left to chance and the hope that a natural incident of geological devastation would occur to facilitate a planned invasion?

"Others (terrorists) are engaging even in an eco-type of terrorism whereby they can alter the climate, set off earthquakes and volcanoes remotely through the use of electromagnetic waves… So there are plenty of ingenious minds out there that are at work finding ways in which they can wreak terror upon other nations… It's real, and that's the reason why we have to intensify our (counterterrorism) efforts." US Secretary of Defence, William Cohen, April 1997 at a 'counter-terrorism conference'

Anyone spot the irony in this statement?

Cohen confirmed that there are in existence, electromagnetic weapons of this nature and there have been for many years, which have been and are being used to initiate earthquakes, engineer the weather and climate, and also initiate the eruption of volcanoes. Several nations now indeed have these weapons.

Following on shortly after the tsunami, on 28th March 2005, there was another earthquake in Northern Sumatra and Indonesia, measuring an almost equally gigantic 8.7 on the Richter scale, according to the National Earthquake Information Centre, US Geological Survey. In addition, there was a total of six giant earthquakes around the World in 2005, all with a magnitude greater than 7.0.

Woods Hole Oceanographic Institution issued a news release entitled *"Major Caribbean Earthquakes and Tsunamis a Real Risk – Events rare, but scientists call for public awareness, warning system."*

Paragraph 2 states that, *"In a new study published December 24, 2004 in the Journal of Geophysical Research from the American Geophysical Union, geologists Uri ten Brink of the U.S. Geological Survey in Woods Hole and Jian Lin of the Woods Hole Oceanographic Institution (WHOI) report a heightened earthquake risk of the Septentrional fault zone, which cuts through the highly populated region of the Cibao Valley in the Dominican Republic. In addition, they caution, the geologically active offshore Puerto Rico and Hispaniola trenches are capable of producing earthquakes of magnitude 7.5 and higher. The Indonesian earthquake on December 26, which generated a tsunami that killed (to date), an estimated 250,000 people, came from a fault of similar structure, but was a magnitude 9.0, much larger than the recorded quakes near the Puerto Rico Trench."*

As previously stated, the significance of this is that there is an active HAARP array situated in Puerto Rico. Coincidence?

There was certainly a sudden inexplicable spate of major earthquakes in 2005. On 8th October 2005, yet another massive quake killed over 25,000 people and injured and displaced many thousands more in the Hindu Kush mountain region of southern Pakistan. Estimates of the quake's magnitude varied from 6.8 to 7.8, with the United States Geological Survey putting the number at 7.6. Its epicentre was roughly 60 miles north of the Pakistani capital, Islamabad, where

20 'significant aftershocks' measuring between 5 and 6.2 magnitude were felt throughout the day, said Dr Qamar-uz-Zaman Chaudhry, director general of the Meteorological Department in Islamabad.

The earthquake sent tremors as far east as New Delhi, the Indian capital, and west to Kabul, the capital of Afghanistan, the 'biggest earthquake to strike the country in a century.'

Two other major, more recent earthquakes are worthy of further mention in this context.

China actually seriously considered going to war with the USA in retaliation for what was seen as deliberate act of terrorism in instigating the 2008 earthquake in China. Indeed there are even suspected HAARP arrays in China and also other ionospheric heaters/transmitters and observatories/receivers in Massachusetts, Colorado, California, Arkansas, Puerto Rico (the Arecibo Ionospheric Observatory, funded by DARPA), Brazil (Sao Luiz Space Observatory), Peru, Russia, Chernobyl, Ukraine (Duga Radar Array/ "Russian Woodpecker"), Norway, Sweden, Wales, Australia, India, Marshall islands, Taiwan and Japan. SuperDARN (Dual Auroral Radar Network) has installations in Canada, Iceland, New Zealand, Newfoundland, Alaska, and Virginia. The suspicion, naturally, is that they can all be used to work together and in conjunction with the main HAARP array in Alaska, (which is still the largest of them all).

Photographs proved that 90 million watt, pulsed long-wave radio waves from the HAARP array in Alaska, combined with pulsed microwaves from a US Military Satellite in orbit, caused Chinese land to 'resonate.' As the land began resonating, its movement tore itself apart, causing the massive earthquake. Photos from Tianshui city, Gansu province, China the city nearest the epicentre of the quake, show strange cloud patterns a full two days before the quake hit. The cloud formations can be seen breaking apart in patterns indicating they were being hit systematically with something from above.

The clouds were being affected by the two sets of waves. As the waves beamed down from the sky, the clouds broke apart in very orderly fashion; proving that something from above, other than the wind,

The Falsification of Science

was affecting this region of China. Some clouds even had their own 'rainbow' signatures, indicative of HAARP activity.

This 'long wave' attack no doubt originated from the HAARP array in Alaska. These extremely long radio waves were pulsed slowly and travelled deep underground when they hit. The much smaller, pulsed microwaves emitted from a US military satellite, pulsed at a faster rate and, because the wavelength is so much smaller, they did not penetrate deeply into the ground. The deep rock began to resonate at one frequency and the surface rock/soil began to resonate at a different, faster rate.

The two separate rates of resonation caused literally billions of tons of dirt and rock to begin subtly, almost imperceptibly grinding against each other within the ground. With all that movement it did not take long for a key geological lock to be crushed, allowing a sudden movement of a large land area underground, which was felt as a massive earthquake at the surface.

If you think that electromagnetic waves cannot cause anything to move, consider your microwave oven. Using as little as 500 watts, microwaves travelling through food 'excite' the molecules within the food. The molecules begin rubbing together and the friction generated by the molecules bouncing off each other causes heat and the heat cooks the food. Now, take that same proven technique and multiply it to ninety million watts, pounding on a particular area for at least two full days. Would it not be possible for that amount of energy, bearing down relentlessly for at least two days, to cause the molecules inside the rocks to start moving? As the first rock molecules start to move, they move against other molecules causing not only heat but also tiny vibrations. As the vibrations spread to other rocks underground, they too start to vibrate.

When several billion tons of earth and rock start to vibrate, even imperceptibly to humans, the enormous weight of that small vibration has a truly devastating impact. All it takes is for that impact to act upon a weak point in the earth's rock structure, crushing it, which then allows a tectonic plate to move a distance possibly as small as one inch. When billions of tons of rock and earth suddenly move one inch, the resulting shockwave is felt as a huge earthquake.

Humans, surrounded by a buffer of air and subjected to local noise from traffic, aircraft, and all other everyday sounds, would not hear or feel the subtle vibration underground, although Chinese seismic sensors had detected the movement. Initially, even meteorologists in China thought this was a natural phenomenon – until the earthquake struck in the exact same area. Then by the simple process of putting two and two together, they deduced that China had been covertly attacked.

At least 90,000 innocent people lost their lives whilst many hundreds of thousands were injured and left homeless. More than half a million buildings collapsed, and the full scale of the disaster is unimaginable, and its effects will be felt for decades to come. While the human cost was steep, the financial impact of this attack totalled in the hundreds of billions, perhaps even trillions of dollars.

I believe that this attack was deliberate, and it was done to show the Chinese who were the masters. Obey us (the Western Elite, 'New World Order') or China will be destroyed with a weapon that cannot be defended against and which leaves no forensic evidence to prove the quake was deliberate. The long-term ramifications of this are yet to fully unfold but rest assured that China has not forgotten and will not just stand by and allow itself to be attacked in this manner. There have already been suspected moves on the part of the Chinese to use their massive financial capabilities to attack the US economy in retaliation and possibly destabilise the dollar and we are just now beginning to see the effects of this policy coming to fruition.

And in another similar event, the 7.0 Earthquake striking Haiti in 2010 destroyed much of the Haitian National Palace and most of Haiti as well. With most living well below the poverty threshold on $1 per day and one in four people living in houses with mud floors, one can only imagine the devastation to a country that was barely surviving after being hit with four hurricanes in quick succession. The

The Falsification of Science

wealthier Haitians were crushed to death in their concrete homes whilst the poor survived in their tents and 'mud huts.'

Haiti is an independent Black nation, a long time, alleged threat to American hegemony and imperialism and has defied the US government control structure for years. It is relatively close to a HAARP array in Puerto Rico, just a few hundred miles away and is a Third world country right in the middle of 'Western civilisation.' Its citizens are little better off than peasants and ripe for exploitation without the rest of the world asking too many questions.

A street camera at the time of the earthquake clearly showed the ground moving up and down only, as if roiling, percolating, or boiling. There was very little sway in the telephone poles. This very salient fact indicates superheated ground water, not crust breaking or crust slippage, which always goes sideways. This has to be the most telling evidence technological interference uncovered so far.

A video, taken above the city http://www.youtube.com/watch?v=tB XXqu6pmk shows dust and what appears to be steam rising all across the quake zone. There also appear to be some HAARP clouds, but this could possibly be just the sunset catching the bottoms of clouds as the quake occurred at 5.00pm.

A strange anomaly is that the earthquake was not actually on a fault line, but very far south of the North American and Caribbean plates. Earthquakes are a particularly rare occurrence there, especially huge ones such as this was. However, HAARP is able to create earthquakes anywhere there is water or trapped moisture, and our world is obviously riddled with underground water and pockets of gas. There was also an excessive number of 5.0 aftershocks. This is highly unusual and may indicate that the superheated ground could not cool off quickly enough, causing a recurrence of the problem. Scalar waves, which are generated from electromagnetic waves, can accomplish this effect, thus providing crucial circumstantial evidence pointing directly to HAARP as the culprit behind this event.

Mainstream news blatantly lied about the earthquake being on the North American and Caribbean fault lines. So, if they told lies about this issue, the pertinent question is 'why is the media deliberately peddling disinformation?' When seemingly unnecessary lies are being propagated, the obvious conclusion is that something is being covered-up.

Actually, Haiti is prime real estate once all the locals are disposed of or marginalised. Due to the 'natural' disaster, the US military is now in control, without having fired a shot and they seem to be doing a pretty good job of bringing about the destruction of the country's infrastructure having been given a good head start by their clandestine technology. The speed with which the 20,000 US troops arrived on the island, proves that plans were already in place before the event. And subsequent delays in the provision of vital food and medical supplies created a desperate situation for the frustrated masses of survivors, which resulted in violence that provided a perfect excuse or justification for the imposition of heavy security measures.

Respect to the Elite is due. They are indeed very, very clever, and audacious. Using HAARP technology to manipulate the environment, they have now perfected the science of taking over a country, murdering thousands of its citizens in cold blood and then actually being thanked by their victims for the 'honour.'

In conjunction with the absorption of metals into our bodies via the high quantities of barium and aluminium present in the air (see the following section on chemtrails), it would appear that we are, if not already under severe attack from HAARP at present, certainly being prepared for some sort of future, sinister purpose. Whether or not this plan actually comes to fruition, only time will tell, by that time however, it will be far too late to prevent it.

And this sinister purpose, apart from mass murder due to weather manipulation, could well be to do with the manipulation of the human brain itself. This excerpt from Elana Freeland's 2014 book, 'Chemtrails, HAARP, and the Full Spectrum Dominance of Planet Earth,' explains just how this is possible...

"HAARP differs from other Ionospheric heaters in that it is a phased array. As an illegal, over-the-horizon phased-array RADAR weapon able

The Falsification of Science

to track hundreds of objects simultaneously, it is capable of tremendous focusing ability due to its pulsed sequential firing. That the Ionospheric waveguide oscillates naturally at 8HZ (cycles per second) means it is an excellent harmonic carrier of low-frequency sound (LFS) waves into the alpha range of human grains. Because LFS waves are so long, they are virtually impossible to detect. In other words, HAARP provides LFS-wave field strengths that can simultaneously affect large geographic swaths of the population."

We will leave this particular section with some further very apposite words of Dr Nick Begich in his essay, 'Star Wars, Star Trek and Killing Politely'.

"This essay is about some of the science being developed and contemplated by military planners and others which could profoundly affect our lives. The intent of this essay is to focus discussion on these new systems by bringing them into the light of day.

Is it possible to trigger earthquakes, volcanic eruptions, or weather changes by manmade activities? Is it possible to create and direct balls of energy at lightning speeds, to destroy an enemy? Is it possible to manipulate the behavior, and even the memories, of people using specialized technologies? The United States military and others believe that this is the case. Many of these systems are well on their way to being used in the battlefield.

If you believe, as I and others do, that we've gone past the realm of probability into the realm of a dangerous reality, especially given the frequency of catastrophic events, isn't it time to send our own beam of energy to the US government, and demand a full investigation into the activities of HAARP, in particular the manipulation of hurricanes and earthquakes? This kind of activity is another form of terror we live with because it seems like it can strike anywhere, anytime, in the guise of natural event.

Perhaps it is just Mother Nature kicking up her heels, perhaps not. In either case, a return to reality is called for in these surreal days. The simple presence of HAARP is a form of psychological terror, threatening not just the weather, but the well-being of man and survival of the planet. And that's something you'll never hear from your media weatherman."

Chemtrails

How often have you seen skies that look like this? Would you believe that prior to around 1996, you would never have seen such a sky? It is totally a modern phenomenon, but we have become so accustomed now to this sight that many believe it has always been like this.

Often on a beautiful summer morning, the sky is a lovely clear, deep blue with no clouds in sight at all. However this will then over the course of the next few hours become criss-crossed with jet 'vapour trails,' which instead of dissipating quickly, as they should (it is only crystalline water vapour after all), remain in the sky several hours after the planes have passed over, slowly spreading until they often obliterate the sun completely and turn the entire sky a watery grey colour.

There is nothing 'normal' about these 'contrails' despite what the controlled mainstream media and government sources would have us believe in their ceaseless and relentless efforts to deflect us from the truth with their blatant disinformation. One needs only to watch the skies intermittently for a few hours to soon notice the bizarre patterns emerging as these planes track back and forth creating deliberate grid type, x-shaped or even sometimes circular patterns, to ensure maximum coverage. If this were a natural event, then it would happen every day in line with regular plane flight schedules, but keen observers will soon notice that there are days when we get none at all, days when we get isolated batches of trails and other days when the entire skyscape is totally obliterated.

Normal contrails are composed of tiny, fragile ice crystals formed by aircraft flying at altitudes of 33,000 feet (10km) or greater. At altitudes below 33,000 feet, normal vapour condensation trails are unable to form behind an aircraft, regardless of its type or design. Whereas above 33,000 feet, normal contrails are formed, and these will appear to an observer on the ground as narrow streaks of white cloud-like material which totally dissipates in seconds, rarely extending for any appreciable distance behind the aircraft.

Chemtrails, in complete contrast to this, may be observed at any height in the sky. They have been seen emitting from aircraft flying at altitudes as low as 8,000 feet but may be seen at all altitudes above this also. Since normal contrails cannot form at these low altitudes, any contrail formation that is observed at these elevations is probably not a contrail at all, but a genuine 'chemtrail.'

Chemtrails usually appear as normal contrails for the first few seconds, but unlike standard contrails do not evaporate almost immediately, but slowly become broader and denser over time. Over periods of several hours, parallel and ninety-degree chemtrail grid formations will eventually spread and merge to form a continuous, cirrus-like cloud formation in the sky.

Shortly after this merging, what just hours earlier was a perfectly clear blue sky will appear as a kind of insipid milky haze, totally unnatural in all respects. Another distinguishing factor between contrails and chemtrails concerns their relative location or position in the local sky, as well as their directional characteristics.

Aircraft emitting normal contrails are constrained by FAA regulations to operate only over designated air routes; therefore, the contrails that they generate will be found consistently only within local flight-corridors, which are easily discerned by an observer at any given geographic location on the ground. In addition, air traffic along these designated routes is always unidirectional. Aircraft flying in opposing compass directions are never permitted to use the same air routes at the same altitudes, for obvious reasons. Therefore, only those contrails that are formed within these restricted air corridors and consistently with the same vector or direction should be considered normal contrails resulting from normal commercial airline traffic.

Another phenomenon observed to be associated with chemtrail formation is the significance of the local surface wind speed over the dispersion area. Chemtrail flights are invariably suspended whenever the ground wind speed reaches or exceeds 20 miles per hour (28 kph). This has been a consistent limiting factor throughout the range of areas where chemtrail activity has been observed and obviously normal contrail activity is never subject to this restriction and can be seen over a wide range of measured surface wind velocities.

So of what exactly *are* these chemtrails composed? Laboratory analysis of their contents, taken from samples that have fallen to Earth, reveal the presence of several biological agents, Pseudomonas Fluorescens, Streptomyces to name but two. A chemtrail researcher, who had been travelling around the country for several years, had a medical test which discovered dangerous pathogens in his body; ones that should only be found in laboratories. In addition, many samples analysed after falling to the ground as a sticky spiderweb type substance have been found to contain unusually high traces of metals, and in far higher concentrations than could be expected naturally, particularly aluminium and barium.

Strange cloud patterns are not the only result of the chemtrail campaign. There is another, far more sinister side to this phenomenon. But whether or not this is a by-product, or the objective of these attacks is yet to be determined. In either case however, the side effects associated with these chemtrails should be the primary focus of our attention, as this ancillary effect could very well pose a particularly serious threat to every human being living on Earth today.

There are no known health or environmental issues associated with normal contrail emissions. They appear at this time to be completely harmless and benign. They are, after all, composed almost entirely of harmless water vapour. This same statement, however, cannot be made in the case of authentic chemtrails. It has been shown by many dedicated researchers and posited extensively on the Internet, that the formation of chemtrails has a direct correlation to the localised, time-synchronised, outbreak of a broad range of primarily, respiratory related or flulike illnesses. In light of this apparent relationship, chemtrails should not be considered inherently benign, but should be treated with caution and indeed concern by those who are either studying them or simply observing them.

Since the regular spraying of our skies commenced in the late 1990s, another disturbing phenomenon has been noted and that is its correlation to the outbreak of a flulike disease, which has been named, 'Respiratory Distress Syndrome.' Disturbingly, the components of this illness, namely the chemical, bacterial and fungal elements have been strongly linked to the bio-chemical footprint of chemtrails.

Once again, we can detect the same pattern unfolding, the denials by government and military agencies and as usual, total, and complete avoidance of the topic by the media. Radio, television, and the press alike have all either ignored it completely or in instances where it has been brought to their attention, dismissed it as mass-paranoia or that old favourite, a *'conspiracy theory!'*

Literally thousands of eyewitnesses, with not only photographic proof, but with the entire evidence manifest in the skies above them, have been either totally ignored or dismissed by *all* the mainstream media organisations. Up to the current time, no individual researcher has as yet been successful in attracting even so much as the passing attention of even the smallest of these media organisations. What is even more interesting and disconcerting is the fact that with all of the widespread discussion surrounding this phenomenon, its link to a near-epidemic outbreak of a serious and debilitating range of illnesses, and the evidence in the skies right above their offices and studios, not one member of the mainstream media has come forward with a story to discredit the data, contradict the evidence, or calm the growing concerns and justified suspicions of the public. Is this not the very minimum response that we have a right to expect from any news organisation that claims to serve 'the people?' Their all-pervading silence speaks absolute volumes.

Another use for chemtrails has been suggested as weather control. This may or may not be wholly or partly true. However weather changes are often observed to follow directly after heavy spray days, most notably an increase in precipitation in that region. It would make sense that an increase in clouds, whether naturally or artificially formed, would increase the chances of rain. Many who even discredit the chemtrail claims, may still be familiar with the practice (which appears to pre-date chemtrails) of 'cloud-seeding,' which makes use of plane aerosols in order to increase cloud formations. And of course it is only logical that blocking out the sun on such a regular basis in so much of the world today, would have some kind of effect upon weather. In some instances artificial cloud cover actually inhibits natural rain patterns. For example in America's southwestern desert regions, the bulk of the annual rainfall happens in summer during monsoon season. Monsoons actually result from intense dry heat from sunny cloudless skies in early summer months where the heat build up of the land attracts the rains. If the sun is blocked at this time of year, the monsoon rains may be delayed or non-existent resulting in dangerous droughts. There is some evidence to suggest that weather control may be happening in conjunction with the HAARP project and chemtrails, with the latter serving to increase conductivity of the atmosphere via sprayed aluminium particles, thus enhancing HAARP's reach and potential. Whilst this theory has some possibilities, it may be a secondary objective only but bears further investigation. It is more than possible that GMO giants are using the technology to devastate farms around the world to enable them to buy the land cheaply and then change back the weather as they like afterwards. It also explains why Monsanto™ (now Bayer AG™) would have invented aluminium resistant seeds, in anticipation of the environmental effect of aluminium contamination of the earth, soil, and water caused by chemtrails.

There has also been much disinformation spread on the Internet as with all these issues, regarding the 'real reason' for chemtrails. One

The Falsification of Science

mainstream explanation is that they exist to counter the effects of 'so-called' global warming, whilst applying the technical sounding label of 'geoengineering' to this application. Some prominent 'geo-engineering' scientists such as David Keith, have openly admitted to experiments being undertaken by our governments and military bodies (whilst only admitting small-scale operations) to cover our skies with aerosolised aluminium nanoparticles as a means to alter weather and combat 'global warming' by creating an artificial cloud cover to literally block out the sun and 'cool' the planet. However this simplistic model is totally flawed since blocking sunlight often results in temperature inversions where warmer air gets trapped in the lower stratosphere under cloud cover, and of course results in unforeseen domino effects of artificial weather alterations. However this conveniently neglects to consider that global warming is a complete fabrication and the Elite know this only too well. Indeed it is they who invented the myth, for other reasons, which are covered in detail earlier in this chapter. However, the most important negative here is that chemtrails have been around since the mid-nineties and the nonsensical pseudoscience of global warming and the made-up branch of science called 'geoengineering' did not actually come to prominence until several years later.

In closing, consider that when Dr Keith was once asked about the environmental and health consequences to these 'geo-engineering' aerosol spraying practices, he calmly stated that it was of no great concern to us as we would simply be 'piggy-backing off our grand-kids,' meaning that it is they who will pay the true cost of these monstrous manipulations.

Fracking

'Fracking' is the process of drilling down into the earth whilst utilising a high-pressure water mixture is directed at underground rock to release the gas inside. Water, sand, and chemicals are injected into the rock at high pressure, which allows the gas to flow out to the head of the well. The process can be carried out vertically or, more commonly, by drilling horizontally to the rock layer, which can create new pathways to release gas or used to extend existing channels.

The term 'fracking' refers to the rock being fractured and torn apart by the high-pressure mixture.

This method of gas extraction has been associated with a growing number of health risks from exposure to the subsequent chemically contaminated water and air surrounding fracking sites. These are not hypothetical concerns either – there are now more than 700 studies looking at risks – and more than 80% of the health studies confirm risks or even actual harm.

It is also important to note that these risks are likely to be seriously *underestimated* because the environmental agencies have been enthusiastically downplaying the risks to the public – as always. A new exposé from investigative journalists takes an in-depth look at the Pennsylvania Department of Environmental Protection's misconduct and negligence, as the DEP studiously ignored citizens' complaints, sometimes not even testing water samples.

Coughing, shortages of breath and wheezing are the most common complaints of residents living near fracked wells and toxic gases such as benzene are regularly released from the rock by fracking. Similarly, a toxic waste brew of water and chemicals is often stored in open pits, releasing volatile organic compounds into the air. These noxious chemicals and particulates are also released by the diesel-powered pumps used to inject the water. An epidemiological study of more than 400,000 patients of Pennsylvania's *'Geisinger Clinic,'* in a joint exercise with Johns Hopkins School of Public Health in Baltimore, USA found a significant association between fracking and increases in mild, moderate, and severe cases of asthma (odds ratios 4.4 to 1.5). Johns Hopkins' Dr Brian Schwartz cautioned that residents should be aware of this hazard as *"some 'pristine' rural areas are converted to heavily trafficked industrial areas."*

Fracking chemicals are very harmful to pregnant women and their developing babies. In West Virginia, USA, researchers found endocrine-disrupting chemicals in surface waters near wastewater disposal sites and these types of chemicals can easily damage the developing foetus even when present in very low concentrations.

Another Hopkins/Geisinger study looked at records of almost 11,000 women with newborn babies who lived near fracking sites and found there to be a 40% increased chance of giving birth prematurely and a 30% risk of having the pregnancy be classified as 'high-risk,' even though they 'controlled' for socio-economic status and other risk factors. Contributing factors likely include air and water pollution and stress from the noise and traffic.

Premature babies accounted for 35% of infant deaths in 2010. In addition to the personal anguish inflicted on the families, premature babies are very expensive for society and prematurity is a major cause of neurologic disabilities in children. The cost of care was more than $26 billion in 2005 alone, or $51,600 per child.

Other studies have found that the noise from the drilling itself, the gas compressors, other heavy equipment, and the site traffic is persistent and loud enough to disturb sleep, cause stress and increase the incidence of high blood pressure in local populations. Longer-term exposure to noise pollution contributes to endocrine abnormalities and diabetes, heart disease, stress, and depression, and has been linked to learning difficulties in children. Sleep deprivation also has negative, pervasive public health consequences, ranging from causing accidents to chronic diseases.

Another epidemiologic study from two US universities compared the hospitalisation rates between a county with active fracking and a neighbouring county without. This study found that fracking well density was significantly associated with higher in-patient hospitalisation for cardiac or neurologic problems and there was also a link between skin conditions, cancer and urologic problems and the proximity of homes to active wells.

With disturbing frequency, new spills or accidents are reported at the same time as the industry tries vainly to reassure us that fracking brings 'safe and clean energy.' Try telling that to the residents of communities who have had their drinking water destroyed, of which there are far too many examples.

Ageing pipelines pose special risks as they deteriorate. For example, an Exxon-Mobil pipeline built in 1947 recently spilled 134,000 gallons of gasoline (petrol) in Arkansas. But new fracking has additional risks, as the conventional pipes often used are unable to withstand the high pressure of the fracking mixture being injected. In fact, new wells were not safer, and 6% of unconventional (fracked) wells drilled since 2000 proved to have problems, confirming more than 100 contaminated drinking water wells.

But of course, the oil and gas industry insists that these health problems are not proven to be caused by fracking. That is partially true, but only because agencies such as the Pennsylvania DEP in the USA have actively hidden complaints or even failed to test the water of residents. Some serious health problems however, such as cancer and neurological problems also take years to develop after exposure. Fracking profits go to private industry, but it is the public, families and communities that bear the costs of the many issues arising from the drilling, in terms of both financial and health.

Since 2009 the people of Dimock, Pennsylvania, USA have insisted that, as natural gas companies drilled into their hillsides, shaking, and fracturing their ground, their water had become undrinkable. It turned a milky-brown, with percolating bubbles of explosive methane gas. Some online video posts have even demonstrated totally flammable tap water, ignitable with a match! Many people have reported it made them sick. Totally unsurprisingly, I would suggest. Their stories, first told through an investigation into the safety of gas drilling by 'ProPublica,' turned Dimock into the epicentre of what would evolve into a national debate about natural gas energy and the dangers of fracking.

But the last word about the quality of Dimock's water came from assurances in a 2012 statement from the U.S. Environmental Protection Agency, the federal department charged with safeguarding Americans' drinking water. The agency actually declared that the

The Falsification of Science

water emanating from Dimock's taps did not require emergency action, such as any kind of clean-up. This stance was widely interpreted to mean that the water was perfectly safe. However, another federal agency charged with protecting public health analysed the same set of water samples and determined that that was definitely not the case.

This finding, from the Agency for Toxic Substances and Disease Registry, a part of our 'old friends' the CDC (the Centers for Disease Control and Prevention) warned that a list of contaminants the EPA had previously identified were indeed dangerous for people to consume. The report found that the wells of twenty-seven houses in Dimock contained, to varying degrees, high levels of lead, cadmium, arsenic, and copper – sufficient enough to pose a grave health risk. It also warned of a mysterious compound called 4-chlorophenyl phenyl ether, a substance for which the agency could not even evaluate the risk and noted that in earlier water samples non-natural pollutants including acetone, toluene and chloroform were detected. Those contaminants are known to be dangerous, but they registered at such low concentrations that their health effects could not easily be evaluated. The water in seventeen of the homes also contained enough flammable gas to pose a great risk of explosion.

The fact that the contaminants were detected in their water had been shared with residents in 2012 but the qualitative assessment of whether those contaminants posed a danger appeared to call into question the EPA's assurances, and could well reignite a smouldering controversy over whether the agency had prematurely abandoned its research into the safety of fracking in Dimock and other sites across the country. ProPublica reported in 2012 that the agency had curtailed investigations it had begun into potential water contamination in Pennsylvania, Texas, and Wyoming.

The findings regarding Dimock also lead to an obvious question... 'How can two federal agencies charged with safeguarding the country's health and environment use the same water samples and reach such different conclusions?' A spokesperson for the EPA offered a seemingly cryptic explanation. The EPA was testing whether the contaminants were *"hazardous,"* whilst the ATSDR was considering whether they were *"safe to drink."* Hmmm, that explains that then. In a statement the EPA sent to ProPublica, it described the ATSDR report as *"useful information"* for Dimock residents and the spokesperson promised that the agency would *"consider the findings."*

Of course, the EPA, in certain respects, had already considered the findings. After collecting and analysing the same water samples used by the ATSDR, the agency detected the same assortment of contaminants. It just ultimately did not describe their presence as a significant health threat. The agency eventually ended its investigation into the groundwater in Dimock. Its detailed water test results were distributed confidentially to homeowners without any evaluation of their meaning, whilst a general report was issued, publicly downplaying the danger.

At the time, ProPublica, which had been leaked a portion of the sample results shared privately with Dimock landowners, wrote that the water contained significant amounts of metals and chemicals and that there appeared to be a disconnect between the EPA's conclusions and the chemical makeup of the water. *"I'm sitting here looking at the values I have on my sheet – I'm over the thresholds – and yet they are telling me my water is drinkable,"* Scott Ely, a Dimock resident, told ProPublica in March 2012. *"I'm confused."* I am too, Scott, I am too!

Pennsylvania environment officials had already determined that a number of drilling violations by a company named 'Cabot Oil and Gas' had likely contributed to some pollution and disturbance underground. It sanctioned Cabot, suspended their drilling permits in the area, and required them, for a time, to supply uncontaminated bottled drinking water to Dimock homes. In 2012, Cabot settled with dozens of families who had sued the company for an undisclosed amount. That contamination existed, and that it may have been caused by the drilling, was not in dispute when the EPA made its judgment call. *"Spills and other releases have been documented…from these drilling activities,"* noted an EPA document. *"There is reason to believe that a release of hazardous substances has occurred."*

The Falsification of Science

The EPA's January action memo went on to describe the contaminants and summarised that *"a chronic health risk exists for most wells and that the situation supports a 'Do Not Use the Water' action."* Beside several of the most concerning substances detected, the agency wrote ... *"note that children reside at this location."*

Still, Superfund law sets several criteria beyond a health concern to qualify for a clean-up, including the possibility of alternative solutions to a clean-up. *"The United States federal Superfund law is officially known as the Comprehensive Environmental Response, Compensation, and Liability Act of 1980 (CERCLA). The federal Superfund program, administered by the U.S. Environmental Protection Agency (EPA) is designed to investigate and clean-up sites contaminated with hazardous substances. Sites managed under this program are referred to as 'Superfund' sites. There are 40,000 federal Superfund sites across the country, and approximately 1,600 of those sites have been listed on the National Priorities List (NPL). Sites on the NPL are considered the most highly contaminated and undergo longer-term remedial investigation and remedial action (clean-ups)."* Wikipedia™

There is also an explicit loophole built into the Superfund law for petroleum and natural gas; neither are allowed to be defined as a 'hazardous substance' requiring clean-up. How those factors weighed on the EPA's characterisations of, and decisions concerning, Dimock's water is unclear. The agency has never explained its decision in detail but just six months later, the EPA issued a carefully worded statement declaring what appeared to be a reversal, saying it *"has determined that there are not levels of contaminants present that would require additional action by the Agency."* It said that it would stop delivering clean emergency water to Dimock homes, and noted that in the homes with the worst contamination residents planned to install water treatment systems. Again, the EPA's 2012 decision was widely interpreted to mean that the water supplies in Dimock had been found to be 'safe.'

"EPA finds Pennsylvania well water safe after drilling," announced a headline in the *Los Angeles Times*. It went on to say *"In a response to ProPublica's questions this week, an EPA spokesperson said the agency had completed a second round of testing and concluded that the worst water*

problems were confined to just four homes. Those homes either already had, or planned to install, water treatment systems that the EPA expected would reduce their exposure risk. The EPA emphasised that ultimately cleaning rural residential water is not its responsibility."

"*Private drinking water well owners are responsible for sampling and maintaining their wells and addressing the operational issues that can affect drinking water quality,*" wrote Michael d'Andrea, the EPA's director of communications for its Mid-Atlantic region, in an email.

The ATSDR, on the other hand, serves to advise residents with regards to their health. It examined the same water samples as the EPA but assumed that the highest levels of contaminants detected were what people would be exposed to. And it simply advises on issues of public health. The ATSDR does not address those uncomfortable political questions that have bogged down the conversations about fracking.

And in the UK, two small earthquakes struck near the coastal town of Blackpool, in the northwest of England. Suspicion immediately fell on, and indeed a subsequent report has now confirmed, that it was fracking that caused the earthquakes. A magnitude 2.3 earthquake occurred on 1st April, followed by a magnitude 1.5 quake on 27th May. Both occurred close to the Preese Hall drilling site, where 'Cuadrilla Resources' was using fracking to extract gas from a shale bed.

Initial studies by the British Geological Survey (BGS) suggested that the quakes were linked to Cuadrilla's fracking activities. The epicentre of the second quake was within 500 metres of the drilling site, at a depth of two kilometres. Less information was available on the first quake, but it seems to have been similar. And the link with fracking has now been confirmed by an independent report commissioned by Cuadrilla, the '*Geomechanical Study of Bowland Shale Seismicity,*' which states that... "*Most likely, the repeated seismicity was induced by direct injection of fluid into the fault zone.*" The two geologists who wrote the report ran detailed models to show that the fracking could, and most likely did, provoke the earthquakes.

The Falsification of Science

There is growing evidence of a variety of geological and health problems being associated with fracking. Common sense dictates that drinking and breathing cancer-causing agents will take their toll on people and the correlation is too strong to ignore, especially when we have other, 'cleaner' energy options than so-called 'natural' gas. For our safety and that of future generations, we should not allow our governments to sell off public lands, nor allow drilling on our land, and fracking should be banned completely.

"Fracking is a nightmare! Toxic and radioactive water contamination, severe air pollution and tens of thousands of wells, pipelines and compressor stations devastating our countryside and blighting communities."
'Frack-Off' The Extreme Energy Action Network

Renewable Energy

The Global Wind Energy Council recently released its latest report, excitedly boasting that, *"the proliferation of wind energy into the global power market continues at a furious pace, after it was revealed that more than 54 gigawatts of clean renewable wind power was installed across the global market last year."*

You may have formed the impression from announcements such as that, and from the obligatory pictures of wind turbines in any typical mainstream 'fake news' story or airport advertisement about energy, that wind power is making a huge contribution to world energy today. However, you would be wrong, very, very wrong. Its contribution is still, after many decades of development, trivial almost to the point of irrelevance.

Rounded down to the nearest whole number, what percentage of the world's energy consumption do you think was supplied by wind power in 2018, the last year for which there are reliable figures? Was it 20 per cent, 10 per cent or 5 per cent? Give up yet? Sorry no, in fact none of those figures are correct, it was a very round number indeed, it was actually 0%. That is, to the nearest whole number, there is still **zero** wind power energy generation on Earth.

Even put together, both wind and photovoltaic solar power are supplying much less than 1% of global energy demand. From the International Energy Agency's 2016 Key Renewables Trends, we can discern that wind provided 0.46% of global energy consumption in 2018, and solar and tide combined, provided 0.35%. Remember this is total energy, not just electricity, which is less than a fifth of all final energy, the rest being the solid, gaseous, and liquid fuels that perform the 'heavy lifting' for heat, transport, and industry.

Such numbers are not too difficult to find, but they do not figure prominently in reports on energy derived from the unreliable lobbies for solar and wind power. Their trick is to hide behind the statement that close to 14% of the world's energy is renewable, with the implication that this is wind and solar. In fact the vast majority – three quarters in fact – is biomass (mainly wood) and another very large part of that is 'traditional biomass,' sticks, logs, peat, and dung burned by the world's dispossessed poor in their homes for heating and cooking purposes. Those people need that energy, but they also pay a huge price in health problems caused by the resultant smoke and frequently, chemical inhalation.

Even in wealthy countries playing with subsidised wind and solar, a huge portion of their renewable energy comes from wood and hydro, the most reliable of the renewables. Meanwhile, world energy demand has been growing at about 2% per year for nearly forty years. Between 2013 and 2018, again using International Energy Agency data, it grew by just under 2,000 terawatt-hours.

If wind turbines were to supply all of that growth but no more, how many would need to be built each year? The answer is nearly 350,000, since a two-megawatt turbine can produce about 0.005 terawatt-hours per annum. That is one and a half times as many as have been built in the world since governments started pouring startling amounts of public funds into this so-called 'industry' in the early 2000s.

At a density of, very roughly, fifty acres per megawatt, typical for wind farms, that number of turbines would require a land area greater than the British Isles, including Ireland, every year. If we kept this up for fifty years, we would have covered every square mile of a land area the size of Russia with wind farms. Remember, this would be just to fulfil the new demand for energy, not to displace the vast existing supply of energy from fossil fuels, which currently supply 80% of global energy needs.

And please do not believe the false notion that wind turbines could eventually become more efficient. There is a finite limit to how much energy can be extracted from a moving fluid, the Betz limit, and wind turbines are already close to it. Their effectiveness (the load factor, to use the engineering term) is determined by the wind that is available, and that varies greatly from second to second, day-to-day, year-to-year.

As machines per se, wind turbines are pretty efficient already; the problem is the wind resource itself, and we cannot change that. It is an ever-fluctuating stream of low-density energy and mankind stopped using it for mission-critical transportation and mechanical power long ago, and for sound reasons. It is simply not very reliable. As for resource consumption and environmental impacts, the direct effects of wind turbines – killing birds and bats, sinking concrete foundations deep into wild lands – is bad enough. But out of sight and out of mind and remaining totally unconsidered by its apologists is the extreme pollution generated in Inner Mongolia by the mining of rare-earth metals for the magnets in the turbines. This generates toxic and radioactive waste on an epic scale, which is why the phrase 'clean energy' is such a sick joke and therefore wind power's 'eco fanatic' proponents should be ashamed every time it passes their lips.

And it gets even worse than this. Wind turbines, apart from the fibre-glass blades, are made mostly of steel, with concrete bases. They need about 200 times as much material per unit of capacity as a modern combined-cycle gas turbine. Steel is made with coal, not just to provide the heat for smelting ore, but to supply the carbon in the alloy and cement is also often made using coal. The machinery of 'clean' renewables is the output of the fossil fuel economy, and largely the coal economy.

A two-megawatt wind turbine weighs about 250 tonnes, including the tower, nacelle, rotor, and blades. Globally, it takes about half a tonne of coal to make a tonne of steel. Add another 25 tonnes of coal for making the cement and you're talking 150 tonnes of coal per turbine. Now if we were to build 350,000 wind turbines a year (or a smaller number of larger ones), just to keep up with increasing energy demand, that will require 50 million tonnes of coal a year. That's about half of the entire EU's coal – mining output.

The point of running through these numbers is to demonstrate that it is utterly futile, on *a priori* grounds, even to consider that wind power can make any significant contribution to the world energy supply, let alone to emissions reductions, without ruining the planet. The arithmetic is firmly against such unreliable renewables.

The truth is, in order to power civilisation with fewer greenhouse gas emissions, then the focus should be on shifting power generation, heat and transport to more naturally sourced gas, the economically recoverable reserves of which are much more abundant than we dreamed they ever could be. We should also accept that wind power is actually counterproductive as a climate policy and, worst of all, shamefully robs the poor to make the rich even richer. That old story again, eh?

The Falsification of Science

Scarcely a day goes by without more evidence to show why the Government's obsession with wind turbines, now at the centre of national energy policy, is nothing more than yet another political hoax. Under a target agreed with the communist agenda-driven EU, Britain was committed at astronomic expense, to generating nearly a third of its electricity from renewable sources, mainly through building thousands more wind turbines, and all within the next ten years. It remains to be seen however, the impact that Brexit will have on this policy, but to be frank we would expect 'not much.'

But relying on the modern day version of windmills simply to keep our lights switched on is a colossal and very dangerous act of self-deception. Take, for example, the 350ft monstrosity familiar to millions of motorists who drive past as it sluggishly revolves above the M4 motorway just outside Reading, Berkshire. This wind turbine performed so poorly (working at only 15 per cent of its capacity) that the £130,000 government subsidy given to its owners was more than the £100,000 worth of electricity it produced last year. Nice work if you can get it!

Meanwhile, official figures have confirmed that during those many cold, yet windless weeks so common in British winters, and when electricity demands are at record levels, the contribution made by Britain's 3,500 turbines is so utterly minuscule as to be undetectable. So, in order to keep ourselves warm in winter we have to import vast amounts of power generated by nuclear reactors in France. It really is just a sick, sick joke – but a very unfunny one indeed.

In fact wind turbines are so expensive and inefficient that Holland recently became the first country in Europe to abandon its EU renewable energy target, announcing that it will slash its annual subsidy by billions of Euros. And so unpopular are wind turbines now that the UK Government has just offered 'bribes,' sorry, incentives, to local communities, in the form of lower council taxes and electricity bills.

And in Scotland, the 800 residents of the beautiful island of Tiree are desperately trying to resist plans to create what will be the largest offshore wind farm in the world, covering 139 square miles off their coast, which they quite rightly say will destroy their community by driving away the tourists who provide so much of their income.

But possibly worse than this, is the health destroying effects that wind farms have on neighbouring communities due to the infrasound created by the turbines. Infrasound is a low frequency form of electromagnetic field pollution, which can travel through the earth up to 15 miles from the source, causing well-documented sleep disturbances, headaches, and a wide range of other disturbing symptoms in humans. If close enough to the turbines, this infrasound, normally not audible, especially at further distances, can be heard by some residents, manifesting as a persistent hum. Those most able to detect the hum can be driven insane by it and forced to move away, if they are able.

So riddled with environmental hypocrisy is the lobbying for wind energy, that a recent newspaper report exposed the immense human and ecological catastrophe being inflicted on northern China by the extraction of the rare earth minerals needed to make the giant magnets that every turbine in the 'West' uses to generate its power.

Here in a nutshell are some of the reasons why people are beginning to wake-up to the horrific downside of the wind business. The case for wind turbines rests on three great lies and the megawatts supplied by our 3,500 turbines is derisory; no more than the output of a single, medium-sized conventional power station...

The first of these lies is the pretence that turbines are anything other than ludicrously inefficient in producing 'value for money' energy.

The most glaring dishonesty peddled by the wind industry and echoed by gullible politicians and those with lucrative vested interests, is to greatly exaggerate the output of turbines by deliberately talking about them only in terms of their 'capacity,' as if this were what they actually produce. Rather, it is the total maximum amount of power they have the capability of producing. The point about wind, of course, is that it constantly varies in speed and therefore power, so that the output of turbines averages barely a quarter of their 'capacity.' This means in effect that the 1,000 megawatts all those

The Falsification of Science

3,500 turbines sited around the country feed on average into the grid is derisory and no more than the output of a single, medium-sized conventional power station.

Furthermore, as they increase in number (the government would like to see 10,000 more in the next few years) it will, quite farcically, become necessary to build a dozen or more gas-fired power stations, running constantly, simply to provide instant back-up for those frequent occasions when the wind ceases to blow.

The second great lie about wind power is the pretence that it is not a preposterously expensive way to produce electricity. No one in their right minds would ever dream of building wind turbines unless they were guaranteed a huge government subsidy. This takes the form of the Renewables Obligation Certificate subsidy scheme, paid for through household bills, whereby owners of wind turbines earn an additional £49 for every 'megawatt hour' of electricity they produce, and twice that sum for offshore turbines.

The third great lie is that this industry is somehow making a vital contribution to 'saving the planet' by cutting our emissions of CO_2 – it is absolutely not. The whole 'climate change by emissions of CO_2' is yet another vast money-making scam (see the relevant, earlier section). What other industry receives a public subsidy equivalent to 100% or even 200% of the value of what it actually produces?

This is why so many people are now realising that the wind bonanza, almost entirely dominated in Britain by French, German, Spanish and other foreign-owned firms, is one of the greatest scams of our age. We may not be aware of just how much we are pouring into the pockets of the wind turbine developers, because our bills hide this from us – but as ever more turbines are built, this could soon be adding hundreds of pounds a year to energy bills in the UK. And of course this situation is not simply restricted to a tiny island off the northwest coast of the European mainland, this same principle applies worldwide.

When a Swedish company recently opened what is now the world's largest offshore wind farm off the coast of Kent in southeast England, at a cost of £800m, we were told that its 'capacity' was 300 megawatts, enough to provide 'green' power for tens of thousands of homes. But what we were not told was that its actual output will average only a mere 80 megawatts, a tenth of that supplied by a gas-fired power station, for which we will all be paying a subsidy of £60million a year, or £1.5billion over the 25-year lifespan of the turbines.

Even if you believe the farcical notion that curbing our use of fossil fuels could change the Earth's climate for the better, the CO2 reduction achieved by wind turbines is so insignificant that one large wind farm saves considerably less in a year than is emitted over the same period by a single jet flying daily between Britain and America. A sobering truth indeed.

Then, of course, the construction of the turbines generates enormous CO2 emissions (according to the climate alarmists) as a result of the mining and smelting of the metals used, the carbon-intensive cement needed for their huge concrete foundations, the building of miles of road often needed to move them to the site, and the releasing of immense quantities of CO2 locked up in the peat bogs where many turbines are built. The whole thing is one hugely unfunny joke.

When you also consider those gas-fired power stations wastefully running 24 hours a day just to provide backup for the unreliability of the wind, any savings will vanish altogether. Yet it is on the strength of these three massive self-deceptions that governments have embarked on one of the most reckless strategies in the history of energy production, the idea that we can rely upon that most fickle of providers, the wind, to generate nearly a third of the electricity we need to keep our economy running, well over 90% of which is still currently supplied by coal, gas and nuclear power.

It is a fact that a target of raising the contribution made by wind by more than ten times in the next nine years was set by the EU. But it is not just Brussels to blame for such an absurdly ambitious target, because no one was more anxious to adopt it than our own politicians, no doubt at the behest of their Elite puppet masters and all their 'friends' within the industry itself who will obviously be the main beneficiaries of the vast profits generated by this utter scam.

In order to meet this target however, the government plans to spend another £100 billion building 10,000 more turbines, plus another £40 billion on connecting them to the grid. Someone, somewhere is definitely going to be making a huge killing out of this, and it is certainly not you or me.

And furthermore, because of this insanity, the UK will soon be facing a colossal energy gap and be dependent on politically unreliable countries such as Russia and Algeria for gas supplies. According to the electricity industry, we will then need to spend another £100 billion on those 'conventional' power stations to provide backup – all of which adds up to £240 billion, or just over £1,000 a year for every household in the land.

And for this our politicians are quite happy to see our countryside and the seas around our coasts permanently blighted by vast arrays of ugly, unsightly giant wind turbines, kill thousands upon thousands of birds and bats and destroy human health, all in order to produce an amount of electricity that could be provided by conventional power stations at one tenth of the cost. This appears to be sheer madness and insanity, unless you factor in, as previously stated, that there are individuals and organisations, no doubt possessing huge lobbying power and very deep pockets that are making billions out of the entire process. Once we realise and accept this, then the agenda becomes much clearer.

All but two of the UK's ageing nuclear power stations are nearing the end of their useful life, with little chance of them being replaced for many years. Six of the large coal-fired stations will be forced to close under an anti-pollution directive, and the government is doing its best to ensure that we build no more. There is no way that it will be possible to make up more than a fraction of the resulting energy gap solely with wind turbines, for the simple and obvious reason that wind is such an intermittent and unreliable energy source.

What we are seeing, in short, is the price we are beginning to pay for the past two decades, during which energy policies have become hopelessly skewed by vested interests and the unrealistic demands of the rabid 'greens' and environmentalists, firstly in persuading politicians to switch from coal and not to build any more nuclear power stations, and then in buying into the romantic vision of gambling the nation's future on the 'free' and 'clean' power of wind and sun. All of which is abject nonsense of course and facilitated ultimately by the systematic falsification of data and of science itself.

Throughout the EU, other politicians are waking up to the dead-end to which this madness has been leading us. The Danes, who have built more wind turbines per head than anyone, have finally realised the idiocy of a policy that has given them the highest electricity prices in Europe, while they have to import much of their power from abroad. And in Spain, their desire for wind and solar power has proved a national disaster. In Germany, having built more turbines than any other country in the world, they are now, of necessity building new coal-fired stations all across the country, whilst in Holland, meanwhile, they have now defied the Marxist EU directives by slashing all their renewables subsidies.

In Europe, it is only in Britain that the political class, still so imprisoned in its money-driven infatuation with wind power, is still prepared to court this dangerously misguided pipedream.

But what about the rest of the 'Western world,' you may ask?

Well, around once a month, according to local reports, a golden eagle is killed by the spinning turbines of the wind farms in Wyoming, USA. Killing these protected birds is a federal crime, one for which the government has pursued oil companies when these birds drown in waste pits, or electric companies when power lines electrocute them.

The *Wildlife Society Bulletin* claimed that more than half a million birds of all species are killed by American wind farms each year, including endangered species such as falcons and eagles. This study was a 'peer reviewed' report.

The Falsification of Science

But critics have claimed that due to the 'green' nature of the wind turbine technology, government officials are reluctant to pursue those responsible, for fear of criticism and as a result, the government has never once fined or prosecuted a wind-energy company, even those that break the law repeatedly.

"It is the rationale that we have to get off of carbon, we have to get off of fossil fuels, that allows them to justify this, but at what cost? In this case, the cost is too high." Tom Dougherty, a long-time environmentalist who worked for nearly 20 years for the *National Wildlife Federation*.

In 2010, BP was fined $100m for killing and harming birds during the Gulf Coast oil spill. *PacifiCorp* paid over $10.5m a year earlier for electrocuting 232 eagles along its power lines and at substations.

"What it boils down to is this; if you electrocute an eagle, that is bad, but if you chop it to pieces, that is OK," said Tim Eicher, a former *US Fish and Wildlife Service* enforcement agent, who helped prosecute the PacifiCorp power line case.

The US government even proposed a rule that would give wind energy companies decades of freedom from prosecution for killing eagles. This is currently under review. The proposal would allow companies to apply for thirty-year permits to kill a set number of bald or golden eagles, whereas previously, companies were only eligible for five-year permits. The government claimed that the longer permit was needed to *"facilitate responsible development of renewable energy while continuing to protect eagles."* What utter claptrap.

Under both the Migratory Bird Treaty Act and the Bald and Golden Eagle Protection Act, the killing of a single bird without a permit is illegal, however it is a fact that the US government continues to overrule experts, and the wind industry, which was part of the committee that drafted and edited guidelines on the matter, continues to 'get away with' almost everything it does.

Is it just me that finds the whole concept of a government 'issuing permits' to kill defenceless birds utterly obscene and totally morally reprehensible? It is unclear whether the wind power companies actually have to pay for this 'privilege' or whether the permits are free. However the former is probably the case, but in whichever event, words just fail me.

Everyone is being brainwashed into believing that alternative energies are the answer to all our energy needs and also the so-called climate change 'crisis.' However, to be clear, alternative energies are not 'alternative' at all. If energy cannot be immediately called upon at will to fulfil power needs, the accurate description would be to refer to energies from wind, solar, waves, etc., as 'supplementary' rather than 'alternative.' Ultimately, the large majority of electricity demand absolutely must be met by reliable sources. Any other policy is simply disastrous.

Wind power will only ever at very best, produce a small portion of the energy that is necessary to design, fabricate, transport, erect, operate, maintain, and decommission it. Therefore, by definition, wind power is unsustainable.

Much of the same objections to wind power also apply to other forms of renewable energy such as wave and tidal power, and solar power. In northerly, colder climates such as that experienced by the UK where sunlight is highly unreliable, even virtually non-existent during the winter months, there are huge shortfalls, which somehow need to be supplemented by more conventional methods of power generation, rendering solar power a fairly useless 'alternative' energy source.

Renewable energy is now 'Big Business' though. The big winners will be developers, landowners, brokerage houses, banks, manufacturers, governments, the 'green' movement, environmentalists, researchers, academia, and the news media – in fact the 'usual suspects.' The big losers will be we, the taxpayers and those that pay energy bills – indeed 99% of us all, and ironically the environment itself.

The only reason that 'alternative' energies are being heavily promoted is because they allegedly offer a solution to 'reducing our carbon footprint' and 'greenhouse gas emissions.' However, once global warming and climate change are exposed for the scams that they most certainly are, and 'alternative energies' are revealed as being unsustainable, the justification for 'alternative' energies simply evaporates into thin air, despite all the claims of this pseudoscience and its deceitful protagonists.

5G and 'Smart' Technology

"The first 5G services will be launched in multiple cities across the UK in 2019, introducing faster download speeds and better responsiveness than current 4G services. The new mobile technology is exciting not only because of the faster speeds for consumers and business users, but also because it enables a broad set of use cases that will benefit a variety of industrial sectors." 5G.co.uk, the website of the company promoting and advocating 5G technology in the UK

5G is (obviously) the fifth generation of mobile networks. It follows the previous mobile generations 2G, 3G and 4G and compared with today's networks (which primarily use 3G and 4G technology), 5G is set to be far faster and more reliable, with greater capacity and lower latency times.

Unlike those previous generations of mobile network, 5G is unlikely to be defined by any single form of technology. It is often referred to as 'the network of networks' for the way it will bind together multiple existing and future standards, including current advanced LTE (4G) networks.

Beyond a simple performance increase, 5G is set to open up a whole new set of usage cases such as ... superfast mobile broadband with no need for land lines, super smart factories, car to car and car to infrastructure communication, driverless cars, holographic technology, the arrival of 5G phones and devices such as 5G TVs and remote/distance health care for patients.

Sounds wonderful does it not? But let us examine it in detail and more closely for a moment. There are some other features of 5G that we need to be aware of ...

Unlike previous generations, 5G utilises millimetre waves rather than microwaves for communication. Millimetre waves (MMWs) do not travel well through buildings and they tend to be absorbed by rain and plants. This interferes with the signal, not to mention the metabolism of the plants themselves. Added to this, high frequency waves like MMWs also have much shorter wavelengths that cannot travel far. In order to counter this problem, 5G will utilise smaller cell stations (and the technology of beam-forming) that will scramble/unscramble and redirect packets of data on a non-interference path back to us. This could mean wireless antennas on every lamppost, utility pole, home and business throughout entire neighbourhoods, towns, and cities. However, having said that, there would appear to be some controversy surrounding this particular point.

The strong rumours persist that preparation for 5G continues apace with the felling of hundreds of trees and bushes in many cities, allegedly in order to prevent the signals from being disrupted in some way by the abundant foliage. But of course we, the general public, are not told officially that this is the reason for the indiscriminate tree-felling bonanza. The city of Sheffield in South Yorkshire, which is fairly local to me, is very active in this respect, in the process generating loud and indignant protests from many so-called 'green' action groups who are quite rightly opposed to this particular form of what can only be described as environmental vandalism.

On the one hand, the industry insists that the cell points need to be fairly closely spaced, yet on the other hand, some researchers, significantly those that are non-industry funded, have proven this to be untrue and insist that 5G proponents are exaggerating the alleged limitations of the technology. The significance of this fact would seem to be that it is the industry itself which is deliberately creating an increased market for the technology but whether the reason for this is financial or to deliberately cause harm – or even a combination of both – your guess is as good as mine.

"When [Verizon] went out in these 11 [5G test] markets, we tested for well over a year, so we could see every part of foliage and every storm that went through. We have now busted the myth that [5G frequencies] have

The Falsification of Science

to be line-of-sight – they do not. We busted the myth that foliage will shut [5G] down... that does not happen. And the 200 feet from a home? We are now designing the network for over 2,000 feet from transmitter to receiver, which has a huge impact on our capital need going forward. Those myths have disappeared." Lowell McAdam, CEO of Verizon

"[Verizon 5G] is really high frequency [28,000 MHz and 39,000 MHz], so everybody thinks it doesn't go very far, but it's a really big pipe and so that's what allows you to gain the super-fast speeds... We are 3,000 feet away from our radio node. The cool thing about this is that we did not move the radio node. It is pointing down to serve the customers in that area... here even 3,000 feet away, we're still getting 1,000 [Megabits per second] speeds... So now we've driven about one third of a mile away [1,760 feet] from the radio node and we are still getting very good speeds even though we have foliage in between [800 Megabits per second]."
Jason L., Verizon Field Engineer

So, this is all quite confusing really. One faction seems determined to facilitate the proliferation of these highly insidious monstrosities (the cell nodes) and allegedly to also facilitate the spread of 5G MMWs by cutting down trees, whilst certain industry insiders, such as those quoted above, imply that this is all totally unnecessary. Could this be a case of certain Elites 'using' the advent of 5G tech to further their own agenda of deliberately harming the population and destroying the environment? Surely not?!!

Current 4G cell towers have about a dozen or so antenna ports to support all communication, whilst the new, smaller 5G cell towers (or bases) will be MIMO (Multiple Input Multiple Output) and carry about a hundred ports. These towers will probably be about 4 feet tall as opposed to the usual 90 feet towers currently erected around us. Cells will be available within a 100 metre range and these smart antennas will be able to differentiate between various mixed-up signals – like radio waves and Wi-Fi signals – in the air and beam them back in an orderly fashion, so to speak.

5G will break down data and send it in smaller sizes to offer significantly reduced transmission times and data will be sent with only a 1 millisecond delay instead of a 50 millisecond delay commonly found with 4G. With communication this fast, it will allow machines to talk to each other with practically no room for error. As Marcus Weldon the Chief Technology Officer of Alcatel Lucent commented, "...*up until now, we've designed the networks for people and their needs, and now we're designing it for things.*"

The USA is currently leading the way with 5G. At the June 2016 press conference where the Federal Communications Commission's (FCC) head, Tom Wheeler, announced the opening up of low, mid, and high spectrums, there was no mention of health effects whatsoever, but nevertheless, the dangers are real.

Tom Wheeler in fact, was the head of the wireless industry's lobbying organisation for more than ten years. He hired George Carlo, a scientist, to head a safety study regarding cell phone exposure, brain tumours and cancers in the 1990s and Carlo actually found that they *do* cause tumours and cancer. This was a $23m funded study over a nine-year period and they had hoped that Carlo, being an 'industry guy,' would find in the industry's favour. However, he did not, and they therefore completely dismissed his research and falsely tarnished his hitherto excellent professional reputation. He even had his house burned down for his pains. No doubt you have heard of these kinds of tactics somewhere before!

Carlo was interviewed a few years ago and related this highly revealing, not to mention, damning information at that time. He was also interviewed in an excellent documentary called 'Generation Zapped.' Yet another prime example of self-regulating industries and 'revolving doors!' That same tired old story again.

When confronted with opposition to the 5G roll-outs without testing and safety standards in place, Wheeler was quoted as saying *"Let the innovators innovate."* And that as it is now the 21st century, they should not have to *"...jump through a bunch of hoops."* Well Tom, as it IS indeed the 21st century now, should we not have learned from the abundant mistakes of the 20th century and understand the exact implications of what happens when industries are unregulated?

The Falsification of Science

Thousands of studies link low-level wireless radio frequency radiation exposures to a long list of adverse biological effects, including...

- DNA single and double strand breaks

- Oxidative damage

- Disruption of cell metabolism

- Increased blood brain barrier permeability

- Melatonin reduction

- Disruption to brain glucose metabolism

- Generation of stress proteins

Let us also not forget that in 2011 the World Health Organisation (WHO) classified radio frequency radiation as a possible 2B carcinogen and more recently the $25 million 'National Toxicology Program' concluded that radio frequency radiation of the type currently used by mobile (cell) phones can cause cancer.

But where does 5G fit into all this? Given that 5G is set to utilise frequencies above existing frequency bands, 5G sits in the middle of all this. But the tendency (it varies from country to country) is for 5G to utilise the higher frequency bands, which brings its own particular concerns, and the following paragraphs summarise exactly how this will adversely affect all our health...

We are going to be bombarded by really high frequencies at low, short-range intensities creating a yet more complicated denser soup of 'electro smog.' To work with the higher range MMW in 5G, the antennas required are smaller. Some experts are talking about as small as 3mm by 3mm. The low intensity is for efficiency and to deal with signal disruption from natural and manmade obstacles.

One of the biggest concerns regarding 5G technology is how these new wavelengths will affect the skin. The human body has between two million to four million sweat glands. Dr Ben-Ishai of Hebrew University, Israel explains that our sweat ducts act like "... *an array of helical antennas when exposed to these wavelengths,*" meaning that we become more conductive. A recent New York study, which experimented with 60GHz waves stated that, "... *the analyses of penetration depth show that more than 90% of the transmitted power is absorbed in the epidermis and dermis layer.*"

The effects of MMWs as studied by Dr Yael Stein, also of Hebrew University is said to also cause humans physical pain as our nociceptors flare up in recognition of the wave as a damaging stimuli. So, we are looking at the possibilities of many skin diseases and cancer as well as physical pain to our skin.

Interestingly, these many new and ubiquitous fungal skin infections enable so-called 'smart dust,' nano-metallic particles, particularly manganese, to adhere to the skin and act as antennae, further exacerbating the problem. Also, significantly the huge, relatively recent, proliferation of body tattoos, especially in the younger generation, but now very common in almost all social groups, further facilitates the harmful effects of 5G upon the skin, as tattoo ink contains high quantities of several heavy metals. And perhaps significantly there is now a patent for a new 'smart tattoo,' by Nokia, which will be coming to a tattoo parlour near you very soon, if it is not there already. This is a tattoo, which can, for example, vibrate when you receive a text on your cellphone! So, all these people, who actually pay good money for this dubious privilege, and no doubt there will be many, are likely to be even more adversely affected by 5G MMWs, than we non-tattooed folks.

A 1994 study found that low-level millimetre wave radiation produced eye lens opacity in rats, which is linked to the production of cataracts and an experiment conducted by the Medical Research Institute of Kanazawa Medical University found that 60GHz *"millimeter-wave antennas can cause thermal injuries of varying types of levels. The thermal effects induced by millimetre-waves can apparently penetrate below the surface of the eye."*

A 2003 Chinese study has also found damage to the lens epithelial cells of rabbits after eight hours of exposure to microwave radiation and a 2009 study conducted by the College of Physicians and Surgeons in Pakistan concluded that EMFs emitted by a mobile phone cause derangement of chicken embryo retinal differentiation.

A 1992 Russian study found that frequencies in the range 53-78GHz (that which 5G proposes to use) impacted the heart rate variability (an indicator of stress) in rats and another Russian study on frogs whose skin was exposed to MMWs found heart rate changes (arrhythmias).

A 2002 Russian study examined the effects of 42HGz microwave radiation exposure on the blood of healthy mice. It was concluded that *"the whole-body exposure of healthy mice to low-intensity EHF EMR has a profound effect on the indices of nonspecific immunity."*

More recently, a 2016 Armenian study observed MMWs at low intensity, mirroring the future environment brought about by 5G. Their study conducted on E-coli and other bacteria stated that the waves had depressed their growth as well as *"changing properties and activity"* of the cells. The obvious concern is that it would do the same to human cells.

The very same Armenian study also suggested that MMW's effects are mainly on water, cell plasma membrane and genome too. They had found that MMW's interaction with bacteria altered their sensitivity to "... *different biologically active chemicals, including antibiotics.*" More specifically, the combination of MMW and antibiotics showed that it may be leading to antibiotic resistance in bacteria. This groundbreaking finding could have a magnum effect on the health of human beings as the bandwidth is rolled out worldwide. The concern is that we develop a lower resistance to bacteria as our cells become more vulnerable – and therefore we become more vulnerable.

One of the features of 5G is that the MMW is particularly susceptible to being absorbed by plants and rain. Humans and animals alike consume plants as a food source and therefore the effects of MMW on plants could leave us with food that is unsafe for consumption. The water that falls from the sky onto these plants will also be irradiated. A 2010 study on aspen seedlings showed that the exposure to radio frequencies led to the leaves showing necrosis symptoms.

Of course, there is already a problem in this regard with foods frequently being sold in shops in close proximity to Wi-Fi and cell towers. Not forgetting too that food is also subject to radiation exposure to a greater or lesser degree in transit, often intentionally with 'irradiation' in order to prolong shelf life or whatever nonsense they wish to convey to us. And, especially in the US in the more remote areas, and to a lesser extent in other countries too, many water tanks now double-up as sites for multiple cell tower arrays, meaning that both our food *and* water is already contaminated and compromised. However, be that as it may, exposure to the far more egregious 5G radiation would worsen this problem exponentially, especially when considering the planned global coverage, and also that it will be used in conjunction with existing 4G proven-to-damage-health microwave technology, but brought down to street level, closer to our homes and communities.

Another Armenian study found that MMWs of low intensity, "... *invoke(s) peroxidase isoenzyme spectrum changes of wheat shoots.*" Peroxidase is a stress protein existing in plants. Indications are that 5G will be particularly harmful to plants – perhaps even more so than to humans.

The Falsification of Science

Since the year 2000, there have been reports of birds abandoning their nests as well as health issues such as *"plumage deterioration, locomotion problems, reduced survivorship and death,"* says researcher Alfonso Balmori. Bird species that are especially affected by this low level, non-ionising microwave radiation are House Sparrows, Rock Doves, White Storks, Collared Doves and Magpies, among many others. But it is not just birds. The declining bee population is also said to be linked to this non-ionising EMF radiation. It reduces the egg-laying abilities of the queen leading to a decline in colony strength and overall bee population numbers, which has an obvious knock-on effect on plant pollination, and thus affecting the entire ecosystem.

A study conducted by Chennai's Loyola College in 2012 concluded that out of 919 research studies carried out on birds, plants, bees and other animals and humans, 593 of them showed impacts from RF-EMF radiations. 5G will be greatly adding to the effects of this electro smog. It will also use pulsed millimeter waves to carry information, but as Dr Joel Moskowitz pointed-out, most 5G studies are misleading because they do not pulse the waves. This is important because research on microwaves already tells us how pulsed waves have more profound biological effects on our body, compared to non-pulsed waves. Previous studies, for instance, show how pulse rates of the frequencies led to gene toxicity and DNA strand breaks.

It may seem difficult to believe, but the 5G frequencies, which will soon be blanketing our towns and cities and indeed the entire countryside, have been under development for a number of years by defence agencies, including the US, Russian and Chinese, as a means of crowd control. The US army uses millimetre waves in crowd dispersal guns called 'Active Denial Systems.' Dr Ben-Ishai explained this as, " ... *if you are unlucky enough to be standing there when it hits you, you will feel like your body is on fire."*

Of course, these devices are claimed to be 'safe' by the authorities because they 'only' make people *feel* like they are on fire, so they (quite understandably) tend to disperse quickly as a result. However, they also admitted that in some 'rare' cases, some people developed boils on their skins and even, yes, burns! But please do not worry, it is still all perfectly 'safe,' we are emphatically told by the industry!! This must be some strange, new definition of the word 'safe,' of which I was not previously aware. To be scrupulously fair though, they are probably 'safe' in much the same way that tasers are described as 'safe,' yet provably cause hundreds of deaths every year.

AT&T have announced the availability of their 5G Evolution in Austin, Texas, USA. 5G Evolution allows Samsung S8 and S8+ users access to faster speeds. This is part of AT&T's plan to lay the 5G foundation while the standards are still being finalised. AT&T is also considering 19 other metropolitan areas such as Chicago, Los Angeles, Boston, Atlanta, San Francisco etc.

Charter, the second largest cable operator in the US, has been approved for an experimental 28 GHz licence in Los Angeles. The outdoor tests will use fixed transmitters with a 1 km or smaller effective radius.

Qualcomm has already demonstrated a 5G antenna system with about 27 decibel gain. According to ABI Research, this is *"about 10 to 12 more decibels than a typical cellular base station antenna."* Not a good sign.

Many more private sector companies such as HTC, Oracle, Sprint, T-Mobile are playing a role in the developing of testing platforms by contributing time, knowledge or money.

In the UK, the 3.4GHz band has been earmarked for 5G use with contracts awarded to O2, Vodaphone, EE and Three. While the 2.3GHz band, awarded to O2, is likely to be used for 5G too in time.

Research and pre-testing is rampant by companies who are interested to tap into the lucrative waters of 5G. But few are willing to research its effects on health. The International Commission on Non-Ionizing Radiation Protection (ICNIRP) guidelines remain essentially unchanged since 1998, not allowing for the recognition of radio frequency microwave radiation and MMWs as harmful unless there is a heating effect. But a few experts are now beginning to speak out against the wholesale deployment of 5G technology...

Dariusz Leszczynski from the University of Helsinki and also a former member of the International Agency Research on Cancer is one of them. He has brought to attention the ICNIRP intention to classify skin as limbs, even though skin is actually our largest *organ* and clearly not an appendage. Limbs are paid lesser attention when classifying exposure levels and yet research indicates that MMWs affect the skin and the eyes most of all. If skin is classified as a limb, this will pave the way for industry giants to introduce even higher exposures and put even more people at risk. Typical underhand tactics of any major industry.

The Global Union Against Radiation Deployment from Space (GUARD) addressed a letter to the FCC in September of 2016, bringing to their attention the harm 5G will inflict. GUARD warned the FCC that 5G violates Article 3 of The UN Declaration of Human Rights which states that "*...everyone has the right to life, liberty and security of person.*" The document is laden with research, information and global support.

Here is what some other experts on the topic have to say about 5G...

"*There is an urgent need to evaluate 5G health effects now before millions are exposed.... We need to know if 5G increases the risk of skin diseases such as melanoma or other skin cancers.*" Ron Melnick, a former National Institute of Health scientist, now retired.

"*Along with the 5G there is another thing coming—Internet of Things. If you look at it combined, the radiation level is going to increase tremendously and yet the industry is very excited about it...they project 5G/IoT business to be a $7 trillion business.*" Professor Girish Kumar, the Electrical Engineering Department at IIT Bombay

"*The new 5G wireless technology involves millimetre waves (extremely high frequencies) producing photons of much greater energy than even 4G and Wi-Fi. Allowing this technology to be used without proving its safety is reckless in the extreme, as the millimetre waves are known to have a profound effect on all parts of the human body."* Professor Trevor Marshall, Director Autoimmunity Research Foundation, California

"The plans to beam highly penetrative 5G milli-wave radiation at us from space must surely be one of the greatest follies ever conceived of by mankind. There will be nowhere safe to live." Olga Sheean former WHO employee and author of 'No Safe Place'

"It would irradiate everyone, including the most vulnerable to harm from radiofrequency radiation…pregnant women, unborn children, young children, teenagers, men of reproductive age, the elderly, the disabled, and the chronically ill." Ronald Powell, PhD

So, in order to protect yourself from 5G radiation…

- Understand the different types of EMFs and how they behave – hence the need to read and share information as much as possible.

- Use an EMF meter to obtain readings and identify hotspots.

- Mitigate your exposure. Which means either eliminate the source, move further away from the source of radiation or shield your body.

This document was submitted to the UN Human Rights Council on 11th February 2019 by the Planetary Association for Clean Energy in Geneva (UNOG,) Vienna (UNOV) and New York (UNHQ)…

"In 1954, the tobacco industry founded the precursor to what is known today as the Council for Tobacco Research. This organization financed hundreds of so-called independent researchers, who published several thousand peer-reviewed studies the goal of which, as we now know, was to create controversy and doubt about a causal link between smoking and a wide array of grave illnesses.

They used arguments which claimed to be 'scientific' although industry insiders knew as early as 1950 that their product was dangerous. In 1969, an internal note from a subsidiary of a leading tobacco firm stated, 'Doubt is our product.'

The Falsification of Science

These techniques are still extensively used today by telecommunications companies. The parallels with the tobacco industry are striking although the tactics subsequently improved with relentless lobbying.

However, this is where the comparison between the tobacco and telecommunications industries stops. Electromagnetic radiation (EMR) has no smell, and you cannot see it. It is everywhere, you cannot escape it, thus the consequences of biased science combined with the impalpable nature of EMR are far more insidious and far-reaching.

Dr Richard Horton, Editor of The Lancet, after a symposium held in April 2015 on the Reproducibility and Reliability of Medical Research, wrote as follows... 'A lot of what is published is incorrect. The case against science is straightforward: much of the scientific literature, perhaps half, may simply be untrue. Afflicted by studies with small sample sizes, tiny effects, invalid exploratory analyses, and flagrant conflicts of interest, science has taken a turn towards darkness.'

Among others, Professor Emeritus Henry Lai, a leading bioengineer at the University of Washington who produced groundbreaking work on the effects of low-level radiation on DNA, faced full-scale efforts to discredit his work when he published it in 1995.

In an internal company memo leaked to a scientific publication, Motorola described its plan to 'war-game' and undermine his research.

After accepting industry funding for continued research from the Wireless Technology Research (WTR) programme, Professor Lai wrote an open letter to Microwave News questioning restrictions placed on his research by the funders. The head of WTR then asked University of Washington president Richard McCormick to fire Professor Lai, which he refused to do.

Professor Lai says that without government funding, most scientific research is funded by private industry and 'you don't bite the hand that feeds you. The pressure is very impressive.'

In 2006, faced with contradictory research, Professor Lai did an analysis of 326 studies on cell phone radiation conducted between 1990 and 2006, and where their funding came from. He found that 56 per cent of the 326 studies showed a biological effect from radio-frequency radiation and 44 per cent did not. But when he looked at their funding, he discovered that 73 per cent of independently funded studies found an effect, as opposed to only 27 per cent of industry-funded studies.

Despite what is being portrayed in the mainstream, wireless radiation has biological effects, and this is not a subject for debate. This was already established more than 60 years ago when the US Department of Defense tested the impact of EMR on animals and human beings under a variety of conditions. These biological effects are seen in all life forms – plants, animals, insects and microbes.

There are more than 10,000 peer-reviewed studies pertaining to the health impacts of EMR and substantial evidence for the cumulative nature and eventual irreversibility of some effects, whether neurological/ neuropsychiatric, reproductive, cardiac, mutations in DNA, or hormonal effects. Some may affect the evolution of the human race.

In humans, there is clear evidence that EMR is causing not only cancer but a wide array of debilitating ailments including cognitive impairment, learning and memory deficits, neurological damage, miscarriage, impaired sperm function and quality, obesity, diabetes, tinnitus, impacts on general well-being, alteration of heart rhythm, and cardiovascular diseases. At the cellular level EMR causes alterations in metabolism and stem cell development, gene and protein expression, increased free radicals, oxidative stress and DNA damage.

Effects in children are important and include some of the above plus autism, attention deficit hyperactivity disorder (ADHD) and asthma. EMR has immediate effects on certain aspects of biology. These may be expressed faster in people already suffering from electro-sensitivity (ES) and electro-hypersensitivity (EHS). Although these are not medical terms, they refer to up to 13% of people globally who have happened to discover what is making them sick in spite of the disbelief of others. Many such people cannot work, are homeless, or have committed suicide because they had nowhere to hide from the radiation.

The impact of wireless telecommunication technologies on humans and their environment was never tested before each and every new generation

The Falsification of Science

was globally deployed. Average adults and their children have been used as experimental guinea pigs without ever being informed or asked for their consent. On the contrary, the public has been actively misled.

Economic interests have prevailed over the precautionary principle and precautionary approaches.

There is no opt-out. With the advent of 5G, everyone is indiscriminately irradiated in ever-increasing doses.

Those responsible for keeping this industry in check, including the World Health Organisation, US Federal Communications Commission and other national and international bodies have not ever been forthcoming about the dangers of radio-frequency radiation. Instead they have protected the industry's interests, with total disregard of known health impacts.

Working groups focused on health impacts of EMR at the International Commission on Non-Ionising Radiation Protection (ICNIRP,) the Scientific Committee on Emerging and Newly Identified Health Risks, the Institute of Electrical and Electronics Engineers, the International Electrotechnical Commission and the International Telecommunication Union, for example, are notoriously plagued by conflicts of interests and/or directly working with the industry.

Despite the unequivocal consequences, the media are still actively misleading the public. All the elements of a scientific experiment gone wrong are present, along with a profit and liability motive for a coverup. Economic interests now worth over $3.4 trillion US in assets have prevailed over public health.

To deploy 5G, not only will the density of antennas be increased by at least a factor of 5 on average, but the current ICNIRP radiation limits will have to be increased by 30 to 40% in order to make its deployment technologically feasible. This won't be enough to ensure total 5G coverage, so thousands of low earth orbit (LEO) satellites will beam the signal from above. This implies radiation impacts not only on our health but also on the earth's atmosphere. LEOs will be emitting modulated signals at millions of watts of effective power straight into the atmosphere, whose nature is inherently electrical.

5G networks will exist alongside previous generations of wireless technology, but unlike them, will pulse millimetre waves from phased-array antennas at levels of EMR tens to hundreds of times greater than those existing today. The idea that the human body can tolerate 5G radiation is based on the faulty assumption that shallow absorption by the skin is harmless.

When an ordinary electromagnetic field enters the body, it causes charges to move and currents to flow. But when extremely short electromagnetic pulses enter the body, the moving charges themselves become little antennas that reradiate the electromagnetic field and send it deeper into the body. They become more significant when either the power or the phase of the waves changes rapidly, and 5G will likely satisfy both criteria.

Shallow penetration of millimetre waves also poses a unique danger to the eyes and skin, as well as to very small creatures. Peer-reviewed studies recently published predict thermal skin burns in humans from 5G and resonant absorption by insects, which absorb much more radiation at millimetre wavelengths than they do at wavelengths presently in use. Since populations of flying insects have declined by 75-80 per cent since 1989, which also coincides with early deployments of cellular networks, 5G radiation could have catastrophic effects worldwide.

PACE believes that 5G, together with previous generations of wireless technology, is an experiment on humanity that constitutes cruel, inhuman and degrading treatment under General Assembly resolution 39/46 of 10 December 1984. The deployment of 5G violates over 15 international agreements, treaties and recommendations, including article 7 of the International Covenant on Civil and Political Rights, which derives from the Nuremberg Code of 1947. It also violates the Declaration of Helsinki of 1964 and its several revisions, as well as other international guidelines that have been translated into national laws in various countries."
Olivier Vuillemin

Pretty scary stuff, you will doubtless agree!?

The Falsification of Science

5G may well instigate a new form of industrial revolution, human connectivity and even a new reality. It offers endless possibilities for the future. But already what is clear, as we have hopefully demonstrated above, is that there are real, tangible and substantial dangers with this technology, which is why it is extremely important to act to protect yourself and your loved ones as much as possible. Please share this crucial information with your friends and family, before 5G takes an unshakeable grip on all of our lives. Indeed, it may well already have done so, with the current (at the time of writing this, December 2020) 'pandemic' crisis, as in every case, including the affected cruise ships, the first and hardest hit 'epicentres' of this 'disease' had populations already exposed to the new 5G technology, and many researchers feel that one of the intended reasons for the invention of this new pandemic myth is to serve as a scapegoat for the anticipated harmful effects of 5G technology.

Overall, please remember that the wireless industry and its many shills and apologists are NOT your friend in any way, and they use deception aplenty as well as downright lies and pseudoscience in order to present a very twisted version of reality for their own financial gain.

Chapter 10

Suppressed Technology

As early as the late 1880s, trade journals in the electrical sciences sphere, were predicting free electricity and free energy in the very near future. Incredible discoveries about the nature of electricity were becoming commonplace and Nikola Tesla was demonstrating 'wireless lighting' and other wonders associated with high frequency currents. There was indeed much excitement about the future, such as had never been seen before.

Within twenty years, there would be automobiles, aeroplanes, motion pictures, recorded music, telephones, radio and hand-held, portable cameras and the Victorian Age was giving way to a truly technologically based future. For the first time in history, the masses were being encouraged to envision a utopian future filled with abundant modern aids, communication, and entertainment devices, as well as plenty of jobs, housing, and food for everyone. It was predicted that all disease would be eradicated as would poverty; life was improving exponentially for the previously dispossessed and everyone was set to benefit from this 'brave new world' of science. So, what happened to shatter this optimistic illusion? Where did all the promises of 'free energy' breakthroughs go? Or was it all simply wishful thinking that science eventually disproved?

Actually, the answer to that question is 'no.' In fact, the opposite is true. Many free energy technologies were developed with all the other breakthroughs and even since that time, multiple methods for producing vast amounts of energy, free or at worst, extremely low cost have been developed. None of these technologies have ever managed to find their way to the bulk of the world's consumers, however. The exact reasons why this is the case, will be covered in finer detail at the end of this chapter, but firstly, here are just a few of the suppressed technologies that are currently in existence and that are proven beyond all reasonable doubt to be effective. The common feature connecting all of these discoveries is that they use a small amount of one form of energy to control or release a large amount of a different kind of energy. Some of them tap into the underlying ether field in some way; a source of energy which is conveniently ignored by 'modern science' and technology. But of course modern science chooses to ignore or attempts to debunk their efficacy, for reasons which will become apparent.

Nikola Tesla

Nikola Tesla was one of the greatest scientific minds the world has ever seen, although, up until the last twenty years or so, he was virtually unknown and forgotten, rather conveniently for some, for sure. Yet I find it HIGHLY suspicious that after almost a century of being virtually ignored, he is suddenly the great unsung hero of our day, right at the point when more wireless technology is being pushed on us and Elon Musk, one of the most evil men alive, has a company named after him!) He was born in Smiljan in what is now Croatia on the 10th July 1856 and immigrated to the USA in 1884, where he eventually became a naturalised citizen.

Albert Einstein was once asked what it felt like to be the smartest man alive and his response was, *"I don't know, you'll have to ask Nikola Tesla."*

He was a prolific inventor and spent most of his time developing and patenting products that he hoped he could bring to market. For example, Tesla conducted a range of experiments with mechanical oscillators/generators, electrical discharge tubes, and early x-ray imaging. He also built a wireless-controlled boat, one of the first ever exhibited. Tesla became well known as an inventor and often demonstrated his achievements to celebrities and wealthy patrons at his laboratory and was noted for his showmanship at public lectures. Throughout the 1890s, Tesla pursued his ideas for wireless lighting and worldwide wireless electric power distribution in his high-voltage, high-frequency power experiments in New York and Colorado Springs and in 1893, he made pronouncements on the possibility of wireless communication with his devices. Tesla tried to put these ideas to practical use in his unfinished Wardenclyffe Tower project, an intercontinental wireless communication and power transmitter, but ran out of funding before he could complete it.

"Today's scientists have substituted mathematics for experiments, and they wander off through equation after equation and eventually build a structure which has no relation to reality." Nikola Tesla

In fact it was JP Morgan and George Westinghouse who withdrew all the financing for Tesla's Wardenclyffe project and Tesla eventually died penniless in a New York hotel room after having been under almost constant supervision and even harassment by the authorities until his death in 1943. Shortly after his death, the FBI classified all Tesla's papers and immediately following WWII, they magically disappeared and now no-one knows what happened to them. Indeed, for many years, scientists and researchers have sought Tesla's missing papers without success. I think it now safe to say however, seventy five years later, that they were probably destroyed by opposing vested interests.

This begs the questions, how could a brilliant inventor like Tesla die penniless, and what motivation did Wall Street bankers have for removing all their funding and keeping him bankrupt until his death?

Maybe it had something to do with the information and technology that Tesla discovered and developed that those in power and with much to lose, did not want the public to know about... such as energy weapons that could be secretly used against the public.

It is definitely the opinion of Shannon and I that not one single thing that Tesla invented has helped humanity. Quite the contrary in fact, it has all done nothing but harm. Certainly, he was a victim in terms of having his inventions and money stolen from him, but we believe that too many people view him as a hero simply because he was a victim in this way. And he was no real friend to the people as is now claimed. His wanting to provide free wireless energy, is tantamount to saying, 'how about giving me some *free* cancer!' Very much like the 'free Wi-Fi' of today which doubtless contributes greatly to the general demise of health we see everywhere. In addition, he wantonly burned down entire forests with his energy weapons without any qualms and doubtless had some idea of how harmful alternating current is, but nevertheless promoted it anyway. He was obsessed with invention simply for the sake of proving just what he could achieve, NOT through some humanitarian or altruistic intent, but more as a kind of 'playing god' type obsession. So we do take some issue with his being presented as some victim/unsung hero in need of our recognition. And from that of course, derives the idea of supporting and implementing all of his inventions, you know, since he was a 'genius'... well, many psychopaths are also geniuses!

Direct Energy Weapons

To get an idea of what these technologies were, let us take a closer look at a telegram printed in *'The New York World'* on 11th July 1935...

"Nikola Tesla revealed that an earthquake which drew police and ambulances to the region of his laboratory at 48 E. Houston St., New York, in 1898, was the result of a little machine he was experimenting with at the time which 'you could put in your overcoat pocket.'"

The bewildered newspapermen pounced upon this as at least one thing they could understand and 'the father of modern electricity' told what had happened as follows...

"I was experimenting with vibrations. I had one of my machines going and I wanted to see if I could get it in tune with the vibration of the building. I put it up notch after notch. There was a peculiar cracking sound. I asked my assistants where the sound came from. They did not know. I put the machine up a few more notches. There was a louder cracking sound. I knew I was approaching the vibration of the steel

building. I pushed the machine a little higher. Suddenly all the heavy machinery in the place was flying around. I grabbed a hammer and broke the machine. The building would have been about our ears in another few minutes. Outside in the street there was pandemonium. The police and ambulances arrived. I told my assistants to say nothing. We told the police it must have been an earthquake. That's all they ever knew about it."

Some shrewd reporter asked Dr Tesla at this point what he would need to destroy the Empire State Building and he replied, *"Vibration will do anything. It would only be necessary to step up the vibrations of the machine to fit the natural vibration of the building and the building would come crashing down. That's why soldiers break step crossing a bridge."*

At a party in 1935, Tesla also claimed that his mechanical oscillator could destroy the Empire State Building with only *"five pounds of air pressure"* if attached on a girder.

"He put his little vibrator in his coat-pocket and went out to hunt a half-erected steel building. Down in the Wall Street district, he found one; ten stories of steel framework without a brick or a stone laid around it. He clamped the vibrator to one of the beams and fussed with the adjustment until he got it. Tesla said finally the structure began to creak and weave and the steel-workers came to the ground panic-stricken, believing that there had been an earthquake. Police were called out. Tesla put the vibrator in his pocket and went away. Ten minutes more and he could have laid the building in the street. And, with the same vibrator he could have dropped the Brooklyn Bridge into the East River in less than an hour."

So, Tesla had developed a pocket sized machine capable of bringing buildings down...Especially steel framed ones and with a weird crackling sound. It is interesting also to note that many eyewitnesses in New York on 9/11 reported a strange crackling sound before each building fell.

(For an example of how vibrational resonance can bring a structure down, I would ask you to 'Google™' it and watch the fascinating

The Falsification of Science

video of the Tacoma Narrows Bridge collapse in 1940 after gale-force winds resonated precisely with the bridge.)

Was, in fact, a direct energy weapon possibly the technology that was used to bring the twin towers and WTC7, down? Many people are unaware of this, but a third tower also 'collapsed' – due to minor fires, or so we are led to believe. In fact the 'collapse' of WTC building 7 looked suspiciously similar to many other controlled demolitions of high-rise buildings.

Strangely, or maybe not(?) the BBC 'on the spot' reporter, Jane Standley reported the collapse of WTC7, live on air at around midday EST and 5.00pm BST, but this was approximately twenty minutes before the event and WTC7 could actually be seen in the background, completely intact. Of course that report was immediately 'pulled' once the truth was realised, never to be seen again!

Most people who have researched the events of 9/11 in any depth will have heard of 'thermite,' which is used to cut metal beams. This was most certainly used to weaken the structure of the towers, as evidenced by the perfect diagonal cuts on the towers' steel beams (such as the one pictured below.)

Then of course, there is also the tell-tale molten iron that was seen running-out of the tower debris after the many small but deadly blasts and also the molten iron at 'Ground Zero' that was still hot and steaming several months after 9/11.

And it should come as no surprise to anyone that Marvin Bush (George W. Bush's younger brother) was a principal in a company called Securacom that provided the security for the World Trade Center complex, United Airlines, and Dulles International Airport. It is also a fact that from 1999 to January 2002 (Marvin and George W.'s cousin) Wirt Walker III was the CEO of the company. Oh, what coincidences.

There were also well-documented power outages and entire swathes of floor shutdowns and evacuations in the weeks leading up to 9/11, under the guise of updating security systems and computer network cables, all of which offered perfect opportunities to wire the buildings for controlled demolition-type explosives. This of course served to fuel what were well-placed suspicions of subterfuge.

After the dust had settled from the event, literally as well as figuratively speaking, the fact that hundreds of emergency personnel developed similar rare forms of cancer at unnatural rates also suggests that some form of covert technology was used on the towers. They also must have been exposed to extremely toxic, burning chemicals considering that the building itself and its contents would necessarily contain things that would be highly toxic if burned. This is further evidenced by the fact that entire steel structures simply disintegrated into billions of tiny Nano-particles of dust – all of which could and would have been inhaled by the largely unprotected rescue workers.

There is little doubt in the minds of anyone who has even a semi-functioning brain, that advanced weaponry of some kind and explosives were used on these three towers. The information to prove this is ubiquitous, and many others, including scientists, architects, doctors of all disciplines, airline pilots, and highly qualified and experienced construction engineers have already covered most of the issues in far greater technical detail than I ever could, so please allow me to bypass any further critique of the events of 9/11 and return to the subject of the suppressed technology of the weapons that were used on that sad day in history. I (and many others) believe that it is very possible that the US government used the particle beam weapon that Tesla developed, in order to bring down the towers.

The Falsification of Science

Only seven miles from the location where Tesla's Wardenclyffe tower stood, in the same neighbourhood on Long Island, is the Brookhaven National Laboratory. This laboratory was formally established in 1947 at the site of Camp Upton, a former US Army base and in 2013 alone, its reported budget was $700 million of US taxpayer money.

However, that figure is probably much, much higher when black project budgets are factored-in. So, it really is quite an important facility.

As for its use, although originally conceived as a nuclear research facility, its mission has greatly expanded, and it now focuses upon...

- Nuclear and high-energy physics

- Physics and chemistry of materials

- Environmental and energy research

- Non-proliferation

- Neurosciences and medical imaging

- Structural biology

And the major facilities at Brookhaven labs include...

The Relativistic Heavy Ion Collider (RHIC) which was designed to research quark–gluon plasma. Until 2009 it was the world's most powerful heavy ion collider and is the only collider of spin-polarised protons.

The Centre for Functional Nanomaterials (CFN) used for the study of nanoscale materials.

The National Synchrotron Light Source (NSLS)

The Alternating Gradient Synchrotron, a particle accelerator that was used in three of the laboratory's Nobel prizes.

An Accelerator Test Facility, which generates, accelerates, and monitors particle beams.

Tandem Van de Graaff, which was once the world's largest electrostatic accelerator.

The New York Blue Gene supercomputer is an 18 rack Blue Gene/L and a 2 rack Blue Gene/P massively parallel supercomputer that involves a cooperative effort between Brookhaven National Laboratory and Stony Brook University. It is the world's 5th fastest supercomputer and the world's 2nd most powerful for open access research. So in a nutshell, it contains some extremely expensive and serious technology, with a huge focus on nano-materials, particle accelerators, and particle beams.

'But what has all this to do with 9/11?' I hear you ask.

Have you ever seen the photographs and videos of the many hundreds of burned-out and melted cars near the twin towers on 9/11? Here are some pictures to remind you of the scene of utter devastation...

The Falsification of Science

Have you never wondered how they all burned and melted in such a uniform fashion when there were no fires whatsoever, reported at street level?

And interestingly, almost all the incinerated cars were on the same two streets: Vesey Street and Church Street, shown in red here...

In order to discover what really happened to these incinerated cars, let us begin by drawing a straight line from the middle of Ground Zero through the point where the two red lines on Vesey Street and Church Street meet.

And then, zooming out, let us see where that takes us...

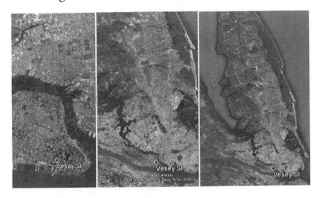

...and then zooming in at the other end of the line, we see that the line from Ground Zero perfectly bisects the Brookhaven particle accelerator facility.

Furthermore, the line passes directly over the two buildings on either end of the ring. So, Brookhaven – built in 1947, is *perfectly* aligned with Ground Zero, 61.16 miles away. What are the odds of that?

This distance between Ground Zero and the far end of the Brookhaven ring – 61.16 miles, is also a 'mirror number,' and encodes 9/11 both upside down and backwards (i.e. 6 is 9.)

If you are still sceptical about all these coincidences, you will no doubt be astonished to see that the twin towers themselves even encoded the date of their destruction in their degree heading of 119.00. Another interesting point to note is that the laboratory has its own Zip (postal) code, which is 11973. As you may have noticed, this Zip code also encodes 9/11 (119) and similarly it also encodes the year that the twin towers were first opened (1973.)

Curiously, the patent for Tesla's 'magnifying apparatus for transmitting electrical energy,' which covered the basic function of the earlier yet similar device used at Wardenclyffe is U.S. Patent 1119732, which also encodes 9/11 (119) and 1973. You may think that this is another astounding coincidence, but personally speaking and as I have stated many times, I do not believe in coincidences.

As you may probably imagine, an absolutely enormous amount of energy would have been required in order to turn thousands of tonnes of steel beams to dust within a mere matter of seconds, but this is precisely what did happen on the morning of 9/11.

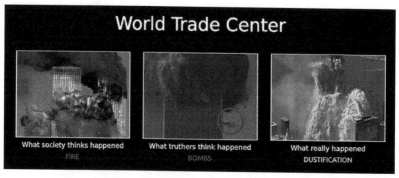

If you are not familiar with the concept of 'grounding' or 'earthing' in electrical engineering, ground or earth is the reference point in an

The Falsification of Science

electrical circuit from which voltages are measured, a common return path for electric current, or a direct physical connection to the Earth. In essence, with any electrical event, there has to be an earthing, so that the electrical energy can follow a controlled path of least resistance to earth, or whatever it is that is the 'ground.'

As you can see in the above images of the cars on Church Street and Vesey Street, those two streets and the cars themselves apparently acted as the grounding for the energy beam on 9/11, either side of the beam itself.

It's also worth noting that only the metal parts of the cars were melted, which makes sense considering that in cars only the metal can conduct electricity. Dr Judy Wood did a remarkable job of outlining this energy weapon and I strongly recommend that you research this for yourself. One particular facet of her research that is particularly astonishing is the behaviour of Hurricane Erin – the storm near the US north-eastern coast on the days around 9/11.

This storm began moving in a straight line towards New York about a week before 9/11, gathering more and more energy (i.e. electrical charge) in the days leading up to 9/11, before suddenly weakening and turning back out towards the Atlantic Ocean immediately after the events of 9/11.

It was almost as though the storm had been 'pulled-In' towards a high energy source and gathered electrical energy, before losing energy once that high-powered energy source released all its power on the twin towers. Or maybe this was some kind or HAARP-induced activity to facilitate the energy beam in some way? Another interesting point to note is that the direction that the smoke from WTC7 and the towers themselves, and also the wind, shifted 180 degrees (attracted directly towards Brookhaven as opposed to the Atlantic Ocean).

From this we may conclude two things:

- A phenomenally huge amount of energy was produced and used on 9/11

- The hidden powers possess technology that can create and move storms

As we have already seen in chapter 9, geoengineering, weather, and geological phenomena manipulation, is very real indeed.

Wireless Electricity

Nikola Tesla, whose huge contribution to subsequently suppressed technologies cannot be underestimated, was the undisputed 'father' of many of the inventions that define the modern electronic era and was the first person to demonstrate this principle, in 1890.

Tesla based his wireless electricity idea on a concept known as electromagnetic induction, which was discovered by Michael Faraday in 1831 and purports that electric current flowing through one wire can induce current to flow in another wire, nearby. To illustrate that principle, Tesla built two huge 'world power' towers that would broadcast current into the air, to be received remotely by electrical devices around the world.

Few believed it could work and to be fair to the doubters, it did not work that well. When Tesla first switched on his 200 foot-tall, 1,000,000-volt Colorado Springs tower, 130-foot-long bolts of electricity shot out of it, sparks played around the toes of the observers and the grass around the laboratory glowed blue. Despite this initial failure, there is little doubt that Tesla eventually succeeded in his quest to produce wireless electricity, but it has remained commercially unused and dormant for more than one hundred years, largely due to the energy cartel's suppression of the technology to protect their own profit-streams.

However, after a gap of greater than a century, several companies are now coming to market with technologies that can safely (allegedly) transmit power through the air, a breakthrough that portends the literal 'untethering' of our abundant electronic devices. Until this development, the phrase 'mobile electronics' has been somewhat of a deception. How 'portable' is your laptop if it has to feed every four hours, like an embryo, through a cord? How 'mobile' is your phone if it shuts down after too long away from a power-source?

The technology about to arrive upon the electronics market is an inductive device, much like the one Tesla designed, but much smaller. It looks like a mouse pad and can send power through the air, over a distance of up to a few inches. A powered coil inside that pad creates a magnetic field, which as Faraday predicted, induces current to flow through a small secondary coil that's built into any portable device, such as a flashlight, a phone, or a portable computer. The electrical current that then flows in that secondary coil charges the device's integral rechargeable battery.

However, please do not misunderstand my motives here. I do not in any way regard this device (or indeed wireless electricity in general) as a positive development. Yes, purely from a 'convenience' point of view it has its merits, but the downsides far outweigh the positives in my view. Just imagine exactly what a worldwide proliferation of wireless electricity will do to the human body and especially to those among us such as your editor and contributor, Shannon who are extremely sensitive to any form of EMF radiation.

Radiant Energy

Nikola Tesla's magnifying transmitter, T. Henry Moray's radiant energy device, Edwin Gray's EMA motor and Paul Baumann's Testatika machine all run on radiant energy. This natural energy form can be gathered directly from the environment via 'static' electricity or extracted from standard electricity by a method known as fractionation. Radiant energy can perform exactly the same functions as ordinary electricity, at less than 1% of the cost. It does not behave exactly like electricity, however, which to be fair, has contributed to the scientific community's misunderstanding of it. However, the Methernitha Community in Switzerland currently has 5 or 6 working models of fuel-less, self-running devices that tap this energy.

Motionless Electromagnetic Generator

Dr Robert Adams of New Zealand has developed several electric motors, generators and heaters that run via the use of 'permanent magnets.' One such device extracts 100 watts of electricity from the source, generates 100 watts to recharge the source and in addition produces over 140 BTUs of heat in two minutes. An American scientist, Dr Tom Bearden has constructed two working models of a permanent magnet-powered electrical transformer which use a 6 watt electrical input to control the path of a magnetic field coming out of a permanent magnet and then by channelling this magnetic field, first to one output coil and then to a second output coil and by repeating this rapidly, the device can produce a 96 watt electrical output with no moving parts. Bearden calls this device a Motionless Electromagnetic Generator, or MEG. Jean-Louis Naudin has also duplicated Bearden's device in France based on principles which were first discovered by Frank Richardson in the US in 1978. Troy Reed, also in the US, has developed working models of a special magnetised fan that heats-up as it spins. It takes exactly the same amount of energy to spin the fan whether it is generating heat or not and in addition to these developments, many inventors have identified working mechanisms that produce motor torque from permanent magnets alone.

Mechanical Heater

There are two classes of machines that are able to transform a small amount of mechanical energy into a large amount of heat. The best of these purely mechanical designs are the rotating cylinder systems designed by Frenette in the US and Perkins, also in the US. Within these machines, one cylinder is rotated within another cylinder with about one eighth of an inch of clearance between them. The space between the cylinders is filled with a liquid such as water or oil, and it is this fluid that heats-up as the inner cylinder spins. Another method uses magnets mounted on a wheel to produce large currents in a plate of aluminium, causing the aluminium to heat rapidly. These magnetic heaters have been demonstrated by several scientists from various countries. All of these systems can produce ten times more heat than standard methods using the same energy input.

Water Powered Car

Water can be broken down into its constituent parts of hydrogen and oxygen using electricity. Mainstream science, however, claims that this process requires more energy than can be recovered when the gases are recombined. However, this is true only in the worst case scenario. When water is bombarded with its own molecular resonant frequency, using a system developed by Stan Meyers, and also by others, it collapses into hydrogen and oxygen gas with very little electrical input. Also, adding electrolytes (additives that make the water conduct electricity better) changes the efficiency of the process dramatically. It is also known that certain geometric structures and surface textures are more effective than others. The implication of this is that unlimited amounts of hydrogen fuel can be made to drive engines e.g. in cars for simply the cost of water. Even more amazing is the fact that a special metal alloy was patented by Freedman in 1957 that spontaneously breaks water into hydrogen and oxygen with no outside electrical input and without causing any chemical changes in the metal itself. This, in effect, means that this special metal alloy can make hydrogen from water free of cost, forever.

Stan Meyers, the American inventor of a working, patented vehicle that ran only on normal, household water, was murdered by poisoning in the late 1990s after successfully demonstrating his prototype which was capable of 100 mph+ and 100 miles on a single gallon of water. He had long been the subject of harassment and threats by Elite 'Big Oil' interests and his death was perhaps the not too surprising culmination of his refusal to cease the project.

Meyer and his brother were invited by a so-called 'potential manufacturer' to discuss funding the bringing of his amazing invention to the market. Over dinner in a restaurant with the 'two investors' and his brother, Meyer was suddenly seized by an incredibly violent pain in his stomach after drinking from his water glass and in panic dashed out into the street where he collapsed and quickly died in extreme agony. After following him outside, his distraught brother could only watch on in shock as the two perpetrators, calmly left the restaurant and walked past Stan's newly deceased body without either saying a word – or even a backward glance.

So what is happening today regarding Meyers' great invention? Absolutely nothing. The patent still exists and is available to view on the Internet, therefore the technology, or at least the wherewithal to recreate the technology, still exists, so why is this invention not being manufactured today? Simply because it would solve all the world's energy problems instantly and almost literally overnight, eliminate oil as an essential fuel and as we now know. That is not how politics and commerce works. It is always money and vested interest that rules, and common sense and the public good, are never the determining factors.

Implosion Engine

All current, major industrial engines use the release of heat to cause expansion and pressure to produce energy, as in a standard internal combustion engine. Nature uses the opposite process of cooling to cause suction and vacuum to produce energy, as in a tornado. Viktor Schauberger of Austria was the first to build working models of implosion engines in the 1930s and 1940s. Since that time, Callum Coats has written extensively on Schauberger's work in his book '*Living Energies*' and subsequently, a number of researchers have built working models of implosion turbine engines. These are fuel-less engines that produce mechanical energy directly from a vacuum. There are also much simpler designs in existence that use vortex motions to tap centrifugal force to produce a continuous, perpetual motion in fluids.

Schauberger's implosion turbines function according to the spiralling rhythmic motions of nature (biomimicry). The vortex inside is like a pump that is expanding and contracting by compressing the scalar wave-fields of the atomic elements centripetally as they pass through them and simultaneously expanding them centrifugally as they exit. These motions produce energy from water or air without chemically (explosively) converting either into inferior products called toxic waste or air pollution. The air and water that pass through the implosion turbines undergo a transformation of size from their normal wave-fields to extremely compressed ones at the turbine's vortex centre (the fulcrum point of the organic vacuum) and back again to normal size as they exit the turbine. This sets up an equal, but unbalanced rhythmic interchange of perpetual generation and regeneration, simulating the processes which (allegedly) power our universe and give it it's form.

The temperature differences due to internal friction (the outer casing) in these implosion turbines and the much cooler vortex centre, sets up more imbalance and the molecular sorting by weight due to centrifugal and centripetal motions adds more still, so that a rhythmical imbalance is continuously seeking to balance itself by sucking-in more air or water to relieve it's excited condition of imbalance. There is also a process of atomic transformation which takes place as the atmospheric and aqueous elements interact with each other in this artificial environment of extreme speed and compression. This is the most energetic process at work and the most difficult to account for according to contemporary academic 'thought,' – if we can legitimately call it that? This process produces enormous amounts of biomagnetic levity (outward push) which streams upward in the opposite direction of the imploding vortices.

The sucking action of an implosion turbine is far more powerful than the explosive action of an internal combustion engine of the same scale. *"The enormous power of suction or the forces of implosion which, according to the research of Professor Felix Ehrenhaft, who helped Viktor Schauberger periodically, are 127 times more powerful than explosive forces."* Living Energies, Page 275

There is no resistance to the implosion turbine's velocity as its speed increases from the internal imbalances which are greatly amplified by its vortex-producing architecture. The now-doomed internal combustion engine is met with a resistance to its velocity which is squared with the increase in velocity and therefore is a failed system as a result of its backward design, which steals from nature to push against her, while she squares her resistance, pushing back, the faster we accelerate our doomed, fuel-exploding contraptions.

The rotational velocity of the longitudinal axis of the implosion turbine can be braked by a generator to produce electricity for conventional power distribution with far greater efficiency than our present wasteful system of exploding fuel to obtain a fraction of its potential energy in the form of usable power. The levity of the flowing biomagnetism created by the implosion turbine can likewise be used to counter gravity's miniscule inward pull and perform a staggering amount of work for mankind without creating any of the various forms of noxious pollution produced by current methods. At the same time, an implosion turbine requires no combustible, 'scarce' and highly priced fuel as an energy source because it uses its medium for the catalyst to produce energy, whether the turbine operates in air or water.

Unlike internal combustion systems which produce less energy than the fuel they consume, implosion turbines require no fuel and therefore are free energy machines by nature. They will be the technological salvation of humanity and the world as a whole, once they are declassified by the NSA, CIA, AEC and the US Department of Defense and released to the public.

Just imagine what we could do with all of the money we are currently wasting on fuel, every time we fill-up our fuel tanks with petrol/gasoline/diesel sold to us at extortionate prices by the Big Oil cartel and their group of energy cohorts, who are currently sucking the life out of this planet, so that they may profit exponentially from this corporate, rigged game.

Cold Fusion

In March 1989, two chemists from the University of Utah announced that they had produced atomic-fusion reactions in a simple tabletop device. The claims were forcefully 'debunked' within six months and

The Falsification of Science

the public subsequently lost interest. Nevertheless, cold fusion is a very real phenomenon and not only has excess heat production been repeatedly documented, but also low-energy atomic element transmutation has been catalogued, involving many different reactions. This technology definitely could produce low-cost energy and prove beneficial in many, many other important industrial processes but of course, we cannot have that now, can we?

In the late 1970s, Rory Johnson, a brilliant inventor in Elgin, Illinois, created a cold fusion, laser-activated, magnetic motor that produced 525 horsepower, weighed 475 pounds, and would propel a large truck or bus 100,000 miles on two pounds (in weight) of deuterium and gallium. Johnson entered negotiations with the *Greyhound Bus Company* in the US, to install this revolutionary motor in several buses in order to demonstrate fuel savings, maintenance reduction and, hence, the possibility of greater profits for Greyhound.

Little did he know, however, that OPEC keeps close track of any potential competition to its oil businesses and that he was number one on its hit list. His first mistake was publicising, in many magazines, his plans to manufacture and distribute his revolutionary motor.

But after a year of hearing nothing but stony silence from Johnson, Greyhound agents tried to contact him, only to be notified that he had passed away suddenly and unexpectedly. This is a particularly troubling part of the story and shades of Stan Meyer are apparent, since at the time he was in his early fifties and in robust health. Later, Greyhound learned that shortly before he died, Johnson had inexplicably moved out of his laboratory in extreme haste in the middle of the night and taken all of his motors and technology to California.

It was then later revealed that the US Department of Energy had placed a restraining order on Johnson's company, Magnetron Inc., prohibiting it from producing the Magnetron engine.

Antigravity Devices

Perhaps the most fanciful of all the possible alternate technologies, antigravity propulsion is today regarded by science as impossible. However, there are reports from several sources that state that 'captured,' so-called 'flying saucers' have been back-engineered in order to produce and take advantage of this supposed 'sci-fi' technology.

Bob Lazar claims to have worked at Area 51 in the late 1980s on a back-engineering programme that he claims began there in 1979. He says that an 'exchange' programme with Extra Terrestrials occurred in the 1970s, which resulted in the acquisition of nine UFOs so that their technologies could be researched. That there were indeed strange craft at Area 51 seems to be corroborated by several other sources spanning several decades.

Here is Bob Lazar's own version of what he discovered...

"Assuming they're in space, they will focus the three gravity generators on the point they want to go to. Now, to give an analogy: If you take a thin rubber sheet, say, lay it on a table and put thumbtacks in each corner, then take a big stone and set it on one end of the rubber sheet and say that's your spacecraft, you pick out a point that you want to go to – which could be anywhere on the rubber sheet, pinch that point with your fingers and pull it all the way up to the craft. That's how it focuses and pulls that point to it. When you then shut off the gravity generators, the stone (or spacecraft) follows that stretched rubber back to its point. There's no linear travel through space; it actually bends space and time and follows space as it retracts. In the first mode of travel, around the surface of a planet, they essentially balance on the gravitational field that the gravity generators put out, and they can ride a 'wave', like a cork does in the ocean. In that mode they're very unstable and are affected by the weather. In the other mode of travel, where they can travel vast distances, they can't really do that in a strong gravitational field like Earth, because to do that, first of all, they need to tilt on their side, usually out in space, then they can focus on the point they need to with the gravity generators and move on. If you can picture space as a fabric, and the speed of light is your limit, it'll take you so long, even at the speed of light, to get from point A to point B. You

can't exceed it, not in this universe anyway. Should there be other parallel universes, maybe the laws are different, but anyone that's here has to abide by those rules."

Interestingly, recent research into the Bose-Einstein condensate has found that by slowing down a body of atoms, to within a fraction of one degree Kelvin (near absolute zero) they coalesce into a 'super-atom' and when suitably excited by an oscillating field this B-EC super-atom propagates matter waves. It is at a very early stage of development at present but it is hoped that one day this technology will produce a tightly focused 'matter wave beam' (much like that of the laser light beam) and what is so interesting, is that elements of the gravity generators described and drawn by Lazar, look exactly like the rings of optical lasers and magnetic traps used in B-EC technology to slow down the atoms. And that these generators emit a beam, one of which is enough for the craft to ride upon, it could mean that the ETs use a system closely related to the propagation of such matter waves.

All highly speculative, granted, but nevertheless strong rumours that this technology exists and is being suppressed by the US government, do not seem to want to go away.

The OTC-X1

In 1925, a certain Otis Carr was an art student and worked part-time at the hotel where Nikola Tesla lived. Over a period of about three years, Tesla passed on to Carr some of his knowledge on 'free energy' generators.

In 1955, Carr started a company, *OTC Enterprises*, in an effort to bring 'free energy' to the world. He claimed to have built six 'flying saucer' prototypes and began trying to announce this to the world following a huge wave of 'UFO' sightings, which began with the Roswell incident in 1947. Then in 1959, Carr announced that he would undertake a demonstration flight of one of his prototypes, but unfortunately for him it was a failure.

That same year however, he hired Ralph Ring, a diver who specialised in testing diving equipment on underwater explorer, Jacques Cousteau's ship, '*The Calypso*' and then immediately commenced his research into electromagnetism.

According to Ring's testimony in 2006, he had taken a 'test flight' in Carr's prototype craft named OTC-X1, which was forty-five feet in diameter, and this had in effect teleported a distance of ten miles without any noticeable, negative effects upon either the craft or upon Ring himself.

However, as a result of this, two weeks later, the FBI physically halted Carr's work, telling him that the reason was *"because of your threat to overthrow the monetary system of the USA."* This statement speaks volumes I believe. The Elite were obviously concerned about the threat posed by 'free energy' to their fraudulent financial system and were terrified of the implications of the possible introduction of free energy into their highly-controlled society. A more comprehensive explanation of this is detailed in the conclusion to this chapter (below).

So Carr's team was disbanded and he was forced to officially abandon all his research. But, Carr was not finished there. He managed to side-step the restrictions imposed upon him by publishing the current state of his research in the form of a patent for 'an amusement similar to a flying saucer.'

The OTC-X1 (below) comprised of two nested, circular frames, one inside the other and rotating in opposite directions. The internal frame included twelve identical patterns with electromagnets. The method of using a coil in the electromagnetic field and then subsequently increasing its distance from the frame created an induced current in the coil which then loads the capacitor on a half-cycle and then discharges during the following half-cycle. Excited at a given frequency, the circuit then resonates.

In 1895 Tesla had first applied for a patent for a 'flat coil' with a very high performance and in 1897 for another, 'conical-shaped' coil and it was these patents that had inspired Carr to take the developments a step further and actually build the machine that had hitherto existed only in Tesla's imagination.

Tesla's (and Carr's) patents still exist and can be searched and examined online.

Patent US512340 – Tesla

Patent US593138 – Tesla

Patent US2912244 – Carr

Conclusion

So why would any technology or invention be suppressed, and science falsified to cover it up? The obvious answer is that those who control existing technologies, primarily the Elite families of this world, have a great deal to lose, financially speaking if other competing technologies are allowed to come to market.

Here are some of the techniques used in technology suppression...

Intimidation and even the murder of inventors

Extreme character assassination

Arson (destroying of equipment, inventions, and premises)

The use of 'expert' de-bunkers to discredit the inventions

Buying-up and shelving of technology

Heavily promoting the 'scientific' principle that free energy is impossible as per the laws of thermodynamics

In addition to the above list of examples, there are dozens of other inventions that have been omitted here due to space constraints. Many of them are just as viable and well tested as the ones listed, but this short list is sufficient to prove the point; free energy technology is here, now. It offers the world pollution-free energy abundance for everyone, everywhere for next to nothing at all and also thoroughly debunks the immovable three laws of thermodynamics, which propound the idea that free energy is 'impossible.'

It is now possible to halt the production of greenhouse gases (if anyone really still believes that to be an issue!) and shut down all of the nuclear power plants (ditto). We can now desalinate unlimited amounts of seawater at an affordable cost and bring adequate fresh water to even the most remote habitats. Transportation costs and the production costs for just about everything could decrease dramatically and food can even be grown in heated greenhouses in the winter, anywhere at virtually zero cost. All these possibilities provide wonderful benefits that could make life on this planet so much easier and better for everyone and yet have been suppressed and covered-up for many decades. Why would this be? Whose purposes are served by this action and what forces are impeding the availability of free-energy?

In the Western world there is a heavily protected money monopoly in place. This money monopoly is solely in the hands of a small number of privately owned banks and institutions, and these institutions are owned by the wealthiest families in the world, the Elite bloodlines. Their future plan is to eventually control 100% of all of the capital resources of the world, and thereby control everyone's lives through the availability (or nonavailability) of all goods and services. Therefore an independent source of wealth (a free energy device) within the reach of every person in the world ruins this plan for world domination of the money supply, permanently. See my book 'Behind the

The Falsification of Science

Curtain,' for a detailed explanation of the relentless and unbridled power of the banking industry.

Currently, a nation's economy can be either slowed down or speeded up by the raising or lowering of interest rates, but if an independent source of capital via free energy were present in the economy and any business or person could raise more capital without borrowing it from a bank, this centralised strangulation of interest rates would simply not have the same effect. Free energy technology changes the value of money, simply put. The Elite do not want any competition, they understandably wish to maintain their current monopolised control of the money supply and so for them, free energy technology is not just something to suppress, it must be permanently deleted from history if at all possible.

Their motivations are their imagined divine right to rule over us all, sheer greed and their insatiable need to control everything except themselves. As already mentioned, the weapons they have used to enforce the permanent suppression of technology beneficial to mankind include intimidation, 'expert' debunkers, the covert buying-up and shelving of competitive technology, and often murder or attempted murder of the inventors, character assassination, arson and a wide variety of financial incentives and disincentives to manipulate possible supporters. They have also promoted and supported the general acceptance of an alleged scientific principle that states that free energy is impossible, as per the laws of thermodynamics.

The second force in operation against free energy technology is national governments. The problem is not so much related to competition in the printing of currency, but in the maintenance of national security. The big wide world out there is a 'dog eat dog' world and humans can be counted upon to be very cruel, dishonest, devious, and sneaky and it is government's job to provide a defence against this behaviour. For this, police powers are delegated by the executive branch of government to enforce the rule of law. Most of us who consent to the rule of law do so because we believe it is the right thing to do, for our own and others' benefit. There are always a few individuals, however, that believe that their own benefit is best served by behaviour that does not voluntarily conform to the generally agreed-upon social order. These people choose to operate outside of the rule of law and are considered outlaws, criminals, subversives, traitors, revolutionaries and often now, terrorists.

Most national governments have discovered, by trial and error that the only foreign policy that really works, over time, is a policy of 'tit for tat.' What this really means to us all is that governments treat each other the way they are being treated. There is a constant jockeying for position and influence in world affairs and the 'strongest' country always wins. In economics, the golden rule states that, 'the one who owns the gold makes the rules.' So it is with politics also, but its appearance is more Darwinian in nature – it is simply the survival of the fittest. However, in politics the fittest has come to mean the strongest party, i.e. the one who is also willing to fight the dirtiest. Absolutely every means available is used to maintain an advantage over the adversary and everyone else is the adversary regardless of whether they are considered friend or foe. This includes outrageous psychological posturing, lying, cheating, spying, stealing, brinkmanship, the assassination of world leaders, proxy wars, shifting alliances, treaties, foreign aid, and the presence of military forces wherever possible.

Like it or not, this is the psychological and actual arena that national governments operate within. No national government will do anything that simply gives an adversary an advantage for free, it is national suicide. An activity by any individual, inside or outside the country that is interpreted as giving an adversary an edge or advantage will be deemed a threat to 'national security.'

Therefore, free energy technology is any national government's worst nightmare. Openly acknowledged, free energy technology sparks an unlimited 'arms race' by all governments in a final attempt to gain absolute advantage and domination. For example, would Japan not feel intimidated if China had free energy? And would Israel sit by quietly if Iran acquired free energy? Unlimited energy available in the current state of affairs on this planet would lead to an inevitable reshuffling of the balance of power. This could become an all-out war

The Falsification of Science

to prevent the 'other' from having the advantage of unlimited wealth and power. Every country will covet it and at the same time, want to prevent every other country from obtaining it.

There is also the credible argument that governments will suppress alternate technology for the simple reason of preserving income streams derived from *taxing* energy sources currently in use. The weapons that they use against this include the preventing of the issuance of patents based on spurious, trumped-up national security grounds, the legal and illegal harassment of inventors via criminal charges, invasive tax audits, threats, phone-taps, arrest, arson, theft of property during shipment and a host of other intimidations which make the business of building and marketing a free energy technology practically impossible.

The third force operating in an attempt to postpone the public availability of free energy technology consists of a group of deluded inventors and outright charlatans and conmen. On the periphery of the extraordinary scientific breakthroughs that constitute the real free energy technologies, lies a shadowy world of unexplained anomalies, marginalised inventions, and unscrupulous promoters. The first two forces have constantly used the media to promote the worst examples of this group, to distract the public's attention and to discredit the real breakthroughs by associating them with the obvious frauds.

Over the last hundred years or so, dozens of stories have surfaced about unusual inventions. Some of these ideas have so captivated the public's imagination that a mythology about these systems continues to this day. There may possibly be real technologies behind these names, but there simply is not enough technical data available in the public domain to make a determination. These ideas remain associated with free-energy mythology however and are cited by debunkers as examples of fraud.

So, the third force postponing the public availability of free energy technology is delusion and dishonesty within the movement itself. The motivations are self-aggrandisement, greed, desire for power over others and a false sense of self-importance. The weapons used are lying, cheating, the 'bait and switch' con, self-delusion and arrogance combined with false science.

The fourth force operating in order to postpone the public availability of free energy technology is 'all of the rest of us.' It may be easy to see how narrow and selfish the motivations of the other forces are, but actually, these motivations are still very much alive in each of us as well. As with the Elite, do we not each secretly harbour illusions of false superiority and the desire to control others instead of ourselves? Also, would anyone of us not 'sell out' if the price was right, say, a million pounds cash, today, in our hands? Or like governments, do we not all wish to ensure our own survival? For example, if caught in the middle of a packed, burning theatre, would we not panic and push all of the weaker people out of the way in a mad, scramble for the door? Or like the deluded inventor, would we not trade a comfortable illusion once in a while for an uncomfortable fact and do we not still fear the unknown, even if it promises a great reward? None of us may wish to believe that this is how we would behave, but it is a fact that self-preservation is a very strong motivator and who among us, if we are being totally honest, would sacrifice our lives for others if the need to do so arose?

All four forces are just different aspects of the same process, operating at different levels in our society. There is really only one force preventing the public availability of free energy technology, and that is the non-spiritual motivations of we humans, but in the final analysis, free energy technology is an outward manifestation of divine abundance. It is the engine of the economy of an enlightened society, where people voluntarily behave in a respectful and civil manner towards each other, where each member of society has everything they need and does not covet his neighbour's possessions, where war and physical violence has become socially unacceptable behaviour and people's differences are at least tolerated, if not actively enjoyed.

The appearance of free energy technology in the public domain would represent the dawning of a truly civilised age. It would be an epochal event in human history and no one individual could take credit for it. No one can become rich on it, no one can rule the world with it, it is simply a gift from the 'gods.' It forces us all to take responsibility for our own actions and for our own self-discipline and self-restraint when needed.

Upon reading Ayn Rand's work, *'Atlas Shrugged'* or the *Club of Rome Report*, it becomes obvious that the Elite families have understood this for decades. Their plan is to live in a world of free energy, but permanently freeze-out everyone else. This is not new, for example, royalty has always considered the general population to be its subjects. However, what is new is that we can communicate with each other now better than at any time in the past. The Internet offers us all an opportunity to overcome the combined efforts of the other three forces preventing free energy technology from spreading.

"The world as it is currently ordered cannot have free energy technology without being totally transformed by it into something else. Unfortunately, those who control us have shown that they cannot be trusted with energy resources; they will only do what they have always done, which is to take merciless advantage of each other and kill each other and themselves in their endless quest for profits." John Hamer, *'The Falsification of History'* (2012)

Chapter 11

The 'Unexplained' Human Mind

"Consciousness is the greatest mystery of science, and it is perhaps the scientific mystery that we most urgently need to solve. And we have at our disposal an array of natural substances called the psychedelics which allow us to switch on and off altered states of consciousness at will and which are superb devices for exploring the mysteries of consciousness. And yet, for ideological reasons the justification for them simply does not exist and we are prevented from doing so. And this is a cover for all sorts of other restrictions and controls on the freedom of the adult over his or her own body and over his or her own consciousness." Graham Hancock

"The day science begins to study the non-physical phenomena; it will make more progress in one decade than in all the previous centuries of its existence." Nikola Tesla

So what do I mean by 'the unexplained mind?' It is quite straightforward really, I am just referring to any phenomenon of the mind for which mainstream science has no explanation and therefore simply refuses to acknowledge its existence, despite copious supporting evidence.

Of course mainstream science makes no attempts to study these phenomena as they clearly do not fit into the proscribed scientific paradigm. But there have been plenty of examples over the past centuries of incidents that although they totally defy conventional belief, have just been ignored or denied by those who 'set the rules' as to what is and is not an acceptable subject for consideration and investigation.

The Falsification of Science

The human mind, as opposed to the physical brain itself, is a very curious entity. Scientists and philosophers alike speculate endlessly and debate about where the mind is actually physically located, but the truth is that no one really knows. In fact the whole subject of mind and consciousness is a vast, unexplored area and mainstream science especially, I believe, does not really want to explore the topic too deeply, as this takes them too far out of the materialistic comfort zone they tend to rigidly inhabit.

In fact the whole topic of 'unexplained phenomena' is so vast that it is virtually impossible to cover it in such a volume as this. However, I feel it is important to at least try to cover some of the issues involved. I think it is safe to say without fear of contradiction though, that what we as a species do NOT know, understand, or can even come close to explaining is by several orders of magnitude, much greater than that which mainstream science **does** claim to understand.

So, in essence, the following chapter is not some feeble, futile attempt to explain the unexplainable, but to merely draw the reader's attention to *some* of those areas of the mind that mainstream science shuns and dismisses out of hand, as nonsense and superstition.

Extrasensory Perception

"Extrasensory perception or ESP, also called 'sixth sense,' includes claimed reception of information not gained through the recognized physical senses but sensed with the mind. The term was adopted by Duke University psychologist J. B. Rhine to denote psychic abilities such as intuition, telepathy, psychometry, clairvoyance, and their trans-temporal operation as precognition or retrocognition." Wikipedia™

However, many people believe that we all possess extrasensory abilities to some greater or lesser extent, and that it is really simply a matter of harnessing those powers through practice, whilst others tend to believe that extrasensory perception is something that is innate and can only be used by 'real' psychics and mediums.

In all, ESP abilities come in nine different forms. Some people have none of these psychic 'gifts,' while others may have one, two, three, or even more. Here below are the 9 different types of extrasensory perception and how each one works.

Clairaudience

Clairaudience translates from French into 'clear hearing.' It is the ability to hear messages or receive information from sounds beyond our ordinary senses.

These messages may come from those who have passed on beyond our life into another dimension. They may also come from the energies of the universe, a spirit or animal messenger, or any other source that exists separate from our physical existence here on Earth.

Most often, these sounds are voices, reaching out with warnings or advice. Other times, a clairaudient may hear music, nature sounds (birds chirping for example), or rhythms. Sometimes it may be akin to white noise or static or even an internal dialogue.

Regardless, clairaudients who keep themselves open and practice their abilities can grow to recognise and understand these messages.

Clairvoyance

Clairvoyance is another one of the 'clair' talents. In French, *clair* means 'clear,' and *voyance* means 'vision.'

So, clairvoyance is the ability to see beyond the Earthly world. Some clairvoyants see visions of the past, present, or future in their mind, similar to viewing a TV show or movie. Seeing the past is known as postcognition, while seeing visions or images of the future is called precognition.

Others see symbols, images, or auras-- the energetic bodies that surround us. And clairvoyants typically receive these visual messages through their 'third eye.' The third eye is the metaphysical gate or portal in the forehead that leads to a higher spiritual plane and often symbolises enlightenment. Seers and mystics use their third eye as a tool to receive their spiritual and psychic messages.

Some of the most famous mediums from history have been clairvoyants. Figures such as Edgar Cayce and Nostradamus, and some believe, Joan of Arc, had clairvoyant powers.

Mediumship

Mediums represent a different type of ESP ability than those with clairvoyance or precognition. And, while mediums may be psychics, psychics are not always mediums.

Mediums are those who speak with the spirits of those who have passed on. They use trances, voluntary possession, seances, or their 'clair' abilities to communicate with the energy of those that have transitioned from this life to the afterlife.

As with all psychic abilities, each medium works slightly differently. They finetune their paranormal talents in their own way. Some mediums can only connect with their own particular spirit guides while others can communicate with anyone's spiritual energy.

Mediums have often been significant figures in history, even advising royalty and bringing solace to the bereaved.

Psychometry

Psychometry comes from two Greek words. One means spirit or soul, and the other means measure. It is sometimes also called psychoscopy or token object reading.

People with a talent for psychometry can read the lore or history of objects through the energetic vibrations stored in those objects. Token object reading is also a form of scrying or divination. It works best with metals, but these vibrations – our vibrations – are stored in everything we touch.

When reading an object, the psychic holds it in his or her hand or presses it to his or her forehead. Emotions, particularly strong ones, leave the most tangible record behind. Some psychometrists can even perceive the object's owner, experience sequences of his or her life, and grasp that person's personality.

Precognition

Those with precognition can divine events and experiences about locations or people that have yet to occur. It is also known as prescience, future sight, future vision, and foresight.

The word precognition comes from the Latin 'pre' and 'cognito,' which together means 'foreknowledge.'

In simple terms, precognition is the sixth sense used for looking into the future. These psychic visions can happen in dreams or during waking hours and often involve the 'clair' talents. While some psychics can use precognition at will, many experience this prescience when reading or practicing their supernormal abilities.

Remote Viewing

Psychics with remote viewing talents can sense accurate details about a person, place, or event that is not local to them or visible through the normal senses. Remote viewing is also known as anomalous cognition or second sight.

Agencies that include the US Government, the Central Intelligence Agency (CIA), and police services have frequently tapped into the talents of remote viewers to help find missing persons, solve crimes, or gain insight into important world events past, present or future.

Remote viewers project their minds to a distant locale and describe details of their 'target' (which could be a person, place, or event in time, given them by any of the aforementioned employers). Some also use their 'clair' talents or telepathy to assist their viewings.

Retrocognition

Retrocognition is the ability to see into the past. Often connected to the sense of déjà vu, retrocognition is the opposite of precognition.

From the Latin, retrocognition means 'backward knowing.' This talent gives the psychic knowledge of past events. Sometimes the psychic picks up information about a living person's past, whilst at other times, their knowledge relates to distant histories.

Psychics often express their past visions through the arts, such as writing, drawing, or painting. Dreams are often a vehicle for retrocognitive viewings.

Telepathy

Telepathy is the ability to read people's minds or communicate with others without speaking.

You may be familiar with the phenomenon of saying the same thing at the same time as someone else? Or thinking of someone and suddenly receiving a call from that person at that exact moment. These are regarded as instances of telepathy at work.

Telepathy can also pick up on thoughts, emotions, and physical needs or desires and may occur during waking moments, as well as via dreams.

Many individuals seem to possess a natural telepathic bond with another individual, and this is often the case with twins. They seem to have an innate sense of what the other is doing or going to do in the future and often mirror those actions, sometimes even at remote distances.

Telekinesis/Psychokinesis

The ability to move distant objects through non-physical means is known as telekinesis. Sometimes the word psychokinesis is used instead, although there is a slight difference in their definitions.

Telekinesis derives from Greek words meaning 'far off movement,' whereas psychokinesis is also from the Greek and means 'soul movement.' So telekinesis is the ability to move objects psychically at a distance, whereas psychokinesis is a more general term.

Further, psychokinesis is an ability to control matter with the mind, whereby objects move or change because of a person's intention. 'Psi' refers to the mind, as in the word 'psyche,' whilst 'kinesis' signifies movement. Webster's dictionary defines psychokinesis as *"the movement of physical objects by the mind without the use of physical means."* So we may say that the concept of psychokinesis describes an ability to consciously influence physical entities using only the power of the mind.

First 'coined' by the Russian psychical researcher Alexander Aksakof, telekinesis can work on a macro or micro level. It includes such phenomena as levitation, psychic healing, and the ability to bend metal, amongst others.

Astral Projection

The term 'astral projection' refers to a process during which the spirit leaves the body for a period of time. Then once the spirit returns to the body, the subject awakens and is able to clearly recall his out-of-body sojourn. Often astral travellers, similar to remote viewers can 'project' themselves to specific places and later recount physically verifiable details of places, people's actions or other events. There are two common types of astral projection . . .

Meditation

Many believe that the astral plane can be visited consciously through deep meditation. This kind of OBE is consciously induced. The subject chooses the time and place to induce his body into a relaxed state of being, thereby allowing his spirit to travel the astral plane. Many astral travel practitioners attempt this just upon awakening when the brain is between 'theta' (light sleep) and 'alpha' (deep meditation), rather than attempt more traditional seated meditation techniques, as the latter tends to be much more challenging than the former method. There are in fact, institutions (such as the Monroe Institute and others), books and online courses, which specialise in teaching astral projection or 'astral travel' techniques to people from all backgrounds, many without any former ESP experience of any kind.

Out of Body Experiences

Sometimes the subject is not consciously attempting to leave his body. It just happens spontaneously, and so have undergone what is known as an 'out of body experience' (OBE). In this state, they find themselves standing outside their bodies or watching from above. A spontaneous OBE is often short-lived because there is a tendency for the subject to panic upon finding himself in such an unfamiliar situation and then he is instantly returned to the physical body.

Ever since ancient times, those who practice astral projection have believed that the spirit journeys outside of the body during sleep.

"In the daylight hours are our feet on the ground and we have no wings with which to fly. But our spirits are not tied to the earth and with the coming of night we overcome our attachment to the earth and join with that which is eternal." From the ancient Essene manuscript, *'Teachings of the Elect'*

Today, the words 'soul' and 'spirit' are often used interchangeably. However, in the ancient texts there was a clear distinction between these two words. The soul was described as the life-force of the body and could not leave without causing death, whilst the spirit was described as being able to come and go at will. Since many astral projection adherents refer to the 'soul' leaving the body, this causes some to be fearful of the practice. So it is important to understand that it is the 'spirit' that makes these sojourns and the life-force (or soul) never actually leaves until actual physical death occurs.

Astral projection has been around since ancient times and has been commonly accepted and used for countless centuries. The old wives tale and superstition of witches flying around on a broomstick is actually merely metaphorical imagery derived from astral projection rather than an actual physical phenomenon. In the past it was quite common for a 'witch' to use a so-called 'flying ointment' to assist them in achieving this state of altered consciousness and awareness. Some of these 'preparations' used herbs or plants which are quite dangerous to the untrained and can be fatal.

A word of warning. It is possible to find supposed 'flying ointment' recipes on the Internet, which include ingredients such as digitalis, hemlock, belladonna, aconite, or cowbane. I must point out that these plants are dangerously toxic, and can be fatal if used incorrectly, and so should never be used for this purpose.

Automatic Writing

"Automatic writing, also called psychography, is a claimed psychic ability allowing a person to produce written words without consciously writing. The words purportedly arise from a subconscious, spiritual, or supernatural source." Wikipedia™

This is a topic that I can at least speak with a little authority about, having some minor experiences of this phenomenon myself during the twenty years or so in which I have been researching and writing articles and books for publication.

Although it is still rare for it to happen to me personally, I have over the years noticed on several significant occasions that the words just seem to flow from my fingertips and onto the keys of my laptop without my previously being consciously aware of them forming within my mind. It is actually quite disconcerting to type, in some instances, a whole, fairly lengthy paragraph in a kind of awake yet trance-like state and then once finished, read it back to myself whilst

The Falsification of Science

experiencing a feeling of reading it for the first time and marvelling at the fact that I had actually written those words myself. How do I explain this? Sadly, I cannot. Maybe it is possibly some kind of connection to the spirit world that is taking place and it is either part of my own spirit essence or some other benevolent 'spirit guide' type entity that is acting on my behalf? I have particularly noticed these occurrences at times when I had been experiencing an episode of 'writer's block' for a while at first.

But what firm conclusion I can draw from all that, would as I say, really be only wild speculation on my part. Maybe it is possible that on occasions, the mind (as opposed to the brain) simply switches into a kind of inexplicable 'automatic mode' which gives the impression of being detached from normal thought processes, but in reality those are working in exactly the same way whilst we are simply not conscious of that happening? No one really knows and maybe never will know.

However, 'automatic writing' is most commonly associated with deliberately attempting to 'channel' messages from spirits and spirit guides by allowing the mind to go blank and simply allowing subconscious (or otherworldly) messages to flow onto the paper.

Automatic writing is writing allegedly directed by a spirit or by the unconscious mind. It is sometimes referred to as 'trance writing' because it is undertaken quickly and without judgement, writing whatever comes to mind, without consciousness, and as if in a trance. It is believed that this allows one to tap into the subconscious mind, where 'the true self' dwells and so, uninhibited by the conscious mind, deep and mystical thoughts can be accessed, or that it comes externally from channelling specific spirits. Trance writing is also used by some psychotherapists who regard it as a way to release repressed memories.

Déjà Vu et al

Déjà Vu

"We have all some experience of a feeling, that comes over us occasionally, of what we are saying and doing having been said and done before, in a remote time – of our having been surrounded, dim ages ago, by the same faces, objects, and circumstances – of our knowing perfectly what will be said next, as if we suddenly remember it!" Charles Dickens

Déjà vu, (translated as 'seen before') as most people know, is the feeling of being certain that we have experienced or seen a seemingly new situation previously – and in which we feel as though the event has already happened or is repeating itself. The experience is usually accompanied by a strong sense of familiarity and a sense of eeriness, strangeness, or just simply downright weirdness. The 'previous' experience is usually attributed to a dream, but sometimes there is a strong, unshakeable sense that it has truly occurred in the past.

Déjà Vécu

Déjà vécu (pronounced vay-koo) (already lived) is what some people are experiencing when they believe that they are experiencing déjà vu. Déjà vu is the sense of having seen something before, whereas déjà vécu is the experience of having seen an event before, but in great detail – such as actually recognising smells and sounds. This is also usually accompanied by a very strong feeling of knowing what is going to come next. Some people have not only known what was going to come next but have been able to tell those around them what is going to come next – and be proven correct. This is a very eerie and inexplicable sensation.

Déjà Visité

Déjà visité (already visited) is a less common experience and it involves having an uncanny knowledge of a new place. For example, knowing the way around a new town or a landscape despite having never been there, and yet knowing that it is impossible to have acquired this knowledge by any conventional means. Déjà visité is more spatial and geographically related, whilst déjà vu and vécu are more related to temporal occurrences. Nathaniel Hawthorne wrote about an experience of this in his book *'Our Old Home'* in which he visited a ruined castle for the first time and yet nevertheless had full knowledge of its entire layout.

Déjà Senti

Déjà senti (already felt) is exclusively a mental phenomenon and seldom remains in the memory afterwards. In the words of a person having experienced it... *"What is occupying the attention is what has occupied it before, and indeed has been familiar, but has been forgotten for a time, and now is recovered with a slight sense of satisfaction as if it had been sought for. The recollection is always started by another person's voice, or by my own verbalised thought, or by what I am reading and mentally verbalise; and I think that during the abnormal state I generally verbalise some such phrase of simple recognition as 'Oh yes, I see,' 'Of course, I remember,' etc., but a minute or two later I can recollect neither the words nor the verbalised thought which gave rise to the recollection. I only find strongly that they resemble what I have felt before under similar abnormal conditions."* In short, it is the eerie feeling of having just spoken, but realising that in fact, you did not actually utter a word.

Jamais Vu

Jamais vu (never seen) describes a usually familiar situation which is not recognised as such. It is often considered to be the opposite of déjà vu and it invokes an equal sense of eeriness. The observer does not recognise the situation despite knowing rationally that they have been there or seen it before. For example when a previously known person, word, or place is inexplicably not recognised. Chris Moulin of Leeds University, England, conducted an experiment whereby he asked 92 volunteers to write out the word 'door' 30 times in 60 seconds. He reported that 68 per cent of his subjects exhibited symptoms of jamais vu, such as doubting that 'door' was actually a real word.

Presque Vu

The term 'presque vu' (almost seen) is best described by 'the answer is on the tip of my tongue' type sensation. It is the strong feeling that you are about to experience an epiphany of some kind, although the epiphany often never arrives. The sensation of presque vu can be very disorienting and distracting.

L'esprit de l'escalier

L'esprit de l'escalier (stairway wit) is the experience of thinking of a clever riposte when it is too late. The phrase can be used to describe a comeback to an insult, or any witty, clever remark that comes to mind too late to be useful, in effect when one is on the 'staircase' ie. already leaving the scene. The German word 'treppenwitz' is used to express the same idea. The closest phrase in the English language to describe this situation is 'being wise after the event.' The phenomenon is usually accompanied by a feeling of deep regret at having not thought of the riposte when it was most needed or appropriate.

Capgras Delusion

Capgras delusion is the phenomenon in which a person believes that a close friend or family member has been replaced by an identical looking impostor. This could be connected to the old belief that babies were stolen and replaced by changelings in medieval folklore, as well as the modern idea of aliens taking over the bodies of people on earth (the 'body snatcher' phenomenon) to live amongst us for reasons unknown. This delusion is most common in people with schizophrenia, but it can also occur in other disorders.

The Falsification of Science

Spirituality

"The world is like a ride in an amusement park. And when you choose to go on it, you think it's real because that's how powerful our minds are. And the ride goes up and down and round and round. It has thrills and chills and it's very brightly coloured and it's very loud and its fun – for a while. Some people have been on the ride for a long time and they begin to question, is this real or is this just a ride? And other people have remembered, and they come back to us, they say, 'Hey – don't worry, don't be afraid, ever, because this is just a ride'... And then we kill those people." Bill Hicks, US comedian

Bill Hicks was a unique comedian. Acid-tongued, uncannily accurate observations of life combined with deep spiritual knowledge and beliefs and a rare understanding of political realities, made him a huge draw with audiences all around the USA and eventually Britain. Of course, he never became a 'super-celebrity,' or a mega-star. For that to happen one must espouse views that totally conform to the mainstream otherwise one will never be allowed a platform for this 'subversive' approach. Hicks sadly died of cancer at the age of 32 and the world lost a true student of spirituality, a deep-thinking intellectual and someone who was able to bring an alternative view of the world to the masses through his insightful comedy shows. He constituted an extreme danger to those who try to enslave us all, as he had a huge following who hung upon his every word. In light of this fact, could his cancer possibly have been natural or is it just feasible that it was given to him deliberately? There is no doubt that the technology to do this exists and is often used against the overt opponents of the New World Order.

Some years ago, the famous actor Larry Hagman, now deceased, took part in an experiment to check-out the effects of the (highly illegal) Class A drug, LSD (**Ly**sergic acid **D**iethylamide). After imbibing the substance and waiting for its effects to manifest, he took an orange from the kitchen and cut it open. Its cellular structure was pulsing, and it looked to him as though the cells were alternating between life and death, which seemed perfectly natural to him in his now highly conscious, altered-state. Looking up from his scrutiny of the pulsating orange, he saw his reflection in a mirror on the wall. He too, was pulsating. Cells were dying, whilst others were in the process of being reborn. An intricate picture of every cell in constant motion became apparent and he realised that he was a constant flow of energy as indeed, was everything else. The scope of this realisation widened with his conviction that: *"I was part of everything, and everything was part of me. Everything was living, dying and being reborn."*

His friend, who had not taken the drug, drove Hagman around Beverly Hills equipped with a sixteen millimetre camera with which he could zoom-in on plants, flowers and people and he found that their cells were also constantly pulsating and changing too. This experience shares a number of similarities with near-death, 'out-of-body,' and certain shamanic experiences. These are usually intensely transforming and empowering, and Hagman's own experience was certainly no exception.

Besides self-insight, he also saw much more deeply into people's emotions and how they were expressed through body and facial language. But most importantly, his view of life and death were profoundly altered by the experience. He realised that so-called 'dying' was actually only a transformation into another expression of the vast creative energy that underlies everything. He concluded that, *"Death is just another stage of our development and we go on to different levels of existence."*

He believed he had an understanding of 'God' consciousness. Fear of manmade concepts of heaven and hell disappeared, and he stopped worrying and indeed felt 'at home' in the universe. It was all so clear and so familiar.

Dangerous knowledge indeed. Can anyone think of a better reason for the illegality of LSD or any other so-called hallucinatory drug such as psilocybin (the active ingredient of 'magic' mushrooms) or Ayahuasca for example, both of which facilitate the expansion of our consciousness, allowing us a brief, tantalising glimpse into 'reality?'

So what exactly is Spirituality? It extends beyond an expression of religion or the practice of religion. The relationship between ourselves

The Falsification of Science

and 'something greater' compels us to seek answers about the infinite. During times of intense emotional, mental, or physical stress, man searches for transcendent meaning, often through nature, music, the arts, or a set of philosophical beliefs. This often results in a broad set of principles that transcends all religions.

While spirituality and religion remain different, sometimes the terms are used interchangeably and this lack of clarity in their definitions frequently leads to debate. Through certain actions, an individual may appear outwardly religious and yet lack any of the underlying principles of spirituality. In its broadest sense, spirituality may include religion for some, but still stands alone without a connection to any specific faith. In my view it is simply a belief that this Earth, this Universe, is not all there is. There is an unseen dimension that may or may not contain our God(s) but most certainly contains a spiritual entity or entities of some form of cosmic super-consciousness.

Whether this entity manifests as a 'oneness' or as individual elements, I believe that one thing is certain; we are indeed all one and the same being, part of the overall 'one-consciousness.' We are all literally 'brothers and sisters' and interconnected at the basic cellular level as is all matter in the Universe, whether sentient or non-sentient.

Consider this; the actual amount of 'solid' matter making up the entire Earth and everything in it, if all the empty space in atoms was removed, would consist of a blob of nuclear material about 1 cm (0.4 inches) in diameter. But more mysterious still, the matter in the nucleus is also not solid as it appears but consists of protons and neutrons with a huge percentage of empty space in-between *them*. And these protons and neutrons are made up of quarks containing yet more space. Quarks consist of neutrinos and even more space, but confusingly this *space* is also 'made of' neutrinos but vibrating at a different rate to the *matter* that makes up the neutrinos.

Therefore the only difference between substance and nothing at all is the vibratory rate of the neutrinos. When experiments were performed to try to understand why the neutrinos would exist in one specific form one moment and another form entirely in the next, it was discovered that neutrinos always become what they are expected to become. So it has now been confirmed by default by mainstream science, what mystics throughout the ages have said all along; that matter is an illusion created by consciousness. The wave-particle duality model shows us that these particles are waves until they are observed. Simply observing the wave/particle changes the waves into particles and vice versa. The whole universe is nothing but a wave pattern that we make real with **our** awareness. Our bodies are our most personal physical creations, which reflect our thoughts, feelings, beliefs, attitudes, choices, and decisions. This phenomenon was first brought to the attention of the world in 1935 by a physicist named Edwin Schrödinger in an hypothesis that has come to be known as 'Schrödinger's Cat' whereby a cat in a box, can be both alive and dead at the same time, depending on the standpoint of the observer.

Every sensory experience, thought and emotion produces an electrical wave, which passes through every DNA molecule in the body. The structure of DNA is a closed spiral, a double-helix, which turns back on itself and also loops back the other way, so any wave that passes through it will travel in both directions simultaneously. The result is a scalar wave, an information wave that has no direction. When this wave is consciously experienced, it moves through the corpus callosum of the brain, which is itself in the shape of a Mobius strip, a strip that is twisted in the middle and coils back on itself, so any wave that passes through it undergoes a 180 degree phase change, which cancels out the wave stored in the DNA coil by destructive interference.

Any thought, feeling or sensory experience that is not fully conscious will remain as a wave within the DNA coil. The electrical wave then draws to itself a melanin-protein complex and forms a crystal. So every suppressed experience, thought or emotion is stored as a crystal in every DNA molecule in the body. So, the body is in effect, a three dimensional hologram, where each part of the DNA affects different parts of the body.

All life is connected, and it is through our DNA that we broadcast and receive information. Our entire body is akin to a giant transmission and reception system being constructed through resonant frequencies. The Sun, Earth and 'heavens' are nothing more than illusions. They are frequencies being generated, from which our subconscious constructs and our conscious mind observes what we euphemistically believe to be 'reality.' And this is the whole essence of the imprisonment of our minds as we are led to believe by our lords and masters that this five-sense universe is all there is and anyone who questions that fact must be deluded in the extreme. All dangerous knowledge such as this must be eradicated at all costs so that they can maintain their vice-like grip on our lives.

Mainstream Religion

A substantial part of the control mechanism used as a tool for centuries by the ruling Elite to facilitate the keeping of the masses in their places has always been mainstream religion (as opposed to spirituality). Through Christianity, Judaism, Islam, Hinduism, Sikhism, and all the other 'isms' and their minor variations and offshoots, we have been managed very effectively and efficiently and thus prevented from gaining access to the real truths of the Universe.

Indeed, through the ages and even today to a great extent, religious dogma has been largely responsible for the suppression of real truth, real history, and even real science in an attempt to prevent the common 'herd' from gaining knowledge of both themselves as spiritual beings and the Universe around them. Of course, that kind of knowledge would be far too dangerous to the control systems and the Elite power-base to allow it to become widespread. The only way that the few can control the many is by deception and deception on an unimaginable scale.

All mainstream religions teach us (broadly speaking) that there are only two courses open to us after death=paradise, or the eternal fires of hell. Allegedly, only those of us who are good citizens and obey the religious edicts and civil laws and believe in the one true God (choose any one of them) will live forever in paradise and the rest will be condemned to a burning pit of fire or similar, for eternity with no hope of redemption. In bygone days when the masses were totally uneducated and illiterate, one can only imagine the impact that this edict had upon the way they conducted their lives. Living in fear of eternal damnation at every turn is a difficult psychological burden to bear, but a wonderfully simple expedient by which the ruling classes, through their thoroughly brain-washed priests and religious hierarchy could keep the 'great unwashed' in check.

I do not mean to condemn or vilify those many genuine, decent, and caring people who follow these religions. Many people find comfort in their religion and there are also caring and charitable people of all religions who are serious about and adhere to their religious values, some of which to be fair, do advocate noble, humanitarian practices. However, it is the imposition of rigidly-structured, restrictive controls by the religious hierarchies that is responsible for the misery and 'imprisoning' of the minds of their adherents that I find problematic.

In order to comprehend the basis upon which all religions were founded, we need to go back in time to the Babylonian and Phoenician eras. All the major religions stem from one great premise – sun worship, the original 'religion' of most of the ancients. If one can understand the original sun worship and its attendant symbolism, then it is possible to understand the basis of all current religious beliefs. All sun symbolism has a basis in the Zodiac, representing as it does the Earth's annual journey around the Sun. As we may see from the diagram below, the Zodiac can be bisected horizontally and vertically and this I believe is the true origin of the cross symbolism in Christianity.

Many pre-Christian deities had their 'birthdays' as 25th December because of this symbolism. The winter solstice occurs on the 21st/22nd December, the point at which the sun is at its weakest in the Northern hemisphere and the point at which the sun had symbolically died according to ancient traditions. By the 25th December each year, three days later, it became observable that the sun had been 'reborn' and had recommenced its climb back to its absolute zenith on 21st/22nd June the following year.

The Falsification of Science

Thus, the ancient peoples regarded the Sun's 'birthday' as 25th December, the day it was born again, Jesus died on Good Friday, which incidentally has no fixed, definitive date and was resurrected (reborn) three days later on Easter Monday. The Christian Christmas festival is purely and simply a Pagan festival reconstituted, as are all Christian events. Below is a short list of other religious 'deities' among many, whose claimed birthday is 25th December...

Horus (c. 3000 BCE)

Osiris (c. 3000 BCE)

Attis of Phrygia (c. 1400 BCE)

Krishna (c. 1400 BCE)

Zoroaster/Zarathustra (c. 1000 BCE)

Mithra of Persia (c. 600 BCE)

Heracles (c. 800 BCE)

Dionysus (c. 186 BCE)

Tammuz (c. 400 BCE)

John Hamer

Adonis (c. 200 BCE)

Hermes

Bacchus

Prometheus

It is a perhaps little known fact that all the major religions tell us the same stories, myths, and legends, albeit using slightly different characters and names which nevertheless are often recycled for mass consumption.

For example, of the above list of commonly known 'deities,' many are attributed similar legends to that of the Jesus 'myth' or even the myth of Moses. They are variously and collectively said to have died and been resurrected three days later, become great teachers at the age of twelve, been purveyors of 'miracles' some of which are also attributed to Jesus, such as the turning of water into wine, walking on water and the healing of the terminally ill. They all had twelve disciples, fasted for forty days and forty nights in the wilderness, were born of a virgin birth and died at the age of thirty three, whilst yet others were found in a basket in water and brought up by royalty as with the Moses story etc. etc. and on and on with broad similarities too numerous to mention.

"There is nothing holy about the Bible, nor is it 'the word of God.' It was not written by God-inspired saints, but by power-seeking priests. Who but priests consider sin the paramount issue? Who but priests write volumes of religious rites and rituals? No one, but for these priestly scribes sin and rituals were imperatives. Their purpose was to found an awe-inspiring religion. By this intellectual tyranny they sought to gain control and they achieved it. By 400 BC, they were the masters of ancient Israel. For such a great project they needed a theme, a framework and this they found in the Creation lore of more knowledgeable races. This they commandeered and perverted—the natural to the supernatural and the truth to error. The Bible is, we assert, but priest-perverted cosmology." Lloyd Graham, *'Myths and Deceptions of the Bible'*

The Christian religion itself was finally shaped into the format with which we are so familiar today at the conference at Nicaea (in what is now modern-day Turkey) in 325AD. This was achieved amongst much conflict, violent disagreement and compromise until eventually 'modern' Christianity was born and evolved from this event. The conference was also characterised by its lies, distortions, and misinformation on a grand scale – all in the name of creating a control mechanism for the masses which was acceptable to all the minor creeds and offshoots of the previous broad base of the religion. In short, Nicaea was a PR exercise and an attempt to sanitise and package for broad consumption, a system of 'belief' that had previously been unacceptable or even unpalatable to so many.

"Real consciousness is as much an anathema to religion as critical thinking is to academia." Robert Bonomo, activistpost.com 13th October 2011

Please understand that the main function of organised religion is to destroy spirituality and to prevent people from making the connection to the consciousness of who we are as a species and as individuals. Given that religions were created in the first place to enslave the population, is it any surprise that they actually attempt to destroy everything for which they purport to stand?

Reincarnation and 'Life between Lives'

"Earth is a training ground for souls." Les Visible, musician and researcher

In the 1970s a prominent psychologist and hypnotherapist in California, Dr Michael Newton, discovered by serendipitous accident, a completely new phenomenon that was to change his life and that of many thousands of individuals worldwide over the course of the next four decades.

In the process of performing hypnotherapy on a patient and whilst regressing him to a past life to attempt to identify the source of his problem (something he had undertaken many thousands of times previously), to his great astonishment, the patient reached the point of his death in that past life and proceeded to describe a most astounding series of events – the transition of the soul after death into the Spirit World and his 'welcome home' by his 'soul mates' and spirit guide.

To say he was stunned is a gross understatement. Although he was obviously aware through his vast experience, that the human soul is reincarnated over and over again through the millennia, he had assumed that the soul's life between each life was simply, in his words, a *"hazy limbo that only served as a bridge from one past life to the next."* Now here before him was 'proof' that the soul has a true 'life between lives' and the stunning implications of this were now apparent, brought home to him by his client in a vividly described yet matter-of-fact related way on his own psychiatrist's couch.

Dr Newton realised that he simply *had* to find a way to uncover any future subject's memories of this 'spirit world' and unlock them as best he could. Eventually after thousands upon thousands of hours of meticulous recording and collating of the experiences of many subsequent patients he was able to construct a theoretical working model of the structure of the 'spirit world' and this is recounted in fine, jaw-dropping detail in his fascinating books, *'Journey of Souls,' 'Destiny of Souls,'* and *'Memories of the Afterlife,'* all of which are highly recommended follow-up reading.

During this long, arduous process, he also observed interestingly, that it did not seem to make any difference whether the subject had passionate religious convictions of any kind, was a 'dyed in the wool' atheist or indeed exhibited any other beliefs in between these two polarities. The outcome was always the same; a clear, coherent, consistent description of the spirit world accompanied by a concise description of the events experienced by souls from death in one life to rebirth in the next.

As Dr Newton relays in his many case histories, collated over the decades, the consistency of description of the experiences of these transitional souls would appear to be proof positive of not only multiple lives, but possibly even more surprisingly, a life between lives. The impact all this has had on Newton's own life may be summed-up succinctly in his observation that as time went on and more and more subjects relayed to him their past experiences, he realised that his own outlook on life had changed substantially. In his own words, he eventually came to the realisation that he had 'lost the fear of death' and in so doing had rid himself of all the unwanted, associated baggage accompanying this most basic of all human fears.

In addition to this, and importantly I feel, Newton also believes he has uncovered the 'meaning of life' *(my interpretation – not specifically his, JH)*. With every passing instance of his numerous 'visits' to the spirit world courtesy of his subjects' vivid descriptions, it became clearer to him that our purpose in living on this Earth (and the many other worlds populated by souls in this and other physical universes) is simply to learn to achieve perfection. Once we reach this state, we apparently cease reincarnating and live out a blissful eternal existence without the need to ever visit this physical realm again.

In essence, Newton believes that Earth is a 'school,' a training academy for souls. Every lifetime we live, we learn valuable lessons in our quest for perfection and our mentors in the soul world, spirit guides if you will, assist us in this quest in any way they can, often pointing us in the right direction when we struggle with any aspect of the lessons we are here to complete. In addition, immediately prior to each reincarnation, we are encouraged by our spirit guides to choose a specific future life from the options they present to us, that will best fulfil our goals and help us learn from our past mistakes and aberrations thus accelerating our 'growth' as an eternal, immortal being and the gradual progression to ultimate perfection.

Towards the end of his working life, Dr. Newton (who sadly passed away in 2016, in his late eighties) founded The Newton Institute in order to train others and provide a tangible platform from which to perpetuate his pioneering work. There are now some 200 practitioners of his methods around the world who through a network of constant communication and training schedules endeavour to keep up-to-date with the latest developments and continue to provide a service to anyone interested in being healed themselves or indeed simply curious about their past lives.

A few years ago, a British TV documentary recounted the story of a small boy in England who could 'remember' a previous lifetime on a remote Scottish island between the two World Wars. His memories were so vivid and persistent and his descriptions of his house and family so clear and consistent that it prompted his family to investigate further. Upon further close investigation they were astounded to discover that the house, the family, the boy and all his siblings as described by their son, actually had existed in the 1920s and 1930s on the island exactly as he had related. The boy himself was too young to have read the story somewhere (and in any case who would have written about an ordinary family living a perfectly ordinary life in a remote outpost of the British Isles, 90 to 100 years ago) and indeed to his parent's knowledge there was absolutely no way that he could have picked up this information without actually having been there at the time. It remains a mystery to this day. However I believe it is a legitimate question to ask whether or not it is possible that 'remembering' past lives is just the accessing of information from the collective subconsciousness. Nevertheless it is all very fascinating.

This story is but one of many similar ones that have begun to emerge since the publicity engendered by the documentary. This is food for thought indeed and yet more circumstantial evidence of reincarnation.

To relate an incident from my own experience, in 2010 and very shortly after becoming aware of the work of Dr Michael Newton and before mentioning it to anyone, I received a phone call from one of my sons, who proceeded to tell me about what had recently been happening with his daughter, Katy (my granddaughter) who was three years old at the time.

He said that Katy, when travelling down a particular road in the car, would regularly come out with statements such as '... *there's the house where I used to live with my other mummy and daddy*' and '... *can I go to my other house today and play with my friends that live there?*'

The Falsification of Science

As you may imagine, I was quite surprised by this revelation and encouraged him to gently question her about it. As a result, he asked her questions such as '*...when did you live at your other house, Katy?*' and was told *'a long time ago.'* Unfortunately, it was all very inconclusive, perhaps not unexpectedly and it eventually ceased altogether, and by the time she attained the ripe old age of four, she had no more memory of these incidents at all! Could it be that we all still retain past life memories as infants but over the course of the next few years lose the ability to 'remember' them as we become more and more conditioned into the five-sense world that we are programmed by society to believe is real?

However, mystery notwithstanding, this short anecdote adds further fuel to the fire and as with all the other examples contributes to the now huge mass of circumstantial evidence in favour of multi-life reincarnation that is growing almost daily.

I believe it is also possible that given the fact that we have all specifically chosen to be here on Earth at this particular time, that we are all here to be part of the battle to ensure the continuity of the human race and fight against the forces of evil that are currently desperately trying to destroy humanity in its present form. There can be no doubt that the next few months and years are going to be critical if we are to survive as a species and some of the greatest challenges the human race has ever had to encounter are just 'around the corner' in our current lifetimes. If we can all 'wake-up' in time and prove ourselves worthy of these challenges and all play our parts in ensuring the survival of the species against the odds, then the karmic lessons we will have learned in our eternal quest for spiritual perfection, will be more than worth the Earthly sacrifices we will all no doubt have to endure along the way.

It is essential though, that we ask ourselves the question; why is all this substantial, circumstantial evidence of life after death and previous lives not even considered worthy of discussion in the mainstream and ignored, let alone properly scientifically investigated? There is far less 'proof' than this of, for example, the truth of the *theory* of evolution, but that particular creed has no trouble in gaining widespread, unthinking acceptance as fact by the 'sheeple' of this world. Unfortunately, we have to draw the same conclusions once more and that is that certain elements of our society have a serious vested interest in keeping this knowledge from general public consumption whilst promoting other huge lies as the truth. Were the masses to lose their fear of death in great numbers, as did Michael Newton and as I have too, in understanding the real point of life, then the implications of this for our controllers could well be critical. The only way they can remain in control is by maintaining our fear of death whilst presenting us with a totally distorted version of reality and history that maintains the status quo in their favour.

There are perhaps 10,000 of 'them' in total and more than 7 billion of 'us.' Just think of the power we would have should we choose to use it, and instead of passively accepting slavery as we do now, simply refuse to bend to their sadistic wills any longer.

Chapter 12

The Great Global Reset 2020

Agenda 21/2030

Let us now examine United Nations' 'Agenda 21,' along with its 'sister' Agenda 2030, the so-called *'Blueprint to Advance Sustainable Development.'* How forward-looking and environmentally responsible, you may think. Who could possibly be against advancing 'sustainable development' and who does not care deeply about Mother Earth, 'Gaia' if you will? Let's 'save the planet' and all that.

Unfortunately, there is rather a large catch, several actually, and one of them is our old friend population reduction, which rears its head once again, preferably by 90-95% of the current total, if the wishes of the Elite are anything to consider – and also according to the 'Georgia Guidestones.'

This Elite plan dressed up as the strategy that will save the world, has been around, for those who do not walk around with their eyes wide shut, since 1992. However in the last four to five years the agenda and timescales have been really cranked-up and we are now starting to see some of these deadly plans beginning to be implemented. Those indispensable allies of the New World Order are continuing all around us as I write. Namely, mass-distractions, the dumbing down of future generations and the poisoning of the planet and its indigenous animal and plant life are all gathering pace. And this is despite the lip service being paid to allegedly saving the planet through such initiatives such as *Agenda 21* which purports to do just that, whilst actually covertly bringing about the destruction of human society as we know it – and putting those of us who survive the 'purge' – under communist, totalitarian control.

The aforementioned 'Georgia Guidestones,' several huge granite slabs, placed by persons unknown, in a meadow in Georgia, USA in 1980 are in effect the 'ten commandments' of the Elite. Whilst creating an impression of being green, eco-friendly, and superficially people friendly, their overt message hides a covert agenda that is far from benevolent, advocating huge population culls and the mass relocation to what in effect would be akin to reservations, of the remaining few 'slaves.' Interestingly they were unveiled on 22nd March 1980, exactly 40 years ago to the day before the first 2020 'COVID-19' lockdown was announced! Coincidence? Maybe.

Indeed, *Agenda 21* is dressed-up in such 'cosy' language and euphemisms that few people seem even remotely concerned about its far-reaching, draconian proposals. *Agenda 21* actually promulgates the total, centralised, global control over human life absolutely in line with the New World Order agenda. It advocates no rights of property ownership, indeed no rights at all, for anyone other than of course the ruling Elite and the methodology by which the planet will be 'sanitised' of its surplus 95% of the population is not made explicit, only that it must happen – and be maintained at that level thereafter.

However, there are enough ongoing policies of the Elite, which currently demonstrate that a pretty good start has already been made on this particular task. I refer here to such delights as the pharmaceutical and medical industries, vaccinations and genetically modified foods and their ilk, a comprehensive list of which, I have compiled below:

Vaccination programmes

Bogus healthcare programmes

Failing healthcare systems

Pharmaceuticals

The Falsification of Science

Contrived epidemics

- Swine Flu
- Bird Flu
- AIDS/HIV
- Ebola
- Legionnaire's disease (flu)
- COVID-19

Sperm count reductions through...

- Radiation (both s0-called ionising and non-ionising)
- Poisons in plastic packaging
- Chemicals in food

Food additives

- Aspartame
- MSG
- Soya Bean Extract
- Refined Sugar
- HFCS

Food restrictive practices

Codex Alimentarius

Food Safety Bill S510

Contrived food shortages

Fluoride in water (tap and bottled) and many other drinks

Chemtrails

Ecological weapons (eg. HAARP) causing man-made...

    Hurricanes

    Tornados

    Earthquakes

    Volcanic activity

    Tsunamis

    Floods

    Famine

    Crop failures

Nuclear power

Ionising radiation (x-rays, radiation 'therapy,' etc.)

Microwave radiation (mobile phones & masts, Wi-Fi, ultrasound, etc.)

Electromagnetic radiation (includes microwave/high frequency and low frequency (alternating current powered electronics, house wiring, and power lines)

Wars – Illegal and contrived

Western-backed political coups

Mass species die-offs/extinctions

Pollution of eco-systems

Pollution of oceans and waterways

Mass deforestation

Economic sanctions

It is a fact that there are tens of millions of family sized plastic coffins stacked-up in huge storage facilities in Georgia, USA and this fact has led to much speculation as to their actual purpose. Could they be there just waiting for the 'great cull of humanity' to commence or is there a more benign reason? Personally, I cannot imagine an innocent reason for the considerable expense to which whoever stockpiled these coffins must have gone, but instead believe that they must be the precursor to some kind of entirely expected and planned-for, cataclysmic event.

It would also appear to be the case that these coffins are airtight and as stated, are capacious enough to comfortably 'lay to rest' an entire family of four people. It has also been reported that no one in the US government is prepared to comment on this phenomenon, leaving the obvious conclusion to be that they must have something to hide.

Some researchers and commentators believe that this is all simply preparation for the Elite to launch their oft-vaunted disease pandemic, which we have all been conditioned to expect over the years with the various bird and swine flu epidemics, SARS, Ebola et all perhaps being used as 'dry runs' and to establish public reaction and also to prepare the ground for forcible mass-vaccination programmes. Has this begun already using the oh-so-fake COVID-19 'pandemic' as a kind of 'dry run' too?

Apparently the US Government has also completed the building of several hundred concentration camps, under the auspices of FEMA, which are, at the moment, standing empty yet fully equipped. These camps are not dissimilar to the Nazi concentration camps in Germany and Eastern Europe during WWII, with barbed wire fences and railway tracks leading to their gates and each are designed to house thousands of families in what are in effect highly-secured units and now it appears that the government is ready to push forward with the bidding for contracts for these camps. Services up for tender include catering, temporary fencing and barricades, laundry and medical services, power generation, refuse collection and other services required for temporary 'emergency environment' camps located in five regions of the United States. (see document below)

> From: Carlton, Bobbi (CED) <Bobbi.Carlton@ky.gov>
> Date: Wed, Nov 16, 2011 at 10:44 AM
> Subject: FW: Potential Subcontracting Opportunity
> To: "Carlton, Bobbi (CED)" <Bobbi.Carlton@ky.gov>
>
> Below is an opportunity for potential contracting.
>
> If interested, please submit a brief description of your services/capabilities to the following: Bob Siefert at bob.siefert@kbr.com. This is all the information that was given to me. Please contact Bob Seifert with any questions.
>
> Kellogg, Brown & Root Services (KBR) is seeking subcontractors on a national basis to provide temporary camp services and facilities as part of its current and future emergency services contracts for the Federal Emergency Management Agency (FEMA), U.S. Army Corps of Engineers (USACE), and state/local government agencies - http://www.aptac-us.org/new/upload/File/RFI%20for%20KBR%2011-16-11.pdf.
>
> **Government, Defense & Infrastructure**
>
> **Project Overview and Anticipated Project Requirements**
> KBR is establishing a National Quick Response Team for our current Federal Emergency Management Agency (FEMA) and U.S. Army Corps of Engineers (USACE) work, and for anticipated future contracts. Upon completion of evaluation, certain subcontractors may be invited to establish a Master Services Agreement (MSA) with pre-established lease rates and terms and conditions.
>
> The Continental US will be broken up into five regions - Services will be required in each State within each region. See map in link above.
>
> **Anticipated Project Requirements:**
> Establish services listed below within 72 hours for initial set-up and respond within 24 hours for incremental services. This is a CONTINGENCY PROJECT and it should be stressed that lead times will be short with critical requirements due to the nature of emergency responses. Subcontractors must be flexible and able to handle multiple, shifting priorities in an emergency environment. Supply lines needed must be short but not necessarily pre-positioned.
>
> The personnel on site to be covered by these services will depend on the size and scope of the recovery effort, but for estimating purposes the camp will range in size from 301 to 2,000 persons for up to 30 days in length.
>
> - The offeror will not have to submit a proposal for each service in each state.
> - Please identify which state and/or region your company can perform the requested services.
>
> *The descriptions of the services are for reference only. Any and all specific requirements will be forthcoming with a detailed Statement(s) of Work in an RFP solicitation.*
>
> **Catering Services**
> This service is open to companies that wish to provide food preparation services only, and to companies that wish to provide the food supplies in addition to food preparation services. Subcontractor shall provide food and food preparation services capable of providing meals per feeding sufficient to meet the prime contractor occupancy levels. All meals may be prepared in accordance with the Army 14 Day Menu program (or equivalent like NIFC Mobile Food Services contract (http://fs.fed.us/fire/contracting/index.htm) and may be enhanced based on individual chef specialties and skills.
>
> **Temporary Fencing and Barricades**
> Subcontractor will mobilize, transport, erect, install and demobilize temporary fencing, barricades, and associated equipment according to federal, state and local laws, codes and manufacturer installation instructions. The Subcontractor shall be able to mobilize and deploy key personnel(s) within four (4) hours of NTP to meet with KBR Site Manager at the Responder Support Camp (RSC) site in order to finalize the site design plan and acquire site specific design requirements and layout.
> Number of linear footage:
> Approximately 2,300LF for a 301 person camp after 36 hours of NTP
> Approximately 3,600LF for 1,000 person camp after 72 hours of NTP

The government contracting organisation, *'Kellogg, Brown, Root'* (KBR)'s tender for FEMA camp service bids came soon after the US Senate passed the National Defence Authorisation Act (NDAA) which expressly permits the military to detain and interrogate supposed 'domestic terror' suspects – in other words, those who overtly disagree with the government, and which is in direct violation of the Fourth Amendment and also Posse Comitatus.

Section 1031 of the NDAA bill declares the whole of the United States as a 'battlefield' environment and allows American citizens to be arrested on US soil and incarcerated at the camps at Guantanamo Bay. This is indeed the stuff of Orwellian nightmares.

Next time you hear anyone say 'it's just not possible' or 'they would never do that,' please show them the above email. The extreme

The Falsification of Science

gravity of Agenda 21 renders it difficult to write about without sounding alarmist, but these Elite psychopaths are obsessed with power and the centralised control over all aspects of human life and innocent-sounding euphemisms such as 'sustainable development,' 'saving the planet,' and the 'green cause,' are often used to mask their true intent. Agenda 21 is indeed worthy of far more public scrutiny than that to which it is currently being subjected.

In fact the whole concept of Agenda 21, when one delves beneath its thin, outer veneer would appear to be as far from true environmentalism as it is possible to get, but as with all things Elite and New World Order oriented, nothing is ever as it would seem. And what is more, it is these very same psychopathic Elite that are actually destroying the very planet about which they pretend concern, in their endless quest for obscene profits and the total control of every aspect of humanity.

Indeed, the same power-mad psychopaths most responsible for the self-defeating 'green movement' are the very drivers of Agenda 21. They falsely claim that the perpetual, colossal consumption of earthly resources, especially energy, cannot be maintained with so many people on *their* planet competing for *their* resources and this is the primary reason that only a deliberately engineered cull of 95% of humanity can safely preserve those Elite 'comfort zones' such as their highly exclusive 'gated communities' liberally sprinkled with sprawling mansions, golf courses, country clubs and helipads. Of course, it takes obscene amounts of starvation, polluted water, destruction of the environment, death, and disease to support even just one billionaire.

So, as the old saying goes, 'something's gotta give' and one thing we can count on for sure is that it most certainly will not be any of the Elite's own home-comforts – if of course they are to have their evil way. Psychopaths have no conscience, no empathy and very little humanity. However, the one thing that really scares them is that we, the 99%, to use a popular current idiom, outnumber them by almost a million to one and the ultimate power belongs to us. Is it that much of a leap to actually organise ourselves to do something about it and stop them in their tracks? Apparently so, judging by the disbelief and apathy I encounter whenever the subject is raised.

But I digress slightly. To wake up the masses to nefarious, crackpot, genocidal schemes such as Agenda 21, how many more loudly ringing alarm-bells do we need? This agenda-from-hell claims to be a "...*comprehensive blueprint of action to be taken globally, nationally and locally by organisations of the UN, governments and major groups in every area in which humans directly affect the environment.*" However, it is really about depopulation, purloining much of the Earth's surface for their own selfish purposes in the name of protecting it, and imposing a rigid control structure that would bring back the Dark Ages for humanity as a whole.

If none of the above makes any kind of sense, perhaps we should ask ourselves this somewhat obvious question...

Why would the very people who are almost entirely responsible for all the pollution, destruction, radiation, poisoning, and devastation of the environment suddenly turn out to be so eco-friendly?

Obviously, this is not the case as Agenda 21 is wholly the work of the bloodline-created and controlled United Nations. It propounds a programme to hijack the world on behalf of its devotees and the ultimate goals include, among many other disturbing facets... (by the way I have not stolen this list from the Communist Manifesto, although you could be forgiven for thinking so)...

- An end to national sovereignty

- State control of all land resources, ecosystems, deserts, forests, mountains, oceans, fresh-water lakes and rivers, agriculture, rural development, biotechnology and ensuring equity (In other words we will all be equally enslaved.)

- State defined roles of business and financial resources

- Abolition of private property

- 'Restructuring' the family unit
- Children raised by the State
- People allocated their jobs
- Major restrictions on movement
- Creation of 'human settlement zones'
- Mass resettlement as people are forced to vacate land where they live
- Dumbing down education (see relevant section)
- Mass depopulation

Indeed contrary to what we are all led to believe, the monsters who instigated this plan are actually communistic in belief and intention. I know this is counter-intuitive in many ways as we are always encouraged to think of them as 'fascists' but nevertheless, that is simply not the case. As I have related in my other books and articles, fascism is not the real problem, whereas communism definitely is. But this is why fascism, past and present is being systematically and constantly demonised – and also why we never hear 'communism' mentioned in this regard. They desperately need you to believe that fascism is the real problem. (For a much fuller, more explicit explanation of this premise, please refer to my book 'The Falsification of History.')

But this is the plan for the entire world and not just for the USA, assuming that the Elite succeed in their nefarious plans.

Harvey Rubin, vice-chairman of the Agenda 21 front operation, the 'International Local Governments for Sustainability,' was asked how all this would affect liberties with regard to the US Constitution and Bill of Rights, private property, and freedom of speech. His reply was short but not very sweet. *"Individual rights must take a backseat to the collective."* Spoken like a true communist – which of course he is, as are they all! (Note the above name and also note the fact that the communist, so-called Russian revolution was 100% instigated and perpetrated by that very same 'tribe' – as were many others!) What does this tell us!?

One of the reasons for the planned systematic dumbing-down of education is encapsulated in this comment in another 'sustainability' document:

"Generally more highly educated people who have higher incomes can consume more resources, than poorly educated people who tend to have lower incomes. In this case more education increases the threat to sustainability."

This of course is utterly terrifying stuff that could have been lifted verbatim, from the pages of 'Brave New World' or '1984.' Have these people no shame at all? That, by the way, is a rhetorical question.

The late Aaron Russo, the award-winning film producer who produced *'Trading Places'* with Eddie Murphy, began to alert people to this conspiracy shortly before he died. He said publicly in 2007 that a member of the Rockefeller family, Nick Rockefeller, had told him that the world population was going to be reduced by 'at least half.'

John P. Holdren, the 'Science Czar' appointed to the Obama administration, is another advocate of 'human culling.' He said that the optimum human population is one billion and he co-wrote the 1977 book, *'Ecoscience,'* which proposes mass-sterilisation by medicating food and the water supply and imposing a regime of forced abortion, government seizure of children born outside of marriage and mandatory bodily implants to prevent pregnancies.

Indeed, there has been nothing of significance omitted from their plan; it guarantees both birth-control and death-control, not by the individual but by the 'State.' It promises the basic necessities of life in return for total and absolute submission to the 'State.' It guarantees the substitution of critical analysis for re-education and brainwashing by the 'State.' It destroys the very fabric of society, culture, and the family unit, proscribing only one way to live, that of the 'State's' own choosing. It 'herds' the population (we 'cattle') into small, easily managed areas in order that we may be more effectively controlled by the 'State' and it creates an environment where we will all be more easily managed via the promise of being taken care of by the 'State.' The Communist Manifesto is alive and well.

Karl Marx, Friedrich Engels, Leon Trotsky, Mao Tse-tung, Vladimir Lenin, and Josef Stalin would be very proud indeed of all of their many, twenty-first century protégés.

COVID-19 – The Real Truth

Once again, I should thank my co-author of this section, Shannon Rowan, for another excellent contribution here. She and I wrote this section jointly and it was originally published on 26th March 2020 on John's Falsification of History website, mere days after the initial 'lockdown' in the UK.

Agenda 21/2030 as described, is the catalyst for the very latest development in this whole, sorry saga. What is probably one of the greatest hoaxes ever perpetrated on mankind took place at the end of 2019 in China and in early 2020 onwards in the 'Western' world.

This totally fake pandemic is already being referred-to by its perpetrators as 'The Great Global Reset' as that is its primary and original function... to facilitate further steps down the road to achieving the initial aims of Agenda 21 by keeping the population of the world in a constant state of fear whilst using the current 'scamdemic' as an excuse to destroy the world economy as we know it and bring about total 24/7 surveillance of our every movement and interaction with each other and society at large.

Indeed, there are several schools of thought amongst those of us who have a 'finger on the pulse,' regarding what is really going on with the COVID-19 'crisis.' But one thing of which we are very sure is that what it is NOT all about is a highly dangerous disease that will kill millions if left unchecked.

Since when did our 'lords and masters' care anything whatsoever about the deaths of we the 'great unwashed,' 'useless eaters,' or 'herd' as they are known to refer to us? Under other circumstances, do our governments ever really care about the welfare of the poor, the sick, the disadvantaged or the elderly? The obvious answer to that is a very resounding, 'no!'

This simple fact alone should be enough to tell us that all the utterly draconian sanctions, restrictions, and loss of our basic freedoms to move around and live freely, right now, are really not at all about protecting 'vulnerable' (or any other) members of society. So, what **are** we to conclude regarding all the COVID-19 hype and propaganda currently being spewed out at us 24/7 through the airwaves and in the written press? Indeed, this veritable, overwhelming cacophony of extremely insidious fear mongering?

We believe that this is a deliberate, mass manipulation of our worst fears and emotions and every available statistic demonstrates that this 'virus,' even if it exists, which it categorically does not, (see chapter 8 and the section on 'Germ Theory') is definitely no more, and in most cases considerably less, 'dangerous' than many other common illnesses. But despite this, the measures now being put in place to allegedly counter this 'dire emergency' are totally disproportionate to any likely threat, either real or perceived.

Indeed, what is currently taking place in every corner of the world now, is akin to martial law. The total lockdown of certain cities and countries, curfews, threats of home confinement for the elderly (and soon everyone else too, no doubt), police and military intervention, closed borders and a moratorium on international travel, could be lifted straight from the pages of some dystopian, futuristic novel or Sci-Fi 'B' movie, except that this is not fiction or the future, this is fact and the present, and we are all living this outrageous and flagrant disregard for our basic human rights and freedoms, right now – and all in the guise of it being for our 'own protection.'

But are 'they' telling us the truth about just how deadly this virus is? Staggeringly, we have incontrovertible proof that they are not!! And

The Falsification of Science

what's probably more significant is that this is not hidden away in some dark corner of the Internet on a crazy 'conspiracy' website but is actually spelled out for us in plain English on the UK government's official website! Don't believe us? Just click on the link below (or key it into a web browser if reading a paper version of this book) and scroll down to 'Status of COVID-19.'

https://www.gov.uk/guidance/high-consequence-infectious-diseases-hcid

In case this link is ever broken or removed altogether, here is a 'screenshot' of the relevant part presented here for your information...

Status of COVID-19

As of 19 March 2020, COVID-19 is no longer considered to be a high consequence infectious diseases (HCID) in the UK.

The 4 nations public health HCID group made an interim recommendation in January 2020 to classify COVID-19 as an HCID. This was based on consideration of the UK HCID criteria about the virus and the disease with information available during the early stages of the outbreak. Now that more is known about COVID-19, the public health bodies in the UK have reviewed the most up to date information about COVID-19 against the UK HCID criteria. They have determined that several features have now changed; in particular, more information is available about mortality rates (low overall), and there is now greater clinical awareness and a specific and sensitive laboratory test, the availability of which continues to increase.

The Advisory Committee on Dangerous Pathogens (ACDP) is also of the opinion that COVID-19 should no longer be classified as an HCID.

The need to have a national, coordinated response remains, but this is being met by the government's COVID-19 response.

Definition of HCID

In the UK, a high consequence infectious disease (HCID) is defined according to the following criteria:

- acute infectious disease
- typically has a high case-fatality rate
- may not have effective prophylaxis or treatment
- often difficult to recognise and detect rapidly
- ability to spread in the community and within healthcare settings
- requires an enhanced individual, population and system response to ensure it is managed effectively, efficiently and safely

In effect then, what they are telling us, utterly incredibly, about COVID-19 is that mortality rates are 'low overall' and that the virus is no longer regarded officially as being an HCID. (High Consequence Infectious Disease). Contrast this with all the media hype and their so obviously staged shots of lines of ambulances, weeping relatives, over-worked hospitals, distraught doctors and nurses and deserted city streets, like something out of a cheap, apocalyptic Sci-Fi movie. What more proof is needed of this than that contained in the above website? Truly shameful!

And even more incredulous is the fact that the nationwide lockdown of its citizens in the UK happened some days after the UK's own official proclamation that COVID-19 is not considered to be an HCID. So why then, after that announcement, was the whole nation confined to their homes (allowing one 'essential' venture outside per day per household), a situation that remained in place for around 2-3 months and was only eventually temporarily lifted in June 2020, slowly eased step by step only for another full lockdown to be announced in November 2020 due to the onset of the proclaimed, yet equally fictitious 'second wave?'

Please check out this astounding quote from the well-respected, mainstream '*British Medical Journal*' ...

"*Politicians and governments are suppressing science. They do so in the public interest, they say, to accelerate availability of diagnostics and treatments. They do so to support innovation, to bring products to market at unprecedented speed. Both of these reasons are partly plausible; the greatest deceptions are founded in a grain of truth. But the underlying behaviour is troubling. Science is being suppressed for political and financial gain. COVID-19 has unleashed state corruption on a grand scale, and it is harmful to public health. Politicians and industry are responsible for this opportunistic embezzlement. So too are scientists and health experts.*" The *British Medical Journal*, 13th November 2020.

And what are we to make of these book excerpts? The first is from a Dean Koontz novel, 'The Eyes of Darkness' written in 1981 – almost 40 years ago! And the second one is from Sylvia Browne's novel, 'End of Days'!

THE EYES OF DARKNESS

"I'm not interested in the philosophy or morality of biological warfare," Tina said. "Right now I just want to know how the hell Danny wound up in this place."

"To understand that," Dombey said, "you have to go back twenty months. It was around then that a Chinese scientist named Li Chen defected to the United States, carrying a diskette record of China's most important and dangerous new biological weapon in a decade. They call the stuff 'Wuhan-400' because it was developed at their RDNA labs outside of the city of Wuhan, and it was the four-hundredth viable strain of man-made microorganisms created at that research center.

"Wuhan-400 is a perfect weapon. It afflicts only human beings. No other living creature can carry it. And like syphilis, Wuhan-400 can't survive outside a living human body for longer than a minute, which means it can't permanently contaminate objects or entire places the way anthrax and other virulent microorganisms can. And when the host expires, the Wuhan-400 within him perishes a short while later, as soon as the temperature of the corpse drops below eighty-six degrees Fahrenheit. Do you see the advan-

...microscopic mites undetectably imported on exotic birds. Known medications and antibiotics will be completely ineffective against this funguslike, extremely contagious disease, and its victims will be dead until it's discovered that the bacteria can be destroyed through some combination of electrical currents and extreme heat.

• In around 2020 a severe pneumonia-like illness will spread throughout the globe, attacking the lungs and the bronchial tubes and resisting all known treatments. Almost more baffling than the illness itself will be the fact that it will suddenly vanish as quickly as it arrived, attack again ten years later, and then disappear completely.

...rides in mental health in the first half of this century will be ex-

The Falsification of Science

"In around 2020 a severe pneumonia-like illness will spread throughout the globe, attacking the lungs and the bronchial tube and resisting all known treatments. Almost more baffling than the illness itself will be the fact that it will suddenly vanish as quickly as it arrived, attack again ten years later, and then disappear completely."

And...

"They call the stuff 'Wuhan-400' because it was developed at their RDNA labs outside the city of Wuhan, and it was the four-hundredth viable strain of man-made microorganisms created at the research center. Wuhan-400 is a perfect weapon. It afflicts only human beings. No other living creature can carry it. And like syphilis, Wuhan-400 can't survive outside a living human body for longer than a minute, which means it can't permanently contaminate objects or entire places, the way anthrax and other virulent microorganisms can. And when the host expires, the Wuhan-400 within him perishes a short while later.. do you see the advantages of this?... the Chinese could use Wuhan-400 to wipe out a city or a country, and then there wouldn't be any need for them to conduct a tricky and expensive decontamination before they moved in and took over the conquered territory."

Is this a case of uncannily accurate predictive programming – or a simple coincidence? Once again, we will let you, the reader, decide. However please be aware that this is a common occurrence and has shades of the 'Titan' novel written in 1898, 14 years before the Titanic disaster and yet which accurately predicted that incident down to the most minute of details.

So, if we are correct in our assertions, what are the purposes and the real root of this apparent worldwide exercise in the control of populations? We can only conclude that there is a hidden agenda of some kind at play, but exactly what that may be, is perhaps still up for debate. However, there are several possibilities, any or all of which could be the case, so we invite you to read on and draw your own conclusions but also please be aware that any attempt to draw this information to anyone's attention on social media will result in the following censorship, which could easily have given Soviet Russia in the 1920s to 1980s, a 'good run for its money!'

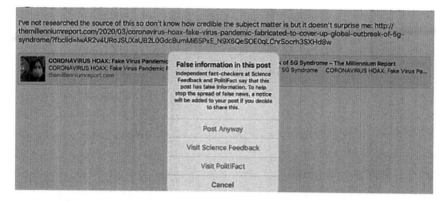

For anyone who doesn't succumb to the media whipped up hysteria or chooses to delve into doing some research of their own, looks like Big Brother is already watching. I tried posting a link earlier and got this pop up, see screenshot. Now THIS is something to be worried about. Looks like it's ok to believe the tripe the newspapers and tv churn out but God forbid you should do some free thinking. Blindfolds back on people, let's be good little boys and girls and lap up all the shite for of course we are only ever fed 100% truth and facts, never false information, right? 🙄 #1984

We are so fortunate are we not, that we have these so-called 'Independent fact-checkers at 'Science Feedback' to correct our gross errors and misconceptions? Go back to sleep everyone. It's all OK, 'they' are taking good care of you!

Mandatory Vaccinations

So, could this 'exercise' be aimed at something as simple as imposing a mandatory vaccine on the entire world, as some free-thinking commentators have suggested? Or is this merely a 'fortunate' (for some)

The Falsification of Science

by-product of what is transpiring throughout the world today? It would certainly generate an absolute mountain of cash for 'Big Pharma,' there is no doubt about that, and it would absolutely not be a precedent either. Think 'Bird Flu,' 'Swine Flu,' 'SARS,' 'MERS,' 'Ebola,' 'AIDS,' and several other fairly recent health 'catastrophes,' all of which had suddenly appeared from nowhere (but mysteriously disappeared, just as quickly) and which also nevertheless desperately 'needed' a new vaccine to combat them. And imagine if you can how many billions (if not trillions) in revenue did these alleged crises generate for the Pharma companies?

Indeed Bill Gates, 'Mr. Vaccine' himself, predicted a deadly 'coronavirus outbreak' as long ago as 2015. And in 2019, several months prior to the 'outbreak' in China, the Pirbright Institute founded by Bill and Melinda Gates, ran a simulation of a coronavirus pandemic which predicted up to 65 million deaths (they wish)! And the lovely Bill and Melinda, just by pure coincidence you understand, also happen to own the patent to this very same coronavirus and are 'working' on a vaccine as we write this, to which they will no doubt own the patent too. But what is really significant is that it is not possible to patent naturally occurring organisms, which should tell us implicitly that the coronavirus is manmade. But by whom and to what ends?

Could this link to what was predicted in 'The Eyes of Darkness' novel? If so, it's interesting to note, that in that (hopefully) 'fictitious' scenario the microorganism referred to as 'Wuhan-400' was not in fact contagious and could not live on surfaces. So if there is any correlation and predictive programming at work and if COVID-19 (as the UK government has recently declared) is not in fact highly infectious (if at all) then why the worldwide lockdown quarantine measures? Is it to keep us in one place as sitting ducks for the directed bio-weapon, and, if so, what is the delivery system? And is 'social distancing' merely meant to keep us all separated in our individual personal spaces in order that their facial recognitions systems can identify us more easily? Or could it be related to the new 5G millimeter wave communications technology currently being rolled out throughout the world? We will explore that aspect in more detail shortly.

Gates is without doubt a Rothschild/Rockefeller 'insider' who is fully aware, and indeed part of the ongoing agenda to subjugate humankind in exactly the way in which it is happening today.

And of course, as most already know, Gates was the founder of Microsoft and who has for some unfathomable reason (unless of course we discount his huge potential fortune) morphed from being a computer software guru into becoming the world expert on health, all without a single medical qualification to his name.

His Microsoft model for generating extreme profits has become the template for his new and soon to be 'magnitudes more profitable' scam, ie. the production of mandatory vaccines which are the alleged cure to a fake pandemic... all designed to be the first of many more to come, as Gates himself openly admits.

The unimaginable profitability of Microsoft was centred upon a monopoly on computer operating systems, the licences of which must be renewed every year otherwise all the admin functions of our computers stop working. And then after a few years, the entire operating systems get upgraded making the predecessors initially obsolescent and eventually obsolete and require the purchase of a new one if we are to prevent our computers from grinding to a sickening halt.

Likewise, his patents on the Coronavirus itself and the soon-to-be-introduced vaccine, allow him to operate a similarly legalised monopolistic scam and he has used his vast wealth and standing to buy-off all the major influencers and world health institutions to facilitate this.

In his new scam, for 'computers,' think 'people' and for 'renewable software,' think 'renewable vaccines.' We are already being conditioned by the complicit media to the effect that the Coronavirus pandemic is simply going to be the first of many more to come. Ask yourself how they know this? The answer will of course be that 'scientists' tell us this with a great degree of certainty. Well maybe these scientists tell us this because they have been told so, too... by certain vested interests who know this to be a fact because it is they who

The Falsification of Science

are actually planning for it to become a reality? It is all so transparent really.

And then if so, how convenient is this for Gates? He has the perfect replacement (plus untold billions more besides) for his former Microsoft profits. Firstly he campaigns and pays out millions to institutions and influential individuals to get them 'on side' and then comes up with a 'miracle' vaccine to stop the non-existent virus in its tracks. 7.7 billion mandatory doses would add up to how much do you reckon? And then of course this is just the beginning.

Once we have all been vaccinated for COVID-19, then oh gosh, along comes the next virus, Covid-21 anyone? Or a mutation of the original one – COVID-19a?!! Well we did warn you that it could mutate and become immune to the original vaccine you know, didn't we? And then maybe next time we will have a completely different virus arrive on the scene which will also need a new, and of course, compulsory vaccine to neutralise it! And so on and on this will go.

The perfect storm in fact! And a perfect money generating machine for Gates. And will he stop there? Oh no, this scam is set to run and run now. Now they have us in their grip with this, they will never let go!!

But, many will argue, is the development of a new vaccine not a 'good thing!?' We strongly believe not. As laid out in no uncertain terms in the chapter on 'health,' vaccines are hyped-up constantly by those to whom we are permanently in thrall and promoted as being the ultimate, effective answer to all disease control. However, surprisingly to most and especially those who are taken in by official propaganda (and we include unknowing health professionals at all levels of expertise in this too), vaccines are definitely NOT what they are widely portrayed and purported to be by those who stand to benefit greatly – and their apologists in the controlled mainstream media. Indeed, vaccines are proven not simply to be totally ineffective, but on the contrary are actually extremely harmful to all, especially, the very young, the very old and those already suffering from an ailment of some kind. Suffice to say that an Internet search for the ingredients of vaccines and their many dangers, will certainly more than prove this point, despite plenty of disinformation designed to portray exactly the opposite.

A Financial Crash?

Some have also suggested that the 'crisis' could be the prelude to and/or designed to facilitate an engineered financial crash. The impact on the economy already is quite devastating and governments are now offering tangible assistance to corporations and businesses in order to help them 'survive' the situation, but of course, there will be little if anything at all to compensate those poor souls whose livelihoods may be curtailed or destroyed completely by the insidious actions of our governments and media.

Again, some may ask how a financial crash could be beneficial to anyone at all, particularly the wealthy, who at first glance would appear to have the most to lose? However, this is definitely not so either! It is a well-established fact that large quantities of money can be generated by financial crashes, particularly by those 'in the know' and who are able to sell their assets at the 'top' of the market and then re-buy those same assets for considerably less after the event. This has been a strategy for centuries and is the foremost reason for all the major financial 'boom and bust' cycles down the ages, all of which are artificially generated in order to enable the already obscenely rich to acquire even more wealth.

And we believe it is likely that this coming devastating financial collapse will be engineered in order to 'reset' society at large and create even greater dependence on government by dint of their 'generous' debt forgiveness and provision of what they are already openly referring to as a UBI – or a Universal Basic Income. At first glance this seems like a desirable outcome for those who would otherwise be penniless and out of work – and there will be tens of millions in this situation. However, before your hopes are raised too high, please forgive us for dashing them prematurely. This proposed UBI will be very basic subsistence level only and even more sinisterly will come with several strings attached.

We will only qualify for this payment upon the acceptance of several stringent conditions. For example it is highly likely that compulsory vaccination on a regular basis will be one of those. As already described in an earlier chapter, the 'cure-all' vaccine will be extremely invasive and insidious and will include such delightful elements as DNA alteration and genetic modification as well as the usual vaccine components, such as mercury and aborted human foetus tissue. Those of us who do consent to it will change – we will in effect become walking 'barcodes' as per the 'transhumanism' agenda. We will gradually become 'plugged into the planned Matrix' and there will be regular 'upgrades' so only those vaccinated and with an 'up-to-date' health passport, will continue to be allowed to travel, by bus, train, or air, etc. And all human rights will disappear since we will no longer be classed as truly 'human,' and as such could potentially be owned by the vaccine manufacturer since we will be carrying patented genetic material belonging to them and not to us.

Already, vaccine trials with young, healthy people, have resulted in serious adverse reactions in 20% of the subjects, with one person paralysed for life and at least one dead (to date – at the time of this book's publication). This will be no ordinary vaccine. It will contain carcinogens, nano-particles, and substances that cause sterility and those who refuse it, may not be allowed to re-enter normal life, and may be shipped off to detention centres and lose their properties. Big Pharma, on the other hand, will make even more money than usual as a result.

1984 is here at last!

But for a truly worrying and unsettling Orwellian slant on the whole affair, allow us to draw your attention to something that, if true, and this is extremely plausible based on what we understand to be going on in the background, whilst we all worry ourselves to death about contracting a probably non-existent virus (or at very worst, one which is less dangerous than influenza)...

According to the website of the MIT Technology Review, a mainstream website and one not naturally given to disseminating so-called 'conspiracy theories,' this is all just the tip of an extremely scary and disturbing iceberg.

The scientists at this 'well-respected' institution, have decreed that in order to put a complete halt to the spread of this 'deadly disease,' society needs to make changes to almost every aspect of our lives as nothing will EVER be the same again! We can only expect this to be the first of a series of further deadly outbreaks and so must be prepared to make many sacrifices and accept even more insidious attacks upon our basic freedoms. Freedoms which of course have been systematically eroded for decades, following closely on the blueprint outlined in George Orwell's terrifying, dystopian novel, '1984.'

And again, according to the MIT Technology Review, here are some or all of the following further draconian restrictions of our freedoms to be actioned imminently (yes we really do have SOME still left at present – but as you will see, probably not for too much longer!)...

1. In order to stop COVID-19 (and all future outbreaks) we will need to radically change almost everything we do; how we work, exercise, socialise, shop, manage our health, educate our kids, and take care of family members.

2. We should expect a permanent 10-person limit on social gatherings to be announced.

3. We also need to impose 'severe' and 'extreme' social distancing.

4. Under this model, the researchers conclude, social distancing and school closures would need to be in force some two-thirds of the time, roughly two months on and one month off, until a vaccine is available. And this is no 'temporary restriction,' this will be a totally radical and different way of life, permanently.

The Falsification of Science

5. Ultimately, however, they predict that the ability to socialise safely will be restored only by developing more sophisticated ways to identify who is a disease risk and who is not and discriminating – legally – against those who are.

6. Israel is already using cell-phone location data, with which its intelligence services track 'terrorists' (for 'terrorists,' read anyone whom they deem to be a threat to their power) to trace people who've been in touch with known carriers of the virus. Singapore does exhaustive contact tracing and publishes detailed data on each known case, all but identifying people by name! This of course will then be extended worldwide – and of course not just to 'terrorists,' but to all of us. And all whilst we continue to slumber in our abject apathy to these machinations.

And amongst other outrages, the report then goes on to state that...

"We don't know exactly what this new future looks like, of course. But one can imagine a world in which, to get on a flight, perhaps you'll have to be signed up to a service that tracks your movements via your phone. The airline wouldn't be able to see where you'd gone, but it would get an alert if you'd been close to known infected people or disease hot spots. There'd be similar requirements at the entrance to large venues, government buildings, or public transport hubs. There would be temperature scanners everywhere, and your workplace might demand you wear a monitor that tracks your temperature or other vital signs. Where nightclubs ask for proof of age, in future they might ask for proof of immunity – an identity card or some kind of digital verification via your phone, showing you've already recovered from or been vaccinated against the latest virus strains."

And perhaps most insidiously of all...

"We'll adapt to and accept such measures, much as we've adapted to increasingly stringent airport security screenings in the wake of terrorist attacks. The intrusive surveillance will be considered a small price to pay for the basic freedom to be with other people."

Well, certainly not by us, it will not and nor we suspect to anyone else who cherishes their rapidly eroding privacy and freedoms! So there we have it, straight from the 'horse's mouth' or at least the mouth of the researchers at Imperial College, London, England, from whence the information in the MIT Technological Review was sourced – and indeed significantly, the source for **all** UK government advice and technical and statistical data on the 'outbreak!' So just to be clear, this is the organisation upon whose every word and recommendation our governments act! It is also highly significant that much of the funding for Imperial college is from the Bill and Melinda Gates foundation.

I make no apologies either for repeating this very important earlier statement regarding any further strict clampdowns on our freedoms. Many people blithely 'parrot' the line, *'if you are not doing anything wrong, you have nothing to fear,'* but this is wrong and irrelevant on so many levels. This glib, oft-quoted statement is misguided in the extreme. Firstly, who decides what is wrong or right? Obviously, one man's terrorist is another man's freedom fighter. Is it wrong for example to peacefully challenge the government if you believe that it is acting illegally or against the best interests of its citizens? Is it wrong to write articles such as this that challenge the status quo and expose what is really happening throughout the world?

You may say 'yes' and you may well be right, we do not have a personal monopoly on the truth, but we do believe strongly that it is everyone's right to peacefully challenge any form of injustice or suppression, as we are attempting to do in writing this piece. But would our own (or any other) governments agree with any sentiments that challenge its very authority to dictate to its citizens what is and is not acceptable? We doubt that very much. In fact the very writing of this article, if it went 'viral' (no pun intended) probably would expose us to covert surveillance, but we strongly do not believe we are doing wrong by writing this and nor we suspect do most free-thinking, fair-minded individuals, whether or not they wholeheartedly agree with the content herein.

And in our view, our privacy and freedoms are not eligible for trading-off against any relaxations in security, whether or not they are randomly or unilaterally deemed to be 'for our own good.' Privacy and freedom are fundamental human rights and not privileges as they would have us all believe – and are certainly not 'currencies' to be used in any kind of barter situation.

Also, do we actually trust any entity that has the power to suppress us if we deviate from their decrees? According to the previous quote above, we would have nothing to fear unless we were 'doing wrong.' But do we trust our governments to make the ultimate decision as to whether we are doing wrong or not? What if it was decided by them arbitrarily that we (or you) were doing something wrong when we patently were not, by any standards? Do we trust governments to always act in the best interests of its individual citizens? We think not. That statement is at best misguided and at worst, a deceit of great magnitude. All governments will do 'whatever it takes' to maintain control and stay in power, without exception and are far from being the benevolent yet bumbling, monolithic institutions that they overtly appear to be and are portrayed to be by the compliant media. Any one of their own citizens who is perceived either as an actual physical or even a passive threat to their supreme power will not be tolerated and will be dealt with severely, either overtly or covertly. This is the very embodiment of extreme communism/ socialism (communitarianism/totalitarianism) and is precisely what we have in place today in our so-called but grossly misnamed 'democratic' societies.

Also, if politicians believe that they have the right to impose any 'law' they wish and police and security forces maintain the attitude that as long as anything is deemed 'lawful,' they will enforce it rightly or wrongly, what is there to prevent complete tyranny? Not the consciences of the 'law-makers' or their legal enforcers obviously, and not even peaceful petitions to the politicians will be effective. Politely petitioning or appealing to oppressors not to be oppressive has a very poor track record of success, historically. When tyrants define what constitutes 'law,' then by definition it is up to the 'law-breakers' to combat tyranny and sometimes the end will justify the means.

Those who are proud to be 'law-abiding' at any cost to their liberties however, may not agree and may even decline to think rationally about this, but what is the alternative? If we do not have the right to resist injustice, even if that injustice is officially called 'the law,' that logically implies that we have an obligation to allow governments to do to us whatsoever they may choose, and also to our homes and families. Realistically, there are only two alternatives; we are either slaves, the 'property' of the Elite, the politicians, and their lackeys, with no rights at all, or we have the right to resist government or Elite attempts to oppress us. There can logically be no other options.

But when 'the law' has deviated from common sense and become an evil tool used for the robbery of others' freedoms, should we blindly and subserviently obey it – or join together in peaceful protest?

So, how do we dissect this truly stunning, not to mention terrifying scenario of the imminent future?

It is our view that THIS is the real overriding reason for the fake COVID-19 scare. Yes, there are several other possible interpretations of all the conflicting and seemingly logical information we receive on an ongoing basis from the lackeys of the media, but reading between the lines of the above report, provides us with very real clues as to the true intent behind this hoax, for hoax it most surely is.

It is well-known amongst the geopolitical researcher fraternity that our 'masters' are constantly seeking new and more insidious ways of deceptively stealing our basic human rights and freedoms in their desperate drive towards a totalitarian state, and so, recognising this to be a fact, what better way to proceed towards their goals than to employ that old favourite of theirs... step forward the good old, reliable Hegelian Dialectic, aka 'Thesis, Antithesis, Synthesis' or 'Problem, Reaction, Solution,' a methodology that was seen to great effect in the 9/11 disaster, as one out of many examples.

For those of you unaware of this particular tactic, it simply goes as follows...

Step 1. Create a **problem** – e.g. The terrorist attacks on several iconic sites (as on 9/11) or the outbreak of COVID-19 in 2020.

Step 2. Await the public **reaction**, which is always going to be as predicted, e.g. 'something must be done!' e.g. 'Attack them in their own

The Falsification of Science

homelands' in the case of 9/11 and 'we need an effective vaccine' and/or 'much more public safety, surveillance and security,' in the case of COVID-19.

Step 3. Offer the ready-made **solution**, which is always rather conveniently, the intended next phase of 'the agenda' anyway.

...and all achieved with not only no protests whatsoever, but with universal approval and gratitude as we consign ourselves into the slavery and servitude we seem to so crave as a species.

So, if our suspicions are correct, this 'virus outbreak' is just the latest in a long, long line of 'false flags,' insidious tactics to further subjugate the human race and we are being subtly manipulated into willingly selling them our freedoms simply to enable us to continue as we were before! Pure psychopathy in action.

In case you still have any lingering doubts about this, here is an excerpt from an online mainstream medical website, offering advice on COVID-19 and its symptoms...

"The symptoms of coronaviruses are similar to any other upper respiratory infection, including runny nose, coughing, sore throat, and sometimes a fever. In most cases, you won't know whether you have a COVID-19 or a different cold-causing virus, such as rhinovirus."

So, incredibly, what they are telling us is that it's next to impossible to discern whether anyone may have this allegedly 'deadly' virus – or just simply a case of common flu or cold – and also that its symptoms are indistinguishable from regular flu. This being the case, how does anyone actually know whether they have COVID-19 or just a dose of regular flu – or even a cold? Quite simply they do not and so this renders all government statistics on deaths and actual transmission rates, null and void. Particularly when patients are reporting being told based on their flu symptoms that they have COVID-19 without being given a specific COVID-19 test at all. These tests, they are told, are being reserved for ICU patients only as there are 'not enough to go around.' But what is not being reported is that there is no specific test to date which can detect this 'new' hypothetical strain of coronavirus, instead they are testing for ANY strain of coronavirus which any of us could 'carry' at any time since we all are capable of housing hundreds of so-called 'coronavirus' strains, sick or healthy. How convenient for them though, to be able to include anyone with COVID-19 symptoms in their official figures, even if they 'only' have a cold or a dose of flu. See images below.

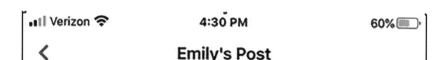

> This 4-hour trip to the ER is the ONLY time I have left home since Thursday, March 12. I had a mild fever March 15 and digestive issues and for a few days I felt like I had a 200-pound weight on my chest, making it difficult to breathe. Then I felt better, then I felt worse, and then I just felt tired and achy. The entire time I had been following all the rules, not going out, calling my primary care doctor, taking Tylenol, etc. We couldn't totally isolate me from the rest of the family because we have three kids under 5 and Steven (who works from home) has to work. Miraculously, so far, no one else in the family has had symptoms. Then on March 24, I was laying on the sofa and felt like someone was kicking or stabbing me in the lungs. This got me to finally go into the ER. The doctors and nurses were so incredibly nice and kept apologizing for putting me in a scary isolation room. I feel very fortunate not to have lung damage. It turns out that pain is just what coronavirus feels like. So, I'm home, I'm ok, everyone is ok. I'm still taking Tylenol and feeling groggy and achy and a little wheezey. Our kids have been total champs through all this. Sage apparently thrives when neglected because she just taught herself how to walk at 10 months old. I feel bad that I got sick, I feel bad that my house is a mess, I feel bad that my kids have been watching too many cartoons. But, I feel happy that I am not more seriously ill and I appreciate all the prayers and the help that has been offered to our family. Much love!

All of which then begs the question, of what does the COVID-19 test actually consist? Interestingly, we assumed that due to Trump's moratorium on incoming flights from Europe (now including the UK and Ireland) to the USA, that all flights would be cancelled. However, not so! Apparently flights are still taking place (at the time of writing anyway) but everyone entering the US is 'screened for the virus' on arrival. When we queried as to what 'screening' actually consisted of, we were told that it would simply be a 'temperature check!' How very scientific then! What they are actually saying in effect is that if anyone has a slightly high temperature then they will be refused entry on the grounds that they have, whether they agree or not, become a

COVID-19 statistic! Of course there is now widespread testing via PCR and other similar tests taking place, which claim 'reassuring' 80% false positive results, and which also come with a disclaimer that they are not to be used for diagnostic purposes! One of the most recent and horrifyingly invasive and clearly sadistic new tests involves sticking a probe up one's nose into one's brain in order to cross the blood-brain barrier. Yet sadly, thousands are submitting to such violating protocols in order to travel or work.

It could also be (or maybe this is just a bonus for them) that they (our masters) are checking the world population's response to a truly worldwide crisis of some kind. And by that I mean, to see how easily they would be able to control us, if they decided to lockdown the entire world (again!) at some future date, for whatever reason? And one thing is almost certain, and that is, that if this 'exercise' goes the way they expect, they will more than likely use it again and again as we have seen constantly with other false flag events.

EMF (Electro-Magnetic Frequency) Sensitivity and a Clear Connection to 5G

The 'virus' may well also be connected to the proposed 5G wireless communications rollout – and quite frankly this is a distinct possibility. The city of Wuhan in China where the outbreak allegedly began in late 2019 was also (by pure coincidence, they tell us!) the epicentre of the large scale 5G rollout in China last year and in that city alone there are over 10,000 5G antennae emitting deadly 5G radiation which is proven to have extremely serious deleterious effects on human (and indeed all life form's) health. These transmitters totally 'bathe' the entire Wuhan area in dangerous electromagnetic frequencies (EMF) and there is literally nowhere anyone can escape being assaulted by this noxious non-ionizing, millimeter wave radiation. Every cubic inch of the city in fact, is being permeated by these damaging waves.

But 'waves' in this case may be a bit of a misnomer since millimeter waves, unlike microwaves, function as 'phased array' antennas, which means they form beams of radiation, much like lasers and thus, like

The Falsification of Science

lasers can be precisely targeted at an object (or person) and so for 5G to reach vast areas, (as is intended), each 'antenna' must contain hundreds of beams of millimeter waves. And since they are claimed to be (there is some debate on this topic) only capable of traveling short distances, the antenna need to be placed every 100 feet (and at street level) in order to form a functional network. Millimeter wave radiation, incidentally, (unlike microwave radiation used for 2G-4G communications tech – which is still a known carcinogen), has up until this point, only been used in weapons' applications such as the 'non-lethal' crowd control weapon called ADS ('Active Denial System') which causes the victims' skin to heat to the point of burning sensations and immediately flee as a result, can in some cases cause 'rare' skin blistering, and never been used or tested for communications applications and never had clinical testing done for human safety. But we should not worry, since we are told by wireless industry leaders that 5G tech is 'perfectly harmless.' Oh good, that's OK then.

Even if the virus exists at all, which we seriously doubt, then prolonged (or even short) exposure to 5G radiation will quickly and severely deplete our natural ability to resist any kind of new assault upon it, leaving the human body entirely vulnerable to any 'virus,' bacteria or disease.

Additionally, if there was some manmade microorganism/bio-weapon (or nanobot?) invented by Bill Gates, (or others), and could act in some lethal (non-contagious) manner, it is possible that millimeter waves (or microwaves) could be used in conjunction with a bio-weapon as a delivery system to 'activate' it in some way by carrying it deeply into the body. Microwaves are already used in hospitals to aid in delivering drug therapies like chemo, since it causes the blood-brain barrier to become permeable and allow the toxin (normally blocked via this protective barrier) into the brain. And it should be noted that 5G technology works in conjunction with 4G (microwave) technology so that each installation of 5G antennas includes the earlier 4G system as well, increasingly populations exposures to harmful microwaves as well as the new millimeter waves.

And it could also be that the 'virus outbreak' was originally designed to cover up the effects that 5G was having on the exposed populations and that they then saw an opportunity to 'use' it to further other agendas?. After all we admit that the releasing of a bioweapon/patented 'virus' upon the world is not easily provable (if at all) and may even be a kind of diversion, giving 'conspiracy theorists' in particular something to sink their teeth into, all the while supporting the likely false mainstream 'viral paradigm' and focusing attention away from the more likely source of illness, the global 5G rollout. By pure 'coincidence' the alleged symptoms of COVID-19 are virtually identical to those caused by 5G! Yes, this is absolutely true, and we would urge you to further investigate this for yourselves.

All illness can be viewed as being tied directly to our environment (i.e. healthy environment = healthy humans). Disease (dis-ease) is a natural response to a toxin (invader) or imbalance of some kind in the body. We react to irritants (dust, pollen, chemicals- which include chemical hand sanitizers and other disinfectant chemicals currently being heralded as **the** solution to protecting from the 'virus') by way of increased mucous membrane secretions intended to push out the irritant/invader such as coughing, sneezing (to clear out sinuses), fever (to raise body temperature to the point which initiates sweating in order to push toxins out via sweat glands in our largest organ, the skin), diarrhoea (to flush toxins from bowels), and vomiting (to remove toxins from the stomach); all supposed symptoms of the 'coronavirus,' and brought on any time we reach threshold levels of toxic exposures in our environment. Thresholds, which will vary from individual to individual, causing some to be more 'susceptible' (reactionary) than others. Some call this reaction an appropriate bodily response to a toxin, indicating a better ability for those bodies to detoxify when needed. Those unable to respond appropriately to toxic overload end up with chronic conditions (often also deadly) like cancers.

For Shannon and millions of others like her (2-18% depending on the population varying by country's levels of pollution—far greater numbers than purportedly affected by COVID-19) who carry the label of living with 'EHS' (Electro-Hyper-Sensitivity, aka Microwave

The Falsification of Science

Sickness/syndrome or EMS, Electro-Magnetic-Sensitivity) and 'MCS' (Multiple Chemical Sensitivity) or simply 'EI' (Environmental Illness) who do respond appropriately to toxic exposures, 'self-isolating' and practicing 'social distancing' have been their only means of coping and surviving in our currently EMR (Electromagnetic Radiation) and chemically saturated world, with levels particularly bad in areas of high population density such as in cities or any place where large numbers of people gather (yes more than the current allowable number of 10!). And often it only takes one person carrying a smart phone or with a Wi-Fi router in his/her home or wearing perfume/detergents or hand sanitizers to bring on flu-like symptoms for sensitive individuals. Wearing a filtered facemask on a regular basis in public, practicing regular hand washing (with non-toxic soaps), fully showering after trips to town, and social avoidance, are the norm, not the exception, for these afflicted people. All practiced, not out of fear of supposed pathogen transmissions, but in order to cope with coming into contact with toxic chemically laden environments and people.

But even with millions living as EI refuges (self-isolating for not just weeks or months but for years, and even decades), have our governments done anything about this particular pandemic of environmental Illness? By their own definitions, more than 2% of a population affected by a condition exhibiting disease symptomology technically earns the epidemic (or pandemic as this is a worldwide problem) label and EHS & MCS both qualify under the World Health Organization as 'functional impairments' (disabilities), yet nothing is done to protect these growing populations of disenfranchised, tormented, suffering people; people who have lost livelihoods, family and friends who disbelieve or do not understand, who are in fact too 'selfish' (a label being tossed about in today's current pandemic-panicked climate for anyone noncompliant with the martial law restrictions to which we find ourselves suddenly subjected) to give up use of wireless devices or chemical cleaners and personal care products in order to protect those afflicted. No one is mandating the disabling of wireless routers and devices in public spaces and replacing Wi-Fi with Ethernet/wired Internet connections, or dismantling/

decommissioning cell towers, or banning toxic petrochemicals in an effort to protect them. Governments and industry leaders (basically one and the same entity) and health protection organizations (such as the WHO) are all perfectly aware of these conditions and the plight of these millions and have been relentlessly petitioned by concerned educators, lawyers, scientists, medical professionals and those suffering with EI to make changes to protect suffering populations and our environment but do nothing. What does this attitude tell you? As we mentioned earlier in this piece, do you think they really care about us and some supposed 'virus' threat?

And as the encroachment of wireless technology increases so too will the number of biologically appropriate reactive people. But these reactions may easily be mislabelled as a viral infection response as they share identical symptoms. (Exposures to both EMF/microwave radiation and/or chemicals can cause headaches and migraines, brain fog, dizziness, nausea, fever, chest pain and coughing, difficulty breathing, insomnia, irregular heartbeat, sweating, disorientation, vomiting, muscle aches and pains, overall malaise, irritable bowel symptoms, and convulsions or seizures.)

Is it really so surprising that the populations most affected by this so-called COVID-19 pandemic have been concentrated in areas of our world with the greatest amount of chemically polluted air and EMF radiation? Is it a surprise that high numbers of people in Wuhan (a neighbouring city to Beijing, infamous for its horrendous air 'quality' so terrible it is common to be unable to see the sun at midday through the black haze) would experience respiratory symptoms? Top that off with the added 10,000 5G antennae in the past year and certainly you will see a rise in the number of symptomatic people.

And this sickness ('virus') is sure to 'spread' as the 5G rollout spreads across our world. Especially as billionaire Elon Musk's tech company 'SpaceX' (an Orwellian name if there ever was one, as it seeks to obliterate space as we know it), continues to launch their dangerous 5G satellites into space. The program started in the autumn of 2019, with the launching of over 120 satellites and plans to add

The Falsification of Science

a total of 30,000 satellites within the coming years, with 'minimal' global service beginning once 420 satellites are in orbit, slotted to commence as early as the end of February 2020. To put this into perspective, prior to 'SpaceX's launches there were approximately 2,000 telecommunications satellites total in the Earth's orbit. When the very first satellites were launched many people around the world fell ill. Scientist, author, and researcher, Arthur Firstenberg, organizer of 'The international Appeal to Stop 5G on Earth and in Space,' gives us a very clear picture of the dangers of this technology and what is in store for us if more satellites are permitted to launch:

"On September 23, 1998, a company called 'Iridium' activated 66 satellites that it had launched into the ionosphere for global cell phone service. On that morning, a majority of electrically sensitive people became suddenly ill, all over the world – so ill that many were not sure they would live. For the next two weeks, birds were not flying in the sky. Homing pigeons got lost by the thousands, and the sport of pigeon racing never recovered. Weekly mortality for the United States rose by four to five percent.

A second satellite phone service, Globalstar, began commercial service with only 48 satellites on February 28, 2000. Again came reports of nausea, headaches, leg pain, respiratory problems, depression, and lack of energy, all over the world, both from "electrically sensitive" people and from "normal" people.

Iridium emerged from bankruptcy and resumed satellite phone service on March 30, 2001. Again, came reports of nausea, flu-like illness and feelings of oppression, as well as catastrophic losses of racehorse foals, all over the world. On June 5, 2001, Iridium added data and Internet to its satellite service. Again, came widespread reports of nausea, flu-like illness, oppression, and hoarseness.

The reason for such a drastic effect from a small number of satellites is not the direct radiation on the surface of the Earth, but the pollution of the ionosphere with millions of pulsed signals. This alters the Earth's electromagnetic environment in which we all live and pollutes the global electrical circuit that passes through every living thing, upon which we all depend for life and health."

'Coincidentally' the flooding of mainstream media with news of the virus outbreak is propped against the backdrop of the massive pushback against 5G technology with the first official global protests (including 200 participating countries) occurring on the same day, 25th January 2020.

And word is quickly spreading and evidence mounting that 5G and biometric systems are being installed in schools while children are forced to stay at home during the COVID-19 lockdown. Many have seen, and even video-recorded towers and antennae being installed on school grounds during the quarantine. Fleets of suspicious white vans with logos linking them to Biometric companies have been parked outside of schools, entering empty buildings under the pretence of disinfection orders.

"My sister is a teacher in TN and they have been told to stay out while maintenance cleans."
-Break the Chains, March 20

"The installation of 5G towers while everyone is being 'quarantined'."
-Phoenix M, March 20 (BC)

"At least 40+ vans. Not including the ones parked in corner of buildings or in front of schools."
-Ivonne J, March 20 (North Texas)

"They are installing 5g around schools here in Richmond, Virginia today-the first day schools are closed due to this fake coronavirus. Trucks are out digging up roads for the fiber installation of 5g."
-Commenter, March 17

"This is a Federal Company, a simple google search leads you to their licensing. They are sub licensed as "Business Radio Licensing", pretty generic name huh? which is a front company for the feds and 5g."
-WTF198, March 19

"Im actually watching these Persons in their altech trucks placing 5G boxes on poles right outside of my apartment building right.now. They started down the whole block actually."
-Wanda F, March 17

"I live across the street from two schools... This morning I saw 2 white trucks and men wearing yellow vest. They seemed to be with the electric company but nothing on their truck or person said so."
-Barbara L, March 18

" One of the shopping malls in downtown Seattle is closed till April 2nd "because of the Virus" – A worker there told my Mom they are installing Five Jee during the closure. All dining establishments and our library are also closed and one of the downtown Starbucks is doing "renovations" – I think they are using the Virus as a cover for the nationwide roll-out."
-Rusty S, March 19

These are not just Internet rumours and 'conspiracy theories.' These claims are being verified and these actions documented.

Josh Del Sol, activist, and filmmaker, famous for his documentary about smart meters, 'Take Back Your Power,' posted the following on his blog, March 21:

"Our concern is that 5G could be installed without our knowledge while we are grappling with the fallout of the COVID-19 pandemic, and that the installation of biometric systems could be a part of a more sinister agenda.

Namely, we are concerned that after the COVID-19 pandemic passes we will still be dealing with the repercussions of newly installed 5, biometric systems, thermal imaging cameras or even temperature guns to detect who MAY have COVID-19.

Even more worrisome is the idea of government-mandated vaccinations; and, for example, that only those who can prove they have received the COVID-19 vaccine (once it's developed) will be allowed back to work, school, public parks, public transportation, etc.

This is even more worrisome when you consider that Bill Gates – a long-time proponent of vaccinations and population control recently stated: 'Eventually we will have some digital certificates to show who has recovered or been tested recently or when we have a vaccine who has received it.'"

The Internet of Things and YOU

'What could this 'digital certificate' possibly mean?' asks Josh, as do we. It may be an electronic Nano-tattoo, which Gates himself is said to have invented, tellingly named 'ID2020,' which is in effect a digital microchip implanted under the skin and was already being pushed forward back in 2018 as a means to ID vaccinated kids in schools.

5G along with AI (Artificial Intelligence) is supposed to power the 'Internet of things,' which effectively enables all of your household appliances to spy on you (and for citizens to be tracked in real time, especially once implanted with microchip IDs). And this microchip RFID tag (similar to what is already being used to track wild animals and household pets), has the very real potential for being implemented as a federal mandate if it can see its way into any of the new 'emergency funding' bills, with any prior resistance easily overridden in the midst of a 'health emergency.' Getting micro-chipped (something which when previously proposed, the average person would have protested with an 'over my dead body' response), will be an easier sell in the current climate of panic, especially if it promises any kind of 'return to normal' for the masses.

And up until now, many citizens have expressed concern over proposed 5G grids due to impacts on health/environment and its enabling of increased surveillance and breaches of our constitutional rights to privacy. For these reasons tech giants have had trouble selling to the public. After all who really cares about remote controlled toilets and fridges? But with the new virus pandemic scare they have the perfect excuse to push forward long planned-for increased surveillance technology on a resistant public, under the guise of enabling contact-tracing programmes, all in the name of public safety and protection.

In Shannon's previous, rural community, her own local utility company is also taking advantage of the pandemic distraction to install the controversial 'smart' meters against which her whole community fought nearly two years. Suddenly the utility workers are on the streets, seemingly untouched by quarantine laws, and when stopped have admitted their plans to smart-grid the whole valley within the next month. (This in April 2020.) The community residents, now heavily caught up in the pandemic panic, fail to notice they are sitting ducks and will not now protest or make much effort against the meter installations as they have 'more pressing' matters at hand. All the while ignoring the thing they should be most afraid of, which can (unlike the virus hoax) actually harm and kill them. And it can be guaranteed that the disease/'virus' is sure to 'come back' and in 'full force' after everyone returns to school and work with the new 5G installed, fully ignorant of reports like the one coming from Ripon, California, where a 5G antenna was removed from a primary school after parents began suspecting a link between the installation and a number of cancer cases at the school.

The 'Spanish Flu' Pandemic of 1918-1919 – Reprise!

Of course, the 'Spanish Flu' is garnering new space in our collective psyche as it is being referenced as an example of just how bad things could get for us now and used as another excuse to add more bars to our growing collective prison. And this 'flu' was the first time any 'flu' (the 'flu' similar to 'AIDS' is a set of symptoms supposedly caused by several strains of different influenza viruses, although it used to be attributed to a bacteria and initially even thought to a be a German bio-weapon during the 'Spanish Flu' outbreak) had taken over large numbers of populations. Previous to the Spanish Flu, flu was practically non-existent. Arthur Firstenberg, connected the dots between the Spanish Flu outbreak of 1918 to EMF radiation exposures in his 2017 book, 'The Invisible Rainbow.' Using epidemiological data, he makes a strong case for the 'flu' coming about as a direct result of exposures to newly introduced manmade (non-native to our environment and thereby also to our bodies) electromagnetic fields via electrification. Prior to that year most of the world did not have an electrical grid and it was at that time when the 'rollout' of the electrical grid really gained momentum. The 'flu' just happened to only 'break-out' in those countries that had just been electrified. This is how it came to be 'pandemic' especially, as Firstenberg astutely points out, where in some cases there were no regular means of travel between affected countries on opposite sides of the Earth and no possible way a pathogen like a virus could have spread so far so quickly. Instead what these countries shared was the fact that their populations were now being ubiquitously exposed to an environmental toxin via the electricity newly introduced to their homes and cities.

And as research has revealed (and as covered in a previous chapter), it was not the 'flu' that actually killed those millions who died but instead the treatment for the flu symptoms in the form of aspirin. As aspirin (manufactured by Bayer) had just come to market shortly before the outbreak and at the time had much laxer safe dosage recommendations than today, because no one had any real idea as to what a safe dosage consisted of. It is a known fact (and is listed under 'side-effects' on aspirin labels) that aspirin can cause pneumonia (amongst several other nasties) and it is pneumonia to which sick patients succumb and die, not actually from the 'flu' itself. During the Spanish Flu outbreak, it was only patients receiving allopathic medical intervention and being given high doses of aspirin who died, none of the patients who opted for treatment by homeopaths (traditional natural medicine doctors) died. But these are the kinds of statistics buried and hidden from general public knowledge. And now, in the wake of the current pandemic scare, it is near impossible to find statistical mortality data related to modern day flu cases. Coincidence? Or does someone not want us to take notice of the fact that regular, annual winter flu claims more lives than COVID-19 has so far?

Ever since the Spanish flu pandemic, the world has not been free from the grip of Influenza, which has accompanied the spread of electrification in the early 20th century and subsequent spread of radar, wireless technology grids throughout our world and satellites orbiting space, up until today. Increasing in strength and reach each time new, artificial non-biological EMFs are added to our environments and cause our bodies to react in kind to this intrusion. The only reason it is seen to be 'seasonal' is because in winter we tend to have less exposure to natural beneficial forms of EMF (from the sun on our bare skins and earth under bare feet) and more exposure to artificial EMF (indoor lighting, electric heating, more time in front of computers, etc.) thereby increasing symptoms. And as we now carry our EMF radiating devices with us everywhere we go, even while outdoors, we see an increase in 'summertime flu' as well.

Now that the world seems to be under quarantine orders, in many cases people are confined to indoor environments with extreme restrictions on outdoor movements, severely limiting their ability to stay healthy and engage in self-care. And we see our masters dictating to us what is considered an 'essential good' or 'medicine.' Amazon™ is now 'prioritising' the sales of 'essential' goods and many other businesses are following suit, not merely because of public and social pressures, but because of government agency mandates. The CDC, using the powers bequeathed them under 'The Model State Emergency Health Powers Act' drafted 21st December 2001 (in the wake of the last big scare used to subjugate the people of the world to new control measures; the World Trade Center attacks, aka '9/11') they are allowed to place restrictions on sales of anything they deem 'non-essential,' are able to enforce quarantines, medication, hospitalisations, surveillance of suspected contagion carriers (very similar to suspected terrorists), road and building closures, dictate disposal of human remains, etc.

You can 'bet your bottom dollar' that natural medicines (herbal, homeopathic, vitamin supplements and maybe even organic foods) will not be listed as 'essential' or 'medical' by our masters. This is another benefit to the pharmaceutical industry (which has strong ties to the CDC) whose profits are being threatened by the renewed interest in 'alternative' (oftentimes actually ancient medicine, long predating the disaster we now call 'modern medicine') holistic non-toxic medicines.

And it is also unsurprising that online blogs and other posts where authors offer alternatives to keeping healthy and 'pandemic-free' are coming under fire by the 'authorities' (who in reality have no authority at all) and being shut down, even when well referenced with valid scientific citations. This information, and any other that goes counter to the establishment, related to the so-called 'crisis' is being labeled as 'fake news' and a 'threat' to public health and safety. It is not hard to see the terrifying implications of this extreme curtailing of free speech and what may happen in the near future to those spreading this information. Will critical thinking and attempts to uncover truth soon become an offense on par with treason and incur the same harsh penalties for treason or 'terrorism'?

As of 25th March 2020 in the UK, the recently voted-in, far-reaching extra powers granted to the government amongst other measures, now allow for the 'sectioning' (compulsory committal to a mental health facility) of individuals deemed to be suffering mental health issues, to be sanctioned by one doctor only, whereas previously the signatures of two doctors were required! Why? And what connection to this 'pandemic' could that possibly have? Well, let's see. Could it perchance be a way of countering the claims of those seeking the real truth? With only one doctor's approval now required, how easy would it be for them to commit anyone who dares to question their actions, by appointing a 'controlled' (by them) doctor to endorse that committal on the flimsy grounds that anyone doubting the government's word on this (or anything) must be insane? It may seem a fantastic notion, but this is exactly the way it worked in the Soviet Union in the 20th century.... But we digress, slightly.

Dr Thomas Cowan, MD, and author of several excellent alternative medicine books, speaking at the 'Human Health and Rights' summit in Tucson, Arizona, 12th March 2020, related that the famous philosopher, Rudolph Steiner, when asked about the cause of the Spanish Flu, responded with, *"Viruses are simply excretions of a toxic cell."* Cowan expounded thus...

"Viruses are pieces of DNA or RNA with a few other proteins. They bug-out from the cell. They happen when the cell is poisoned. They are not the cause of anything. And the first way I would encourage you to think about this is if you were a famous dolphin doctor and you've been studying dolphins in the Arctic Circle for hundreds of years, or at least a long time, and they call you up saying, 'Fred, all the dolphins, or a lot of the dolphins are dying in the Arctic Circle, can you come and investigate?' And you have one question to ask, how many of you would say 'I want to investigate the dolphin to see the genetic makeup of that dolphin'? Nobody. Because that is stupid. How many of you would say 'I want to see if this dolphin and that dolphin has a virus because it might be contagious and that's why all these dolphins are getting sick.'? How many of you would say, 'somebody put some shit in the water'...like Exxon Valdez? Anybody? Everybody. Because that's what happened. And the cells get poisoned, they try to purify themselves by excreting debris, which we call 'viruses.'...I had a dramatic example of this when I was growing up. Right outside our house there was a wetlands and it was full of frogs and the frogs kept me up at night and in the spring they made a big racket, and over time the frogs were all gone. How many think the frogs had a genetic disease? How many think the frogs had a virus? How many

think somebody put DDT into the water? That's what happened. [DDT in the water] Diseases are poisonings. So what happened in 1918 [with the Spanish Flu]? With every pandemic in the last 150 years there was a huge, quantum leap in the electrification of the Earth. In 1918 there was the introduction of radio waves around the world. Whenever you expose any biological system to a new electromagnetic field, you poison it, you kill some and the rest go into a kind of suspended animation, so interestingly they live a little bit longer and sicker."

A Pandemic of Testing

Researcher and author, David Crowe's 14th March 2020 'GreenMedinfo' article, 'Does the 2019 COVID-19 Exist' shares compelling evidence which calls into question not just the existence of the supposed COVID-19 virus but the reliability of the RNA testing being used to determine the number of infected cases. He makes the point that when countries stop testing for COVID-19, then there will be no more new cases. *"The definition, which assumes perfection from the test, does not have the safety valve that the definition of SARS did, thus the scare can go on until public health officials change the definition or realize that the test is not reliable."* So really this pandemic can go on as long as 'they' wish. It is a classic case of 'the Emperor has no clothes.'

Does the virus even exist? David shares his viewpoint:

"What I learned from studying SARS, the previous big coronavirus scare, after the 2003 epidemic, was that nobody had proved a coronavirus existed, let alone was pathogenic. There was evidence against transmission, and afterwards, negative assessments of the extreme treatments that patients were subjected to, the nucleoside analogue antiviral drug Ribavirin, high-dose corticosteroids, invasive respiratory assistance, and sometimes oseltamivir (Tamiflu)."

We should definitely be much less worried about being infected with an unproven virus than we should be afraid of the treatments to which those displaying symptoms may be (and are being) subjected. There are reports circulating of experimental anti-viral drugs being used on hospital quarantined patients (and when they die from treatment they can claim they succumbed to the virus) and now rumours in the mainstream media that controversial (quite dangerous) antimalarial drugs (Hydroxychloroquine) may be effective in treating COVID-19. In fact, President Trump came under fire on 23rd March 2020, after three people were hospitalised in Nigeria from drug overdoses of Hydroxychloroquine, having self-medicated with the drug after Trump publicly and repeatedly touted, without evidence, the possibility of using the antimalarial drug for treatment of COVID-19. The Nigerian government is now urgently warning its citizens to not self-medicate with unproven treatments.

The Trouble with RNA Tests

David Crowe opines that the world is *"…suffering from a massive delusion based on the belief that a test for RNA is a test for a deadly new virus, a virus that has emerged from wild bats in China, supported by the western assumption that Chinese people will eat anything that moves."*

We are so conditioned to be lured in by fairy tales that we quickly fall for stories of evil, deadly viruses lurking deep inside bat caves, just like the stories we are told about the origins of life emanating from 'deep sea vents,' and other such fantasies, which currently pass as 'science.'

The bat cave origin for the COVID-19 story we found in an 11th March 2020 article from the 'respected' (read – 'industry-funded'), 'Scientific American,' titled 'How China's "Bat Woman" Hunted Down Viruses from SARS to the New COVID-19' with the tagline reading 'Wuhan-based virologist Shi Zhengli has identified dozens of deadly SARS-like viruses in bat caves, and she warns there are more out there.'

Oh no, there are 'more out there!' Be very scared everyone, this is not the end to the threat; the bat caves are harbouring more sinister viruses which could be released on an unsuspecting public at any time! And in fact at the time of writing, mere days before publication of this book, the UK government has just announced that an 'even

The Falsification of Science

more dangerous, mutant strain of Covid-19' has now struck, and which is 'terrifyingly' 70 times more transmissible than the original strain. How they can possibly know this, is just utterly unbelievable yet so typical of the fear-porn emanating from whatever passes as 'science' in these dark days!

But not only can viruses originate in deep, dark bat caves, we are told they are also being birthed in the 'wet markets' of China as presented in the Netflix 'documentary' November 2019 film 'Pandemic,' the narrator engages our imagination with the following...

"This is a wet market in the Lianghua, China. Unlike markets in much of the West, where animals are already dead when they arrive, this wet market sells meat that's very fresh. It's killed on sight. That's what makes it a disease factory...All the while their viruses are mixing and mutating, increasing the odds that one finds its way to humans."

Wow! 'Mixing and mutating' sounds very scary and oh so scientific! It is totally illogical and certainly plays on our prejudicial attitudes about those weird bat-eating Asians whilst preserving our belief in Western ways as being superior and more civilised than those in the East.

But back to the trouble with testing, Crowe continues with:

"If the virus exists, then it should be possible to purify viral particles. From these particles RNA can be extracted and should match the RNA used in this test. Until this is done it is possible that the RNA comes from another source, which could be the cells of the patient, bacteria, fungi, etc. There might be an association with elevated levels of RNA and illness, but that is not proof that the RNA is from a virus. Without purification and characterization of virus particles, it cannot be accepted that an RNA test is proof that a virus is present.

This strange new disease, officially named COVID-19, has none of its own symptoms. Fever and cough, previously blamed on uncountable viruses and bacteria, as well as environmental contaminants, are most common, as well as abnormal lung images, despite those being found in healthy people. Yet, despite the fact that only a minority of people tested will test positive (often less than 5%), it is assumed that this disease is easily recognized. If that was truly the case, the majority of people routed for testing by doctors should be positive.

The COVID-19 test is based on PCR, a manufacturing technique. When used as a test it does not produce a positive/negative result, but simply the number of cycles required to detect genetic material. The division between positive and negative is an arbitrary number of cycles chosen by the testers. IF positive means infected and negative means uninfected, then there are cases of people going from infected to uninfected and back to infected again in a couple of days."

One reader's response to Crowe's article succinctly summed up the situation at hand:

"Everyone is operating in fear.

- *The doctors are afraid of public opinion if they just send symptomatic people home to drink hot tea and eat garlic.*

- *The lab techs are worried that if they don't count even negatives as positives, then they might get sued or lose their jobs if someone dies.*

- *The medical establishment is afraid to step away from the status quo 'medical remedies' of hand sanitisers/steroids/supplemental oxygen/inhalers,...because if they DO create doubt in the public mind about these 'sanctioned' treatments, people might just get too smart...and then they might sue their doctors for aggressively and invasively treating something that might have been better served with chicken soup.*

- *Nobody wants to be a high profile, public naysayer (and good on David Crowe's head for speaking out!) because they are afraid of public opinion, litigation, and censure from those in high places.*

It's a huge panic bubble, and the principalities of this world must be having a heyday watching everyone scramble around in out-of-control paranoia."

The Falsification of Science

During the infamous Spanish Flu pandemic, Dr Joseph Goldberger working for the US public health services, and wishing to establish proof of contagious human-to-human transmission for the 'Spanish Flu' conducted a large-scale experiment involving one hundred volunteer prisoners. The prisoners participating in the experiment were promised reduced sentences in exchange for their help in the study. The healthy prison volunteers were subsequently exposed to mucous secretions of extremely sick, terminally ill prisoners. Secretions were swabbed into their nostrils and in their throats. They were exposed to coughing and sneezing in their faces. They were even injected with bacteria from the sick patients. Yet not a single one of them fell ill in the slightest and thus Dr Goldberger had to admit failure. There are in fact many other such examples of similar experiments with all of them resulting in failure to prove diseases to be contagious leading many researchers in the field (such as Crowe) to disbelieve the 'infectious myth' and 'Germ Theory' (that vastly over-used word 'theory,' again!) that all of modern 'science' is now based upon. Take that myth away and the house of cards that is our current medical establishment falls to the ground.

And yet even though this supposedly 'new' coronavirus displays none of its own unique symptoms and 'fever' is open to interpretation depending on the beliefs of individual health practitioners. There is no universal definition for 'fever,' and as Crowe points out in his article, for SARS *"a fever was defined as 38°C even though normal body temperature is considered to be 37°C (98.6°F)"*, *temperature is being used to signal out suspected viral cases."* (This in spite of the fact that some people will run higher normal temperatures than others especially if they have medical conditions such as hyperthyroidism.) And in spite of these problems with using body temperature as a litmus test, airports all over the world have now installed or are installing infrared body temperature scanners or airport personnel are individually taking passengers temperatures. If your temperature happens to be a little high you may be subject to medical detention for further testing, or denied passage, deported, or quarantined.

Yet President Trump was quoted as stating that the virus should 'burn itself out' by early-mid April because there will be warmer weather and viruses cannot survive the increased heat. (Trump had also first declared the virus a 'hoax' before he was presumably pressured politically to change his attitude and 'play ball' if he had hoped to win re-election.) And some 'experts' have backed up this claim, stating we may see a drop-off in cases with a change in seasons but also stating that this means the virus could come back again in the fall or winter and with a vengeance (as with the shark in the 'Jaws' movies, 'this time it's personal!'). But this begs several questions; if the virus can be killed off by heat/increased temperatures shouldn't we welcome and encourage fever in sick patients? Also, should we not then simply treat patients or even those among us who could be carriers with sauna therapy? And how does this explain the spread of the 'virus' throughout the world, in countries like Australia where it was still summer when the outbreak occurred or in the tropics where it is hot year round?

'Hidden in Plain Sight' . . . .

. . . reads the headline of one of the latest *New York Times* Op-ed propaganda pieces (they are nothing if not persistent, not a day goes by that our lords and masters do not dispense their fear mongering), tells us that 'new studies show' more asymptomatic people could be carriers of the virus and responsible for spreading it 'more than was previous supposed' . . . no facts or data or even names of scientists and their mysterious studies to back up these claims.

However even with the media *infecting* us with their lies about asymptomatic carriers (used as excuses to be sure none of us walk the streets a free people during this worldwide lockdown), hospitals offering free testing for COVID-19 will only test those demonstrating symptoms. This is to ensure that the public does not become suspicious of the test results. With a 70-80% likelihood of false positive test results it would soon become apparent to the general public that something is amiss when growing numbers of healthy people test positive yet never display symptoms and never infect anyone who comes into contact with them. With sick people testing positive (even if a small percentage) they can support their viral paradigm (or 'theory,' which is all it is – another insubstantial, unproven theory) that the virus causes the sickness even though the virus was never isolated (or purified) and no clinical trials were performed, and Koch's postulate was ignored.

(Koch's postulate (criteria set by German bacteriologist Robert Koch in the late 1800s, and which is supposed to be used today in order to establish disease including viral disease) requires that the virus/pathogen be purified and that clinical trials are conducted (on animals) by exposing them to the pathogen and verifying that the same illness is produced in each case. But this postulate has been completely ignored.)

It has also recently (again, just prior to publication), come to light that an FOI (freedom of information) request to both the UK and Republic of Ireland governments asking whether or not the COV-SARS2 (Covid-19) virus has ever been isolated, yielded the honest response in both instances, 'no.' In other words, and for clarity's sake, it is therefore impossible to test for it and more importantly and significantly, therefore... IT DOES NOT EXIST. So whatever all these alleged tens of thousands of people are dying of, it is certainly NOT Covid-19.

But, the key here is to keep the belief in the public's mind that the virus is real, and very contagious. The conclusion that this particular virus is so very contagious derives from assumptions based on a need to explain why people who did not come in close contact with one another became infected. 'Because there was no close contact and the individuals displayed similar symptomology and we believe the virus is the cause, therefore the virus must then be very contagious, can live on surfaces for days or weeks, etc.' This is circular logic and thus belief-based, not hard scientific fact.

If the general public ever gets wind of this incredible hoax to which we have all been subjected in the past weeks and months, imagine the backlash and the loss of faith in our institutions and leaders? And the incredible anger and upset over loss of jobs, livelihoods, income, in some cases homes, shelter and lives due to the panic and insane decisions our leaders have made in order to 'protect us' from an unproven non-existent threat. The real threat then is not the fake virus but our rulers. The truth simply must not come to light and must be obfuscated at all costs.

Robot, Good--- Human, Bad

The COVID-19 outbreak is at its core, also part of a transhumanist agenda serving to influence humanity into viewing itself as a pathogen, a plague upon the earth, in much need of correction and control. Transhumanists believe that it is inevitable that humans merge with machine/technology (which they call 'the singularity') and that this transformation is somehow necessary for our own good and survival as well as that of our planet and also part of a 'natural' process they call 'evolution.' The pandemic scare helps to further this agenda by removing obstacles to the 'necessary' global 5G grid, pushing forward human microchip IDs and use of emergency relief funding to further develop and implement AI technology and robots to meet the rising demand for aid to hospital workers and in the workplace at large. In other words, while we stay sequestered in our homes the robots can take over.

SpaceX founder/futurist tech guru, Elon Musk, while standing to profit immensely from Artificial Intelligence and related 5G technologies (as previously mentioned, being the driving force behind the irradiation of our atmosphere via many thousands of 5G satellites), oddly is known for his outspokenness against AI, continually warning us of its dangers. His motivation, it would appear, in igniting our fear of AI, is also linked to transhumanist agendas. Musk warns us that AI will soon surpass human intelligence thereby making humans obsolete in many job markets and then presents us with his 'solution' to this dilemma. Namely, that humans connect themselves with artificial intelligence (via the AI-powered 5G 'cloud' enabled in large part by Musk's own 'SpaceX') by way of microchip implants into our brains! We are really not making this up!

The Falsification of Science

Humans are a Virus

Our elite masters do not view the masses as 'human' at all. They believe themselves to be the only true humans and the rest of us cattle or some kind of genetic mutation. However, the reverse is closer to the truth since our leaders do not demonstrate truly human traits such as empathy and love. And they themselves are responsible for designing and implementing the ecologically destructive systems in which we find ourselves imprisoned, though they love to cast the blame for ecological disasters on 'we the people.'

Alien 'Agent Smith' from 'The Matrix' movies, possibly representational of our non-human (or sub-human) psychopathic leaders, tells us exactly what our leaders think of humanity; nothing more than a virus or scourge.

"I'd like to share a revelation that I've had during my time here. It came to me when I tried to classify your species and I realised that you're not actually mammals. Every mammal on this planet instinctively develops a natural equilibrium with the surrounding environment but you humans do not. You move to an area and you multiply and multiply until every natural resource is consumed and the only way you can survive is to spread to another area. There is another organism on this planet that follows the same pattern. Do you know what it is? A virus. Human beings are a disease, a cancer of this planet. You're a plague and we are the cure."

And at another point in the film he says:

"I'm going to be honest with you. I hate this place, this zoo, this prison, this reality, whatever you want to call it. I can't stand it any longer. It's the smell if there is such a thing. I feel saturated by it. I can taste your stink, and every time I do I fear that I have somehow been infected by it. It's repulsive! Isn't it?"

So, we the unwashed masses, 'stink' and are 'infectious' and 'carriers' of all manner of disease and pathogens. We carry this stigma, but machines do not. (However, it is totally overlooked that the use of these machines cause illness and even death due to the toxic EMFs and chemicals to which they expose humanity and that the manufacturing of them causes severe degradation to our environment.)

In fact, these 'benign' and 'intelligent' machines are portrayed as 'helping' us as can be seen in videos of robots admitting patients in Wuhan hospitals found on YouTube™. And technology (especially wireless) is aiding us in carrying on with work, school and socially while under quarantine.

'We Live in Zoom Now'

With this new forced social isolation our right to assembly under constitutional law has been stripped away and now meeting only in the virtual reality is becoming the accepted form of human-to-human interaction.

A New York Times 22nd March 2020 article announced, *"We Live in Zoom Now – Zoom is where we work, go to school and party these days."* We are being pushed into 'meeting' on a virtual platform (which has the added benefit for our masters of recording all of our conversations during these meetings) as our only means of socialising. Even some in isolated rural communities such as the one in which Shannon lived until recently are heralding the advantages of 'meeting on Zoom' and suggests we all 'meet' there in lieu of any in-person meetings, even when we are still legally permitted to gather in groups with ten persons or less.

Increasingly (under the NWO orders) we are being denied physical contact with other humans, which is a basic human need (and thus right), but this need is treated as if it can be replaced with virtual contact and that virtual contact is its equal when it is most definitely not. And now under the new COVID19 'laws,' gatherings of first more than 500, then 200, then 30 and now, incredibly, 10 (with numbers changing daily it would seem), is considered and treated as a criminal offence and as such carries harsh penalties and criminal punishments.

But nary a mention of 5G or other carcinogenic hazardous EMF radiation as being criminal, (even though many millions, including your author, argue that this vast experiment being carried out on the world's population is in fact criminal in the extreme) and (as mentioned previously) certainly no mandates are ordered to halt or reduce exposures to wireless radiation while it is considered a criminal offence instead for victims of forced 'smart' meters and soon to be forced 5G antennas on their homes and properties to refuse the above, even and especially on the grounds of health concerns.

A New Type of War

The 'War on Terror'™ has been a failure in the eyes of our elitist masters. It has fallen short of the mark in realising their New World Order wet dreams and it has failed to instil enough fear in the people of the Earth and bring them to give up all of their liberties and submit to one-world governance. Hence the need for a new threat, one that is not limited to small groups of people or demographics as was the case with 'terrorists.'

And with a 'contagion' anyone could be a threat (carrier), anyone-- infants, elderly, or your neighbour's dog. It is indiscriminate, can affect any race, ethnicity, or gender. (And as of 23rd March 2020, we are told 'might' even be airborne!) However discrimination against marginalised people (homeless, and others without means to comply with quarantine/social distancing orders), defiant people (those who refuse to buy into the NWO propaganda, or who understand the science to be faulty), and the elderly (those who are under stricter containment laws for 'their own protection') will prevail.

And now in the UK, police are being given 'greater powers' to enforce lockdown because too many people are 'flouting' the 'social distancing' guidelines. The new powers allow them to set up roadblocks and checkpoints, and to arrest lockdown offenders. One of the more ludicrous headlines on the BBC homepage at the start of the 'pandemic' stated, *"Coughing on Emergency Workers to Be Punishable by Jail."* And on the other side of the 'pond' in the United States, Oregon state's governor announced that she did not want *"...to HAVE to arrest people"* when addressing the issue of the people not complying with quarantine measures. All this is happening right now in our laughably, so-called 'free' and 'democratic' nations!

Never before, certainly not since WWII, have world governments had this much carte blanche to restrict the freedoms of the people. (Curfews, severe travel restrictions, quarantines, military roadblocks, etc.) While we are forced indoors, our invisible wireless radiation prison will continue to be erected with extra billions in 'emergency' funding from the US government allocated for 'medical surveillance' technology and 5G being covertly installed in now empty schools as students are forced to study from home, and 5G boxes are placed outside our homes to spy on and sicken us.

During the 9/11 (so-called) 'terrorist attack' scare, flights were only grounded for one day and even that rocked the world and had the people in shock and awe. But with this latest 'pandemic' scare, flights worldwide were grounded for weeks and months. (However it is interesting to note that this moratorium on flights does not seem to have curtailed the constant 'chemtrail' aerosol spraying from jets crisscrossing our skies on a daily basis. And now that COVID-19 is being described as a possible 'aerosol,' could the virus be used as a cover-up for when more people become sick from toxic aerosol spraying of our planet?)

Depopulation Agenda

On social media COVID-19 has been nicknamed the 'boomer remover' because the segment of our population of over 60s known as 'baby-boomers' is being singled out as 'high-risk' for the virus. Under the guise of 'protecting' the more vulnerable, those over 70 are being shut-in, removed from human contact, and denied freedom to go out of doors. Shannon's friend's mother at a nursing home in Missouri is an example of one of the countless such shut-ins. Now a prisoner, confined to only her room over a week (at the time of writing), she cannot have visitors or go out of doors, or even leave her room, but she is permitted to use a cell phone to make calls to family members. With nothing else to do but use cell phones and watch TV in

The Falsification of Science

chemically toxic dangerously high-EMF environments our seniors are sure to succumb to the effects of both social isolation (social contact often being the only thing previously keeping nursing home residents alive) and increased toxic chemical and EMF exposures. And when they do succumb, the authorities can say it was to the virus and that our 'efforts' were not enough, and next time need to be more stringent! If there is a targeted depopulation agenda aimed at removing the most 'useless' of we 'useless eaters,' aka those living on pensions and social security, it will be successful, not due to a fake virus, but because of increased and forced exposures to EMF radiation, chemicals, and social isolation.

In Shannon's former rural community of about 100 people spread out over a distance of 20 miles with most residents living on acreage (4-150 acres each) with the nearest neighbour often up to a mile away, still community members are panicking. This, in large part, because the mean resident age is over 60 and because they trust the authorities and 'official' sources of news. This has led to serious discussions on the email list-serve about everything from hoarding toilet paper (one member bought a case of 96 rolls online to share with neighbours after finding the local Wal-Mart shelves empty), the need to use hand sanitiser liberally (and the recent rash of thefts of bottles of such hand sanitisers from cars in town), getting flu shots, availability of free testing (for which everyone seems to be clamouring), self-imposed quarantines and social-distancing (cancelling all local gatherings even though they rarely see more than 20 in attendance), setting up checkpoints along the main road (this later was said to be a joke but many members had believed it and thought it a good idea), and using the community centre to house sick patients. Shannon did her best to dispel the mainstream myths and fear, pointing to lack of evidence of veracity of COVID-19 claims, but her voice was drowned out amidst the cacophony of establishment-backed hysteria. One would think that in such an isolated rural wilderness community 25 miles from the nearest town, there would be some sense of safety, calm, security, and distance from the panic, but this just proves how powerful massive amounts of propaganda can be, particularly in our digitally connected world. Measures to control the entire world's population to this level would not have been possible without our current level of constant connectivity.

"First there will be a planned outbreak of an engineered bioweapon. This will be done on a global scale—an engineered bioweapon will be released. There will be massive calls for government funding of the vaccine industry to come up with a vaccine and miraculously they will have a vaccine in record time. Everyone will be required to line up and take this vaccine shot, except the shot itself will be formulated to kill people. The kill-switch for humanity, a hard kill. Laws will be passed demanding that everyone line-up to be injected. People who are not vaccinated will be disallowed from participating in public transportation, you will be denied a bank account or banking services.

In this injection they will of course feed more bioengineered weapons, there will be viral strains in there with long latency that are programmed to activate perhaps months later and once the payload begins to express itself you will see people dying in the streets. Once that begins, the media will begin to push a 'second round' of vaccines which will be a shorter duration 'kill-switch' that will kill people in weeks not months, half-days in some cases. If you line up to be injected you will likely find yourself dead before very long. Vaccines have become weaponised. They are a kill-switch vector for humanity. Remember, the globalists don't want humans around—at least not most of us!" Mike Adams, writing several years ago in 'Natural News'

Those of you already familiar with our work will also know that we are very suspicious indeed of Mike Adams and believe him to be a peddler of propaganda disguised as real information. However, this statement definitely seems to hit the nail firmly on the head. If as suspected, he is a COINTELPRO operative (or similar) then maybe this is simply a classic example of 'revelation of the method?'

The Real Death Toll

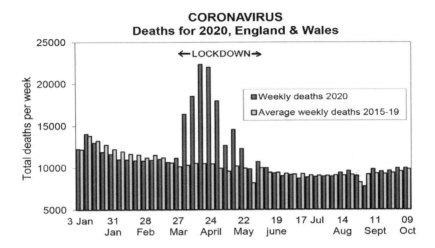

These figures also prove that when the UK lockdown eased-off in early June, cafes, parks, restaurants, pubs, and other leisure amenities re-opened) and deaths plummeted once again to normal levels. For the subsequent **three months,** deaths were 1-2% below the average again, for that time of year.

So why additional deaths during lockdown? Older people were killed off by the lockdown, through fear, stress, and loneliness. And in addition to this it is also difficult to calculate correctly exactly how many people died through being unable to receive treatment from hospitals for ailments other than so-called COVID-19.

In every country for which reliable death-data exists, the surge in deaths occurred AFTER the lockdown, *not* before. In autumn 2020, the UK Government and the compliant media began terrorising people again, with talk of a 'second wave' and lockdown over Christmas, and *in consequence of this* we notice a slight increase in deaths during October 2020.

Talk of 'cases' is totally misleading

All the rhetoric regarding 'cases' or 'infections' is just meaningless waffle. All this means is that someone has tested 'positive' on their 'PCR test,' which obviously artificially boosts the numbers of those allegedly infected creating a totally distorted picture of its spread. This deceptive test is promoted as being genuine science, but it is absolutely not!

Its inventor Kary Mullis, the American epidemiologist clearly stated that it should never be used for virus diagnostic purposes and according to the *Centers for Disease Control* (CDC) in the US, the virus has 'never been isolated' which renders accurate tests absolutely impossible.

In any event the so-called test produces 90-100% false positives and many people who placed untested swabs in the return envelope have received back positive test results nevertheless! In addition a goat and a papaya tested positive when 'given' the test. Indeed one lady in the UK visited her local test station, gave her name and address upon entry but then was kept waiting for so long that she got tired of waiting and left without taking the test. Sure enough, two days later, she received a 'positive' result in the post!

It is also worth mentioning here that the test is constant, but viruses (allegedly) mutate!

A Coronavirus is Nothing to Fear

Having coronaviruses in your system is absolutely normal and your healthy body can cope with them perfectly well. In fact, politicians have talked themselves into a dangerously dishonest corner where they can no longer now admit that the virus is fake (https://tinyurl.com/yytxae8a), that the lockdown measures are not protecting people and indeed are causing immense damage to the fabric of society and the economy. All as planned of course.

The Falsification of Science

A 'coronavirus' death on a death certificate neatly and conveniently disguises deaths from other causes such as diabetes, a heart attack, cancer, old age or even falling under a bus. All that the doctors need is a positive test result on the PCR test and 'coronavirus' is automatically recorded as the cause of death – all as per government decree.

As millions 'raid' grocery stories, clamouring to hoard food and supplies in anticipation of increased lockdown measures and prolonged quarantines, it is the masses of impoverished, homeless or otherwise disenfranchised, unable to afford food stockpiling or even have access to a storage space, who will suffer the most and possibly die from lack of access to basic life necessities. There may be mass starvation of millions, and death by exposure in extreme climates for those unable to find adequate shelter when forced to 'shelter in place.'

In India, a population of 1.3 billion people were given less than four hours' notice for a three-week lockdown, leaving millions stranded, without jobs or income and without access to food. Some reported being physically beaten for doing their jobs under lockdown and thousands of migrant workers fled the cities attempting to make their way back to their home villages, with at least one person dead from attempting to walk the 168 mile journey to his home.

And furthermore, what of the loss of life due to heightened suicide rates as the social isolation, loss of employment and hope for a brighter future take its toll? When the dust settles, if it is ever allowed to settle (and with 28th March 2020, BBC headlines reading 'The Birth of Quarantine Culture' and '10 TV Shows to Watch in Isolation' and the like, we can get a very accurate idea of what our masters have in store for us). As shown in the graph above, statistics prove that far more deaths occurred due to extreme police state lockdown measures than any supposed 'COVID-19' ever did.

What does all this tell us about the true intent behind this massive hoax, when measures being taken to 'protect us' kill more people than any 'pandemic' ever could? And when we are being fed messages from mainstream media sources 'urging couples to *"rethink getting pregnant"* during the Coronavirus pandemic'? It seems likely that the 'pandemic' could be nothing more than a ruse to further implement (on a massive scale), the infamous 'Agenda 21' depopulation goals.

Big Brother is Watching

Social pressure to properly perform 'social distancing' is creating a truly Orwellian nightmare with constant messages circulating on 'what to do if your friends and family are not 'on board.' 'Big brother' manifests as friends and family reporting your 'misbehaviour'/non-compliance for your own good. Again, shades of the Soviet Union c. 1920-1990 methinks, not to mention present day China!

Online daily articles pop-up on our home pages geared at 'helping us' learn 'social distancing.' In essence they serve to further social engineering agendas by teaching the scripted behaviour modification necessary for the transformation of a democratic society into a dictatorship.

Here is but one example out of many such nauseating pieces, this from the ahem, 'respected' Globalist propaganda machine and far left 'rag' otherwise known as 'The Guardian' newspaper entitled, 'An Expert Guide to Social Distancing – and What to Do If Friends and Family Aren't Onboard.' 'Experts' (in what exactly, is not specified – maybe experts in complete tyranny?) say avoiding close contact is the key to slowing COVID-19 – but what if you live with someone who's throwing caution to the wind?'

The author of this utter drivel lists many quotes from these so-called 'experts' dispensing their sage advice. One of the most offensive quotes comes from Dr Arthur L Caplan:

"Right now, mental health has to take a secondary place. Stay indoors, stay away from others as much as you can, and get your food delivered. Minimise your trips out for medicine or to the grocery store. Bring your hand sanitiser and use it all the time. Don't shake hands. Don't use paper money. Get ready to do a lot of television watching. You should call your shut-in parents and so forth."

The Falsification of Science

The agenda is therefore rendered crystal clear. In other words, our emotional and mental wellbeing do not matter (nor has it ever mattered) to our masters. They would like nothing better than for us to stay indoors away from life giving energies from our sun, air, and earth. They want us to constantly poison ourselves with toxic petrochemicals, eat lots of take-away junk foods and binge-watch television (to keep the programming in place) and only maintain contact with loved ones via the use of electronic technologies. And, perhaps most significantly, to dispense with the use of paper currency, which has been an unmet goal of the global Elite for some time now.

But with this new excuse, now backed by the WHO also suggesting the end of paper currency (since they have decreed it can be contaminated with viruses!), this goal may be nearing fulfilment, further enslaving the masses by cutting off any chances of using untraceable, tax-free currency. But more importantly, once all forms of money are digital only, we lose our ability to directly possess and control our own wealth, making it simple for any one of us to lose everything with one swipe of a keystroke if we disobey orders from our leaders, exactly as has been witnessed with the 'social credit' system which has been recently implemented in China. Fully digitised currency via credit and debit cards also has the highly desirable benefit (for them) of leaving a conspicuous digital paper trail, recording our every purchase and associated movement. How convenient that is, for tyrants!

Caplan doesn't stop there, when asked what to do if 'my partner or roommate is not practicing social distancing,' he continues to lay out the agenda with the following statement...

"Do your best to socially distance. Maybe don't sleep in the same bed? Minimise sexual contact. Don't share toothbrushes. Try to use separate things. You don't want to be hugging and kissing.... You might not want to share the same forks and knives...that kind of thing...If they won't go along, get away."

Non-compliant individuals are to be socially punished and loving close relationships and family come second to the demands of the state. Caplan and his ilk do not want we peasants hugging and kissing. But who is this Caplan anyway?

Interestingly Art Caplan, a 'bioethicist' (whatever the hell that is supposed to be?!) was written about in 'The Atlantic,' in a prophetic 20th December 2016 article 'How a Pandemic Might Play Out Under Trump.'

The 'article' (in reality, nothing more than another sickeningly blatant propaganda piece) begins with listing all of the major 'pandemic' outbreaks in recent decades (aka other less successful hoaxes than our current one; Swine Flu, Bird Flu, MERS, Ebola, and the Zika virus) and how US presidents responded to those crises, then goes on to speculate about a possible future pandemic under Trump's administration. *"As Donald Trump prepares to become America's 45th President in January 2017, the question isn't whether he'll face a deadly outbreak during his presidency, but when? And more importantly how will he cope?"*

And here is where our friend, the benevolent Dr Caplan shares his amazingly clairvoyant predictions:

"Bioethicist Art Caplan from the New York University School of Medicine envisages a quick slide towards isolation and authoritarianism. In a blog post that can only be described as pandemic fan-fiction, he imagines that a lethal mutant strain of H7N9 flu emerges in China and spreads to America. A hypothetical President Trump responds with a quick succession of moves: he seals the borders with Canada and Mexico; he quarantines sick Americans: he declares martial law, builds detention-style camps for quarantine-defiers, and uses epidemic conspiracies to launch a trade war with China."

As we have witnessed in the last weeks, this nightmare dystopian scenario has nearly played out in full. Let us just hope he was not completely accurate in *every* aspect and that all is not yet lost. But that is a pretty good assessment of the situation we now find ourselves, leading to the obvious conclusion that dear Mr. Caplan was probably privy four years ago to the intended agenda.

The Falsification of Science

Science (or more accurately, pseudoscience) is the new religion of the NWO (New World Order), and its dogma is dictating international policy, which is policed by authorities and the people themselves. (The people self-policing often even without any external prompting, is a very telling symptom of a true police state.) Those who are non-compliant and non-believing are seen as a threat to the rest of the population, much like the 'witches' of old but today are labelled as 'domestic terrorists.' And accompanying anger and outrage is being directed at those who choose to simply hug one another and not spray themselves and their homes with chemicals, such is the strength of the superstitions driving the prejudice; a prejudice which in Ecuador has led to locals setting up their own roadblocks and asking police to perform immigration checks on the 'Gringo' expat community of Vilcabamba. They are outraged that the more enlightened foreigners are not taking the 'pandemic' threat more seriously and tensions are rising. Apparently a simple act such as someone sitting in a park with no apparently urgent business being attended to, or not wearing a mask or gloves, is causing offence, and igniting fear. Everyone is being asked to stay at home and even to stay inside unless they absolutely have to go out.

But most alarming is the latest news of this escalating situation related in this Vilcabamba forum posting from 22nd March 2020...

"The locals are now FURIOUS with foreigners who are not complying with the quarantine laws. They are saying many foreigners are doing the following:

-walking in town barefoot

-not wearing masks

-walking their dogs

-riding their bikes

The locals see those people as a potential carrier of COVID-19 and infecting the entire community (even though it is high probability that is not happening).

The locals are now responding with the following:

1- Taking pictures and videos of anyone in town

2- Submitting those pictures to the authorities

3- Requesting the authorities arrest and imprison anyone who is out

4- They are now profiling people and taking notes, pictures and videos so they can take matters into their own hands when the time is right.

The tensions are now at a point where they can easily get out of control and it could become directed to all foreigners.. even those of us who are trying to cooperate.

PLEASE COMPLY WITH THE LOCAL PEOPLE'S DESIRE FOR EVERYONE TO NOW REMAIN IN THE HOUSE AT ALL TIMES, INCLUDING THE LEGAL CURFEW STARTING AT 7PM."

Why would sitting outside be dangerous to anyone if 'social distancing' is being practiced, and if you live on acreage and near wilderness (which is actually the case for that small mountain Ecuadorian village)? Can we no longer indulge in simple pleasures; are we all under house arrest, having become criminals overnight; enemies of the state? In the fast approaching dystopian future will our every movement be monitored and assessed not just by authorities, but also by our own neighbours? Will enjoyment of the natural world, leisure activities, contact with our friends, families and pets all be luxuries, privileges we must earn through servitude to our lords and masters instead of basic human rights? This seems a very real possibility if the predictions highlighted in the MIT Review article we referenced earlier in this chapter come to pass.

Panicked and fearful people are easy to control. They panic-buy, increasing profit margins for some businesses a hundredfold whilst others have to 'fold' as they are forced to close their 'non-essential' services. Fear makes us compliant, easily led, and ready to sacrifice our liberties for any promise of protection. We will use more drugs to ease our anxiety (both street drugs and pharmaceuticals), we will watch more television (as Carson dictated, I mean *predicted*, and

The Falsification of Science

apparently according to Amazon Prime viewing statistics, the popular films of the day tend to depict fictitious pathogen pandemic scenarios, with titles like 'Contagion,' 'Parasite,' 'Flu,' with Zombie apocalypse movies taking a firm 2nd place, as if our real lives aren't enough 'fear porn' for the masses currently), we will distance ourselves from life-giving, life-affirming people, surroundings and activities, and increasingly tether ourselves to life-destroying technologies. With one foot teetering over the gap between our natural and virtual worlds, we will allow ourselves to slip into a brave new 'smart' world, one our technocratic masters have envisioned for us for a very long time and fall into that virtual abyss. And if we are fortunate enough to re-emerge, it will be as part-human, part-robot, ready and willing to hook ourselves up to the new AI-driven 5G control grid by any means 'they' deem necessary in order to be permitted the crumbs of contact with our natural world and other humans they will let fall to our feet.

So, we are very sad to conclude, COVID-19 is simply an extremely elaborate and meticulously planned worldwide psy-op (psychological operation – one of many over the past few decades), a prelude to Agenda 21/2030, facilitated by the Elite powers in collaboration with all governments and is absolutely NOT what it is purported to be. The incredible over-the-top response to what should be just a minor health scare (if real) should really tell us that. They are placing the entire world into a lockdown situation for something that has killed far, far less people in the months since the scare began (even by their grossly over-hyped statistics), than the common flu and indeed many other diseases, none of which are assigned the same importance as COVID-19. (In the 2017/2018 Flu season in the US alone, 80,000 people died from 'flu', and as previously stated, it is estimated that up to 800,000 die annually in the United States from hospital and doctor 'error' thus rendering the Rockefeller-controlled medical industry the leading cause of death in the US.)

And whatever you do, please do not worry about contracting this so-called virus. (However you may want to minimise your EMF exposures and rally together with your communities against the real threat of 5G encroachments and continue to educate yourselves and investigate into real causes of illness not covered by mainstream media outlets.) They rely on the fact that the symptoms are also stated to be very similar to normal colds and flu and so it's quite easy to convince someone (and that includes medical professionals) that one has the virus, especially with all the media hype and propaganda that is currently 'doing the rounds.' We believe that the entirety of the world's health services have been propagandised by these monsters in human guise and so we cannot even rely on our health professionals to tell us any different from what we hear in the 'ever-compliant and eager to serve their masters' every bidding,' media.

Furthermore, on the 29th October 2020, just prior to the announcement of a second nationwide lockdown in the UK, the British government responded to a private individual's Freedom of Information (FOI) request by stating that there were no records in existence that proved that so-called 'COVID-19' existed. Just let that sink-in for a moment! In case you do not comprehend the implications of that, if this is the case (and why would they lie about something so damaging to their agenda?) then obviously it is impossible to test for it. So, what are they actually testing for? The answer is that the PCR test is simply to determine the presence of 'coronavirus' in an individual – ANY coronavirus! And given that around 80% of us have 'coronaviruses' (meaning presence of genetic material, not 'virus' in the true sense of the world) of various kinds within our bodies most of the time (such as linked with colds and flu, but not necessarily the *cause* of either), and given that PCR tests are not positive/negative tests but subject to interpretation by the test administrator, it is surely not difficult to understand – even for the 'hard of thinking,' that the ongoing testing regime will fulfil their insidious purposes of 'proving' how widespread their allegedly 'deadly' disease, actually is! Does that all make sense now? IT IS ALL ONE HUGE HOAX with a very sinister underlying purpose, allied to the onset of Agenda 21 and Agenda 2030.

Please take good care everyone, look after your friends and family wherever and whenever you can, give them hugs, pet your animals, connect with the natural world, and stay *truly* safe and well in these truly scary and unpredictable times! And try to maintain a positive

The Falsification of Science

outlook and sense of humour, as reason and logic are in about as short supply as toilet paper was at the beginning of this farce, leaving the collective shelves of the human psyche empty, as it is continually bombarded with messages of fear, uncertainty, and doubt.

And please do spread the message contained within this chapter as far and wide as is humanly possible! The world at large needs to understand fully what is going on – before it is too late!

** Footnote **

John and Shannon's forthcoming book on the COVID-19 hoax should be published and available sometime during the first half of 2021.

The Epilogue

All branches of science today are predicated upon several unsubstantiated and at best, extremely flimsy theories – specifically those already covered in this book, such as the Big Bang theory, the theory of evolution, Germ Theory etc. However one other such theory that I have not covered in full detail within these pages is Einstein's Theory of Relativity, although I did allude to it briefly. This, perhaps needless to say, is another highly dubious body of work, which has been challenged by more than a few well-respected scientists over the years, but of course, all to no avail as the scientific establishment 'protects its own' with the same zealousness and ferocity as a mother bear protects her cubs.

The highly esteemed English physicist, Professor Herbert Dingle (1890-1978), one-time president of the Royal Astronomical Society, actually discovered a major flaw in Einstein's Special Theory of Relativity (1905) in the 1950s, rendering the whole work null and void, and spent the last twenty years of his long life, fighting the intransigent scientific establishment in a vain attempt to have his voice heard on the topic. His excellent book *'Science at the Crossroads'* is a detailed account of his long, valiant struggle to bring his findings to the attention of the wider scientific community, only to be met with the usual tactics of disbelief, obfuscation, obstruction, ignorance, ridicule and ostracisation. This sad story is unfortunately so typical of many others, who found themselves at severe odds with established science.

Unfortunately, 'Relativity' is an intrinsic cornerstone of modern theoretical physics and if as I believe, Dingle's findings are absolutely correct, then this would have brought the entire 'house of cards'

The Falsification of Science

tumbling down. And as we are all aware, anything that challenges the ongoing 'gravy train,' deeply entrenched, vested interests and the foundations of many a lucrative career, is doomed to failure from the outset, thus bringing into question the entire integrity of the scientific establishment.

And yet another flimsy and totally un-provable 'theory' that has gained popular traction in recent years, is 'string theory.' This is another highly implausible 'logic patch' that attempts to cover up the current mathematical deficiencies in science's feeble attempts to justify its 'theory of everything.' As described in the opening chapter, the 'science' around gravity is totally inadequate in explaining how the universe hangs together, it being by many magnitudes a much weaker force than electromagnetism – even assuming its existence as an independent, quantifiable entity – which of course is extremely doubtful. And so of course, science has to fill that knowledge gap once again with yet another entirely invented construct to plug the gaping mathematical holes, whilst maintaining the illusion of 'wholeness' and appeasing those who deem it necessary to create a plausible 'theory of everything.'

String theory is actually an attempt to unite the two main pillars of 20th century physics, quantum mechanics and Einstein's Theory of Relativity, with an overarching framework that can explain all of physical reality. It attempts to do so by positing that particles are actually one-dimensional, string-like entities whose vibrations determine the particles' properties, such as their mass and charge.

This utterly counterintuitive idea was first developed in the 1960s and 70s, when strings were used to model data emanating from subatomic colliders in Europe and was initially proposed by those great 'propaganda machines,' the University of Oxford and the Royal Society. According to their diktats, 'strings' provide an 'elegant mathematical way' of describing the strong nuclear force, one of the alleged four fundamental forces in the universe, which holds together atomic nuclei. But in order to make this highly theoretical and totally un-provable concept work, this means that we must accept the fantastic notion that the universe consists of up to 26 dimensions. Please excuse my scepticism here!

Modern science also absolutely believes and heavily promotes the fact that the entire universe is made up of purely material, physical entities, and nothing else. According to their deceptive edicts, there is no consciousness, no spirit, no mind, and nothing other than mechanical and chemical processes at work.

This also neatly explains science's obsession with finding smaller and smaller particles using the Large Hadron Collider at CERN. Many scientists actually believe that if the very smallest portions of their alleged 'mechanical universe' are finally identified and labelled – because of course labels are really important to the materialistic worldview – then the entire cosmos will finally be completely mapped and understood and the 'delusion' of a creator or architect can finally be dismissed forever, in their narrow view of everything. Their goal is the ultimate in pessimism…to forestall or preferably destroy any belief in a higher intelligence and to doom we humans to living pointless lives that end in their total destruction at the moment of death. All of which could be lifted almost verbatim, directly from the Marxist creed and is an attempt to bring the masses further under the control of the Elite.

But perhaps the most astonishing delusion of all in 'modern science' is the fact that most scientists do not believe that they are themselves conscious beings.

The huge assumption made by science is that we humans are nothing more than biological robots and that animals are not conscious either. They literally do believe that consciousness is simply an illusory artefact of the wholly chemical and physical brain. Not surprisingly, they also do not believe that plants and other living systems are conscious or animated by spirit or a life-force. Even further, the idea that certain inanimate objects such as minerals or crystals might have some sort of consciousness, is considered heresy by most modern scientists – rather than as it should be in any kind of rational scientific paradigm, a subject for discussion and lively debate.

The Falsification of Science

This denial of consciousness is merely an unfounded assumption, however. There is no evidence whatsoever, supporting this premise. In fact, first-person evidence of the human experience appears to directly contradict the false assumption that humans are not conscious.

These casual, unfounded pronouncements of modern science are especially suspicious given that even conventional cosmologists readily admit that 96% of the universe has not yet been detected at all. That is the 'dark matter/dark energy' portion of the universe as described in chapter one of this book, and to my certain knowledge, neither dark matter nor dark energy have ever been directly measured or even detected, least of all seen, by human scientists.

Except for the risible and wholly theoretical Big Bang, there is no phenomenon by which modern scientists believe the totality of matter and energy can either come into existence – or exit our known universe.

This assumption is especially ridiculous considering the totally theoretical framework of the Big Bang, which utterly bizarrely claims that all the known matter and energy in the entire cosmos spontaneously appeared without cause, all without any intention or reason and totally contradictorily to the First Law of Thermodynamics which asserts that matter, under *natural* circumstances, can neither be created nor destroyed. The Big Bang theory and its accompanying theory of cosmological inflation (the ongoing expansion of the universe) are by any definition an utterly illogical kind of material mysticism that goes to great lengths to deny the existence of any kind of creator/designer/engineer/intelligent advanced civilisation etc.

The assumption that the laws of nature are absolutely rigid and fixed appears to have already disintegrated thanks to the efforts of a few more freely-thinking modern-day scientists themselves. As a simple example, multiple physics experiments are now being conducted all over the world, and widely replicated, which demonstrate 'faster than light' teleportation of information via quantum entanglement, totally contradicting Einstein's heavily flawed and basically useless, Theory of Relativity, which frankly is yet another in the long line of 'theories' seemingly unfit for purpose, yet also heavily promoted as the absolute de facto truth, regardless.

As just one small, largely unreported example of this, an international research team has recently achieved quantum teleportation over a record-breaking distance of 143 kilometres through free space. The experiment saw the successful teleportation of quantum information -- in this case, the states of light particles, or photons -- between the Canary Islands of La Palma and Tenerife. The breakthrough is a crucial step toward quantum communications.

Theoretically, instantaneous quantum teleportation could take place over unlimited distances. The distance makes no difference whatsoever to the principle. Quantum teleportation ignores the apparent laws of physics, including the 'cosmological speed limit' known as the speed of light – another by-product of the failed Theory of Relativity. According to classic laws of physics, such quantum teleportation is impossible. In fact, all quantum computing should be impossible, and even transistors should not function either, if they adhered rigidly to mainstream physics' immutable laws, but nevertheless they (quite obviously) do so.

Yet the far stronger argument for challenging the false assumption that nature's laws are fixed, and immovable is to be found in the 'multiverse theory' which states that our known cosmos is just one of an infinite number of other universes, each with its own 'personalised' version of the laws of physics. Only in a small fraction of all universes we are told, is for example, the strength of the weak nuclear force set at precisely the right number to result in the formation of stars, planets, and carbon-based life. But because, according to this premise, there are an infinite number of universes, it logically follows that there must also therefore be an infinite number of universes where the laws of physics exactly equal our own... and even ones where 'other' human civilisations almost perfectly reflect our own.

The Darwinian, evolutionary theory framework of biological science assumes that nature achieves highly complex biological structures, social structures, mechanical engineering, and behavioural cultures simply through the process of natural selection. And whilst natural selection is constantly taking place throughout nature, it alone is not sufficient to explain the ability of plants, animals, humans and possibly even universes to 'evolve' purely through random chance and inheritance.

There appears to be an inexplicable creative force behind much of what we observe in nature, including in animals and humans. This creative force, appears to have a deep, intrinsic connection with spirit, a non-physical 'mind' if you will, which bestows consciousness upon physical beings of all kinds.

What we see in the natural world, in ecosystems, plants, animals and even humans, is inexplicable wholly through natural selection. Intention, consciousness, and a desire to achieve complex goals by taking fantastic 'evolutionary' leaps which modern science cannot explain, also play an important role in the advancement of a species.

As a simple example of this, consider the fact that although many thousands of humanoid-like fossils have been unearthed in the last two centuries, there are still no fossils that record any of the thousands of theoretical 'missing links' and especially those which are alleged to link humans to other primates. Why have no such fossils ever been found? Almost certainly because they do not exist.

The assertion that DNA controls our bodies, and thereby our lives, is now an ancient myth. Only in the materialistic circles of old school 'science,' do people still believe that DNA alone controls our health, our behaviours and all of our inherited attributes.

Today we know that there are epigenetic factors beyond DNA, which strongly influence the development of all biological beings. We also now know that environmental factors (i.e. exposure to chemicals, heavy metals, nutrients, etc.) strongly influence either the suppression or the hyper-activation of genes. Vitamin D, for example, is one of the most powerful gene activators in human biology, turning on 'healing genes' like microscopic 'light switches.'

Furthermore, consciousness and freewill overrides DNA. Whilst we may have an inherited tendency towards a particular type of behaviour, we can override that behaviour as a matter of choice, when it suits our purposes. In other words, the mind rules the mechanics, and especially so if the mind is sufficiently trained, for example through meditation.

Personally, I find it bewildering that most mainstream scientists still do not acknowledge the existence of the 'mind,' a non-material awareness/presence/consciousness that coexists with the brain but is not derived from or apparently connected to the mechanics and chemistry of the brain.

Ironically, many scientists actually use their minds to attempt to disprove the existence of the mind. They would like us to believe that self-awareness is an illusion or that terms such as 'mind' or 'consciousness' are just semantics used in discussing brain chemistry, not actual concepts that really exist.

But all to no avail. They have utterly failed to date, to prove scientifically the inexplicable notion that consciousness does not exist or that the mind is not present in a conscious being, it is simply yet another unrealistic assumption that this is not the case. Science cannot disprove these things because all the tools of modern-day science are materialistic by definition and therefore incapable of proving or disproving nonmaterial phenomena. It is almost akin to attempting to measure the speed of a moving object with a thermometer or a wristwatch!

Modern scientists actually believe that memories are stored chemically, using the brain as some sort of complex, biological hard drive, and that if they could only find the location of the brain in which these chemical memories are stored, then they could not only literally read your mind but also download it onto a memory stick!

This crazy assumption is so far divorced from reality. Personally, I believe that memories are holographically stored in not only brain matter itself, but also in a non-material spirit matrix of some sort, which interacts with the physical brain. This explains why the physical location of memories in the brain can never be located by scientists and is also why some people are still found to be fully functional in our world even though they have virtually no brain matter at all. For example, there was a recent *New Scientist* article about a man who had almost no brain matter whatsoever, but yet still possessed an average IQ and functioned as a normal part of society.

He also had normal memories, too. So if memories are stored somewhere in the brain as modern-day scientists falsely believe, then how could this man have memories if he had virtually no physical brain to store them in? How could he even function at all? (And this story is just one of many similar ones.) But of course, science has no interest in anything whatsoever that could shatter any of its insistent and persistent myths. Far too many vested interests of various kinds would be shaken out of their puerile delusions by those kinds of actions.

Modern-day sceptics go to great lengths to try to disprove anything that even hints of telepathy or other inexplicable functions of our mysterious brain. But it is impossible to rationally refute the scientific work of people such as Dean Radin, author of *'The Conscious Universe: The Scientific Truth of Psychic Phenomena.'*

Radin has, repeatedly, demonstrated strong, convincing scientific evidence for low-level telepathy and other phenomena such as premonition. Explanations for such phenomena are entirely consistent with quantum non-locality and quantum entanglement, which Einstein referred-to as 'spooky action at a distance.'

The most likely explanation for all this is that the human brain, being a holographic, hybrid physical/nonphysical computational and awareness engine, is also intertwined with all matter in the universe at a quantum level. The brain/mind seems to be both a transmitter and receiver of quantum information that is continually and instantly rippling across the known universe. Tuning into that information is very similar to slow tuning into the correct radio station and gradually realising that the music is becoming crystal clear.

So-called 'sceptics' who attempt to refute the scientific work of people like Dean Radin eventually end up declaring something desperate or ridiculous along the lines of, *"if that were true, we would already know it"* which is a classic example of failed circular reasoning bordering on self-delusion or at least self-congratulatory dogma.

Mechanistic, Rockefeller medicine is an utter failure by any standards. Most scientists and medical professionals today, do not believe that any vitamin, any mineral, or any food has any biological effect whatsoever on the human body other than providing calories, sugars, proteins, fibre, and fat. This utterly delusional belief is enshrined in the FDA's regulatory framework and is practiced throughout hospitals and health clinics across the planet.

But logically, how can vitamin D have no effect on the human body when nearly every organ in the body has vitamin D receptors? How can minerals play no role in human health when elements like magnesium and calcium are necessary for the most fundamental chemical processes of muscle neurology?

The physical part of any human being obviously requires physical building blocks. Those building blocks are nutrients, plant-based chemicals, minerals, proteins, and water. They are NOT statin drugs, blood pressure medications, chemotherapy or radiation, as the scientists and marketeers of Big Pharma would fervently wish us to believe. The mechanistic model of medicine is an utter failure for human civilisation and health, but it has obviously proved a huge success in generating profits for drug companies however, which is exactly why this failed system is so desperately and unethically defended by those who seek to profit from it.

And the medical mechanistic model, similar to the scientific model of the universe, seeks to find answers to disease by dissecting human 'machines' down to the smallest component or particle. It is this mechanistic view and accompanying use of high-powered microscope technologies which has led us to the absurd viral paradigm responsible for today's imprisoned world, which points to the smallest observable particle (in this case a 'virus') as the cause of disease. If viewed in the proper framework, of seeing the world as a whole, alive, vibrant interdependent, symbiotic orchestra, a 'virus' (or bacteria or

The Falsification of Science

cancer cell, etc.) would be viewed as a symptom or effect of disease process and not the cause. As it is, because modern medicine and science refuse to acknowledge life force and symbiotic relationship in rigidly adhering to the mechanistic model, its view is skewed, fully upside down, inside out and backwards and will never succeed in enabling humanity to achieve perfect health.

That 'hero' of the scientific mafia, Albert Einstein is famously quoted as saying, *"We cannot solve our problems with the same thinking we used when we created them."* Yet that is what much of modern science is trying to do... let us cure cancer by finding a chemical that kills cancer cells instead of the more logical approach of discovering the cause of the generation of those cancer cells in the first place. Hmmm. And of course that highlights another issue with mechanistic medicine. It is not in any way shape or form interested in preventative medicine. Of course alleged 'cures' are far more lucrative than natural preventatives such as vitamins and enzymes. Many of these cures are also, to use an analogy, similar to duct taping over a red light on your car dashboard so that it no longer shows, rather than trying to determine the actual cause and fixing it.

But none of these approaches will ever succeed in answering the really important questions because they are rooted in materialistic, 19th century assumptions, which we now know to be false. There is far, far more to our universe than mere materialism, just as there is more to human consciousness than simple brain chemistry and there is also much, much more to biology than genetics and natural selection.

There are thousands of videos and blogs on the Internet now offering 'proof' that the Earth is not an oblate spheroid (a globe) as we are constantly told by the controlled media, mainstream science and NASA and its armies of apologists and shills. However, one aspect of this 'conspiracy' that rarely receives a mention is 'why?' Why the deception? What would be the point of creating, maintaining, and sustaining such an illusion? And *cui bono* (who benefits)? This indeed is the acid test of any question.

It is a commonly accepted premise in crime detection that whoever has the means, the opportunity, and the motive to commit that crime, is very likely to be the actual perpetrator. In others words the motive to commit the crime (or who is most likely to benefit) will usually lead us directly to the source. And there is little doubt, in using this tested principle, as to where that ultimately leads us, in this case.

Why have we been conditioned from cradle to grave, for the past 500 years (and more) to deny the overwhelming evidence of logic and our own eyes and bodies? I strongly believe the answer is to be found in one word... Freemasonry.

I appreciate that that is once again, a somewhat glib answer, so please allow me to expand on that a little...

Freemasonry is an all-pervasive force, an adjunct of Illuminism, totally dedicated to the absolute control of the world's populations and most importantly the destruction of religion and any form of spiritual belief and the spherical Earth and evolutionary model, is merely just one way to achieve those ends.

All early advocates of the spherical Earth were Freemasons. Copernicus, Galileo, and Newton to name just a few of the main protagonists, conspired with the governments of the day to present and promote the totally illogical yet strangely compelling model of the universe that placed the Sun as the focal point, relegating the Earth to a mere rocky outpost on the periphery and thereby virtually eliminating its 'importance' in the grand scheme of things.

The only way that this could be achieved logically was by creating a model whereby the universe consisted of spherical bodies (of varying sizes and importance) all of which were created from nothing in the seemingly areligious, yet almost biblically miraculous 'Big Bang' which we are assured happened exactly as our so-called scientists tell us.

Of course, this principle was to be further reinforced by other elements such as evolution and the 'discovery' that the Earth is actually

4.5 billion years old (we are told!) and which in turn was reinforced by the very first discoveries of dinosaur bones in the 19th century, totally coincidentally, around the time that evolution had first been mooted as an answer to the questions surrounding life's beginnings.

It is no small wonder that the ancient civilisations believed in a flat and motionless Earth. All of our senses and experiences would tend to point to this fact and when we cease to believe our own eyes, we are in fact bowing in reverence to these pseudoscientists and their false science.

So, to repeat myself, why is this deception perpetrated?

By turning the Earth into an insignificant spinning ball, one of several trillion others in a vast, unimaginably large universe which came about by cosmic accident in an explosion of nothing, and declaring that humankind evolved from pond slime, renders us as a species and the entirety of creation, totally irrelevant. It is a form of trauma-based mind control, or a form of mental manipulation, if you will?

NASA is the biggest, black-budget, black hole in existence. It consumed around $30bn of US taxpayer money for the fake moon landings alone and several more billions every year in continuance of the gigantic fraud it perpetuates on behalf of the Elite. It is all just a ploy to devalue human spirituality and render our existence pointless.

The Big Bang-induced, heliocentric, globe-Earth, chance-evolution paradigm controls humanity mentally and spiritually, by removing a creator or any form of intelligent design from the equation and replacing it with haphazard cosmic coincidence. But, by removing the Earth from the centre of the Universe, the Elite have removed us physically and metaphysically from a place of supreme importance to one of complete nihilistic indifference. And if the Earth is just one of trillions of other planets revolving around trillions of other stars in trillions of other galaxies, and we its inhabitants simply 'stardust,' then the ideas of creation, intelligent design, and a specific purpose for humanity, all instantly become highly implausible.

This could legitimately be said to be the greatest hoax and cover-up in history. And believe me, once again, there is some pretty strong competition for that particular accolade. Of course, I cannot put my hand on my heart and state categorically that the Earth is indeed flat. Without the ability to project ourselves high enough above the ground, no one can. (However those ancients and present-day astral travel and OBE-ers may beg to differ. And it is interesting to note that ancients who regularly practiced out-of-body experimentation did not claim a spherical Earth model.) But neither can they say categorically that the Earth is spherical either. You may argue that NASA has this ability and has proved photographically that the Earth is indeed a spinning ball, but given its considerable motivations for lying to us and what we have learned in the previous pages about how NASA and its many apologists, have deceived us all for many decades, faked photographs (and the fact that alleged satellite images of Earth do not show same or even similar sized continents dated 30 years apart) and notwithstanding the fact that a flat Earth is far more logical than an 'oblate spheroid,' how can we be absolutely sure of that? I submit that we cannot.

There is enough evidence for a 'flat Earth' to make it highly plausible or at the very least render the globe Earth model highly questionable. And if the Earth is the centre of the universe, then the ideas of intelligent design, creation and a purpose for human existence are suddenly transformed into supreme importance – and of course this would never 'do' as far as our deceitful masters are concerned.

By surreptitiously indoctrinating us into this 'scientific,' materialistic sun-worship, not only do we as a species lose faith in anything beyond the material, but we also gain absolute, total faith in materialism, superficiality, hedonism, and consumerism. If there is no intelligent design and we are a simple cosmic accident, then all that really matters is me, me, me. But this 'me' is a faint husk of what could be a fully embodied, deeply rooted, consciously aware, fully realised, self-empowered human being, where the potential of 'me' would have a totally different meaning. And this is why materialist self-centeredness is never truly satisfying and has never resulted in anything remotely resembling 'happiness.'

The Falsification of Science

The speed at which the world has become 'industrialised' is absolutely staggering. Only just over one hundred years ago, we were still largely dependent on horse-power for transport, and steam engines powered by coal and water were the only machines in existence capable of challenging the horse's superiority in this regard. Indeed, as a species we progressed from steam power and horses to jet engines and space travel in only around 60 years. A monumental technological feat that had it been used for the benefit of humanity as a whole instead of a tiny minority, would have led to a world of peace and plenty for all instead of the truly appalling state of affairs we see today.

The last two hundred years have seen us shift from being the servants of Mother Nature to being its masters. We have behaved in the past and are still today behaving in ways that seriously threaten the very existence of the Earth upon which we depend, whilst over 100,000 people die of starvation each day. Starvation caused directly by Elite profiteering.

Our so-called economic system depends wholly on the fact that we need to consume the Earth's resources at an ever-increasing rate simply to maintain and increase the profit margins of the corporations. Economic growth, we are constantly told by the Elite-controlled economists, media pundits and financial analysts, is the only meaningful measure of a strong economy and as such is portrayed forcibly to be inherently desirable despite the fact that it is this very 'economic growth' that results in the wholesale destruction of the planet we see today and the constant widening of the gap between those who 'have' and those who do not.

"Somehow, we have come to think the whole purpose of the economy is to grow, yet growth is not a goal or purpose. The pursuit of endless growth is suicidal." David Suzuki

How obvious does it have to be? There is FAR MORE to discover about our universe, if we only would set ourselves free from the mental shackles of dogmatic, vested interest 'science' as practiced today in our Westernised, wholly materialistic culture.

The mega wealthy, all-powerful corporations with their sun cult logos and ultra-slick marketing hype sell us idols and celebrities to worship instead, and slowly but surely bring the entire world under their influence whilst we tacitly believe their false science and not-so-subtle propaganda. We vote for their controlled politicians, read, watch, and listen to the lies and garbage emanating from their controlled media 24/7, buy their heavily promoted, often useless products, gizmos, and gadgets, listen to their dreary, repetitive, propaganda-filled, satanic 'music' and watch their sick, dysfunctional TV and movies and generally sacrifice our own souls upon the altar of materialism.

And that dear reader, is precisely the kind of world that suits their insidious agenda and why and how they are able to maintain control over *all* of our minds and spirits.

Either the world is completely insane, or I am – and I am genuinely not sure which. But should it turn out to be me after all, then to be frank, I much prefer insanity, to what passes as 'sane' in today's world.

Please do not allow this sad state of affairs to continue unchallenged.

END

Made in the USA
Middletown, DE
11 September 2024

60784249R00400